T0177629

Martin Folkes (1690–1754)

Martin Folkes (1690–1754): Newtonian, Antiquary, Connoisseur

by

ANNA MARIE ROOS

OXFORD
UNIVERSITY PRESS

OXFORD
UNIVERSITY PRESS

Great Clarendon Street, Oxford, OX2 6DP,
United Kingdom

Oxford University Press is a department of the University of Oxford.
It furthers the University's objective of excellence in research, scholarship,
and education by publishing worldwide. Oxford is a registered trade mark of
Oxford University Press in the UK and in certain other countries

© Anna Marie Roos 2021

The moral rights of the author have been asserted

First Edition published in 2021

Impression: 1

All rights reserved. No part of this publication may be reproduced, stored in
a retrieval system, or transmitted, in any form or by any means, without the
prior permission in writing of Oxford University Press, or as expressly permitted
by law, by licence or under terms agreed with the appropriate reprographics
rights organization. Enquiries concerning reproduction outside the scope of the
above should be sent to the Rights Department, Oxford University Press, at the
address above

You must not circulate this work in any other form
and you must impose this same condition on any acquirer

Published in the United States of America by Oxford University Press
198 Madison Avenue, New York, NY 10016, United States of America

British Library Cataloguing in Publication Data
Data available

Library of Congress Control Number: 2020950769

ISBN 978-0-19-883006-1

DOI: 10.1093/oso/9780198830061.001.0001

Printed and bound by
CPI Group (UK) Ltd, Croydon, CR0 4YY

Links to third party websites are provided by Oxford in good faith and
for information only. Oxford disclaims any responsibility for the materials
contained in any third party website referenced in this work.

Acknowledgements

I first became interested in Martin Folkes upon the realization that he was the only person to have been President of both the Society of Antiquaries of London and the Royal Society. What would being President of a society dedicated to the material past have to do with leading a society dedicated to natural philosophy? Answering that question has taken me on a six-year odyssey.

To assist me on my scholarly peregrinations, I would like to thank the John Rylands Library at The University of Manchester, The Huntington Library, All Souls College, Oxford, and the Beinecke Library at Yale University, who provided me with fellowships for archival work and time and space to think. I have also had the great privilege of visiting some of the finest libraries and archives in the world, including the Bodleian Library; British Library; Clare College, Cambridge; London Metropolitan Archives; National Art Library at the Victoria and Albert Museum; Royal Society Library; and the Society of Antiquaries of London. My thanks to them and their staff for their invaluable assistance.

Special gratitude goes to the Royal Society Library team, particularly Keith Moore, for his suggestions, insights, and sense of humour. Rupert Baker, Ellen Embleton, Louisiane Ferlier, and Ginny Mills have also been super, and, as a journal editor for the Royal Society, I have had the great and good pleasure to work with Tim Holt, Phil Hurst, Jennifer Kren, and Stuart Taylor, as well as the rest of the Publishing Board.

My home institution, the University of Lincoln, and my colleagues in the School of History and Heritage, the School of Mathematics and Physics, the Lincoln Medical School, and the School of Philosophy are also to be thanked for their kindness and patient support. A special shout-out goes to my head of school, John Morrison, Pro Vice Chancellors Libby John and Abigail Woods, and Vice Chancellor Mary Stuart, who accommodated my eccentric love of writing. My thanks also to Andrei Zvelindovsky and Fabien Paillusson who taught me how to think another way.

My thanks as well to Malcolm Baker, Chiara Beccalossi, Jim Bennett, Paul Cox, Nicholas Cronk, Surekha Davies, Christian Dekesel, William Eisler, Elizabeth Einberg, Patricia Fara, Mordechai Feingold, Christopher Foley, Sietske Fransen, Vera Keller, Rebekah Higgitt, Michael Hunter, Rob Iliffe, Dmitri Levitin, Arthur MacGregor, Anna Maerker, Sir Noel Malcolm, Scott Mandelbrote, Sheila O'Connell, Cesare Pastorino, Heather Rowland, Charlotte Sleigh, Liam Simms, Kim Sloan, Aron Sterk, Mary Terrall, Simon Werrett, Charles and Carol Webster,

Dustin Frazier Wood, and Beth Yale for their insights and conversation. Benjamin Wardhaugh very kindly read the entire manuscript, and his advice was invaluable. The members of the Council of the Society for the History of Natural History have been consistently kind. Philip Beeley, June Benton, Helen and Bill Bynum, Tim Birkhead, Andrew, Joseph and Ruth Bramley, Isabelle Charmantier, Yupin Chung, Robert Fox, Chris and Uta Frith, Rainer Godel, Anita Guerrini, Jo Hedesan, Miranda Lewis, Antonella Luizzo-Scorpo, Gideon Manning, Will Poole, Lori Rausch, Katie Reinhardt, Leon Rocha, Richard Serjeantson, Jade Shepherd, Lisa Smith, Anke Timmermann, and Alexander Wragge-Morley cheered me on when Folkes became irritating or difficult.

Most of all, I would like to thank my husband Ian Benton. Ian, this is the eighth book I have edited or written. I cannot fathom how you put up with it, or me, but I am surely glad you do. I love you.

A Short Chronology of the Life of Martin Folkes

29 October 1690	Born
1699–1706	Privately tutored by James Cappel and Abraham de Moivre
February 1705	Death of Folkes's father, Martin
July 1706	Admitted to Clare College, Cambridge as a Fellow Commoner
December 1713	Proposed as Fellow of the Royal Society
July 1714	Elected Fellow of the Royal Society
October 1714	Marriage to Lucretia Bradshaw, St Helen's Church, Bishopsgate, London
November 1714	Admitted Fellow of the Royal Society
November 1716	Elected to Council, Royal Society
December 1716	Presented first substantial astronomical observation to the Royal Society
1717	Observation of *aurora borealis*, published in *Philosophical Transactions* in 1719
October 1717	Received MA from Cambridge
1718	Birth of first child, Dorothy
1720	Birth of only son, Martin
	Election to the Society of Antiquaries of London
1721	Birth of second daughter, Lucretia
1723	Service on the Repository and Library Committee, the Royal Society
	Appointment as Vice-President of the Royal Society
1724–5	Deputy Grand Master of Grand Lodge
March 1727	Death of Sir Isaac Newton
1727	Folkes's retelling of the Newtonian apple tree parable is in print
November 1727	Folkes defeated in presidential election of the Royal Society
1728	Publication of Newton's *Chronology of Ancient Kingdoms, Amended*, which Folkes co-edited with Thomas Pellett
1733	Publication of Newton's *Observations Upon the Prophecies of Daniel and the Apocalypse of St. John* (1733) which Folkes co-edited with Thomas Pellett
1733–5	Grand Tour
1736	Proposed Copley Medal Prize
	Publication of Folkes's *A Table of English Gold Coins*
1739	Elected Governor, Coram Foundling Hospital
1739	Tour of France
1741–3	Elected Member of the Egyptian Society
1743	Elected corresponding member of the Spalding Gentlemen's Society
1741–52	Served as President of the Royal Society
1742	Elected to *Académie Royale des sciences*
1745	Publication of Folkes's *Table of English Silver Coins*
1746	Given honorary doctorate of laws at Oxford and Cambridge
1747	Elected President of the Royal Society Dining Club
1749–54	Served as President of the Society of Antiquaries
September 1751	Suffered first stroke
November 1753	Suffered second stroke
28 June 1754	Death
1763	Posthumous publication of Folkes's *Tables of English Silver and Gold Coins* by the Society of Antiquaries
1792	Erection of Folkes's memorial, Westminster Abbey

Contents

Acknowledgements   v
List of Illustrations   xi

1. Martin Folkes: 'That Judicious Gent'   1
   1.1 Introduction   1
   1.2 'Scientific' Antiquarianism   2
   1.3 The Vibrancy of the Royal Society   6
   1.4 Newtonianism, and the Heritage of the Royal Society   10
   1.5 Folkes's Archival Lives and Afterlives   16

2. Nascent Newtonian, 1690–1716   19
   2.1 Early Life   19
   2.2 Education   24
   2.3 The Royal Society   38

3. Lucretia Bradshaw: Recovering a Wife and a Life   49
   3.1 A Childhood on Stage   50
   3.2 A Rising Star   55
   3.3 Mrs Bradshaw: A Seasoned Actress   59
   3.4 'Not a More Happy Couple': Mrs Lucretia Folkes   69

4. Folkes and His Social Networks in 1720s London   73
   4.1 Introduction   73
   4.2 Folkes and Freemasonry   75
   4.3 The Order of the Bath   90
   4.4 The Royal Society in the 1720s   96
   4.5 Antiquarianism and the Royal Society in the 1720s   101
   4.6 Bidding for the Royal Society Presidency   112
   4.7 Folkes's Religious Beliefs and his Editions of Newton's
       Chronology and *Observations*   124

5. Taking Newton on Tour   141
   5.1 The Grand Tour   141
   5.2 Metrology   150
   5.3 Venice and Newtonianism   156
   5.4 Venice and Celsius   167
   5.5 Folkes as Newton?   176

6. Martin Folkes, Antiquary   183
   6.1 Numismatics   183
   6.2 The Egyptian Society, 1741–3   208
   6.3 The Egyptian Society Minute Books   215

7. Martin Folkes and the Royal Society Presidency: Patronage,
   Biological Sciences, and Vitalism   237
   7.1 Folkes as Administrator   237
   7.2 Patronage of Gowin Knight   247
   7.3 Folkes and Benjamin Robins   251
   7.4 Folkes, Madame Geoffrin, and Abraham Trembley   257
   7.5 Henry Baker: Gentlemen's Microscopist, Antiquary,
       and Opportunist   274
   7.6 Baker, Folkes, and Preformation   280
   7.7 The Plant–Animal Continuum   281

8. Martin Folkes and the Royal Society Presidency: The Electric
   Imagination   289
   8.1 Keeping Current: Folkes and Electrical Research in the
       Royal Society   289
   8.2 Patronage of Benjamin Wilson   292
   8.3 Illusion and Reality: The Rembrandt Craze   300
   8.4 Public Upheavals   303
   8.5 Private Upheavals   309
   8.6 The Constancy of Friendship: Charles Lennox, the 2nd Duke of
       Richmond and Goodwood   320
   8.7 Friendship Beyond the Grave: Folkes and John, 2nd Duke of
       Montagu   324

9. 'Charting' a Personal and Institutional Life   333
   9.1 President of the Society of Antiquaries of London   333
   9.2 Last Years   341
   9.3 The Legacy of Martin Folkes   348

Afterword: Folkes and Voltaire   353

Bibliography   363
Index   399

List of Illustrations

2.1 Letters patent under the Great Seal, 1676 Dec. 1, granting to
William Chaffinch and Martin Folkes (agents for Nell Gwyn and
Henry Jermyn, Earl of St Albans) reversion and inheritance of the
property then occupied by Nell Gwyn, MS 179A, Pierpont Morgan Library,
Department of Literary and Historical Manuscripts.   21

2.2 Hillington Hall, 1907, Frederic Dawtrey Drewitt, *Bombay in the days of
George IV: memoirs of Sir Edward West, chief justice of the King's court
during its conflict with the East India company, with hitherto unpublished
documents* (London: Longmans, Green, and Co., 1907), pp. 17–19,
Wikimedia Commons, Public Domain.   23

2.3 Jacques Antoine Dassier, *Abraham de Moivre*, 1741, Bronze Medal,
Obverse, 54 mm diameter, Gift of Assunta Sommella Peluso, Ada Peluso,
and Romano I. Peluso, in memory of Ignazio Peluso, 2003, accession
number 2003.406.18, Metropolitan Museum of Art, New York, Public Domain.   27

2.4 David Loggan, etching of the inner court of Clare Hall as part of a larger
broadsheet of the Cambridge Colleges, Gift of C.P.D. Pape, object number
RP-P-1918-1767, Rijksmuseum, Amsterdam, Public Domain.   33

2.5 Edward Dayes, *Queen Square in 1786*, Watercolour with pen and black
ink over graphite on thick, smooth, cream wove paper, 17 1/16 × 23 1/2 in.,
Yale Centre for British Art, Paul Mellon Collection, Public Domain.   40

2.6 Jonathan Richardson, *Martin Folkes*, Graphite on Vellum, 1735,
179 mm × 134 mm, AN327311001, © The Trustees of the British Museum,
CC BY-NC-SA 4.0.   41

2.7 John Smith, *Christopher Wren*, mezzotint after Godfrey Kneller,
1713 (1711), 350 mm × 263 mm © The Royal Society, London.   45

2.8 John Simon, *John Vaughan, 3rd Earl of Carbery* after Godfrey Kneller,
early 18th century, 357 mm × 258 mm, © The Royal Society, London.   45

2.9 John Smith, *Thomas Herbert, 8th Earl of Pembroke*, mezzotint after
Willem Wissing, 1708, 346 mm × 250 mm, © The Royal Society, London.   46

2.10 John Smith, *Charles Montagu, 1st Earl of Halifax*, mezzotint after
Godfrey Kneller, 1693, 336 mm × 247 mm, © The Royal Society, London.   46

3.1 Lucretia Bradshaw's Spoken Epilogue in Mary Pix's *The Deceiver Deceiv'd*
(London: R. Basset, 1698), pp. 47–8, call number 147237, The Huntington
Library, San Marino, California.   55

3.2 Theatre ticket design for *The Mock Doctor*, a benefit at the 'Theatre Royal',
designed by William Hogarth (?), *c.* 1732. 5 13/16 × 5 11/16 in., Metropolitan
Museum of Art, New York, Public Domain.   58

3.3   The Italian Opera House at the Haymarket before the fire of 17 June 1789,
       original drawing by William Capon, etching by Charles John Smith,
       ca. 1837. Charles John Smith, *Historical and literary curiosities,
       consisting of facsimiles of original documents* (London: H.G. Bohn, 1852),
       p. 294, archive.org, University of California Libraries.   61

4.1   List of Freemasons' lodges and meetings, engraved by John Pine,
       © The Royal Society, London.   82

4.2   William Hogarth, *The Mystery of Masonry Brought to Light by Ye Gormagons*
       [sic], Dec. 1724, 8.5 × 13.5 in., Metropolitan Museum of Art, New York,
       Public Domain.   85

4.3   William Hogarth (attributed to), *Examining a watch; two men seated at
       a table, the older (Martin Folkes) looking through his eyeglasses at a watch,
       a paper headed 'Votes of the Commons' (?) on the table.* Pen and brown (?) ink
       and wash, over graphite, *c.* 1720, 125 × 185 mm, Image 1861,0413.508,
       © The Trustees of the British Museum.   87

4.4a, b Ottone Hamerani: *Martin Folkes*, 1742, bronze, 37 mm., reversed, collection
         and photo of the author.   88

       Enlarged view of Folkes in the Gormogon print, Metropolitan Museum
       of Art, New York, Public Domain.   88

4.5   1751 English armorial binding for Martin Folkes (1690–1754) with his
       crest of the golden fleece. Horace, [*Epistulae. Liber 2. 1. English & Latin*]
       *Q. Horatii Flacci epistola ad Augustum. With an English commentary
       and notes.* (London: W. Thurlbourn, 1751), PA6393 E75 1751 cage,
       Folger Shakespeare Library.   89

4.6   Jonathan Richardson the Elder, Martin Folkes, oil on canvas, 1718,
       H 76 × W 63 cm, LDSAL 1316; Scharf Add. CVIII, by kind permission
       of the Society of Antiquaries of London.   92

4.7   Knights of the Order of the Bath, with elaborate plumage. John Pine,
       *The Procession and Ceremonies Observed at the Time of the Installation of the
       Knights Companions of the Most Honourable Military Order of the Bath*
       (London: S. Palmer and J. Huggonson, 1730), GR FOL-353, Bibliothèque
       nationale de France, département de l'Arsenal, Public Domain.   94

4.8   Portrait of Folkes, Thomas Hill, and Matthew Snow, the 'Knights
       Companions' for Charles Lennox, the Duke of Richmond for the Installation
       Ceremony for the Order of the Bath. John Pine, *The Procession and Ceremonies
       Observed at the Time of the Installation of the Knights Companions of the
       Most Honourable Military Order of the Bath* (London: S. Palmer and
       J. Huggonson, 1730), GR FOL-353, Bibliothèque nationale de France,
       département de l'Arsenal, Public Domain. Folkes is portrayed
       gesturing to the right towards the procession.   95

4.9   Bernard Lens (III), miniature of *Martin Folkes*, watercolour and bodycolour
       on ivory, ca. 1720, 83 mm × 64 mm, © National Portrait Gallery, London.   98

4.10 *The appearance of the total solar eclipse from Haradon Hill near Salisbury,*
       May 11 1724, by Elisha Kirkall after William Stukeley, 1984–453,
       © The Board of Trustees of the Science Museum.   101

4.11 Enoch Seeman, *Mohammed Ben Ali Abgali,* oil on canvas, ca. 1725,
       60 × 40 inches (152.5 × 101.5 cm), Private Collection, courtesy of
       Ben Elwes Fine Art.   103

4.12 Figure of 'petrified man', from 'An Account of some human Bones incrusted
       with Stone, now in the Villa Ludovisia at Rome: communicated to the
       Royal Society by the President, with a Drawing of the same', *Philosophical*
       Transactions 43, 477 (1744), p. 558, Image from the Biodiversity Heritage
       Library. Contributed by Natural History Museum Library, London.
       www.biodiversitylibrary.org.   106

5.1   John Smith, *Martin Folkes* after Jonathan Richardson Senior, mezzotint,
       1719, 339 mm × 247 mm, 1902,1011.4540AN121298001, © Trustees of the
       British Museum, Creative Commons BY-NC-SA 4.0 license.   142

5.2   Map of Folkes's Grand Tour, keyed to Martin Folkes' memorandum book,
       1733, (NRO, NRS 20658), Norfolk Record Office, Norwich.   143

5.3   Canaletto, etching of Piazza San Marco, *c.* 1707–58, 144 mm × 211 mm,
       object number RP-P-OB-35.577, Rijksmuseum, Amsterdam, public domain.   152

5.4   Measuring standard observed by Folkes in the Villa Caponi, Rome,
       Diary of Andrew Mitchell in Rome, Add MS 58318, f. 109r, © The British
       Library Board.   154

5.5   Gerard Van der Bucht, *Nicholas Saunderson,* engraving after John
       Vanderbanks's 1719 portrait painted for Martin Folkes, 1740,
       221 mm × 144 mm, Wellcome Collection, Attribution 4.0 International
       (CC BY 4.0).   163

5.6   Watercolour sketch of Anders Celsius's discovery of runes on the Malsta stone in
       Rogsta, Hälsingland, Sweden, which was engraved and published in the
       Philosophical Transactions, CLP/16/51, © The Royal Society, London.   171

5.7   Stereographic projection by Martin Folkes of the Farnese Globe in
       Richard Bentley, ed., *M. Manilii Astronomicon* (London: Henrici Woodfall;
       Pauli et Isaaci Vaillant, 1739), p. xvii, 25.9 × 52.1 cm. The Library, All Souls
       College, Oxford.   174

5.8   Martin Folkes's diagram of Antonio de Ulloa's Triangulation for the
       Peruvian Expedition, Letters and Papers, Decade 1, Volume 10A,
       5 December 1745 to 1 May 1746, MS no. 479, © The Royal Society, London.   175

5.9   Louis François Roubiliac, Sculpture Bust of Sir Isaac Newton, Marble,
       1737–8, S/0018, © The Royal Society, London.   177

5.10 John Faber Jr, *Martin Folkes,* mezzotint after John Vanderbank,
       1737 (1736), 353 mm × 253 mm, Yale Center for British Art, Paul Mellon
       Collection.   178

5.11   John Vanderbank, *Martin Folkes*, oil on canvas, 1739, private collection,
        courtesy of Christopher Foley FSA, Lane Fine Art.   179

5.12   John Vanderbank, *Three-quarter-length portrait of Sir Isaac Newton, aged
        eighty-three years*, oil on canvas, 1725–1726, 1270 mm × 1016 mm,
        © The Royal Society, London.   180

6.1   Ottone Hamerani, Martin Folkes, 1742, bronze, 37 mm, M.8467,
        © The Trustees of the British Museum, Creative Commons BY-SA 4.0.   185

6.2   Christoph Jacob Trew (1695–1769), friendship album, MS 1471, fol. 83r.,
        University Library Erlangen-Nürnberg.   192

6.3   Martin Folkes' memorandum book, coinage weights pasted inside
        back cover, (NRO, NRS 20658), Norfolk Record Office, Norwich.   193

6.4   Plate 1.69: Engraving of a Standard of Weights and Measures, 1497,
        Vetusta Monumenta (1746), vol. 1, Getty Research Institute, archive.org.   199

6.5   George Vertue, engraved coins featuring Charles I, in Martin Folkes,
        Tables of Silver and Gold Coins, (London: Society of Antiquaries 1763–6),
        plate XIII, New York Public Library, Hathi Trust, Public Domain.   202

6.6   Plate XXX, New England Shillings, figures 4, 5, 9, 11–14 in Martin Folkes,
        Tables of English Silver and Gold Coins (London: Society of Antiquaries
        1763–6), New York Public Library, Hathi Trust, Public Domain.   203

6.7   Louis François Roubiliac, Bust of Martin Folkes, 1747, Marble. Earl of
        Pembroke Collection, Wilton House, Image: Conway Library,
        The Courtauld Institute of Art.   205

6.8a,b Jacques-Antoine Dassier, Medal of Martin Folkes, Bronze,
        1740, 54 mm, obverse and reverse. Accession Number 2003.406.17,
        Gift of Assunta Sommella Peluso, Ada Peluso, and Romano I. Peluso,
        in memory of Ignazio Peluso, 2003, Metropolitan Museum of Art,
        New York, Public Domain.   206

6.9   Henry Baker, *Employment for the Microscope* (London: R. Dodsley, 1753),
        facing p. 441, plate IX. Columbia University Libraries, Archive.org,
        Public Domain.   207

6.10   Andrew Miller, *Lebeck*, mezzotint after Sir Godfrey Kneller, mezzotint,
        1739, 353 mm × 250 mm. Medical Historical Library, Harvey Cushing/
        John Hay Whitney Medical Library, Yale University.   210

6.11   Sketch of a camisa, in a letter from William Stukeley to Maurice Johnson,
        16 June 1750, SGS No 80, MB5 fol. 59A, by kind permission of the Spalding
        Gentlemen's Society, Spalding, Lincolnshire.   214

6.12   Sistrum drawing from the Egyptian Society Minute books with
        opening poem, BL MS Add. 52362, f. 1r., © The British Library Board.   216

6.13   Friendship album page from the Egyptian Society meeting minutes,
        with Folkes's motto taken after Newton, BL MS Add. 52362, f. 4r,
        © The British Library Board.   221

6.14   Joseph Highmore, *John Montagu, 4th Earl of Sandwich*, oil on canvas, 1740,
      48 in. × 36 in. (1219 mm × 914 mm), © National Portrait Gallery, London.   222

6.15   Mummified Ibis from Egyptian Society Minute Book, BL MS Add. 52362,
      f. 18r, © The British Library Board.   226

6.16   *A View of the Fire-Workes and Illuminations at his Grace the Duke of
      Richmond's at White-Hall and on the River Thames*, on Monday 15 May, 1749.
      © Victoria and Albert Museum, London.   228

6.17   Page of Coins and Medals from Egyptian Society Minute Book with
      identifying initials for various numismatic collections, BL MS Add. 52362,
      f. 35r, © The British Library Board.   235

7.1   Pierre Louis Surugue, *L'Antiquaire*, etching and engraving, 1743,
     305 mm × 230 mm, after painting by Jean Siméon Chardin, *Le Singe de la
     philosophie*, shown at the Louvre in 1740 and now in the Musée du Louvre,
     Paris, Metropolitan Museum of Art, New York, Public Domain.   283

7.2   Giacomo [James] Poro. 1714, engraving, Wellcome Collection,
     Attribution 4.0 International (CC BY 4.0).   286

8.1   Francis Watkins, *A Particular Account of the Electrical Experiments Hitherto
     made publick* (London: By the author, 1747), Wellcome Collection, CC-BY.   291

8.2   Benjamin Wilson, *Bryan Robinson, M.D. aetatis suae 70*, engraving, 1750,
     32.7 × 26.6 cm, Wellcome Collection, CC-BY.   295

8.3   Rembrandt van Rijn, *St. Jerome Reading in an Italian Landscape*, Etching,
     Engraving and Drypoint, *c*. 1653, 26.2 × 21.4 cm, Bequest of
     Mrs. Severance A. Millikin 1989.233, Cleveland Museum of Art,
     CC Open Access Licence.   302

8.4   Copy of Entry Record about baby 53, written by Martin Folkes, with an
     indication where the letter was pinned, A/FH/A09/001, Billet Book, 1741,
     March-May. © Coram.   312

8.5   Portrait of Emanuel Mendes da Costa, LMA/4553/01/06/001,
     London Metropolitan Archives, City of London Corporation.   322

8.6   Monument of John, Second Duke of Montagu by Louis-François Roubiliac,
     St Edmund's Church, Warkton, Northamptonshire, 1752. Photograph by
     Ian Benton.   326

8.7   Louis-François Roubiliac, Monument of John, Second Duke of Montagu
     by Louis-François Roubiliac, St. Edmund's Church, Warkton,
     Northamptonshire, 1752. Detail of Lady Mary. Photograph by Ian Benton.   327

8.8   Louis-François Roubiliac, Monument to Lady Mary, Duchess of
     Montagu (detail), St. Edmund's Church, Warkton, Northamptonshire, 1752.
     Photograph by Ian Benton.   328

9.1   Tomb of Martin Folkes, Church of St Mary, Hillington, Norfolk,
     Photograph by the Author.   344

9.2   Monument, Lucretia Folkes Betenson, St George's Church, Wrotham, Kent,
     Photograph by the Author.   346

9.3   Monument, Martin Folkes, Westminster Abbey, Image © 2019 Dean and
        Chapter of Westminster.   349

9.4a   Michael Dahl, *Joseph Addison*, oil on canvas, 1719, 40 1/2 in. × 31 1/4 in.
        (1029 mm × 794 mm), © National Portrait Gallery, London.   357

9.4b   Godfrey Kneller, *Joseph Addison*, engraved by John Faber the Younger,
        c. 1733, Image: 12 9/16 × 9 11/16 inches (31.9 × 24.6 cm), Yale Centre for
        British Art, Paul Mellon Collection.   357

9.5a   Nicolas de Largillière (1656–1746), *Voltaire,* oil on canvas, *c.* 1718–24,
        80 cm × 65 cm, Musée Carnavalet, Histoire de Paris, CC Open Access
        license.   359

9.5b   Nicolas de Largillière, *Voltaire*, oil on canvas, *c.* 1724–5,
        © Château de Versailles, Dist. RMN-Grand Palais/Christophe Fouin.   359

9.5c   Close-up of Hogarthian Sketch in Figure 4.3, reversed horizontally.   359

Indeed I never looked upon Martin Folkes Esq. as a very fashionable man, or a man who would do any thing merely to comply with the fashion

—Archibald Bower

…but we are all citizens of the world, and see different customs and tastes without dislike or prejudice, as we do different names and colours

—Martin Folkes to Emanuel Mendes da Costa, 1747

1

Martin Folkes

'That Judicious Gent'

1.1 Introduction

Martin Folkes (1690–1754) was Sir Isaac Newton's disciple, his unique distinction being his presidency of both the Royal Society and the Society of Antiquaries. Astronomer, mathematician, art connoisseur, Voltaire's friend, and William Hogarth's patron, his was an intellectually vibrant world. Folkes served as a governor of the newly established Coram Foundling Hospital and, as an expert in scientific instrumentation, was on the Board of Longitude. A notorious Freemason and Enlightenment freethinker who frequently violated social conventions—his personal motto was '*qui sera sera*'—Folkes was one of the first members of the gentry to marry an actress, Lucretia Bradshaw, whose debut performance in 1696 was in *The Royal Mischief*. He commissioned works by Louis-François Roubiliac, whom his brother William then supervised on the creation of John, Duke of Montagu's monumental family sepulchre at Warkton, Northamptonshire, the gestures of the grieving widow adopted from theatrical poses and tropes.[1] When Lewis Theobald was preparing his definitive edition of Shakespeare's works, Folkes lent him his First Folio from his vast library for editorial concordance and answered his queries about Elizabethan coinage, as Folkes wrote one of the first treatises on England's numismatic history.[2]

As a member of the Académie Royale des Sciences, a frequent participant in French salon culture, and member of the Florentine Masonic Lodge, Folkes also continued the Newtonian programme in mathematics, optics, and astronomy in the Royal Society and on the Continent, literally 'taking Newton on tour'. Folkes's Grand Tour from 1732/3 to 1735 established his reputation as an international

---

[1] David Bindman and Malcolm Baker, *Roubiliac and the Eighteenth-Century Monument: Sculpture as Theatre* (London: Paul Mellon, 1995); Tessa Murdoch, 'Roubiliac as an Architect? The Bill for the Warkton Monuments', *Burlington Magazine* 122, 922 (January 1980), pp. 40–6. Philip Lindley, 'Roubiliac's Monuments for the Second Duchess of Montagu and the Building of the New Chancel at Warkton in Northamptonshire', *The Volume of the Walpole Society* 76 (2014), pp. 237–288.
[2] John Nichols, ed., *Illustrations of the Literary History of the Eighteenth Century*, 8 vols (London: For the author, 1817), vol. 2, pp. 618–20 and p. 732. The First Folio is now in the John Rylands Library in Manchester. For Folkes's work on coins see Hugh Pagan, 'Martin Folkes and the Study of English Coinage in the Eighteenth Century', in R. G. W. Anderson et al., *Enlightening the British: Knowledge, Discovery and the Museum in the Eighteenth Century* (London: The British Museum Press, 2003), pp. 158–63.

Martin Folkes (1690–1754): Newtonian, Antiquary, Connoisseur. Anna Marie Roos, Oxford University Press (2021).
© Anna Marie Roos. DOI: 10.1093/oso/9780198830061.003.0001

broker of Newtonianism as well as the overall primacy of English instrumentation to Italian virtuosi, his large number of far-flung social contacts engendering scientific creativity in the Royal Society. His work challenges the long-standing, mistaken impression among scholars that the organization was in decline in the eighteenth century. Analysing Folkes's activities as a 'scientific statesman' both abroad and in London as Royal Society Vice President and President refines our definition of Newtonianism and its scope in the early eighteenth century, elucidating and reclaiming the vibrant research programme that he promoted in the period of English science least well understood between the age of Francis Bacon and the present.

Although Folkes had instructed that upon his death his papers should be destroyed, this eradication of material was only partially achieved.[3] Folkes's travel diary, extensive (if scattered) manuscripts, correspondence with fellow antiquaries such as William Stukeley, and numismatic works permit a reconstruction through Folkes's eyes of what it was like to be a collector and patron, a Masonic freethinker, and an antiquary in the period before natural philosophy became sub-specialized. It was not until Folkes's death in 1754 that the Antiquaries, with their interests in material culture and history, and the Royal Society as a 'scientific' institution began to go their separate ways. This book's analysis of his life and letters thus forces us, in a larger sense, to re-examine the disciplinary boundaries between the humanities and 'sciences' in early Georgian Britain and to reassess to what extent the separation of these 'two cultures' developed in this era of 'Enlightenment'. This biography thus places Folkes's work in the larger historical context of his interinstitutional presidential roles, analysing the intellectual, social, and cultural relationships between natural philosophy, antiquarianism, collecting, and connoisseurship.

1.2 'Scientific' Antiquarianism

Although Folkes was perhaps the best-connected and most versatile natural philosopher and antiquary of his age, an epitome of Enlightenment sociability, he has been a surprisingly neglected figure in the literature. The long shadow of Newton—Folkes's patron and hero—has tended to obscure those who followed him. Folkes also promoted an admixture of antiquarianism and natural philosophy in the Royal Society, subsequent scholars considering his actions 'to be an adulteration of science, a body of knowledge based on new things, not old

---

[3] George S. Rousseau and David Haycock, 'Voices Calling for Reform: The Royal Society in the Mid-Eighteenth Century: Martin Folkes, John Hill and William Stukeley', *History of Science* 37, 118 (1999), pp. 377–406, on p. 380.

words'.[4] What would being President of a society dedicated to the material past have to do with leading a society dedicated to natural philosophy? In the eighteenth century, the ability to observe nature was thought to make natural philosophers better suited to understanding the empirical details of ancient artefacts and how they were created. Artefacts were not only empirically observed to a high degree of detail but, in contrast to the practices of modern conservation, sometimes reused in experiments to better understand their original purpose, or even refined to incorporate new knowledge.[5] Naturalists, experimentalists, chymists, and physicians of the early modern British world were also archivists and antiquaries, and their work in the latter sphere was central to prosecuting their work in the former. The development of this 'scientific' antiquarianism which Folkes encouraged, one of the major themes of this book, with its emphasis on measurement, careful observation, and consideration of historical context, anticipated the development of archaeology.

Indeed, in his own work, Folkes also represented a characteristic paradox of his era: an 'ancient' in his cultural sympathies and affinities, yet a 'modern' in the analytical and scientific techniques he applied to antiquarian and natural philosophical study. This seeming paradox requires a subtler interpretation of his interests and of antiquarianism than one solely relating his work and administrative priorities to the battle of the books or the respective virtues of ancient versus modern learning. Both ancient and modern approaches had their utility; for example, we will see in chapter five that Folkes used elements of his collection of antiquities to measure the Roman 'foot' in building and to assess the fidelity of Renaissance architect Jacopo Sansovino (1486–1570) to the Vitruvian Canon. Folkes incorporated techniques and affinities from antiquarianism— natural history and landscape—as well as the 'new science'—engineering principles, metrology, and empiricism.

As Woolf has shown, two impulses or practices were at the heart of early modern English antiquarianism.[6] The first stemmed from the humanist tradition, inherited from Continental philologists like Guillaume Budé (1467–1540) and their Italian predecessors such as Lorenzo Valla (1406–57). English intellectuals in this group, such as John Leland (1503–52), analysed the etymology of words

---

[4] Vera Keller, Anna Marie Roos, and Elizabeth Yale, 'Introduction', *Archival Afterlives: Life, Death, and Knowledge-Making in Early Modern British Scientific and Medical Archives* (Leiden: Brill, 2017), p. 1. We might think, perhaps, of Walter Houghton's attempt to 'winnow the "genuine scientist" from those who encouraged only the "dilution and distortion of the scientific mind"'. For a discussion of Houghton, see Craig Ashley Hanson, *The English Virtuoso: Art, Medicine, and Antiquarianism in the Age of Empiricism* (Chicago: University of Chicago Press, 2009), p. 8.

[5] See Simon Werrett, *Thrifty Science: Making the Most of Materials in the History of Experiment* (Chicago: University of Chicago Press, 2019).

[6] David Woolf, *The Social Circulation of the Past: English Historical Culture 1500–1730* (Oxford: Oxford University Press, 2003), pp. 142–50. For other studies of early modern English antiquarianism see Angus Vine, *In Defiance of Time: Antiquarian Writing in Early Modern England* (Oxford: Oxford University Press, 2010).

and sought linguistic and verbal remains to understand the historical record. The second form of antiquarianism, which became more prevalent by the end of the seventeenth century, considered the landscape in analysis of ancient objects and buried artefacts.

The early Royal Society was also involved in projects that integrated natural history and antiquarianism, particularly before the establishment of the Society of Antiquaries in 1717. Antiquarianism was 'cognate with natural history at the very least in the sense that the type of activity involved (field work, collection, display, classification) was of a similar character'.[7] Fellows of the Royal Society applied the methods of the antiquary to natural history, and in particular to the study of fossils and geology. For Robert Hooke the work of a natural historian was similar to that of an antiquary in studying man-made objects. For instance, he remarked,

> There is no Coin can so well inform an Antiquary that there has been such or such a place subject to such a Prince, as these [fossil shells] will certify a Natural Antiquary, that such and such places have been under the Water, that there have been such kind of Animals, that there have been such and such preceding Alterations and Changes of the superficial Parts of the Earth. And methinks Providence does seem to have design'd these permanent shapes. As Monuments and Records to instruct succeeding Ages of what past in preceding [ages].[8]

As Miller has indicated, a prosopographical account of Royal Society Council members in 1750 at the end of Folkes's presidency shows 'an affinity between antiquarian interests and natural historical ones. There is also a marked tendency for those with such interests to provide accounts of earthquakes, comets, weather, agitation of waters, etc. to the Royal Society, a number of which were published in the *Philosophical Transactions*'.[9] Folkes himself, in addition to his work on metrology and Roman antiquities, wrote a series of reports on *aurora borealis*, interested in their relationship to electricity.

Several members of the Royal Society were also physicians, and the doctor's habit of interpreting symptoms seems to have made them sensitive to visual evidence; after all, 'like artists, they were trained observers'.[10] Visuality was also intimately related to *autopsia*; not only in the post-mortem examination, but also

---

[7] David Miller, '"Into the Valley of Darkness": Reflections on the Royal Society in the Eighteenth Century', *History of Science* 27, 2 (1 June 1989), pp. 155–66, on p. 160.

[8] Robert Hooke, *The Posthumous Works of Robert Hooke . . . containing his Cutlerian Lectures, and other discourses read at the Meetings of the Illustrious Royal Society . . . Publish'd Richard Waller* (London: Samuel Smith and Benjamin Walford, 1705), p. 321.

[9] Miller, 'Into the Valley of Darkness', p. 166, footnote 24.

[10] Peter Burke, 'Images as Evidence in Seventeenth-Century Europe', *Journal of the History of Ideas* 64, 2 (2003), pp. 273–96, on p. 294.

in the production of visual images that 'served well for material description'.[11] In the post-Baconian world, a high trust was placed 'in the efficacy of observing the so-called advancements of learning—and of recording these observations in pictorial form'. Reading clues to make diagnoses also ripened early modern physicians' abilities to understand and contextualize the empirical details of ancient artefacts and the processes by which they were created. There were thus multiple interconnections between viewing, understanding, and knowing.

At the same time, the trained eye of physician-collectors meant that they could 'anatomize' and question visual interpretation, as visuality and seeing were also debated topics. Although in Aristotelian philosophy vision was considered one of the most secure means into knowledge (*opsis* and *autopsia*), seeing could be associated with deception, 'mere' appearance, and false perception. Training the eye was thus paramount for the physician-collector and for the connoisseur. Although not a doctor, Folkes, with his keen visual sense developed as both a natural philosopher and a patron of the arts, was fascinated with fake antiquarian and aesthetic objects, optical illusions, and false self-fashioning in portraits, even at one point having himself painted by the portrait artist John Vanderbank to look like his hero Newton (see chapter five). Thus, although recent analysis of English virtuosity claims that antiquarianism and nascent archaeology were fundamentally bound up with an appreciation of the classics, the empiricism, visuality, and *autopsia* of the physician-collector represents a fundamentally different set of skills from those used in philology.[12] Sweet has also drawn attention to the contribution of eighteenth-century antiquaries and classicists to the development of archaeology, but it seems that we should also look among the early modern physicians and natural philosophers to trace the discipline's origins.[13]

Physician-collectors in the Royal Society, like Hans Sloane, also had long developed extensive antiquarian interests. The antiquarian interests of physician-collectors served not only as a means to allow physicians to deal with the contents of already extant collections or to distinguish themselves from less learned collectors. They may also have offered philological tools informing new collecting practices. Via the Baconian discipline of the history of letters (*historia literaria*), antiquarianism offered a quickly developing body of theorizing about categorization, standardization, citation, and information management. Physicians developed newly standardized and professionalized techniques of categorizing and organizing collections and connecting the study of the natural world with antiquity.

---

[11] Benjamin Schmidt, *Inventing Exoticism: Geography, Globalism, and Europe's Early Modern World* (Philadelphia: University of Pennsylvania Press, 2015), p. 127.

[12] See Hanson, *The English Virtuoso.*

[13] Rosemary Sweet, *Antiquaries: The Discovery of the Past in Eighteenth-Century Britain* (London: Hambleton, 2004).

Use of material culture to reconstruct the past was thus inherent to antiquarianism and to natural philosophy at the Royal Society. Fellows attended experimental demonstrations from which axiomatic principles were formulated via discursive practice; members could also access this material by the reading of reports of experiments.[14] Societies allied to the Antiquaries and the Royal Society, such as the Egyptian Society (1741–3), to which Folkes and other Fellows belonged, had the same methodology: to discover principles of sociocultural and religious practices in Egypt, as well as to provide 'object biographies' of artefacts that belonged to members, for instance ancient enamels and pigments, which they examined with speculations on their manufacture and use, as well as the practice of mummification.

Natural philosophy and antiquarianism were thus seen as intimately related; the Society of Antiquaries and the Royal Society had many common members and held their meetings on the same day. As we will see, there may have been an attempt to unite the two groups into one organization; amalgamation would have made sense financially to the mathematically minded Folkes, who kept a tight rein over the Royal Society's finances.[15] As naturalist and detractor John Hill, famous for his attacks on Folkes and the Royal Society, stated in his 'Of Antiquities Commemorated in the Transactions of the Royal Society':

> The Members of this illustrious Society would perhaps have been larger on this Head, if they had intended to have set themselves up as Rivals to a certain other Society; an Event which perhaps nothing could have frustrated but the judicious Contrivance of making the same People Members of both.[16]

Papers on antiquarian and archaeological subjects indeed comprised between six and seven per cent of articles in the *Philosophical Transactions*, particularly as the Antiquaries' own journal *Archaeologia* was not launched until 1770. Consolidation reflected the simple reality that natural philosophy and antiquarianism were perceived to be of the same intellectual enterprise until Folkes's death in 1754.

1.3 The Vibrancy of the Royal Society

Scholarly analysis of attempts at consolidation of the two societies, however, has tended to focus upon Folkes's critics such as William Stukeley, who censured him due to his lack of belief in organized religion, as well as members of the

---

[14] Mark Thomas Young, 'Nature as Spectacle: Experience and Empiricism in Early Modern Experimental Practice', *Centaurus* 59 (2017), pp. 72–96, on p. 91.

[15] RS/MS/645, Royal Society Library, London; Rousseau and Haycock, 'Voices Calling for Reform'.

[16] John Hill, *A Review of the Works of the Royal Society* (London: R. Griffiths, 1751), part 2, p. 47, as noted by Miller, 'Into the Valley of Darkness', p. 159.

Antiquaries who did not belong to the Royal Society and who feared being marginalized.[17] These contemporary critiques, in turn, led to the mistaken impression among some historians that the Royal Society in the first half of the eighteenth century was moribund, a premise this book will strenuously challenge.[18] As a result, the eighteenth-century Presidents of the Royal Society serving between the two major naturalists Sir Hans Sloane (1660–1753) and Sir Joseph Banks (1743–1820) are little understood, and few modern biographical studies have been written about them.[19] Fara has noted that the central years of English Science from 1740 to 1770 'remain an under-researched period, with very few studies of social networks of experimenters'.[20] Sorrenson has noted that these unedifying impressions of the Royal Society during this period also stemmed from the belief that the institution failed to match the standard that Newton had set in the *Principia*; it also stemmed from negative opinions voiced by the Society's critics such as John Hill, and from the popularity of Babbage's *Reflection on the Decline of Science in England* (1830).[21] As an example, Rousseau's survey of English science in the eighteenth century relies heavily on satirists such as Swift,

[17] The connections between the two organisations were analysed by Rousseau and Haycock, although my interpretation of Folkes in this book will be substantially different. See Rousseau and Haycock, 'Voices Calling for Reform', pp. 377–406.

[18] See for instance, John Heilbron's *Elements of Early Modern Physics* (Berkeley, Los Angeles and London: University of California Press, 1982) or his *Physics at the Royal Society During Newton's Presidency* (Los Angeles: William Andrews Clark Memorial Library, 1983), or for a critique of Heilbron's view see Miller, 'Into the Valley of Darkness', pp. 155–66.

[19] For instance, for Folkes, there is not a full biography, but only a series of articles. For a brief treatment of Folkes, see Charles Weld, *A History of the Royal Society*, 2 vols (London: J. W. Parker, 1848). See also: W. Johnson, 'Aspects of the Life and Works of Martin Folkes (1690–1754)', *International Journal of Impact Engineering* 21, 8 (1998), pp. 695–705; David Boyd Haycock, '"The Cabal of a Few Designing Members": The Presidency of Martin Folkes, PRS, and the Society's First Charter', *Antiquaries Journal* 80 (2000), pp. 273–84; Howard M. Lenhoff and Sylvia G. Lenhoff, 'Abraham Trembley and his Polyps, 1744: The Unique Biology of Hydra and Trembley's Correspondence with Martin Folkes', *Eighteenth-Century Thought* 1 (2003), pp. 255–80; William Eisler, 'The Construction of the Image of Martin Folkes (1690–1754) Part I', *The Medal* 58 (Spring 2011), pp. 4–29; William Eisler, 'The Construction of the Image of Martin Folkes (1690–1754) Art, Science and Masonic Sociability in the Age of the Grand Tour Part II', *The Medal* 59 (Autumn 2011), pp. 4–16; George Kolbe, 'Godfather to all Monkeys: Martin Folkes and his 1756 Library Sale', *The Asylum* 32, 2 (April–June 2014), pp. 38–92; Anna Marie Roos, 'Taking Newton on Tour: The Scientific Travels of Martin Folkes, 1733–35', *The British Journal for the History of Science* 50, 4 (2017), pp. 569–601. Folkes's role in the Copley medal is discussed here: Rebekah Higgitt, '"In the Society's Strong Box": A Visual and Material History of the Royal Society's Copley Medal, *c*.1736–1760', *Nuncius: Journal of the Material and Visual History of Science* 34, 3 (2019), pp. 284–316.

[20] Patricia Fara, '"Master of Practical Magnetics": The Construction of an Eighteenth-Century Natural Philosopher', *Enlightenment and Dissent* 14 (1995), pp. 52–87, on p. 56. This lacuna has been partially remedied with the scholarship of Larry Stewart, and by Rebekah Higgitt concerning Rev. Dr Nevil Maskelyne. See, for instance, Rebekah Higgitt, 'Equipping Expeditionary Astronomers: Nevil Maskelyne and the Development of "Precision Exploration"', in F. MacDonald and C. W. J. Withers, eds, *Geography, Technology and Instruments of Exploration* (Basingstoke: Ashgate, 2015), pp. 15–36.

[21] Richard Sorrenson, 'Towards a History of the Royal Society in the Eighteenth Century', *Notes and Records of the Royal Society of London* 50, 1 (1996), pp. 29–46.

Steele, and Hill to conclude that the Society 'from 1680 to 1780 fared ingloriously, except during Newton's presidency'.[22]

It seems, however, that the Royal Society in the eighteenth century manifested considerable strengths in technological innovation, setting new standards in experimental precision and refined instrumentation, with Folkes at the centre of such activity.[23] Even Heilbron, who termed Folkes's presidency as 'an antiquarian-dominated nadir', did admit that 'experimentation at the Society revived somewhat with the popularity of electricity', with Stephen Gray's work creating a new field of electrical conduction.[24] In 1727, Stephen Hales published *Vegetable Staticks*; inspired by Newton's speculations in the *Opticks* concerning attractive and repulsive forces between particles, Hales investigated the air that was 'fixed' in plants, helping to lay the foundations of pneumatic chemistry. Folkes encouraged the electrostatic work of Benjamin Wilson and William Watson in the Royal Society, embracing new investigations in static electricity, primarily with the use of insulators. These experiments were usually in a Newtonian cast, as Folkes was told by Newton that the future of natural philosophy was in the investigation of electrical forces.

Folkes also promoted the study of applied natural philosophy, such as the work of Benjamin Robins in ballistics and research into navigation. As Fara commented,

> Eighteenth-century natural philosophers perceived navigation to be a key area for establishing credibility in a commercial community whose wealth depended on maritime trade. Many of them turned their attention to inventing and promoting devices for making ships more seaworthy, and navigational techniques more reliable. The Royal Society had close links with Christ's Hospital, founded as a mathematical school for training naval officers. As merchants and manufacturers increasingly demanded a new style of education, many lecturers in natural philosophy shared classrooms with teachers of navigation and commerce.[25]

---

[22] George Rousseau, 'The Eighteenth Century: Science', in Pat Rogers, ed., *The Eighteenth Century*, The Context of English Literature Series (London: Holmes & Meier, 1978), p. 158.

[23] Sorrenson, 'Towards a History', pp. 30–1. See also Geoffrey Cantor, 'The Rise and Fall of Emanuel Mendes da Costa', *English Historical Review* 116 (2001), pp. 584–603. For other scholars that have challenged this declinist narrative, see: Andrea Rusnock, 'Correspondence networks and the Royal Society, 1700–1750', *British Journal for the History of Science* xxxii (1999), pp. 155–69; Richard Sorrenson, 'George Graham, Visible Technician', *British Journal for the History of Science* xxxii (1999), pp. 203–21; Palmira Fontes da Costa, 'The Culture of Curiosity at the Royal Society in the First Half of the 18th Century', *Notes and Records of the Royal Society of London* 56, 2 (2002), pp. 147–66; Anita Guerrini, 'Anatomists and Entrepreneurs in Early 18th-Century London', *Journal of the History of Medicine and Allied Sciences* 59, 2 (2004), 219–39; Palmira Fontes da Costa, *The Singular and the Making of Knowledge at the Royal Society of London in the 18th Century* (Newcastle-upon-Tyne: Cambridge Scholars Publishing, 2009).

[24] Heilbron, *Physics at the Royal Society*, p. 4, p. 40, note 114 and Miller, 'Into the Valley', p. 165, note 12. This is also a point made by Lenhoff and Lenhoff, 'Abraham Trembley and his Polyps', p. 266.

[25] Fara, 'Master of Practical Magnetics', pp. 55–6.

For these reasons, Folkes was a fervent patron of astronomers and navigational experts like James Bradley, who became Astronomer Royal, having in 1748 'published the results of nearly 20 years of nightly observations with a highly precise zenith sector made by the mathematical instrument maker and horologist George Graham that established the nutation of the earth's axis'.[26] In 1721, Folkes, when on the Royal Society Council, had supported Bradley for the Savilian Professorship of Astronomy at Oxford. One of the reasons Folkes succeeded in Bradley's election was that William Wake, Folkes's Uncle and Archbishop of Canterbury, also endorsed Bradley for the post.[27] Bradley was probably introduced to Folkes and Wake, as well as Newton and the powerful Whig factions that promised future openings to various offices of state, through his wealthy uncle the Reverend James Pound.[28] Folkes was also a champion of the horologist George Graham (FRS 1720), inventor of the dead-beat escapement, possessing several of his timepieces that he gave as presents to foreign dignitaries. Graham reported the 'results of his thrice-daily, year-long observations that tracked the behaviour of the magnetic dipping needle, even while he was also in the midst of a three-year daily investigation of his new heat-invariant pendulum', that proved so crucial to Maupertuis's expedition to measure the circumference of the Earth.[29] As we will see, Folkes helped secure the instruments that made this expedition possible.

Folkes also could spot and support significant talent in those natural philosophers from humble backgrounds. The mathematician Colin Maclaurin discovered the mechanical abilities of James Ferguson (1710–76), the son of a Scottish tenant farmer, and introduced him to Folkes; Ferguson eventually became a Fellow of the Royal Society. Although his formal education was limited to three months in grammar school, Ferguson was a skilled horologist and designer of clocks, astronomical devices, and planispheres, becoming a renowned public speaker and promoter of Newtonianism and publishing *Astronomy Explain'd upon Sir Isaac Newton's Principles* (1756), which went through numerous editions. Along with eleven other members of the Royal Society that were on the Board of Longitude, Folkes also recommended that John Harrison be encouraged in his attempts to build his chronometer that would secure the Longitude Prize,

---

[26] Sorrenson, 'Towards a History', p. 30.

[27] Martin Folkes to William Wake, 4 September 1721, Wake Letters 16, no. 82, Christchurch Archives, University of Oxford.

[28] John Fisher, 'Conjectures and Refutations: The Composition and Reception of James Bradley's Paper on the Aberration of Light with Some Reference to a Third Unpublished Version', *British Journal for the History of Science* 43, 1 (March 2010), pp. 19–48, on p. 46. As Fisher noted, 'Pound had been a wealthy man with two lucrative church livings at Wanstead and Burstow, a private income independent of these sources and a large dowry that came with his second wife Elizabeth who was the sister of Matthew Wymondsold, a wealthy speculator in South Sea stock who had sold all of his holdings one day before the bubble burst.'

[29] Sorrenson, 'Towards a History', p. 30.

and discouraged those submitting other solutions to the Royal Society. When Folkes presented Harrison with the Royal Society's Copley Medal in 1749, an award that Folkes himself created to reward scientific endeavour, Folkes stated:

> Mr Harrison, who, before he came to this Town lived in a place called Barrow in the County of Lincoln, not far from Barton upon Humber, was not originally brought up to the business he now professes, tho' he was afterwards directed to it, by curiosity and inclination, and by the strong impulses of a natural and uncommon genius, but such a one, as has been sometimes found, instances also, to carry those who have been possessed of it, much further than they could have been led by the most elaborate precepts and rules of art.[30]

Although the uneducated Harrison must 'have seemed utterly foreign to the admirals, churchmen, natural philosophers and mathematicians who were interested in the longitude problem', Folkes recognized Harrison's rare ability.[31] Even William Stukeley who visited Harrison in 1740 at his home in Barrow to see his clocks' 'movements to a perfection surprizing', (probably H1 or H2) called him a 'wonderful genius' who would 'carry the prize of longitude found out'. Harrison and Stukeley, both Lincolnshire men, had met fifteen years previously when Stukeley was carrying out excavations to discover the remains of Humber Castle. Stukeley, like Folkes, evinced a combined interest in natural philosophy and antiquarianism.[32] In addition to his more well-known work on Stonehenge and the ceremonial landscape, Stukeley presented papers at the Royal Society on gout, electricity, and his invention of a new type of carriage.

1.4 Newtonianism, and the Heritage of the Royal Society

Most importantly, this book will analyse the significant degree to which antiquarianism in the Royal Society and Antiquaries, due to Folkes's influence, was simply a continuation of the Newtonian programme of the 'physical sciences' in the Royal Society. Newton was one of Folkes's mentors, and their close relationship in the last years of Newton's presidency in the Royal Society meant that there is no extant correspondence between them. Instead, Folkes was a go-between and emissary for Newton, and then later for Newtonianism (see chapters two and five).

---

[30] RS/JBO/10/184, Royal Society Library, London.

[31] Jim Bennett, 'James Short and John Harrison: Personal Genius and Public Knowledge', *Science Museum Journal* 2 (Autumn 2014), http://dx.doi.org/10.15180/140209.

[32] William Stukeley, *Memoirs of the Royal Society*, vol. 2, ff. 30–1, Spalding Gentlemen's Society, Spalding, Lincolnshire.

In his early career, Folkes was known for his mathematical and astronomical work, switching to antiquarianism and archaeology in his later intellectual life. Folkes's contributions to mathematics and astronomy have been overlooked due to Babbage's claim that mathematical sciences were not in evidence in the Royal Society in the eighteenth century. There was in fact a nexus of important mathematicians, including Abraham de Moivre and Colin Maclaurin, as well as several of their students. Folkes accomplished innovative work in mathematical optics and probability and made significant observations of solar eclipses and the *aurora borealis*. Folkes and Philip, 2nd Earl of Stanhope together studied higher plane curves and gravitational attraction of ellipsoids, of interest to Newtonians, and algebraic analysis, of increasing relevance to eighteenth-century mathematicians.[33]

Folkes's Newtonianism extended to his co-editorship of Newton's *Chronology of Ancient Kingdoms Amended* (1728). Newton's work came from his interests in antiquity that bore on the origin of civilization; he used astronomical records in Hipparchus' *Commentary* to create his timeline of human history. In particular, he used the periodicity provided by the precession of the equinoxes to date historical observations of the heavens. The most recent large-scale publication about the *Chronology* by Buchwald and Feingold, *Newton and the Origin of Civilization*, only briefly analyses Folkes's role as co-editor, making this topic of current historiographic interest.[34] Most of Newton's chronology was based upon an attempt to re-date the history of the Greek, Egyptian, Assyrian, Babylonian, and Persian Empires. For example, in reworking data that he took from the Hellenistic Commentary of Hipparchus, Newton sought to identify historical colures. (A colure is the meridian that passes through the poles of the celestial sphere and cuts the sun's apparent path through the heavens at the points that it has reached at the solstices (or, alternatively, the equinoxes).) The region of the heavens identified by such a colure shifted over time due to the precession of the equinoxes, and thus its description could in theory generate precise dates.

Folkes, as a Masonic freethinker, furthered Newton's interests in chronology to pursue his own interests in astronomy, timekeeping, and antiquarianism. Folkes used the same Newtonian rationale when he produced a stereographic map of the Farnese Globe, the oldest surviving depiction of this set of Western Calculations.[35] Richard Bentley published Folkes's map in his edition of Manilius's *Astronomia* in

[33] David Bellhouse, 'Lord Stanhope's Papers on the Doctrine of Chances', *Historia Mathematica* 34 (2007b), pp. 173–86. The papers are 2nd Earl Lord Stanhope–Folkes letters U1590/C21, Kent Record Office, Canterbury.
[34] Jed Buchwald and Mordechai Feingold, *Newton and the Origin of Civilization* (Princeton: Princeton University Press, 2013). But see Cornelis J. Schilt, 'Manuscripts and Men: The Editorial History of Newton's Chronology and Observations', *Notes and Records: The Royal Society Journal of the History of Science* 74, 3 (2020), pp. 387–408.
[35] Kristen Lippincott, 'A Chapter in the "Nachleben" of the Farnese Atlas: Martin Folkes's Globe', *Journal of the Warburg and Courtauld Institutes* 74 (2011), pp. 281–99.

1739. Folkes measured the Globe when he was in Rome and made a plaster cast which he exhibited to the Society of Antiquaries on 7 July 1736, arguing that the declination of the Arctic and Antarctic Circles corresponded to a particular latitude for the observer whose observations were adopted by the sculptor; a detailed analysis of the Globe would reveal the latitude and epoch for the observations incorporated in the globe. Folkes concluded the Globe dated to the time of the Antonine emperors.

Folkes's interest in chronology was later a topic of great import as he, as Royal Society President, and James Bradley, by this time Astronomer Royal, checked the legislative draft bills for the Gregorian Calendar. As the Royal Society was also heavily involved in the 1752 reform of the Calendar, we will also argue that Folkes's penchant for mathematics, astronomy, and antiquarianism simply found its métier during his presidency, following Newtonian precepts. In fact, Folkes's antiquarian publications on numismatics, such as his *Tables of English Silver and Gold Coins*, and his pronounced interest in metrology could be seen as a reflection of Newton's interest in the reform of the coinage when an officer of the Mint.[36] Fellows of the Royal Society were also deriving metrological standards and tables of specific gravities of metals, so these concerns dovetailed in Folkes's work on the English coinage.

As mentioned, Folkes's natural philosophical work also had ties with intellectual traditions of Newtonianism and natural philosophy in Italy and France that have been little explored.[37] Folkes made two continental tours, and left a detailed travel journal of his 'Journey from Venice to Rome' in the 1730s.[38] In Venice, he recreated Newton's experiments in optics for an invited and sceptical audience at the Palazzo Giustiniani, and he made contact with Anders Celsius (1701–44) and arranged for Celsius to obtain precision scientific instruments used in his expedition with Pierre Louis Maupertuis (1698–1759) to Lapland to measure the shape of the Earth. Their purpose was, again, Newtonian: to confirm Newton's determination by calculation that the Earth was slightly flattened or oblate. During his time in Italy, Folkes also demonstrated Newtonian optics to Florentine virtuosi such as Antonio Cocchi (1695–1758), and was made a member of the Florentine Academy. These ties with Italian natural philosophers were deepened when Folkes assumed the vice presidency and presidency of the Royal Society and Society of Antiquaries, and were mediated by Sir Thomas Dereham (1680–1739), a Jacobite Catholic, and the Royal Society's chief correspondent in Italy. Although Folkes was avowedly a Whig, and he supported his younger brother William in his unsuccessful 1747 run as a Whig against Sir

---

[36] Martin Folkes, *A Table of English Silver Coins from the Norman Conquest to the Present Time* (London: Society of Antiquaries, 1745).

[37] Roos, 'Taking Newton on Tour', pp. 569–601.

[38] Bodl. MS Eng. misc. *c.* 444, Bodleian Library, Oxford.

John Turner, Folkes mixed with those from a variety of political backgrounds, more interested in the merits of the individual than their political allegiance or social standing.[39]

Folkes's second tour included a visit to France in May 1739, where he met Madame Geoffrin, hostess of an illustrious Parisian salon where he was exposed to Abraham Trembley's (1710–84) work on the regeneration of hydra or polyps. Work on polyps challenged the Aristotelian Chain of Being, which had dominated understanding of the natural world and raised questions of vitalism, materialism, and deism.[40] For instance, did the polyp have a soul, and could that soul be split into two? Was the ability of the polyp to regenerate support for preformation, the idea that all living things pre-existed as invisible germs? Folkes's correspondence about hydra ranged far and wide, from exchanges with Georges-Louis Leclerc, Comte de Buffon, René Antoine Ferchault de Réaumur, and Trembley, as well as in heretofore unpublished letters to his friend and patron, Charles Lennox, 2nd Duke of Richmond and Duke of Aubigny (1701–50), who was busily dissecting polyps of his own.[41] Hydra took centre stage in discussions of biological taxonomy and metaphysics, and the exchange of specimens served as a form of patronage and scientific gift giving, establishing intellectual networks.

Folkes and microscopist Henry Baker (1698–1774)[42] discussed similar questions about vitalism and preformation with specific reference to parasitic twins arising from birth defects, which lived in a primitive fashion on the host's body. Under Baker's influence, Folkes also made the study of microscopy dominant in the Royal Society, writing a treatise about the bequest of microscopes to the Royal Society by the sister of Antonie van Leeuwenhoek.[43] Folkes wrote the treatise to help preserve the Society's history, part of his vision as an antiquary and natural philosopher to restore the Society's Repository Museum Collection, Library, and early letters and manuscripts.

In Folkes's era, we see an increasing interest in preserving private papers for posterity by depositing them in public archives and publishing them posthumously, and antiquaries were central to this process.[44] The natural philosopher

---

[39] MS/MC/50/6, 503 x 3, 1745–7, Letters to William Folkes, Norfolk Record Office, Norwich.

[40] Mark Ratcliff, *The Quest for the Invisible—Microscopy in the Enlightenment* (Aldershot: Ashgate, 2009).

[41] RS/MS/865/32, Royal Society Library, London.

[42] Baker was elected FRS in 1740 and to the Society of Antiquaries of London, and was a recipient of the Copley Medal in 1744 for his microscopic observations of salts and a frequent contributor of papers to the *Philosophical Transactions*.

[43] 'Some Account of Mr Leeuwenhoek's curious microscopes lately presented to the Royal Society', by Martin Folkes, 1723, RS/Cl.P/2/17, Royal Society Library, London, also published in *Philosophical Transactions* 32, 380 (1723), pp. 446–53.

[44] Some of this material is adapted from Vera Keller, Anna Marie Roos, and Elizabeth Yale, 'Introduction', *Archival Afterlives*, pp. 1–28. For changing attitudes towards manuscript from the sixteenth until the eighteenth century, see Richard Serjeantson, 'The Division of a Paper Kingdom: The Tragic Afterlives of Francis Bacon's Manuscripts', in *Archival Afterlives*, pp. 29–71.

emerges as a public individual through his instantiation in posthumously printed editions and in an archive, particularly through the transfer of private collections into public hands. The sense that some private papers are worth preservation in public institutions (such as the Ashmolean Museum and the Royal Society Archive) and even preservation by the State, emerged in the history of science in the eighteenth century—before it arose in the nineteenth and even the twentieth centuries in the history of literature.

As they collected, transferred, discussed, and published documents, virtuoso antiquaries found ways of authenticating their documents and thus rendering them, if not authoritative sources, at least starting points for further discussion and experimentation. At the beginning of the seventeenth century, manuscript collectors sought signs of the reliability of a text rather than a document. Frequent destruction of original manuscripts illustrates what little regard the holograph manuscript attracted, other than as an indication of the text's authenticity. But by the time of the eighteenth century, this attitude was changing. Alongside their antiquarian interests, Fellows of the Royal Society during Folkes's presidency also engaged in collecting the scientific works of their own recent past. Naturalists applied bibliographic tools developed for the study of antiquity and more distant history to papers of the relatively recently deceased, increasingly allocated attention to handwriting (looking for holograph materials), to provenance, and to copying and recopying marks of authenticity, such as a particular writer's monogram. There was also an increasing emphasis on the recording of life events of Fellows and ritual remembrance. The Anniversary meetings of the Royal Society record and announce the names of Fellows who died over the past year, a custom introduced by Folkes on 30 November 1743.

Folkes also actively encouraged Fellows to pass along manuscript materials in their possession that elucidated the history of the Royal Society. The naturalist Reverend Dr Henry Miles (1698–1763), in a letter to Henry Baker of 13 March 1744/5, heeded Folkes's call by searching for the papers of early Fellows Henry Oldenburg (1619–77) and John Beale (1608–83).

> I am particularly obliged to our most worthy president [Folkes] for his complements…let him know that I have not forgot my promise of preparing some letters of Mr Oldenburgh's &c to be lodgd in the Library of the R.S. those of Mr Oldenburg's being long since pasted in a Folio Chart book and would have been sent before the meeting of the Society in October but that I had unhappily mislaid a paper, without which I could not finish my design and a variety of interruptions have prevented my adding those of Dr Beale and some others: many being writ on all sides, and so near the edge, that I have found it very difficult without taking up more time than my necessary affairs would allow, to make any considerable progress: however I have taken care that they will not be

lost to the Society, if I should die, and I hope soon to finish and present them my self.[45]

Oldenburg was of course the most eminent secretary of the Royal Society in the seventeenth century, and Beale, a Somerset divine and early Fellow of the Royal Society, was elected in 1663. Miles was a Fellow of the Society of Antiquaries of London and Royal Society with some pronounced scientific interests, including microscopic work with rotifers, and he was one of the first to discover that an electric spark could kindle certain substances (most notably phosphorus). Miles is, however, primarily known as the editor and transcriber (with Thomas Birch) of Robert Boyle's manuscripts that were purchased from Boyle's apothecary. Indeed, the Oldenburg–Beale correspondence is in the present Royal Society Library not an 'official' part of the Early Letters Collection, but an additional guard-book, so it seems that Miles had made good on his promise to Baker and Folkes.[46] Baker also noted to Miles that Folkes was 'obliged for your Care in preserving the Letter to Mr Boyle'.[47]

Lest it be thought that past Royal Society manuscripts and memorabilia were completely totemic during Folkes's presidency, Miles and Birch were also known for throwing away many of Boyle's private papers after his death, having dismissed them as inconsequential. This reflects a common eighteenth-century editorial practice of eliminating materials that could be construed as non-scientific, embarrassing, or too personal. Folkes himself in his role of editor of the biblical chronology of Newton also largely sanitized it of any heretical content, and 'the material that Newton had been preparing to publish was deliberately shorn of most of its context in the history of ancient religion'.[48] Folkes was more interested in preserving Newton's and his own public reputation and avoiding any taint of anti-Trinitarianism, than in any personal religious scruples (see chapter four).

In the spirit of what Werrett has called 'thrifty science', even older objects from the Royal Society's earliest heritage could also be repurposed, such as Christiaan Huygens's 210-foot aerial telescope, which he presented to the Society in 1692.[49] James Bradley, in describing the 'Glass and old Furniture of Mr Huggen's [sic] large Telescope', asked the Royal Society 'to Accept of Such new additions and improvements which his Uncle [Mr James Pound] had made to the Furniture &

[45] John Rylands Library, Henry Baker Correspondence, Eng MSS 19, vol. 2, ff. 19–20.
[46] Email from Keith Moore, Royal Society Librarian to Anna Marie Roos, 14 July 2016. The letters between Beale and Oldenburg are separately catalogued as Early Letters OB (Oldenburg Beale). The letters have been arranged chronologically and indexed as stated in a note by Henry Miles dated 20 August 1745.
[47] Henry Miles to Henry Baker, 26 March 1745, John Rylands Library, Henry Baker Correspondence, Eng MSS 19, vol. 2, f. 33.
[48] Scott Mandelbrote, 'Duty and Dominion, Footprints of the Lion: Isaac Newton at Work'. Online exhibition, Cambridge University Library, https://wwwe.lib.cam.ac.uk/CUL/exhibitions/Footprints_of_the_Lion/duty_dominion.html [Accessed 16 July 2016].
[49] Werrett, *Thrifty Science.*

apparatus whilst he was using it Viz. A Curious Micrometer contrived and made by Mr [George] Graham, a new Eye Glass, a new Director to the Sight & a new Tin Tube to carry the Object Glasse'.[50] Likewise, Henry Baker re-engraved the original copperplates for Hooke's iconic *Micrographia* (1665) to create a new edition, the *Micrographia Restaurata* (1745). The microscopes that Leeuwenhoek left behind were locked in the Society's Council Room as they were utilized for experimental observation (see chapter eight). As Werrett has shown, the specialization of instruments and spaces in early modern natural philosophy existed side by side with the ethos of household economy. The idea that 'second-hand' or 'repaired' was inferior was only a nineteenth-century construct, even in natural philosophy, as was the totemic treatment of scientific heritage. So while objects, books, and papers from the Society's early history were preserved by the antiquaries in the Royal Society during Folkes's presidency, sometimes this preservation was for experimental reuse and republication. Antiquarianism in the first part of the eighteenth century had its own distinctive ideas about conservation of the heritage of natural philosophy that mediated between use, reuse, and preservation.

1.5 Folkes's Archival Lives and Afterlives

Ironically, some of the scholarly neglect of Folkes was also due to the fact that a portion of his *own* manuscripts and books were not preserved in one place. While it is impossible to know how many of his own papers or correspondence were destroyed on his instruction, we do know a good proportion of them were scattered due to the sale of the Macclesfield Library at Shirburn Castle in 2004.[51] In 1872, the Royal Commission on Historical Manuscripts documented the papers of Folkes, consisting of four portfolios of letters and five folios of scientific correspondence. The papers survived at Hillington Hall in Norfolk for another sixty years, at which time they were sold, dispersed, and in some cases presumed lost.[52] Scholars have thus assumed that here is little extant material.[53] This assumption is ill founded, particularly due to recently recovered, rich, and largely unexamined materials, the majority to do with Folkes and Royal Society business, intermixed with some personal correspondence.

---

[50] RS/JBO/14/235, 20 June 1728, Royal Society Library, London.
[51] George Parker, 2nd Earl of Macclesfield, became President of the Royal Society from 1752, after Folkes had become debilitated and confined to his house in Queen Square due to a series of strokes.
[52] Alfred J. Horwood, 'The Manuscripts of Sir Wm. Hovell Browne Ffolkes, Bart., at Hillington Hall, co. Norfolk', in *Third Report of the Royal Commission on Historical Manuscripts* (London: HMSO, 1872), appendix, pp. 247–8.
[53] See, for instance, Rousseau and Haycock, 'Voices Calling for Reform', p. 380.

The Royal Society managed to acquire four lots out of a total of fourteen at this 1932 Sotheby's sale.[54] The successful bids were on the items of personal correspondence that now form MS/250 in the Society's archives, while the major loss was the five sets of scientific papers mentioned by the Royal Commission. These contained papers read before the Royal Society during Folkes's presidency and represent a major break in the record sequence known as 'Letters & Papers' within the Society's archives. Manuscript content that would form the printed *Philosophical Transactions* is entirely lacking at the Society for the period from May 1746 to March 1749/50. Fortunately, the Wellcome Trust managed to acquire ninety of these manuscripts from Sotheby's and from subsequent sales, so that the loss to historians was not absolute.[55]

By 1949, R. E. W. Maddison (1901–93) had recovered two folios of Martin Folkes's correspondence from Mr G. Jeffrey's book stall on Farringdon Road.[56] These were remnant or unsold lots from the Sotheby's sale.[57] Maddison eventually donated them to the Society after producing a description and brief listing for *Notes and Records*.[58] Maddison's cache of letters, MS/790 in the Society's collections, contains a broad range of domestic and European scientific correspondence.

These Folkes letters have been recently joined by a third group of closely related documents which originated with another aristocratic patron and close friend, Charles Lennox.[59] There are thirty-six of Lennox's letters to Folkes, corresponding exactly to Sotheby's lot 118. The existence of an older numbering system in ink common to these letters (beginning at 4 and ending at 105), and others of the 'loose' letters from Sotheby's suggests that they belonged to a single original portfolio, much of which can now be reconstructed, thanks to this donation.[60] Physical evidence from the stub of an old guard on one letter in the Lennox sequence matches precisely the type of guard in the two original folio bindings preserved by Maddison, confirming their common origin.[61]

---

[54] Sotheby and Company, London, sale catalogue 27 June 1932, lots 111–23A. I would like to thank Keith Moore for letting me use material from his unpublished report, 'The Richmond-Folkes *hydra* Letters and a Unique Survival of Early Freemasonry'.

[55] The manuscripts were at Sotheby's, Stevens's, and Glendinning's sales, including Stevens's sale 14–15 August 1934, lots 283–94. The manuscripts are now Wellcome MSS 1302, 2391–2, and 5403.

[56] J. A. Bennett, 'Obituary R. E. W. Maddison (1901–93)', *Annals of Science* 52, 3 (May 1995), p. 306.

[57] Royal Society provenance file, RS/MS/790. Letter, R. E. W. Maddison to the secretary of the Historical Manuscripts Commission, 21 June 1952.

[58] R. E. W. Maddison, 'A Note on the Correspondence of Martin Folkes, P. R. S', *Notes and Records of the Royal Society of London* 11, 1 (1954), pp. 100–9.

[59] T. J. McCann, 'Lennox, Charles, Second Duke of Richmond (1701–1750)', in Sidney Lee, ed., *Oxford Dictionary of National Biography*, 63 vols (New York: Macmillan and Co.; London: Smith Elder & Co., 1894), vol. 33, pp. 359–61.

[60] For example, the letters of Colin Maclaurin in MS/250, Sotheby's lot 115, and a single letter of Sarah Lennox, the Duchess of Richmond, in MS/790.

[61] The guards were made from a printed book; Letter number XX has residue from the same book. My thanks to Keith Moore for this information.

The Lennox–Folkes letters, now designated MS/865 in the Royal Society's archive collections, are personal, but with intermittent material on Royal Society business. This usually takes the form of election recommendations, for example Lennox's request for Folkes's support on behalf of the French chemist Charles Dufay (1698–1739) and the naturalist Abraham Trembley.[62] The author was also extremely fortunate to be given photographs and complete transcriptions of letters between Folkes and Trembley in the Trembley Family Archives of Geneva by Sylvia G. Lenhoff who, with her late husband Harold, edited a critical edition of Trembley's work on hydra: *Mémoires Concerning the Polyps*.[63] Lastly, a series of Folkes's accounts, records of the publication of the *Philosophical Transactions*, and a draft history in his hand of the Royal Society came to the Library through the collections of Folkes's son-in-law, Richard Betenson, 4th Baronet of Wimbledon, Surrey.[64]

Folkes has thus been mischaracterized due to a variety of historiographic and archival reasons. His virtuosic sensibility and the efforts towards the unification of the Society of Antiquaries and the Royal Society tells against the current historiographical assumption that his was the age in which the 'two cultures' of the humanities and sciences split apart, never to be reunited. In this period, antiquarianism and natural philosophy were instead two sides of the same coin, antiquarianism's presence in the Royal Society not a symbol of its supposed moribund nature, but of its strength. In a larger sense, Folkes was part of what Antti Matikkala has termed 'The Chivalric Enlightenment', a movement 'essentially rhetorical, learned, antiquarian and eclectic'.[65] Although the term 'Enlightenment' itself is contested, with so-called 'Enlightened historians' often scorning antiquarian pursuits, others, such as Folkes, had enlightened interest in the past that was informed by the technology and empiricism of natural philosophy. It was a distinctive view of the past and a view of natural philosophy which he, the Antiquaries, and the Royal Society Fellows promoted.

---

[62] RS/MS/865/5, Richmond–M.Folkes 4 May 1729 (Dufay) and RS/MS/865/28, Richmond–M. Folkes 15 February 1743 (Trembley), Royal Society Library, London.

[63] Sylvia G. Lenhoff, Howard M. Lenhoff, and Abraham Trembley. *Hydra and the Birth of Experimental Biology, 1744: Abraham Trembley's Mémoires Concerning the Polyps* (Pacific Grove, CA: Boxwood Press, 1986); the Lenhoffs were given permission by the late Jean Gustave Trembley to make the correspondence available for scholarly study.

[64] RS/MS/702, An Account of the Royal Society from its First Institution; RS/MS/704, Index of Papers Read Before the Royal Society, 1716–38; RS/MS/213, Table of Contents of Journal Book, 1741–8, Royal Society Library, London.

[65] Antti Matikkala, *The Orders of Knighthood and the Formation of the British Honours System 1660–1760* (Cambridge: Cambridge University Press, 2008), p. 43.

2

Nascent Newtonian, 1690–1716

2.1 Early Life

Martin Folkes was born on 29 October 1690 in Little Queen Street, Lincoln's Inn Fields, in the parish of St Giles-in-the-Fields, Westminster. He was baptised in the parish church the following month on 18 November as the son of Martin Folkes, Esq. and his wife, Dorothy Hovell.[1] His father, Martin Folkes Sr (1640–1705), had made his fortune serving as barrister-at-law and attorney general and an estate manager for the Crown, administering the affairs of Queen Catherine Braganza, wife of Charles II.[2] Martin Folkes's grandfather had served as a steward at the Earl of St Albans's estate, Rushbrooke Hall in Suffolk, and his father followed in his footsteps, serving as trustee for the Earl of St Albans's immediate heir, Thomas Lord Jermyn, administering his estates and helping with the selection of vicars on his holdings.[3]

The Folkes and Jermyn families were close. Thomas Lord Jermyn's father, Henry, had been granted the fields of St James's Westminster by King Charles II, which he developed into St James's Square. Henry Lord Jermyn had also commissioned Sir Christopher Wren to design the church of St James, Westminster. 'On 13 July 1684 the Bishop of London went to Henry Jermyn's house in St. James's Square where he received the title-deeds of the site from "Master Fowke" (Martin

---

[1] Christening Book. Parish of St Giles-in-the-Fields, London: 'November 1690/ 18th November/ Martin of Martin Folks Esq. &/ Dorothy his wife born October 29th'. On 20 January 1705, there was a 'Treasury reference to Mr. Travers, Surveyor General of Crown Lands, of the petition of Martin Folkes for a further term in a house in Little Queen Street, a stable, 2 coach houses and 11 other small houses there'. See 'Warrant Book: January 1705, 11–20', in William A. Shaw, ed., *Calendar of Treasury Books, Volume 19, 1704–1705* (London: His Majesty's Stationery Office, 1938), pp. 461–68. *British History Online*. http://www.british-history.ac.uk/cal-treasury-books/vol19/pp461-468 [Accessed 29 May 2019].
[2] BL Add MS 22067, Miscellaneous letters and papers, mostly addressed to, or concerning, Martin Folkes, Attorney-General to Katherine Queen Dowager, 1639–1704, British Library, London.
[3] 'St. James's Square: No 8', in Francis H. W. Sheppard, ed., *Survey of London: Volumes 29 and 30, St James Westminster, Part 1* (London: London County Council, 1960), p. 115, footnote 1. *British History Online*. http://www.british-history.ac.uk/survey-london/vols29-30/pt1/pp115-118 [Accessed 28 May 2019]. Martin Folkes Sr was son of the steward of the Rushbrooke Estate. See Sydenham H. A. H. Hervey, *Biographical List of Boys Educated at King Edward IV's Grammar School (Bury St. Edmunds England)* (Bury St Edmunds: Paul and Matthew, 1908), p. 141. See also MS/MC/50-1-8, Genealogy of Martin Folkes Senior and his parents in the Norwich Archives: 'An Extract, taken out of the Registry book, of Births, Marriages, and burials, kept in the parish of Rushbrooke, in the county of Suffolk', Norfolk Record Office, Norwich. See also BL Add MS 22067, 'Letter to Martin Folkes, 24 July 1689, from Thomas Jermyn on a church living', f. 9, British Library, London.

Martin Folkes (1690–1754): Newtonian, Antiquary, Connoisseur. Anna Marie Roos, Oxford University Press (2021).
© Anna Marie Roos. DOI: 10.1093/oso/9780198830061.003.0002

Folkes), acting on behalf of Thomas, Lord Jermyn; he then went in solemn procession to the church for the consecration service'.[4] Accompanied by Wren, they then proceeded up Duke of York Street and into the church to attend the first service held there. Martin Folkes Sr also had family members living in the Jermyn Estates. Part of Jermyn's lease was 23 Golden Square in Warwick Street, London; the rate books show that Jermyn never lived in the house, but in 1708 the latter assigned the lease to Elizabeth Fowke (Folkes), widow, who was Martin Folkes Sr's daughter.[5] Elizabeth lived there until 1722, after which the Portuguese envoy took up residency.[6]

Folkes Sr also was involved in administering the estates for Nell Gwyn, mistress of Charles II. According to a contemporary account, her house (number 79 on the south side of Pall Mall) 'was given by a long lease by Charles II to Nell Gwyn, and upon her discovering it to be only a lease under the Crown, she returned him (the King) the lease and conveyances saying she had always conveyed free under the Crown, and always would; and would not accept it till it was conveyed free to her by an Act of Parliament'.[7] In other words, her house had been leased to her until 1720, but was not given to her by right. Perhaps as a result of Gywn's displeasure, on '1 December 1676 Charles II granted the freehold of the house to William Chiffinch, one of his confidential servants, and Martin Folkes, a trustee of the Earl of St Albans' estate, and they in turn conveyed it to Nell Gwynne's trustees on the following 6 April'[8] (see figure 2.1). By the deed of conveyance, the property was settled on Nell Gwyn for life, then upon her younger son, James, Lord Beauclaire (Beauclerk) and his heirs, with the remainder to her

---

[4] Guildhall Library, MS 9531/18, ff. 15–19, as quoted in 'St. James's Church, Piccadilly', in F. H. W. Sheppard, ed., *Survey of London: Volumes 29 and 30, St James Westminster*, Part 1 (London: London County Council, 1960), pp. 31–55. *British History Online.* http://www.british-history.ac.uk/survey-london/vols29-30/pt1/pp31-55 [Accessed 3 November 2020].

[5] Elizabeth Folkes (born 1694) who later married Thomas Payne.

[6] 'Golden Square Area: Warwick Street', in Francis H. W. Sheppard, ed., *Survey of London: Volumes 31 and 32, St James Westminster, Part 2* (London: London County Council, 1963), p. 167, footnote 1. *British History Online.* http://www.british-history.ac.uk/survey-london/vols31-2/pt2/pp167-173 [Accessed 28 May 2019].

[7] J. P. Malcolm, ed., *Letters between Rev. James Granger, M.A., Rector of Shiplake, and Many of the Literary Men of his Time: Composing a Copious History and Illustration of his Biographical History of England, with Miscellanies and Notes of Tours in France, Holland and Spain, by the Same Gentleman* (London: Printed by Nichols and Son, for Longman, Hurst, Rees, and Orne, 1805), p. 308, quoted in 'Pall Mall, South Side, Past Buildings: No 79 Pall Mall, Nell Gwynne's House', in Francis H. W. Sheppard, ed., *Survey of London: Volumes 29 and 30, St James Westminster, Part 1* (London: London County Council, 1960), pp. 377–8. *British History Online.* http://www.british-history.ac.uk/survey-london/vols29-30/pt1/pp377-378 [Accessed May 28, 2019].

[8] MS MA 179A, Pierpont Morgan Library, New York; 'Pall Mall, South Side, Past Buildings: No 79 Pall Mall, Nell Gwynne's House', in Francis H. W. Sheppard, ed., *Survey of London: Volumes 29 and 30, St James Westminster, Part 1* (London: London County Council, 1960), pp. 377–8. *British History Online.* http://www.british-history.ac.uk/survey-london/vols29-30/pt1/pp377-378 [Accessed May 28, 2019]. See also J. H. Wilson, 'Nell Gwyn's House in Pall Mall', *Notes and Queries* 194 (2 April 1949), pp. 143–4.

Fig. 2.1 Letters patent under the Great Seal, 1676 Dec. 1, granting to William Chaffinch and Martin Folkes (agents for Nell Gwyn and Henry Jermyn, Earl of St Albans) reversion and inheritance of the property then occupied by Nell Gwyn, MS 179A, Pierpont Morgan Library, Department of Literary and Historical Manuscripts.

eldest son Charles, Earl of Burford.[9] The Earl of St Albans was given, in compensation for releasing his rights to Nell's House, 'the Lay Soule [Laystall?] Veseys Garden, and Watts Close'.[10] Nell was given absolute rights to the property until 1740, and 'reversion and inheritance in perpetuity after that date', her house 'conveyed free'.[11] Bringing this process to a successful conclusion must have taken enormous tact, restraint, and diplomacy on the part of Martin Folkes Sr—lessons passed onto his eldest son, who was known for his amiability.

Martin Folkes Sr also used his position and skills at negotiation to make some shrewd real estate investments; estate books held in the Norfolk Record Office

---

[9] MS MA 176E, Deed of Conveyance, House of Nell Gwyn, Pierpont Morgan Library, New York.
[10] Wilson, 'Nell Gwyn's House in Pall Mall', p. 144.
[11] Wilson, 'Nell Gwyn's House in Pall Mall', p. 144.

show that he left an income to his son of £2,589 (approximately 28,766 times the day's wages of a skilled labourer in 1710), with estate holdings in Norfolk, Lincolnshire, Yorkshire, Middlesex, and London, including the chief lordship of South Lynn and Southmere and Docking.[12] Under the terms of one of his property interests, he held the leasehold of thirteen houses in Queen Square in London. Folkes Sr used his considerable position and influence as attorney general to get lands revalued and rents decreased, sometimes by half.[13] There are no extant portraits of him, as those listed at the family estate of Hillington Hall were scattered to the winds; from 1824 to 1830, the hall was incorporated into a new building designed by W. J. Donthorne, before the house was finally demolished in the 1940s and its collections sold. Despite these losses, there are descriptions of Folkes Sr as a respectable figure dressed conservatively in shades of brown: full brown eyes, brown curly wig, brown robe, 'full shirt sleeves, tight at the wrists, short cross over cravat, black shoes, silver buckles'.[14]

Dorothy Hovell (1655–1724), Folkes's mother, was from well-established gentry stock in Norfolk who also invested well, bringing into the marriage the architecturally striking Hillington Hall with its red brick and later neo-Gothic crenulations, its splendour reflected in a serpentine ornamental lake (see figure 2.2). Samuel Lewis, editor of *A Topographical Dictionary of England* (1848) described it thus:

> Hillington Hall, the seat of the lord of the manor, is a stately mansion, beautifully situated in a richly-wooded park; it was originally erected in 1627, and has been much enlarged and improved by its present proprietor [Sir W. J. H. B. Folkes, Bart., a descendant of Dorothy], who has added to it a noble hall, staircase, and library.[15]

At Hillington, there was a portrait of her at the age of thirty with Martin aged three. Dorothy was described with 'hazel eyes, full, brown hair rather low, and a

---

[12] National Archives Currency Converter, https://www.nationalarchives.gov.uk/currency-converter [accessed 20 August 2020].

[13] Crown Lease Book VI, p. 449. His son Martin later had to petition for a reversionary lease. See 'Warrants, Letters, etc.: 1744, January–March', in William A. Shaw, ed., *Calendar of Treasury Books and Papers, Volume 5, 1742–5* (London: His Majesty's Stationery Office, 1903), pp. 544–56. *British History Online*. http://www.british-history.ac.uk/cal-treasury-books-papers/vol5/pp544-556 [Accessed 28 May 2019].

[14] 'Hillington Hall', in Prince Frederick Duleep Singh, *Portraits in Country Houses of Norfolk* (Norwich: Jarrod and Sons, 1928), p. 244.

[15] *Christ Church College Newsletter*, Trinity Term 2008. Folkes's younger brother William also had extended it in the 1760s. Hillington Hall was demolished in 1946 and a neo-Georgian house was constructed on the site in 1997. The Gate House is still extant. See http://www.heritage.norfolk.gov.uk/record-details?MNF3508-Site-of-Hillington-Hall-gardens-and-dovecote&Index=3254&RecordCount=57339&SessionID=b22f3d1d-758a-4474-9079-d32c1d85b6fa [Accessed 28 May 2019]. The designs for the house's 1820 neo-Gothic refurbishment, designed by William Donthorne, are in the Royal Institute of British Architects' Archive, RIBA65475.

A. E. Purdy, King's Lynn, Photo.

HILLINGTON HALL, NORFOLK, 1907.

Fig. 2.2 Hillington Hall, 1907, Frederic Dawtrey Drewitt, *Bombay in the days of George IV: memoirs of Sir Edward West, chief justice of the King's court during its conflict with the East India company, with hitherto unpublished documents* (London: Longmans, Green, and Co., 1907), pp. 17–19, Wikimedia Commons, Public Domain.

lock of it on the left shoulder' dressed in a long green gown with pearls and black trimmings that must have set off her eyes, her left elbow bent hand down, her right holding the hand of her son who had long and fair hair, the angel-hair of toddlers.[16]

The Hovells also descended from William Chichele, brother of Henry Chichele, Archbishop of Canterbury (1414–43), founder of All Souls College, Oxford.[17] These connections to church preferment may have been behind the marriage of Dorothy's formidable sister Ethelreda to William Wake, Archbishop of Canterbury (1657–1737). Wake's account of their courtship indicates it was a protracted, though ultimately happy affair, involving the consent of the Folkes family. Wake wrote:

I was now in good earnest in my thoughts of marrying; nor did my new settle-ments at Gray's Inn put a stop to them. Mrs Sharpe, the then Dean of Norwich's wife, being with us at Highgate the Xmas last past, had often been speaking of a person whom both she, & Mrs Clagett thought would make me a very good wife, and wish'd they could get Her for me. But they feared mr Folkes who had

[16] 'Hillington Hall', in Prince Frederick Duleep Singh, *Portraits*, p. 245.
[17] One Richard Chichele, for instance, signed Dorothy Hovell Folkes's will on 28 May 1719. MS/ MC/50-1-6, Norfolk Record Office, Norwich.

marry'd Her sister would oppose it; and here the matter still ended as often as they spake of it. After Dr Clagetts death, mrs Sharpe frequently fell upon the subject of my marrying, & seriously persuaded me to it: she was an excellent and prudent woman, and every thing she sayd had a very great weight with me. Making a visit one day at mrs Folkes's, she was willing to try what she would think of a match between us; and was so far encouraged in the discourse, that she resoloued plainly to propose it to the young Lady, mss Hovell; and by consent, it was agreed, that Dr Sharpe should propose it to her Brother in Law, Mr Folkes.[18]

Due to these familial connections with William Wake, Martin Folkes Jr, when he came of age, was involved in the choosing of vicars on the Wake estates, an ironic task for a freethinker and deist, but one which was governed more by networks of patronage and family than religious scruple. No stranger to nepotism, Wake had stipulated in his will (though probably without any legal right to do so), that 'William Wake (the second son of my cousin Doctor Edmund Wake) should receive the next vacancy of a living at Owermoigne [now Overmoigne], in Dorset'.[19] It was tricky for his wish to be fulfilled, as the intended incumbent had not even finished his Oxford BA. In a series of letters with the latitudinarian Bishop of Winchester, Benjamin Hoadly, Folkes would use his connections to support William, the candidacy of William's younger brother, Charles, and one Reverend Conant, the husband of another of the late archbishop's relatives.[20]

2.2 Education

The dice in England are wrought with a care and exactness that might be much better employed in some useful piece of mechanism. In a word, the English mathematicians continue at this day to be very busy in calculating of probabilities.

—Le Blanc, *Letters on the English and French Nations* (1747),
vol. 2, p. 309.

---

[18] MS 541a, Archbishop Wake's autobiography, Christchurch Archives, University of Oxford, p. 22.
[19] TNA PROB/682/5, Will of William Wake, Lord Archbishop of Canterbury, 1 March 1737, The National Archives, Kew.
[20] RS/MS/250, vol. 1, letters: 3, 9, 17, 19, 20, 23, 30–32; vol. 2, letters: 18, 58, 60, 61, 72, 42, Royal Society Library, London. For Folkes's querying Wake about his degree status, see MS 250, vol. 2, letter 15. See also John Dearnley, 'Patronage and Sinecure: Examples of the Practice of Bishop Hoadly at Winchester (1734–61)', *Proceedings of the Hampshire Field Club of Archaeology* 65 (2010), pp. 191–201, esp. p. 195. Dearnley did not seem to realize Folkes was involved in the sinecures due to his family connection with Wake.

Although Martin Folkes Sr and Dorothy's second son, William Folkes, followed his father's trade, serving as a Gray's Inn barrister and legal factor for the good and great, including John, 2nd Duke of Montagu and Charles Lennox, Duke of Richmond, Martin, as the eldest son and heir, did not have to turn his mind to a practical living.[21] His was the education of a wealthy gentleman, with private tutors in arts and letters, and he was allowed to indulge his talents in mathematics and astronomy.

Talented Huguenot refugees who had escaped to the Dutch Republic and then attempted to get to England due to the Revocation of the Edict of Nantes in 1685 shaped Martin Folkes's education. 'The settlement of French immigrants in the Netherlands and their links with Dutch printers and booksellers' made the situation ripe for a 'dissemination of English ideas to Europe through the channel of French-speaking commentators', such as those of Newtonianism.[22] As Baillon indicated,

> Huguenots…contribute[d] to the success of Newtonian science both at home and abroad, but they actually participated in the making of Newtonianism itself, or—to be quite specific—in the elaboration of a distinct variety of Newtonianism which happened to be authorized by Newton himself. One of the characteristic features of that variety was the centrality of its physico-theological dimension. From another perspective, the contribution of several Huguenots to that task can also be considered as subservient to a wider purpose, namely the further integration of the Huguenot community into British society at a time when Britain was increasingly posited as a Protestant haven and possibly a 'deuxiè mepatrie' by the exiled French Protestants in London and elsewhere.[23]

Huguenot John Theophilus Desaguliers (1683–1744) was one of these Newtonians, with a career as a public lecturer in London and a demonstrator at the Royal Society, writing a popular introduction to Isaac Newton's (1643–1727) natural philosophy, the two-volume *A Course of Experimental Philosophy* (1734–44). In 1715, the year of Dutch Newtonian Willem 's Gravesande's visit,

---

[21] Folkes had one sister, Elizabeth Folkes (born 1694), who later married Thomas Payne, and two brothers, William Folkes, also baptised in the church of St Giles-in-the-Fields on 20 July 1704 (+ 10.04.1773) and elected as a Fellow of the Royal Society in 1727, and Henry Folkes (b. 1704). William Folkes married Mary Browne, daughter of Sir William Browne who was President of the Royal College of Physicians and lived in King's Lynn, Norfolk. One Martin Browne, a Rotterdam merchant (d. 1713), invested on behalf of the Folkes family, and William was the recipient of his largess when he died. Martin served as his guardian upon his mother's death until William reached the age of majority. See MS/MC/50-1-6, Will of Dorothy Hovell Folkes, 28 May 1719, Norfolk Record Office, Norwich. See MC/50-1-4, Martin Folkes's copy of Martin Browne's will, Norfolk Record Office, Norwich.

[22] Jean-François Baillon, 'Early Eighteenth-century Newtonianism: The Huguenot Contribution', *Studies in History and Philosophy of Science* 25 (2004), pp. 533–48, on p. 534.

[23] Baillon, 'Huguenot Contribution', p. 534.

Desaguliers performed experiments on colours before members of the French Académie Royale des Sciences at a Royal Society meeting, resulting in greater acceptance of Newton's *Opticks* by French natural philosophers such as Pierre Varignon. Folkes's later adherence to a Newtonian programme when he was Royal Society President may have had its origins in his tutelage under Huguenot intellectuals with similar sympathies, such as James Cappel and Abraham de Moivre.

Contrary to some extant scholarship, Folkes was not educated at the Academy of Saumur, an institution that represented a more liberal side of French Protestantism, as it was closed in 1694 upon Louis XIV's end to religious toleration. From 1699 until 1706, Folkes was, however, privately tutored in his household by the Huguenot prodigy James Cappel (1639–1722), who had been appointed Saumur's Professor of Hebrew at the age of nineteen and was then forced to flee France. James Cappel was the son of another distinguished Hebrew scholar, Louis Cappel (1585–1688), also from Saumur, who compared ancient Hebrew with the language of Masoretic text in the Old Testament, concluding (correctly) that vowel points and accents were not original to the ancient language. James Cappel taught Folkes Greek, Latin, and Hebrew in the humanist tradition, as well as conversational French; this instruction would serve Folkes well when he assisted in editing Newton's work on biblical chronology. Mr Cappel wrote to M. Le Clerc from Hillington Hall in February 1707 (n.s.) that his pupil was 'a choice youth, of penetrating genius, and master of the beauties of the best Roman and Greek writers'.[24]

Cappel and Folkes had a close relationship even after this arrangement ended. Folkes later served as executor to Cappel's will, in which his teacher left to 'his honoured friend Martin Ffolkes Esq. Who was for 7 years my excellent and only pupil…as a surere token of my gratitude the Italy, the Sicily and the German of Cluverin four volumes in ffolio which I desire may be taken from amongst my Books and presented to him'.[25] These were the works of Philipp Clüver (1580–1622), a professor at Leiden University, an antiquary and cartographer best remembered for his edition of Ptolemy's *Geographia*, as well as for his study of the geography of antiquity. He published a lavishly illustrated study of the Rhine in the Roman Era, the *Germania Antiqua*, referred to as the 'Germany' in Cappel's will, as well as the *Siciliae Antiquae libri duo* and the *Italia Antiqua* (1624), the 'Sicily and the 'Italy'. The Elzevir volumes were in Folkes's library until he

---

[24] Rev. David C. A. Agnew, *Protestant Exiles from France, Chiefly in the Reign of Louis XIV. Or, The Huguenot Refugees and Their Descendants in Great Britain and Ireland*, 3 vols, 3rd ed. (London: Private Circulation, 1886), vol. 2, p. 229.

[25] TNA PROB 11/583/91, Will of James Cappel or Cappell of Saint John Hackney, 6 May 1720, Middlesex, The National Archives, Kew. The will was originally written in French and it avers Cappel's Huguenot beliefs; Folkes was the only person bequeathed anything from the estate outside of Cappel's family.

died.[26] It is tempting to think that Folkes's antiquarian interests and his later study of the Farnese Globe (see chapters four and five) may have been inspired, in part, from his perusal of these works with Cappel's guidance. After all, at the auction of his collection after his death, Folkes had sixty-four 'ancient Maps, finely coloured' in addition to thirteen other cartographic lots.[27]

Before he was a tutor to Folkes, Cappel was an itinerant teacher, and became a colleague of the mathematician Abraham de Moivre (1667–1754) while working in a dissenting academy in London. De Moivre would subsequently serve as Folkes's maths tutor on Cappel's recommendation (see figure 2.3).

De Moivre was known for his work connecting trigonometry and complex numbers, as well as on probability and statistics, particularly the existence of regular distribution patterns in the form of a normal or bell curve in games of chance.[28] He ascribed the orderliness of seemingly random probability as due to the ordered plans of God: 'tho' Chance produces Irregularities, still the Odds will be infinitely great, that in process of Time, those Irregularities will bear no proportion to recurrency of that Order which naturally result from ORIGINAL

Fig. 2.3 Jacques Antoine Dassier, *Abraham de Moivre*, 1741, Bronze Medal, Obverse, 54 mm diameter, Gift of Assunta Sommella Peluso, Ada Peluso, and Romano I. Peluso, in memory of Ignazio Peluso, 2003, accession number 2003.406.18, Metropolitan Museum of Art, New York, Public Domain.

---

[26] Item 2182, Cluverii Italia, Sicilia & Germania Antiqua, 4 Tom (1624), 'Seventeenth Day's Sale', in *A Catalogue of the Entire and Valuable Library of Martin Folkes, Esq.* (London: Samuel Baker, 1756).

[27] Item 128, 'First Day's Sale, 2 February 1756' in *Catalogue of the Entire and Valuable Library*.

[28] De Moivre's theorem in modern notation: For a real number θ and integer n we have that $\cos n\theta + i \sin n\theta = (\cos \theta + i \sin \theta)^n$.

DESIGN'.[29] Alexander Pope, in his *Essay on Man,* wrote in reference to God, De Moivre, and the geometry used when a spider spins its web:

> Who taught the nations of the field and wood
> To shun their poison, and to choose their food?
> Prescient, the tides or tempests to withstand,
> Build on the wave, or arch beneath the sand?
> Who made the spider parallels design,
> Sure as Demoivre, without rule or line?[30]

De Moivre was also an intimate friend of Newton, 'who used to fetch him each evening, for philosophical discourse at his own house, from the coffee-house [probably Slaughter's], where he spent most of his time'.[31] De Moivre's *The Doctrine of Chances* (1718) was dedicated to Newton when Newton was President of the Royal Society, and he acted as 'Newton's spokesperson' during the priority quarrel with Leibniz over the development of the calculus.[32] It was reported (probably apocryphally) that 'during the last ten or twelve years of Newton's life, when any person came to ask him for an explanation of any part of his works, he used to say: "Go to Mr De Moivre; he knows all these things better than I do".[33]

De Moivre usually took students when they were approximately sixteen, but Folkes was tutored by him at an earlier age to prepare him for Cambridge.[34] Through De Moivre, Folkes was exposed to Continental ideas of natural philosophy and an intellectual network that would serve him well when he became a council officer, Vice President, and later President of the Royal Society. 'There were grounds for believing that De Moivre fostered a sense of connection between his pupils, as he evidently brought them together at social evenings and later kept them together as a kind of clique'.[35] Peter Davall, John Montagu, 2nd Duke of Montagu, Charles Cavendish, Peter de Mangueville (a colleague of Bernoulli), Charles Cavendish, and George Parker, 2nd Earl of Macclesfield were all students of De Moivre. Other Huguenots who Folkes would come to know through De

---

[29] Abraham de Moivre, *The Doctrine of Chances* (London: W. Pearson, 1718), p. 243.

[30] Alexander Pope, 'Essay on Man', in William Roscoe, ed., *The Works of Alexander Pope,* 10 vols (London: J. Rivington, 1824), vol. 5, p. 130, Epistle Three, lines 99–104.

[31] A. M. Clerke, 'Abraham de Moivre', in Sidney Lee, ed., *Oxford Dictionary of National Biography,* 63 vols (New York: Macmillan and Co.; London: Smith Elder & Co., 1894), vol. 38, pp. 116–17, on p. 116.

[32] A. Rupert Hall and Laura Tilling, eds, *The Correspondence of Isaac Newton,* 7 vols (Cambridge: Cambridge University Press, 1959–77), vol. vii, p. xxxi.

[33] Charles Bossut, *A General History of Mathematics from Earliest Times to the Middle of the Eighteenth Century* (London: J. Johnson, 1803), p. 388.

[34] David Bellhouse, *Abraham de Moivre: Setting the Stage for Classical Probability and its Applications* (Boca Raton, FL: CRC Press, 2011), p. 147.

[35] Christa Jungnickel and Russell McCormmach, *Cavendish: The Experimental Life* (Lewisburg, PA: Bucknell University Press, 1999), p. 53.

Moivre included Pierre Coste (1668–1747), who translated Newton's *Opticks* (1704) into French, which was 'revised by Desaguliers before going to press'.[36] An analysis of De Moivre's subscriber lists indicated that 'politics, landed interests and the Royal Society, both within De Moivre's network and more generally, [were]...closely interconnected'.[37]

These relationships thus developed into lifelong friendships and were formative to Folkes's intellectual development and the creation of his social circle. In 1730, De Moivre dedicated his *Miscellanea Analytica* to Folkes, Folkes keeping his presentation volume until his death and soliciting his brother William and uncle Thomas Folkes (d. 1731) to subscribe; Folkes owned the 1725 and 1743 editions of his teacher's *Annuities upon Lives*, the 1718 and 1738 editions of *The Doctrine of Chances*, as well as a manuscript by De Moivre.[38] In 1747, Folkes and Edward Montagu dined with De Moivre to celebrate his eightieth birthday, and there is one piece of correspondence from De Moivre to Folkes in French asking if a short visit in the afternoon would be a bother.[39] De Moivre later dedicated his 1750 edition of his *Miscellanea Analytica de Seriebus et Quadraturis* ('Analytical miscellany concerning series and quadratures') to his star pupil, 'spectatissimo viro Martino Folkes armigero' ('To an outstanding man, Martin Folkes, Esquire'); the work was the successor to the *Doctrine of Chances* containing the first formulation of De Moivre's theorem which determines a normal approximation to a binomial distribution.[40]

De Moivre also shaped Folkes's conception about what made an interesting mathematical problem. What did Folkes learn in his early mathematics lessons? As Bellhouse indicated,

A gentleman, to live as a gentleman, must therefore be versed in the ideas of the new sciences as well as the practical aspects of his day-to-day life related to a landed estate. The study of natural philosophy included topics in mechanics, hydrostatics, pneumatics, and optics. As a foundation for this study, the student would have to know arithmetic, algebra, Euclidean geometry, and conic sections...according to one eighteenth-century classification of mathematics, topics in the field divided themselves into two general areas: pure and mixed

---

[36] Margaret E. Rumbold, *Traducteur Huguenot: Pierre Coste* (New York: Peter Lang, 1991), p. 75.

[37] David Bellhouse, Elizabeth Renouf, Rajeev Raut, and Michael A. Bauer, 'De Moivre's Knowledge Community: An Analysis of the Subscription List to the Miscellanea Analytica', *Notes and Records of the Royal Society* 63 (2009), pp. 137–62, on p. 145.

[38] Bellhouse, Renouf, Raut, and Bauer, 'De Moivre's Knowledge Community', p. 145.

[39] David R. Bellhouse and Christian Genest, 'Maty's Biography of Abraham De Moivre, Translated, Annotated and Augmented', *Statistical Science* 22, 1 (February 2007), pp. 109–36, on p. 126, note 81. RS/MS/250/30, De Moivre to Folkes, 27 February, Royal Society Library, London. [Vous me fère un sousille plaisir de me faire savoir par le poste de cette lettre et une courte visite aujourd huy sur feu midi, me vous incommoderait pointe. je suis avec la plus parfaite consideration Monsieur].

[40] Agnew, *Protestant Exiles from France*, vol. 2, p. 213.

mathematics. Arithmetic, algebra, geometry, trigonometry, and conic sections belong to pure mathematics; topics such as geography, astronomy, optics, and mechanics belong to mixed mathematics. De Moivre taught his students topics in pure mathematics.[41]

We can know a bit more specifically the nature of Folkes's tuition from some of De Moivre's lecture notes repurposed to tutor a German student in Hanover, George Friedrich Von Steinberg.[42] After a course in basic arithmetic, there was a course in basic algebra with up to three unknowns and quadratic equations. Then, as now, algebraic word problems were a popular tool for learning, and De Moivre gave a number of these to his students.

Two people, A and B, are 59 miles apart. Person A travels 7 miles in 2 hours, while B goes 8 miles in 3 hours. B sets out one hour after A. How far, in miles, does A go before he meets B? To answer the question, the student must equate the lengths of time until A and B meet. Let x be the required distance in miles travelled by A. Then the distance travelled by B is $59-x$ miles. The length of time travelled by A is then $2x/7$ and the length of time travelled by B is $3(59-x)/8$. Then equating the times until they meet yields $2x/7 = 1 + 3(59-x)/8$, since B starts one hour later than A. The solution is $x = 35$ miles.[43]

This problem had its origins in print in Newton's *Arithmetica Universalis*, which was based on his Cambridge lectures (1673–83), from a section on how 'worded questions can be expressed as equations'.[44] It was quite distinctive that, in a period when most schools were still setting and solving this type of problem purely verbally, De Moivre's curriculum made a rudimentary use of algebraic symbols; it may 'be inferred that De Moivre was tutoring mathematics at least at a level corresponding to a student in his early years in Cambridge'.[45]

Another area of emphasis was calculations in compound interest and life annuities, both areas of necessity for a landed gentleman like Folkes engaged in land leases and estate management. De Moivre's book, *Annuities upon Lives* (1725), was published in response to the popularity of a new financial instrument in which a buyer put down a lump sum to either the government or a private firm to receive an annual payment until the death of a named individual purchaser. In the reign of William and Mary, the English government began to sell life

[41] Bellhouse, *Abraham de Moivre*, p. 145.
[42] Menso Folkerts, 'Eine Algebravorlesung von Abraham de Moivre', in Rudolf Seising, Menso Folkerts, and Ulf Hashagen, eds, *Form, Zahl, Ordnung. Studien zur Wissenschafts-und Technikgeschichte. Festschrift für Ivo Schneider zum 65. Geburtstag* (Stuttgart: Franz Steiner, 2004), pp. 269–275; Bellhouse, *Abraham de Moivre*, p. 149.
[43] Bellhouse, *Abraham de Moivre*, p. 149.   [44] Bellhouse, *Abraham de Moivre*, p. 149.
[45] Bellhouse, *Abraham de Moivre*, p. 149. My thanks to Benjamin Wardhaugh for this point.

annuities, called exchequer annuities, to raise money for expenses for the Nine Years' War.[46] With the advent of statistical analysis of the bills of mortality by natural philosophers and physicians such as William Petty, John Graunt, and Edmond Halley, it was also possible to construct life tables.[47] De Moivre simplified Edmond Halley's life tables based on the *Bills of Mortality* for Breslau, making an assumption that after the age of twelve, the mortality table adhered to a linear function which led to an accurate enough calculation.[48]

Although there is little in De Moivre's own notes to elucidate his method for teaching in these areas, it is clear from Martin Folkes's correspondence with Philip Stanhope (1714–86) that the mathematics of annuities was an area of interest. As Bellhouse has noted, 'Philip Stanhope was prevented from studying mathematics by his uncle and guardian, Philip Dormer Stanhope, 4th Earl of Chesterfield. The young Stanhope's studies in mathematics began only when he became independent from his uncle'. Stanhope subsequently enjoyed exchanging mathematical problems with Folkes, mainly derived from De Moivre's *Annuities on Lives*.[49] Having been well instructed under De Moivre's tutelage, Folkes answered a query of Stanhope about annuities by referring him to the set questions in De Moivre's book, such as, 'What is an annuity of 125328 Livres for 50 years worth at 3 7/8 percent Interest. Answer r. v[ide] qu[ery] 1'.[50]

As the advent of mathematical probability in the early eighteenth century did 'very little to sever the connections between gambling and other forms of aleatory contracts like annuities and insurance', Folkes's correspondence with Stanhope also revolved around the statistics of card games.[51] In this sense, Folkes took after his teacher, as De Moivre apparently gave advice at Slaughter's coffee-house on gambling; Le Blanc in his *Letters on the English and French Nations* (1747), commented:

I must add that the great gamesters of this country, who are not usually great geometricians, have a custom of consulting those who are reputed able calculators upon the games of hazard. M. de Moivre gives opinions of this sort every day at Slaughter's coffee-house, as some physicians give their advice upon diseases at several other coffee houses about London.[52]

---

[46] Bellhouse, *Abraham de Moivre*, p. 156.
[47] For Graunt and Halley and the mathematics of life insurance, see Anders Held, *A History of Probability and Statistics and Their Applications Before 1750* (Hoboken, NJ: John Wiley and Sons, 2003), pp. 81–143.
[48] Abraham de Moivre, 'Preface' to *Annuities Upon Lives* (London: W.P, 1725), p.v.
[49] Bellhouse, *Abraham de Moivre*, p. 148; David R. Bellhouse 'Lord Stanhope's papers on the Doctrine of Chances', *Historia Mathematica* 34, 2 (May 2007), pp. 173–86, on pp. 175–6.
[50] MS U1590/C21, f. 18a, undated letter from Folkes to Stanhope, Centre for Kentish Studies, Maidstone, Kent.
[51] Lorraine Daston, *Classical Probability in the Enlightenment* (Princeton: Princeton University Press, 1995), p. 138.
[52] Jean-Bernard Le Blanc, *Letters on the English and French Nations*, 2 vols (London: J. Brindley, et al., 1747), vol. 2, p. 309.

In an undated letter (*c.* April 1744) from Stanhope to Folkes headed 'Monday morning', Stanhope posed the following to Folkes:

> It is disputed at White's [chocolate house and gentleman's club composed of 'the most fashionable exquisites of the town and court' on St James Street with a reputation for gambling], whether it be an equal wager to lay that the Dealer at whist will have four Trumps. Some think it disadvantageous to lay on the Dealer's side, because he has 12 cards left wherein to find Trumps, when all the others have 13 apiece for them. Others say but I don't understand how they can prove it that the advantage to lay on the Dealers side amounts to 25 per cent. If you are at leisure, I should be glad to know your opinion.[53]

Although we do not have Folkes's answer, we do know it was a point of interested discussion. George Lewis Scott FRS (1708–80), another pupil of De Moivre's and editor of Ephraim Chambers's *Encyclopaedia*, wrote Folkes a letter in April 1744 concerning 'the question you mentioned the other Day about Whist'.[54] As 'matters of chance are fertile in paralogismes', Scott did his own set of calculations, finding the advantage for the dealer closer to 40 percent, with Folkes correcting his addition and writing on the side 'or 17 to 12 nearly'.

Folkes would later answer similar questions about annuities and lotteries for aristocratic patrons and friends such as Lennox, and possessed books in his private library about lotteries, such as *Critique Historique…sur les lotteries* by Gregorio Leti (1697).[55] In his work, Leti stated that 'one hears talk only of lotteries among all manner of persons, up to and including the most poor and wretched maidservants, not only in Holland, but in the whole of Europe'.[56] Leti was correct. When Folkes later travelled to Venice in 1733, even Italian natural philosophers such as Cristino Martinelli asked him about the odds of winning the lottery. Folkes wrote in his travel diary on 1 September 1733,

> Mr. Martinelli brought me a scheme from the Resident of the Popes Lottery which he desired me to tell him the disadvantage of. It is rather a game than a Lottery but instead has the royal oak or ace of hearts, there is 26 percent in the

---

[53] RS/MS/790, Royal Society Library, London; William H. Ukers, *All About Coffee* (New York: Tea and Coffee Trade Journal Company, 1922) http://www.web-books.com/Classics/ON/B0/B701/15MB701.html [accessed 2 June 2019].

[54] RS/250/3/26, George Lewis Scott to Martin Folkes, April 1744, Royal Society Library, London.

[55] *Catalogue of the Entire and Valuable Library* (London: Samuel Baker, 1756). These queries continued throughout his life, one Charles Labelye admitting in a letter to Folkes of 22 March 1741/2 that he was 'greatly mistaken in the Case of having 2 honours in 2 cards in the same Hands…as to the Case of 2 Honours in 13 Cards the Result of my Calculation intirely agrees with the Result of yours'. Labelye went on to ask Folkes to check his results. See RS/MS/790/80, Royal Society Library, London.

[56] Gregorio Leti, *Critique historique, politique, morale, économique & comique, sur les loteries, anciennes & modernes, spirituelles & temporelles* (Amsterdam: Chez les Amis de l'Auteur, 1697), p. 1.

best cases against the player and in the others 40 and 70 per Cent yet are all the people at Rome we are told mad after it.[57]

Folkes had apparently learned his lessons well from De Moivre, his mathematical knowledge further shaped as he entered Clare Hall, Cambridge shortly after the death of his father in February 1705[58] (see figure 2.4). Folkes was formally admitted on 31 July 1706 as a Fellow Commoner, his supervisor Richard Laughton

Fig. 2.4 David Loggan, etching of the inner court of Clare Hall as part of a larger broadsheet of the Cambridge Colleges, Gift of C.P.D. Pape, object number RP-P-1918-1767, Rijksmuseum, Amsterdam, Public Domain.

[57] Bodl. MS Eng. misc. c. 444, Martin Folkes, 'Journey from Venice to Rome', f. 39r., Bodleian Library, Oxford. Martinelli owned a Hauksbee-type air pump, which was bought by the Republic of Venice in 1738–9 for the new chair of experimental philosophy at the university; the pump was used by Giovanni Poleni for his lectures on experimental philosophy. See Terje Brundtland, 'Francis Hauksbee and His Air Pump', *Notes and Records of the Royal Society* 66 (2012), pp. 1–20, on p. 15. The pump is still extant at the Museo di Storia della Fisica in Padua. Sir Robert Brown, First Baronet (died 5 October 1760) was a politician and merchant. Son of William Brown and Grisel Brice, he served as King George II's Resident in Venice.

[58] Clare College was founded in 1326 but by 1339 was being called Clare Hall. This name continued to be used until 1856 when the then master, Edgar Atkinson, made several changes including changing the name to Clare College. A subsequent endowment by the College in 1965/6 led to the foundation of Clare Hall as a graduate college. The two institutions were closely linked at first, but now remain as separate institutions (communication by Elizabeth Stratton, Archivist of Clare College, Cambridge). In this book we will refer to the current Clare College as Clare Hall where appropriate to context. My thanks to Christian Dekesel for this information.

(proctor at Clare) receiving his MA on the occasion of King George I paying a visit to the university on 6 October 1717.[59]

At Clare, Folkes quickly became part of a community of Newtonian natural philosophers and mathematicians. As Ellis stated, 'contrary to what is often argued, it was not continental scholars who did most to spread Newton's ideas, but rather his colleagues and pupils from Cambridge. Indeed, the university possessed a number of leading mathematical scholars in the eighteenth century quite apart from Newton: Roger Cotes, Robert Smith and Edward Waring.'[60] Several professorships in natural philosophy were founded in the late seventeenth and early eighteenth centuries, and latitudinarian clergy prevalent at Cambridge such as John Tillotson and Thomas Tenison disseminated Newton's ideas as 'they saw in his theories a powerful scientific justification for the existence of God and proof of his role in forming the universe.'[61] In turn, the proteges of Tillotson and Tenison, such as Laughton (proctor of Clare) and William Whiston, promoted Newtonianism at Clare Hall; in 1710, Whiston, Laughton, and Folkes, along with several other members of the College, waited on William Wake to gain his permission to print his *Translation of the Smaller Epistles of Ignatius*.[62] Whiston also authored some of the earliest Newtonian textbooks to be used at Cambridge and in his *Emendata Academica* of the curriculum stated that 'no uncertain systems of philosophy to be recommended [a dig at the Cartesian system]; but mathematics and experiments to be prefer'd.[63] In 1694, Laughton, 'a tutor celebrated for the great number of the nobility and gentry education under his care, and a person eminent for his learning', persuaded Samuel Clarke to 'defend in the schools a question on physical astronomy taken from the *Principia*, and in the same year the Cartesian theory was ridiculed in the tripos verses.'[64] Laughton's friend, Charles Morgan, who became master of Clare in 1726, was much admired for his mathematical prowess and contributions to Clarke's textbook editions, and was also a close colleague to Samuel Clarke's brother, John, whose *Enquiry into the Cause and Origin of Evil* (1720)

---

[59] John Venn and John Archibald Venn., *Alumni Cantabrigienses, A Biographical List of all Known Students, Graduates and Holders of Office at the University of Cambridge, from the Earliest Times to 1900, Part I: From the Earliest Times to 1751, Volume II: Dabbs-Juxton* (Cambridge: Cambridge University Press, 1922), p. 155; Clare College Admissions Register.

[60] Heather Ellis, *Generational Conflict and University Reform: Oxford in the Age of Revolution* (Leiden: Brill, 2018), p. 27.

[61] Ellis, *Generational Conflict*, p. 28.

[62] William Whiston, *Memoirs of the Life and Writings of Mr William Whiston*, 3 vols (London: the author, 1753), vol. 1, p. 219.

[63] Whiston, *Memoirs of the Life*, p. 44; Ellis, *Generational Conflict*, p. 29.

[64] Walter William Rouse Ball, *A History of the Study of Mathematics at Cambridge* (Cambridge: Cambridge University Press, 1889), p. 75. For Laughton as an exemplary master, see John Nichols, ed., *Illustrations of the Literary History of the Eighteenth Century*, 8 vols (London: the author, 1817), vol. 6, p. 788 and p. 791.

'used Newtonian natural philosophy for apologetical ends'.[65] As Gascoigne noted, 'Whiston, Laughton, Morgan and Folkes all helped to make Clare a centre for the study of Newtonian natural philosophy'.[66]

Robert Greene (1678–1730), Fellow and tutor at Clare, although having religious reservations about Newtonian philosophy, still introduced students to Newton's works and those of his scientific contemporaries in tutorial sessions.[67] Greene was a whimsical character, his will directing 'that his body be dissected and his skeleton hung in King's College Library and monuments erected in the chapels of Clare and King's, in St Mary's and at Tamworth, long and extravagant descriptions of the deceased being supplied for the purpose by himself'.[68] His pamphlet, *ΕΓΚΥΚΛΟΠΑΙΔΕΙΑ (Encyclopaedia) or method of instructing pupils* (1707), however, was indicative of high pedagogical ability.

Although the content of early modern university teaching is notoriously elusive, Greene's published syllabus of his four-year 'circle of studies' well approximated what Folkes studied at Clare. It was a rigorous programme within a larger Newtonian programme, representing the fruitful interaction between natural philosophy, mathematics, and mechanics. Greene's general curriculum in the trivium is similar to that of Daniel Waterland's *Advice to a Young Student* drawn up in 1706 for his pupils at Magdalene College, Cambridge, though it deviates as expected, with its emphasis on Newtonian natural philosophy and mathematics.[69]

Some of Folkes's lessons at Clare would have repeated those he had with De Moivre. But then, in addition to the basic curriculum, he had studied more advanced mathematics with the 'pupil monger' Laughton, and purportedly surpassed his tutor by his 'great genius'.[70] For the more typical student, after a year of classical studies in the first year, mathematics was begun in the second year with *Elements of Geometry* (textbooks included Euclid, Sturmius, Pardies, and Jones), moving onto arithmetic and algebra with study of works by John Wallis, Newton, and Seth Ward, as well as William Oughtred's *Clavis Mathematicae* (1631), which treated fractions, irrationals, decimal expansions, and logarithms all as 'numbers',

[65] John Gascoigne, *Cambridge in the Age of the Enlightenment: Science, Religion and Politics from the Restoration to the French Revolution* (Cambridge: Cambridge University Press, 2002), p. 158.

[66] Gascoigne, *Cambridge in the Age of the Enlightenment*, p. 158.

[67] Gascoigne, *Cambridge in the Age of the Enlightenment*, p. 159.

[68] Douglas McKie and Gavin Rylands de Beer, 'Newton's Apple', *Notes and Records of the Royal Society of London* 9, 1 (1951), pp. 46–54, on p. 51.

[69] Daniel Waterland, *Advice to a Young Student. With a Method of Study for the Four First Years* (London: John Crownfield, 1740), pp. 23–4. The work was created in 1706 and first published in 1740. My thanks to Benjamin Wardhaugh for alerting me to Waterland's work.

[70] Nichols, *Illustrations of the Literary History of the Eighteenth Century*, vol. 3, p. 322. Sir William Browne made the anecdote in a speech recommending mathematics as the paramount qualification for the presidency of the Royal Society. Considering Brown was descended from Folkes's brother and the topic of the speech, there may have been some exaggeration of Folkes's abilities.

in 'contrast with other mathematicians at this time who treated these as distinct concepts'.[71]

Greene, in his tutorials, had little time for scholastic logic, claiming 'the Elements of Geometry they are certainly the best practical Logic that can be, for they inure [habituate] and accustom, as it were, the Mind to true and exact Reasoning…they are the Foundation of the most valuable Natural Knowledge'.[72] To those that objected there were not enough 'Classicks' in his curriculum, Greene noted 'it must be consider'd that this is a course of University Studies, which always suppose Classicks already taught, and begins where the School ends'.[73] The third year was dedicated to conic sections and curves, with studies of unspecified works of De Witte, De la Hire, Sturmius, de l'Hôpital, Newton, Milnes, and Wallis. In the fourth year, pupils learned about infinite series and calculus using the Newtonian method of notation, and finished with logarithms and trigonometry (with readings *inter alia* from Sturmius, Mercator, Briggs, Newton, and Caswell).[74] Mathematics was not only studied for itself, but to build character in Cambridge students: 'Newton's doctor, Sir John Arbuthnot, praised mathematics for giving "a manly vigour to the mind"; a tonic Newton took to such good effect that he mastered all "the noble and manly sciences" and became "the greatest man that ever liv'd"'.[75]

Folkes was also exposed to a heavy dose of natural philosophy, studying the corpuscular philosophy in his second year (with readings from Descartes, Jacques Rohault's *Traité de Physique* (1671), and works of Robert Boyle). On the surface, the emphasis on Cartesian physics seems surprising, as it was a system irrevocably opposed to that of Newton's. However, the Newtonian Greene commented in reference to Rohault's Cartesian textbook, 'Metaphysics and Corpuscular Philosophy are likewise necessary to be known, because most of the exercises in the University depend upon them, tho' the [Cartesian] Corpuscular Philosophy it self seems in most respects to be ridiculous and trifling'.[76] It was simply that there was no comprehensible or comprehensive textbook drawing on Newton's *Principia* to replace Descartes's oeuvre, not even at Cambridge. In 1711, Roger Cotes wrote to Welsh mathematician William Jones (1675-1749) concerning the 'state of Mathematicks' in Cambridge: 'We have nothing of Sir Isaac Newton's that

---

[71] 'William Oughtred', MacTutor, University of St Andrews, http://mathshistory.st-andrews.ac.uk/Biographies/Oughtred.html [Accessed 8 December 2019].
[72] Robert Greene, *ΕΓΚΥΚΛΟΠΑΙΔΕΙΑ (Encyclopaedia) or Method of Instructing Pupils* (Cambridge: n.p., 1707), p. 6.
[73] Greene, *ΕΓΚΥΚΛΟΠΑΙΔΕΙΑ*, p. 8.   [74] Greene, *ΕΓΚΥΚΛΟΠΑΙΔΕΙΑ*, p. 4.
[75] John Heilbron, 'The Measure of Enlightenment', in Tore Frangsmyr, John Heilbron, and Robin E. Rider, eds, *The Quantifying Spirit in the Eighteenth Century* (Berkeley: University of California Press, 1990), p. 210, quoted from, respectively, John Arbuthnot, *Miscellaneous Works*, 2 vols (Glasgow: J. Carlisle, 1751), vol. 1, pp. 1, 9, 36, and James Jurin, 'Dedication to Martin Folkes, Esq; Vice-President of the Royal Society', *Philosophical Transactions* 34, 392 (1727), p. i.
[76] Greene, *ΕΓΚΥΚΛΟΠΑΙΔΕΙΑ*, p. 6.

I know of in Manuscript at Cambridge, besides the first draught of his Principia as he read it in his Lectures; his Algebra Lectures which are printed; and his Optick Lectures the substance of which is for the most part confined in his printed book'.[77]

Until Newton had an interpreter, Rohault's work on physics was taught, albeit in a Latin translation and annotation by English theologian and Newtonian Samuel Clarke (1675-1729). Clarke subsequently annotated Rohault to argue for the primacy of Newtonian physics, giving a 'full answer to such objections made against the author as seem not to have any just foundation, and a great many things in natural philosophy, which have been since found out by the pains and industry of later philosophers'.[78] In this rather strange course of events, Rohault's work 'happened to become the chief vehicle for the dissemination of Newton's physics among his British countrymen in the first half of the eighteenth century'.[79] We also see the emphasis on Rohault in, for example, Waterland's *Advice to a Young Student*.[80] Charles Morgan, who served as Master of Clare from 1726-36, also helped annotate Clarke's editions of Rohault. For instance, Morgan wrote explanations of the rainbow and its colours and size; these were later published as a separate set of 'philosophical dissertations' for students in 1770.[81] Morgan's papers at Clare show he was also taking careful notes on elements of Newton's *Principia* to prepare them for presentation to students, including directions for how to calculate elliptical orbits and atmospheric refraction.

Folkes may have learned from Clarke's fourth edition of 1710, 'furnished increasingly with critical comments on Rohault's text', as well as direct quotations from Newton's *Opticks* and the *Principia*.[82] For instance, Rohault simplified Descartes's theory that explained the colour spectrum kinematically. Particles collided with the corpuscles of matter that composed the prism, causing their deflection from their path and causing a rotary motion. When a pulse of light

[77] Letter from Roger Cotes to William Jones, 30 September 1711, f. 92r., Cambridge University Library, Cambridge. http://www.dspace.cam.ac.uk/handle/1810/194188 [Accessed 3 November 2020]. Elected FRS in 1711, Jones left his collection of mathematical books to George Parker, 2nd Earl of Macclesfield; it was considered to be the most valuable mathematical library in England. In 1749, Jones would later serve as one of Folkes's Vice Presidents, along with Charles Cavendish. See CMO/4/8, 21 April 1749, Royal Society Library, London; Ruth Wallis, 'William Jones (c. 1675-1740), mathematicians', *Oxford DNB* (Oxford: Oxford University Press, 2004).

[78] 'Preface', in John and Samuel Clarke, eds, *Jacobi Rohaulti Physica* (London: Jacobi Knapton, 1697; trans. John Clarke, London, 1723).

[79] Volkmar Schüller, 'Samuel Clarke's Annotations in Jacques Rohault's *Traité de Physique*, and How They Contributed to Popularising Newton's Physics', in Wolfgang Lefèvre, ed., *Between Leibniz, Newton and Kant: Philosophy and Science in the Eighteenth Century* (Dordrecht, Springer, 2001), pp. 95-110. See also Michael A. Hoskin, '"Mining all within": Clarke's Notes to Rohault's *Traité de Physique*', *The Thomist* 24 (1961), pp. 353-63.

[80] Waterland, *Advice to a Young Student*, pp. 23-4.

[81] Schüller, 'Samuel Clarke's Annotations', p. 108; Charles Morgan, *Six Philosophical Dissertations...Published by Dr Samuel Clarke in his Notes Upon Rohault's Physics* (Cambridge: Fletcher and Hodson, 1770).

[82] Schüller, 'Samuel Clarke's Annotations', p. 100.

brushes against the prism corpuscles, the friction imparts spin: if they rotate faster than usual, they cause the sensation of redness, and if they spin slower, they give rise to blueness. Clarke annotated this argument by presenting a simplified version of Newton's theory of white light composing the spectrum. Clarke stated that different colours arose as a result of the differing velocities of light particles which, in turn, explained why 'different colours were refracted to different degrees when passing through a medium'.[83] This comprehensive treatment and comparison of Cartesian and Newtonian optics presented in Clarke's edition of Rohault would have later served Folkes in good stead, when he was called upon in the 1730s to demonstrate the experiments in Newton's *Opticks* to Italian virtuosi and to answer their objections informed by Cartesian theories (see chapter five). Folkes had the two-volume 1723 edition of Rohault in his personal library.[84]

Greene then noted 'after having given an account of the Philosophy which thrust out the Old, it will be necessary to lay the Grounds of a true and adequate Knowledge of nature as far as we are capable of attaining it, which cannot be better accomplish'd than by having recourse to Experiments' and provided the curriculum for a course of 'Experimental Philosophy and Chymistry of Minerals, Plants and Animals', followed by 'Anatomy and Philosophy'.[85] The latter featured articles in the *Philosophical Transactions*, the *Course of Chymistry* by Nicolas Lémery,[86] Giovanni Alfonso Borelli's *De Motu Animalium*, works on embryology and blood circulation by William Harvey, and Nehemiah Grew's *Anatomy of Plants*, as well as Hookes's *Micrographia* and experimental reports by Antonie van Leeuwenhoek. The third year also had a series of lectures on optics, the nature of colour, and catoptrics, featuring works by David Gregory, Isaac Barrow, Newton, Descartes, Christiaan Huygens, and Johannes Kepler. The fourth year was dedicated to the mechanical philosophy, percussion and sound, gravitation, and astronomy (spherical, hypothetical, practical, and physical) for 'Notions confes'dly noble and sublime'.

2.3 The Royal Society

Folkes so distinguished himself at mathematics, that he was elected a Fellow of the Royal Society when he was still a student at Clare. The Royal Society secretary

[83] Schüller, 'Samuel Clarke's Annotations', p. 106.
[84] Item 18, 'First Day's Sale, 2 February 1756' in *Catalogue of the Entire and Valuable Library*.
[85] Greene, *ΕΓΚΥΚΛΟΠΑΙΔΕΙΑ*, p. 3 and p. 6.
[86] Nicolas Lémery, *Course of Chymistry: Containing an Easie Method of Preparing those Chymical Medicines which are used in Physick; With Curious Remarks and Useful Discourses upon each Preparation, for the Benefit of such a Desire to be Instructed in the Knowledge of this Art*, 3rd ed. (London: Kettilby, 1698).

Thomas Birch in his notes and amendments for his English translation of Pierre Bayle's *Dictionnaire Historique et Critique* (1730) noted in his (rather hagiographic) entry about Folkes that:

> The progress he made there [at Cambridge], & after he left the University, in all parts of Learning, & particularly Mathematical & Philosophical, distinguish'd him at so early an age, that when he was but three & twenty years old, he was esteem'd worthy of a seat in the Royal Society into which, having been propos'd as a Candidate on the 13th of December 1713, he was on the 29th of July following elected and on the 11th of November admitted a Fellow...He had not been much above two years a Member, when, on account of his known abilities and constant attendance at the Meetings of the Society, he was at the Anniversary Election November 30th 1716, chosen one of the Council.[87]

After he left Cambridge, the Royal Society Archives indicate that in 1718 he lived on Southampton Street, Covent Garden; on the south corner with Tavistock Street was the Bedford Head Tavern, where Folkes went to Masonic meetings (see chapter four). Folkes later took residence in Queen Square, settling there for the rest of his life, with occasional journeys to Hillington to oversee his estates.[88] Although on the outskirts of London, Queen Square and its residents were at the centre of intellectual and cultural life in the first half of the eighteenth century (see figure 2.5).[89] Between 1713 and 1725, surveyor Thomas Barlow built the first houses on the east and south sides and then expanded to the west. The elegant Georgian terraces surrounded the square, and the northern side opened out to the countryside with views of Hampstead and Highgate.[90] On the south side was the Church of St George the Martyr, built in 1706; William Stukeley, best known for his work on Stonehenge, was rector from 1747 until his death there of a stroke in 1765. Also in the Square was Folkes's friend and near neighbour, the artist Jonathan Richardson, best known for his spirited portraits of immediate sensibility.[91] Richardson sketched Folkes, Sir Hans Sloane, surgeon William Cheselden, Alexander Pope and his mother, and his fellow artist Sir Godfrey Kneller, who

---

[87] BL Add MS 4222, Birch's Biographical Collection, Collection of biographical notices and memoranda of literary, scientific, and other illustrious persons, made by or for Thomas Birch; c.1730–66, vol. 2, D.-L., ff. 22r.–v., British Library, London.

[88] See RS/LBO/16/195, Letter of Henry Bradbury, Hillington Hall to Martin Folkes, at King's Street near Bloomsbury of 11 March 1720, Royal Society Library, London.

[89] Formerly Devonshire Square, but renamed in honour of Queen Anne, it was developed by Sir Nathaniel Curzon, 4th Baronet of Kedleston.

[90] Simon Shorvon and Alastair Compston, *Queen Square: A History of the National Hospital and its Institute of Neurology* (Cambridge: Cambridge University Press, 2018), pp. 51–2.

[91] See Carol Gibson-Wood, 'Jonathan Richardson as Draftsman', *Master Drawings* 32, 3 (Autumn 1994), pp. 203–29.

Fig. 2.5 Edward Dayes, *Queen Square in 1786*, Watercolour with pen and black ink over graphite on thick, smooth, cream wove paper, 17 1/16 × 23 1/2 in., Yale Centre for British Art, Paul Mellon Collection, Public Domain.

often came 'across from his lodgings in Great Queen Street to spend a quiet evening at Richardson's house'[92] (see figure 2.6).

Cromwell Mortimer, the secretary of the Royal Society (1730–52) who made a fortune selling an early form of medical insurance, was also a neighbour as well. Although Folkes was a dedicated attendee of Royal Society meetings, the Society's Journal Book entries reveal his first few years there to be relatively inconsequential, limited to his nominating friends like William Whiston for membership. On 30 June 1715, Folkes amusingly noted that he 'tried the experiment of bleeding a man that had eaten plentifully of asparagus and had found the blood had not the least smell of it, as it is observed in the urine of those that have eaten sparingly thereof'.[93]

Shortly after he was elected to the Royal Society Council, however, he became much more active, contributing observations of scientific merit. In a meeting of 6 December 1716, he presented a paper about the relatively rare event on 21

---

[92] Edward Walford, 'Queen Square and Great Ormond Street', in *Old and New London: Volume 4* (London: Cassell, Petter and Galpin, 1878), pp. 553–64. *British History Online*. http://www.british-history.ac.uk/old-new-london/vol4/pp553-564 [Accessed 26 April 2019].

[93] RS/JBO/12/71, Royal Society Library, London.

Fig. 2.6 Jonathan Richardson, *Martin Folkes*, Graphite on Vellum, 1735, 179 mm × 134 mm, AN327311001, © The Trustees of the British Museum, CC BY-NC-SA 4.0.

November 1716 of the eclipse of a fixed star, α Gemini (Castor), by the body of Jupiter, seen by his use of a good '16-foot tube'. Folkes's observation was corroborated by Stephen Gray; Gray had been observing the times of passage of Jupiter's four larger satellites into and out of the face of the planet since at least 1699 in a perennial attempt to solve the longitude problem, as had James

Pound.[94] Folkes, with the assistance of his uncle William Wake, would later support Pound's nephew James Bradley for the Savilian Professorship of Astronomy at Oxford.[95] As we have seen, Bradley was probably introduced to Folkes and Wake, as well as Newton and the powerful Whig factions that promised future openings to various offices of state, through Pound.[96]

Folkes's observation was quickly followed by another in 1717 concerning the *aurora borealis*, an interest he shared with Edmond Halley, along with tracking meteors, resulting in his first published paper. The early part of the century saw the end of the Maunder minimum, and London and King's Lynn near Hillington Hall were ideally situated to observe aurora.[97] On 30 March 1716/17, Folkes wrote,

> being in the Street, between 8 and 9 a Clock on Saturday last (30 Martii) I perceived a Light over the Houses to the Northwards, little inferior to that the Full Moon gives when the first rises. Upon this suspecting such Meteors as we saw the last year, I made all the hast I could into the Fields, where I immediately found my Conjecture verified; and was for some time agreeably entertain'd with the sight of an Aurora Borealis, attended with most of the Phaenomena that have been describ'd in that very remarkable one of the 6th of March, 1715–6.[98]

The subjects of *aurora borealis* and meteors were prominent ones at the Royal Society. In investigating the origins of these phenomena, Halley considered the work of Aristotle in his *Meteorology* and Pliny in his *Natural History* who had traditionally attributed meteors or fireballs, along with thunder and lightning strikes, to the ignition of sulphurous vapours in the atmosphere. The source of the vapours came from subterranean caves which generated minerals. Early modern German and Swedish mining literature also implicated ore exhalations due to

---

[94] RS/JBO/12/135–6; Robert A. Chapman, 'The Manuscript Letters of Stephen Gray, FRS (1666/7–1736)', *Isis* 49, 4 (December 1958), pp. 414–23. In a committee meeting of 11 Jun 1714, Newton discussed different methods for determining longitude at sea: 'by a watch to keep time exactly'; 'by the eclipses of Jupiter's satellites'; and 'by the place of the Moon'. By 16 Jan 1741/2, it was agreed at the Royal Society that the first method would be the best, as agreed by: Martin Folkes; Robert Smith; James Bradley; J. Colson; George Graham; Edmond Halley; William Jones; George, Earl of Macclesfield; James Jurin; Charles Cavendish; Abraham de Moivre; and John Hadley.

[95] Martin Folkes to William Wake, 4 September 1721, Wake Letters 16, no. 82, Christchurch Archives, Oxford. We do know, for instance, that Pound introduced Bradley to Edmond Halley. See John North, 'The Satellites of Jupiter: From Galileo to Bradley', in Alwyn van der Merwe, ed., *Old and New Questions in Physics, Cosmology, Philosophy, and Theoretical Biology* (Boston: Springer, 1983), pp. 689–718, on p. 712.

[96] John Fisher, 'Conjectures and reputations: The Composition and Reception of James Bradley's Paper on the Aberration of Light with Some Reference to a Third Unpublished Version', *British Journal of the History of Science* 43, 1 (March 2010), pp. 19–48, on p. 46.

[97] Mike Lockwood and Luke Barnard, 'An Arch in the UK: Aurora Catalogue', *A & G: News and Reviews in Astronomy and Geophysics* 56 (August 2015), pp. 4.25–4.24.30, on 4.26.

[98] Martin Folkes, 'An Account of the Aurora Borealis, seen at London, on the 30th of March last, as It Was Curiously Observ'd by Martin Folkes', *Philosophical Transactions* 30, 352 (1719), pp. 586–8, on p. 586.

these vapours (called *witterung*) as causes of thunder and lightning; Martin Lister (1639–1712), naturalist and FRS, even went even further and in his 1684 work *De Fontibus* attributed the sulphurous steams to iron pyrites, or ferrous sulphide.[99] Robert Boyle argued that the subterranean heat observed during the winter was caused by steams of sulphur underground that were copious around mines, but he speculated 'there may be probably very many [other places] that may supply the air with a store of mineral exhalations, proper to generate fiery meteors and winds'.[100] Some writers like natural historian John Woodward (1665–1728) 'argued that when these vapours saturate the air to a critical level', they constitute a kind of 'Aerial Gunpowder' which was the cause of lightning, thunder, and other atmospheric effects.[101] Lightning does produce a smell of ozone which is similar to sulphur, sulphur is an excellent insulator, and static electricity accumulated on it discharges in electrical sparks towards proximate objects, all things which Halley considered when speculating about the origin of meteors and aurorae being due to the lighting of atmospheric vapours.

Halley thought the process might have been akin to one of the demonstrations he saw Ashmolean Keeper John Whiteside (bap. 1679–1729) perform in his course of physics at Oxford; Whiteside ignited gunpowder *in vacuo* in the receiver of an air pump, in the darkened basement of the Ashmolean Museum. Halley noted, 'the Vapours of Gun-powder, when heated in Vacuo to shine in the Dark...ascend to the Top of the Receiver though exhausted'.[102] Halley speculated whether the glow left in the receiver after the explosion was a sort of incorporeal fire. As Beech noted, 'it was presumably the demonstration that convinced Halley that flammable vapours could gather in the Earth's upper atmosphere, and thereby cause the appearance of bright meteors (and possibly aurora)'.[103] Folkes apparently took these observations into consideration, as he noted in his observations of the aurorae that he remembered 'last Year a great many People said there came a ill smell [presumably from sulphur exhalations], which I did not at all perceive; however as I remember it to be the very same Appearance, I thought it might not be improper just to take notice of it'.[104]

[99] Christian Berward, *Interpres phraseologiae metallurgicae* (Frankfurt: Johann David Zunner 1684); Olaus Magnus, *Historia Olai Magni Gothi Archiepiscopi Upsalensis...* (Basel: Henric Petrina, 1567), book six, chapter eleven. See also Frank Dawson Adams, *The Birth and Development of the Geological Sciences* (New York: Dover Publications, 1938), pp. 301–3.

[100] Robert Boyle, 'An Examen of Antiperistasis', in Robert Boyle, *The Works*, ed. Thomas Birch (Reprint, Hildesheim: George Olms Verlagsbuchhandlung, 1966), vol. 2, p. 663, quoted in Vladimir Jankovic, *Reading the Skies: A Cultural History of English Weather, 1650–1820* (Manchester: Manchester University Press, 2000), p. 25. See also: Anna Marie Roos, 'Martin Lister (1639–1712) and Fools' Gold', *Ambix* 51 (2004), pp. 23–41.

[101] Jankovic, *Reading the Skies*, p. 27.

[102] Angus Armitage, *Edmond Halley* (London: Thomas Nelson, 1966), p. 172.

[103] Martin Beech, 'The Makings of Meteor Astronomy: Part VI', *WGN, The Journal of the International Meteor Organisation* (1994), pp. 52–4, on p. 52.

[104] Folkes, 'An Account of the Aurora Borealis', p. 586.

Folkes was active as a natural philosopher and a shrewd observer of human nature, offering a gift to the Royal Society shortly after he was elected to the Council to flatter Newton, who was the sitting President. On 20 December 1716, the Journal Book recorded that 'Mr Folks presented the Society with the Pictures of all the Presidents of the Society that were to be found in Mezzo Tinto. Viz. Sir Christ. Wren, The Earls of Carbery and Pembroke and the Lds Halifax for which he had the thanks of the Society and they were ordered to be hung up in the Meeting Room [at Crane Court]'.[105] Folkes's gift demonstrated his developing connoisseurship of engravings and mezzotints, and he later was often called upon to commission engravings for noble patrons or to buy objects of vertu at auctions. For instance, Lennox wrote about Folkes obtaining some engraved portraits of past Freemason Grand Masters:

> I have too great a reguard, you may say, to the Dukes of Montagu & Bucclugh who were my Predecessor, to have my print done first, butt after they have gott theirs, the D: of Whartons, & the three that go before them viz; Ant: Sawyer, Geo: Payne, & Dr. Dessgr, for I insist upon theirs being done first, then I will consent to your lending my picture, butt possitively, not before those six are finish'd.[106]

Lennox asked Folkes again, via his estate manager Thomas Hill, to go to 'the Duke of Ancaster's auction, and buy him some of the old stones or marbles that have inscriptions on them…at the same time if you can meet with any antique urns that are in good taste'.[107] In 1737, Folkes also arranged for Lennox to be painted by Enoch Zeeman, noting 'I will be by him when he puts in the wording to avoid mistakes, and I think I shall get the coats of arms all made out to compleat our account of it'.[108]

The portraits that Folkes gave to the Royal Society for Newton's delectation were also highly collectable. Three of the four mezzotints were by John Smith, 'the greatest native born British printmaker' of the seventeenth and early eighteenth centuries; his work sold for high prices, with most of his customers among the Whig aristocracy.[109] (The fourth portrait of John Vaughan, 3rd Earl of Carbery was by John Simon who served as Godfrey Kneller's mezzotinter for a brief period in 1708–9.)

The auction catalogue of Folkes's art collection indicated that he collected a number of images by Smith, who created mezzotints of Charles Montagu, 1st Earl

---

[105] RS/JBO/12/139, Royal Society Library, London.

[106] RS/MS/865/4, Charles Lennox, 2nd Duke of Richmond and Duke of Aubigny, to Martin Folkes, Goodwood Tuesday [1728–9], Royal Society Library, London.' Dessgr' refers to Desaguliers.

[107] RS/MS/790/61, Thomas Hill to Martin Folkes, undated, Royal Society Library, London.

[108] Goodwood MS 110, Folkes to Lennox, 23 July 1737, West Sussex Record Office. Zeeman also painted Mohammed Ben Ali Abgali, George II and Queen Caroline, and he was in high demand as an artist in court circles.

[109] Antony Griffiths, *The Print in Stuart Britain, 1603–1689* (London: British Museum Press, 1998), pp. 239–40.

of Halifax; Thomas Herbert, 8th Earl of Pembroke; and Sir Christopher Wren after the original oil portraits painted by Godfrey Kneller and Willem Wissing. 'Smith worked from rooms at the sign of the Lion and Crown in Russell Street, close to Kneller's studio in Covent Garden piazza, and he came to enjoy a virtual monopoly in the production of prints after the artist, engraving 113 of his works'.[110] His skills as engraver and publisher meant 'for the first time, prints made in Britain were taken seriously abroad'.[111] The engraver George Vertue recorded in one of his notebooks that Smith's 'Fame was universally known to all Ingenious persons of his Time in England…his works was of no less esteem abroad…his picture was painted by Sr Godfrey Kneller—whose works he excellently immitated'.[112] Owning his portraits was the mark of an informed connoisseur and a mark of political allegiance; Folkes's choice of the artist may have had symbolic importance to Newton who was politically and personally tied to the Whigs, having served short terms as a Whig Member of Parliament for the University of Cambridge in 1689–90 and 1701–2.

The mezzotints are still extant in the Royal Society Collections (see figures 2.5–2.8). At some point in the past, the mezzotint portraits were taken out of their

Fig. 2.7   John Smith, *Christopher Wren*, mezzotint after Godfrey Kneller, 1713 (1711), 350 mm × 263 mm © The Royal Society, London.

Fig. 2.8   John Simon, *John Vaughan, 3rd Earl of Carbery* after Godfrey Kneller, early 18th century, 357 mm × 258 mm, © The Royal Society, London.

[110] John Smith, Mezzotint Print Maker, National Portrait Gallery, London. https://www.npg.org.uk/research/programmes/early-history-of-mezzotint/john-smith-mezzotint-printmaker-biography [Accessed 20 June 2019].
[111] Griffiths, *The Print in Stuart Britain*, p. 24.
[112] 'Vertue's Note Book B.4 [British Museum Add. MS 23, 079: Volume III]', *The Volume of the Walpole Society* 22 (1933–4), pp. 87–142, on pp. 113–4.

Fig. 2.9 John Smith, *Thomas Herbert, 8th Earl of Pembroke,* mezzotint after Willem Wissing, 1708, 346 mm × 250 mm, © The Royal Society, London.

Fig. 2.10 John Smith, *Charles Montagu, 1st Earl of Halifax,* mezzotint after Godfrey Kneller, 1693, 336 mm × 247 mm, © The Royal Society, London.

wooden frames and torn around the edges (the versos have acid wood marks), but the rich gilt dedications and descriptions that Folkes had an artisan apply to their surface are still intact, promoting them from being just another set of prints into esteemed Royal Society memorabilia.[113] Not only a gift to Newton, the prints were also a reflection of Folkes's friendships; he was associated with Thomas Herbert, Earl of Pembroke, and assisted in the publication of the 1746 collection of plates depicting the Earl's famous coin collection.[114]

During these years, Folkes was often called upon to witness and confirm experiments performed by Desaguliers to confirm Newtonian principles. For example, on 27 April 1719, Desaguliers timed the descent of pasteboard, glass, and lead balls from the upper gallery of St Paul's Cathedral onto the tiled pavement below, their fall measured by a special chronometer made by George Graham which had a quarter of a second accuracy.[115] As Carpenter has indicated, 'When further experiments were carried out on 27 July using as well, for their lightness, carefully contrived balloons made of hogs' bladders, Folkes again recorded the times', as did Newton, Halley, Jurin, and Graham.[116] In 1726, Newton

[113] The mezzotints have since been restored by Sarai Vardi for the Royal Society's collection.

[114] Kolbe, 'Godfather to all Monkeys', p. 42.

[115] Audrey T. Carpenter, *John Theophilus Desaguliers: A Natural Philosopher, Engineer and Freemason* (London: Bloomsbury, 2011), p. 67; John Desaguliers, 'An Account of Experiments Made on 27 April 1719 to Find How Much Resistance of Air Retards Falling Bodies' and 'Further Account of Experiments Made on 27 April Last', *Philosophical Transactions* 30, 362 (1717–19), pp. 1071–8.

[116] Carpenter, *John Theophilus Desaguliers,* p. 67.

would describe these experiments in detail in the third edition of the *Principia*.[117] His talent and diligence recognized and rewarded, Folkes subsequently served on the council for nine years during the presidency of Sir Isaac Newton, and in 1722/23 he was chosen by Newton as the Society's Vice President. No longer the nascent Newtonian, Folkes's career as a natural philosopher seemed assured.

[117] Carpenter, *John Theophilus Desaguliers*, p. 67.

3

Lucretia Bradshaw

Recovering a Wife and a Life

After securing his career as a natural philosopher in the 1710s and 1720s, Folkes continued to expand his social and intellectual network, becoming a Freemason, serving as a Knight's Companion to Lennox in the inaugural ceremony of the Order of the Bath in 1725, and becoming a fellow of the Society of Antiquaries of London. Folkes also formed his most significant personal relationship. Free to make his own decisions after his father's death, on 18 October 1714 Folkes married the Drury Lane actress Lucretia Bradshaw.

It is not definitively known how they met, although before Folkes inherited his father's townhouse on Queen Square, he lived on Southampton Street in Covent Garden, where several actors and actresses were rate-paying occupants, including Anne Oldfield (1709–13), Robert Wilks (1709), and Colley Cibber (1714–20).[1] Folkes's good friend Lennox also knew the impresario Owen Swiney, who helped to establish Italian Opera at the Queen's Theatre in London; Lennox later served as Deputy Director of the Royal Academy of Music.[2] The meeting of the young mathematician and the actress, dancer, and comedienne may also have its clue in the individual who nominated Folkes to the Royal Society in the first place, the unprepossessing Fettiplace Bellers (1687–1750). After his election, Bellers appears to have soon lost interest in the Society; by the late 1710s his name was not on the list of fellows, probably as he did not pay his subscription. Fettiplace had a reputation as a rogue and a wastrel, and an embarrassment and disappointment to his father, John Bellers (1654–1725), a respected Quaker and social reformer.[3] At the time Fettiplace nominated Folkes for the Royal Society, he was, however, an aspiring playwright, engaged in a campaign to get his work staged at Drury Lane. He

---

[1] 'Southampton Street and Tavistock Street Area: Southampton Street', in Francis H. W. Sheppard, ed., *Survey of London: Volume 36, Covent Garden* (London: London County Council, 1970), pp. 207–18. *British History Online*, http://www.british-history.ac.uk/survey-london/vol36/pp207-218 [Accessed 21 November 2019].

[2] Elizabeth Gibson, 'Owen Swiney and the Italian Opera in London', *The Musical Times* 125, 1692 (February 1984), pp. 82–96, on p. 83.

[3] Geoffrey Cantor, 'Quakers in the Royal Society, 1660–1750', *Notes and Records of the Royal Society* 51, 2 (1997), pp. 175–93, on p. 180. In 1685, John Bellars contributed to the 'purchase of 10,000 acres of land in Pennsylvania for the resettlement of French Huguenots displaced by the revocation of the edict of Nantes'. See Tim Hitchcock, 'Bellers, John (1654–1725), Political Economist and Cloth Merchant', *Oxford Dictionary of National Biography*. https://doi.org/10.1093/ref:odnb/2050 [Accessed 10 June 2019].

Martin Folkes (1690–1754): Newtonian, Antiquary, Connoisseur. Anna Marie Roos, Oxford University Press (2021).
© Anna Marie Roos. DOI: 10.1093/oso/9780198830061.003.0003

was eventually successful in 1732, with the performance of his tragedy titled *Injur'd Innocence*, derived from Sir William Davenant's *Unfortunate Lovers* (1649). One wonders if it was Fettiplace that introduced Martin and Lucretia, perhaps with some reservations. After all, his play's epilogue declared:

> Well, Ladies, now our preaching play is over,
> What say you to this philosophic lover?
> Who boasts he'd spend his life in admiration
> Of every part of the whole fair creation:
> For me; I fear, in spight of all his flights.
> He will want power to please you many nights.[4]

3.1 A Childhood on Stage

Lucretia Bradshaw is an enigmatic figure. We know little about her origins, other than that she was purportedly the daughter of a Drury Lane boxkeeper, born sometime between 1680 and 1685.[5] Her education is a matter of surmise, though she must have been able to read her lines, probably learning the rudiments of literacy at a petty dame school. It was typical for an aspiring actress to follow a 'career path from dancer or singer to actress', and we do know that Bradshaw became an accomplished dancer, indicating she might have been sent to a girls' school to learn a bit of French, history, or geography, along with music and dancing.[6] In later life, others were generally not impressed with her conversation, calling it 'nonsense', but her education was one of accomplishment rather than learning.[7] Lucretia became a protégé of the 'incomparable Mrs Barry', who, as a theatrical shareholder in her own right, wielded considerable clout. Barry was known for the unaffected nature of her performances, and the talented Lucretia learned to make 'her self Mistress of her Part', leaving 'the Figure and Action to Nature'.[8] With Elizabeth Barry's backing, Lucretia became the leading actress of Drury Lane; her contemporary, Charles Gildon, remarked rather cuttingly (or jealously), 'Mrs Bradshaw... if she be not the best Actress the Stage has known, she

---

[4] [Fettiplace Bellers], *Injur'd Innocence: A Tragedy. As it is Acted at the Theatre-Royal in Drury Lane* (London: J. Brindley, 1735).

[5] Philip H. Highfill, Jr, Kalman A Burnim, and Edward A. Langhans, *A Biographical Dictionary of Actors, Actresses, Musicians, Dancers, Managers and Other Stage Personnel in London, 1660–1880*, 12 vols (Carbondale: Southern Illinois University Press, 1973–), vol. 2, p. 283.

[6] Moira Goff, *The Incomparable Hester Santlow* (Farnham and Burlington, VT: Ashgate, 2007), p. 24.

[7] Goff, *Hester Santlow*, p. 24.

[8] Charles Gildon, *The Life of Mr Thomas Betterton: The Late Eminent Tragedian* (London: Robert Gosling, 1710), p. 41.

has hindered Mrs Barry from being the only Actress'.[9] Nonetheless, Thomas Betterton's short *History of the English Stage*, published in 1741, related that 'It was the opinion of a very good Judge of Dramatical Performers' that Lucretia 'was one of the greatest, and most promising *Genij* of her Time'.[10]

Lucretia's first recorded appearance on stage was in April 1696, when she spoke an epilogue in Mary de la Rivière Manley's *The Royal Mischief* at the Lincoln's Inn Fields Theatre. Built in 1661, Lincoln's Inn was the reconverted Lisle's Tennis Court, a relatively small theatre with one gallery. It was under the direction of England's foremost actor, Thomas Betterton. The other rival theatre, Drury Lane, was under the control of Christopher Rich, a lawyer and ruthless manager who in the 1690s humiliated and replaced more expensive senior actors like Betterton with younger and cheaper ones.[11] Within a year, the senior actors were in 'open rebellion', and they moved to Lincoln's Inn.

Lincoln's Inn's artistic production was of high quality at the time Lucretia made her debut. It was typical for an aspiring Georgian actress to gain stage experience with small spoken parts, gaining larger roles as she progressed under the tutelage of an older member of the troupe. Manley's play was a sensationalistic tragedy of thwarted passion, set in the fictitious and libidinously named Oriental locale of 'Libardia'. The main character, the princess Homais, entered the stage to inform the audience of her 'Charms unequalled' before pausing to let the audience assess for themselves the truth of her assertion. Homais was one of Elizabeth Barry's title roles, as she was known for her 'emotionally charged interpretation' of tragic characters.[12]

Desperately in love with the married Levan, the Prince of Colcheis, Homais herself is yoked to a jealous and impotent husband, Libardian. When Homais suggests that husbands are not always objects of a wife's desire, and when the handsome Levan comes back from battle to a hero's welcome, it is evident that the mischief of the title was sexual.[13] Although Homais' lust means she comes to a

[9] Gildon, *Betterton*, p. 41. But for another view about the relationship between Barry and Bradshaw, see Susan Martin, 'Actresses on the London Stage: A Prosopographical Study', PhD Dissertation (Pennsylvania State University, 2008), p. 80. Martin stated, 'Any junior actress, particularly one who wished to move up the ranks, might reasonably try to imitate the most successful actress in the company and would have had the opportunity for constant close observation of that actress's style whether or not they were friends or even on speaking terms'.

[10] Thomas Betterton, *The History of the English Stage from the Restauration to the Present Time* (London: E. Curll, 1741), p. 62. By this time Lucretia was married to Martin Folkes and of higher social standing, so Betterton was effusive in his praise.

[11] Judith Milhous and Robert D. Hume, *Vice Chamberlain Coke's Theatrical Papers, 1706–15* (Carbondale and Edwardsville: Southern Illinois University Press, 1982), p. xvii.

[12] David Thomas, 'The 1737 Licensing Act and its Impact', in Julia Swindells and David Francis Taylor, eds, *The Oxford Handbook of the Georgian Theatre 1737–1832* (Oxford: Oxford University Press, 2014), pp. 91–106, on p. 102.

[13] Review of the 'Royal Mischief', Rose Playhouse, London, 16 September 2016. https://www.thereviewshub.com/southwest-template-draft-3-2-5/ [Accessed 15 October 2018].

sticky end, stabbed by Libardian with sword rather than phallus, she retorts by calling her husband 'impotent in all but Mischief'. Even considering the ribald nature of the Georgian stage, Manley was criticized for the warmth of Homais' fierce response, which intimated that the source of socio-political order was firmly due to the patriarchal control of women by forced or arranged marriage, which ultimately had little benefit for either gender.[14] *The Royal Mischief* also served as the basis for *The Female Wits*, an anonymously authored play which ridiculed Manley and two other female playwrights for their scandalous publications.[15]

Lucretia debuted as an ingénue with an epilogue to this bloody bedroom drama, written by Manley not to titillate the audience with recollections of the 'fierce desires made fit for raging Love' they had seen in the play, but with youthful beauty, innocence, and sexual promise. Lucretia spoke,

> Our Poet tells me I am very pretty,
> Have Youth and Innocence to move your pity.
> A few Years hence perhaps you might be kind,
> The Tallest Trees bend in the rustling Wind;
> Then spare me for the good which I may do.
> Early bespeak me, either Friend or Foe:
> Nor think those Youthful Joys I have in store,
> Far distance Promises, unripen'd Oar,
> Meer Fairy Treasure, which you can't Explore:
> The Play-House is a Hot-Bed to young Plants,
> Early supplies your Longings and your Wants.
> Then let your Sunshine send such lively Heat,
> May stamp our Poet's work, and Nature's too Compleat.[16]

Although a number of scholars have rightly challenged the idea that 'actresses were passive erotic objects and victims of male desire', and there has been 'disproportionate scholarly attention given to [their] sexual representation', there is little doubt that in this instance Montague Summers's assessment in his work, the *Restoration Theatre* was roughly true.[17] He noted,

---

[14] Mrs [Delariviere] Manley, 'To the Reader', in *The Royal Mischief; A Tragedy. As it is Acted by His Majesties Servants* (London: R. Bentley, S. Saunders, and J. Knapton, 1696), p. iii.

[15] Rose Reifsnyder, 'Mary de la Rivière Manley', *Salem Press Biographical Encyclopedia*, 2013.

[16] Manley, *The Royal Mischief; A Tragedy*, p. 29.

[17] Helen E. M. Brooks, 'Theorizing the Woman Performer', in Julia Swindells and David Francis Taylor, eds, *The Oxford Handbook of the Georgian Theatre 1737–1832* (Oxford: Oxford University Press, 2014), pp. 551–67, on p. 555.

The poets often endeavoured to give an extra spice and savour to their prologues and epilogues by entrusting the delivery of the addresses to a young girl. It appears …especially piquant that wanton rhymes should be pronounced by lips which if not innocent were at any rate tender and bland, and a smutty jest was winged with far livelier point if given with seeming simplicity and ingenuous artlessness.[18]

As Crouch has noted, the social position of eighteenth-century actresses was problematic, as 'actresses and the theatres in which they performed were obvious targets for criticism from both moral reformers intent on chastising the stage, and satirical authors interested in relating scandal'.[19] On the one hand, either sort of author could criticize an actress like Lucretia as a 'whore', immodestly displaying herself on the stage, but her own independent income and self-fashioning might allow her to emulate the social elite and purchase the accoutrements of gentility.[20] It was a delicately liminal state for Lucretia, requiring her shrewd management of her choice of roles and determined self-fashioning to maintain her public reputation as well as audience acclaim.

Lucretia would have faced a crowd of mixed gender, mostly from upper social groups but with some servants and bourgeois. The pit in front of the stage was where the wits, critics, and gallants stood, 'the sphere of rowdiest licence and quarrels that led to brawls and duelling'.[21] The boxes were reserved for persons of quality, chosen for their more reserved atmosphere. As Dryden noted,

> This makes our Boxes full; for men of Sense
> Pay their four Shillings in their own defence:
> That safe behind the Ladies they may stay;
> Peep o'er the Fan, and Judg, the bloudy Fray.[22]

Although the boxes often had a restricted view of the stage, they were the best place to see the audience and to be seen—important to members of the Georgian beau monde for whom the theatre was an extended drawing room salon, members' club, fashion show, and marriage market. Ladies in their illegal vizard masks added to the air of social intrigue.[23] The tumultuous middle gallery was most popular, with a variety of attendees, and the upper gallery, where prostitutes and

---

[18] Montague Summers, *The Restoration Theatre* (New York: Macmillan, 1933), p. 178.
[19] Kimberly Crouch, 'The Public Life of Actresses: Prostitutes or Ladies?', in Hannah Barker and Elaine Chalus, eds, *Gender in Eighteenth Century England: Roles, Representations and Responsibilities* (London: Taylor and Francis, 1997), pp. 58–79, on p. 58.
[20] Crouch, 'Public Life', p. 58.   [21] Summers, *The Restoration Theatre*, p. 67.
[22] John Dryden, 'Epilogue to the Duke of Guise 1682', in Montague Summers, ed., *Dryden the Dramatic Works*, 6 vols (London: Nonsuch Press, 1931), vol. 5, p. 265, quoted in Summers, *The Restoration Theatre*, p. 67.
[23] *London Gazette*, 17–20 1703/4, Queen Anne's 'Proclamation against vice and immorality in the theatre' stated: 'That no Woman be allow'd or presume to wear a Vizard-Mask in either of the Theatres'.

orange wenches plied their humble wares, was least fashionable.[24] The anonymous satirical tract, *Character of a Beaux*, described the 'Nice, Affected Beaux' at the theatre, flaunting himself:

> he very soberly enters the House; first in one side Box, then in t'other; next in the *Pit*, and sometimes in the *Galleries*, that the *Vulgar* sort may as well behold the *Magnificence* of his *Apparel*, as those of *Quality*. Before the *Play's* half done, whip he's at t'other House, and being in the Pit, between every *Act* leaps upon the benches, to show his Shape, His Leg, his Scarlet Stockings, his Meen and Air; then out comes a *Snuff-Box* as big as an *Alderman's Tobacco Box*, lin'd with a bawdy Picture and the Hand's very gracefully lifted to the Nose, to shew the length of its Fingers, its whiteness, its delicacy, and the Diamond Ring; and having played a few Monkey Tricks, the Musick ceases, and the Gentleman descends, bowing this way, that way, and t 'other way, that the Ladies in the Boxes may take notice of him, and think him a Person of *Quality*, known and respected by every body.[25]

Audiences in the Georgian theatre also had a fairly interactive relationship with stage performers, with intermittent applause, laughter, heckling, or even riots; in 1737, for example, Charles Fleetwood, then manager at Drury Lane, attempted to deny footmen free admission to the top gallery, and it took a guard of fifty soldiers to subdue them.[26] It was not uncommon for audience numbers to watch the same play several times, and they 'prided themselves on their familiarity with the material, and their ability to spot when actors changed a piece of stage business or missed a line'.[27] There is also evidence to suggest that performers—and not plays—were the reason people attended the theatre: stars such as Anne Oldfield, Elizabeth Barry, or, later, David Garrick, were all bound to draw an audience.[28]

The audience reaction to Lucretia must have been reasonably favourable, as 'Miss Bradshaw' spoke another epilogue to Pix's *The Deceiver Deceiv'd* the following year, a 'small Embassadress for Grace, if there was power in such a Childish Face' (see figure 3.1). She was soon launched into her theatrical career.

[24] Summers, *The Restoration Theatre*, p. 67.
[25] Anonymous, *The Character of a Beaux in Five Parts* (London: s.n., 1696), pp. 14–15.
[26] Rev. J. Genest, *Some Account of the English Stage, from the Restoration in 1660 to 1830*, 10 vols (Bath: H. E. Carrington, 1832), vol. 2, p. 499, as quoted in Rebecca Wright, *The Georgian Theatre Audience: Manners and Mores in the Age of Politeness, 1737–1810*, MA Thesis (University of London, 2014), p. 49.
[27] Cecil Price, *Theatre in the Age of Garrick* (Oxford: Wiley-Blackwell, 1973), p. 3, as quoted in Wright, *The Georgian Theatre Audience*, p. 21.
[28] Wright, *The Georgian Theatre Audience*, p. 21.

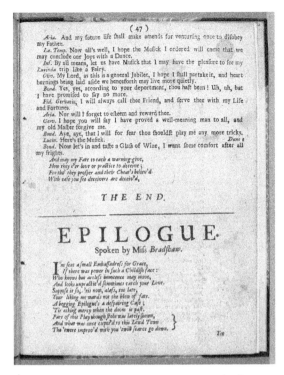

Fig. 3.1   Lucretia Bradshaw's Spoken Epilogue in Mary Pix's *The Deceiver Deceiv'd* (London: R. Basset, 1698), pp. 47–8, call number 147237, The Huntington Library, San Marino, California.

3.2   A Rising Star

Her next role in June 1697 was as the child of Hercules in a farcical masque of the same name by dramatist and librettist Peter Anthony Motteux (1663–1718); Motteux's libretto was a masque for John Eccles to perform in the intermissions of Motteux's play *The Novelty*. At the heart of the masque is the volatile and muscular Hercules, essentially a buffoon who rants about the disasters that ensue when 'Women rule', and a 'comic victim' of the humiliation they inflict on him.[29] For instance, when Hercules' attention is diverted by the lovely Omphale, he tries to please her by offering any favour. She replies, 'all you Men, when Love is new, Promise much, but little do'; she then asks him to learn to spin, initiating a spinning scene in which Hercules is stripped of his lion's skin and clothed in a white hood, a night rail, and a white bib apron.[30] This is then an excuse for a diverting

[29]   Richard Rowland, *Killing Hercules: Deianira and the Politics of Domestic Violence, from Sophocles to the War on Terror* (New York: Routledge, 2016), p. 161.
[30]   David Ross Hurley, 'Dejanira, Omphale, and the Emasculation of Hercules: Allusion and Ambiguity in Handel', *Cambridge Opera Journal* 11, 3 (1999), pp. 199–214, on p. 201.

pantomime and dance featuring two women and two men, and two spinning wheels turned with choreographed postures and motions.[31] Shortly afterwards, Dejanira, Hercules' wife, enters with their children and upbraids Hercules for his flirtatiousness with other women, whereupon the children begin to weep. In this scene, Lucretia first displayed her considerably useful ability to sing; this decade in the English theatre was one of experiment which linked drama and music to satisfy the tastes of a fickle audience 'familiar with Italian violin sonatas by Arcangelo Corelli, French ballet airs by Jean-Baptiste Lully, and Henry Purcell's dramatick operas, as well as street-ballads and the musical drolls at Bartholomew Fair'.[32]

With the nine-year-old James (Jemmy) Larouche, Lucretia sang a comic plaint 'Hee, oh! pray Father' (composed by Eccles with verse by Motteux) for Hercules to return home and see them and their mother Dejanira. His absence (in a bit of double entendre) 'at night makes her Moan, as she cannot lie alone'. Following the song, Hercules attempts to leave, but agrees to put on the shirt Dejanira made for him, and Lucretia apparently dances for joy at the thought of her 'father' returning. Unfortunately, the shirt was dipped into the poisoned blood of the centaur Nessus and was no 'love charm, but an instrument of revenge'.[33] Motteux also wrote a similar juvenile dialogue—'Pretty Miss, let's talk together'—for the play *Love's a Jest*, sung again by Lucretia and Jemmy; it was 'definitely designed to compete' with Drury Lane Theatre, which was offering similar fare sung by Letitia Cross and Jemmy Bowen.[34] Throughout her career, Lucretia would sing at the theatre and at subscription music concerts, performing, for instance, the favourite air 'From Grave Lessons and Restraint' by John Weldon on 26 April 1705.[35]

Benefit concerts on Lucretia's behalf soon followed, the first held on 15 December 1698 at York Buildings on Villiers Street off the Strand, the first proper concert hall in London; on 8 May 1700, she shared a benefit concert at the same place.[36] As the lawyer and amateur musician Roger North (1653–1734) noted in his memoirs, 'in York buildings, a fabrick was reared and furnished on purpose for publick musick. And there was nothing of musick valued in towne, but was to be heard there. It was called the Musick-meeting; and all the quality and *beau mond* [sic] repaired to it'.[37]

[31] Kathryn Lowerre, *Music and Musicians on the London Stage, 1695–1705* (London and New York: Routledge, 2016), p. 53.

[32] Lowerre, *Music and Musicians*, p. 3.   [33] Lowerre, *Music and Musicians*, p. 54.

[34] Lowerre, *Music and Musicians*, p. 45, footnote 53.

[35] Lowerre, *Music and Musicians*, p. 363.

[36] Julie Anne Sadie, ed., *Companion to Baroque Music* (Berkeley: University of California Press, 1998), p. 267.

[37] Edward F. Rimbault, ed., *Memoirs of Musick by the Hon. Roger North* (London: George Bell, 1846), pp. 112–14.

Performers could generally expect 170 days of wages per annum to accommo-date the theatre season, which ran early September to June and often were short-paid; benefit concerns were thus a lucrative technique to boost annual earnings, with actors sometimes earning half their annual wage.[38] The benefit season would open in March, after the theatre owners had had their chance to make as much income as possible during the regular season. Actors would typically sell their own tickets, out of their homes or from a coffee-house, and the theatre owner generally only kept a nightly benefit charge of approximately £40, enough to cover operating costs for keeping the theatre open one night (see figure 3.2). At least, this was the way it was supposed to work. The actors did not always receive their due. In March 1705, an anonymous, unacted comedy called *The lunatic* was pub-lished, with a dedication to The *Three* Ruling *B——s* of the New-House in Lincoln's Inn-Fields [Thomas Betterton, Anne Bracegirdle, and Elizabeth Barry, the trium-virate management of that theatre]. The dedication accused them of 'stopping all the Pay of the Under Actors on Subscription-Nights, when you were allow'd forty or fifty Pound a Night for the House, besides the Benefit of the Galleries.'[39]

In the early eighteenth century, subscription 'musick' concerts also arose among the nobility to fund all-sung operas in the Italian style on the London stage.[40] These were exclusive events, the 1703–4 subscription 'musick' concerts pricing gallery seats to non-subscribers at 2/6 and 1/6: more expensive than prices charged for new plays.[41] Lucretia performed regularly in these benefits when she became a member of the troupe of the new Queen's Theatre at the Haymarket, moving from Lincoln's Inn Fields. John Vanbrugh, the young play-wright and architect of Queen's, had reckoned correctly that actors and actresses like Lucretia wished to perform in a new and elegant theatre.[42] Betterton's Lincoln's Inn Fields had since been in disorder, so Vanbrugh built his own theatre on speculation, asking 'twenty-nine aristocrats and fellow members of the Whig Kit-Cat Club' to contribute one hundred guineas each towards construction.[43] The theatre was constructed on a former stable yard, which led to much public comment, Daniel Defoe noting in his satirical newspaper, *A Review of the Affairs of France,*

[38] Philip Highfill, Jr, Kalman A. Burnim, and Edward A. Langhans, eds, *A Biographical Dictionary of Actors, Actresses, Musicians, Dancers, Managers and Other Stage Personnel in London, 1660–1880,* 16 vols (Carbondale: Southern Illinois University Press, 1973–1993), vol. 2, p. 284.

[39] Olive Baldwin and Thelma Wilson, 'The Subscription Musick of 1703–4', *The Musical Times* 153, 1921 (Winter 2012), pp. 29–44, on pp. 32, 34.

[40] Baldwin and Wilson, 'The Subscription Musick of 1703–4', p. 29.

[41] Baldwin and Wilson, 'The Subscription Musick of 1703–4', p. 44.

[42] Milhous and Hume, *Theatrical Papers,* p. xvii.

[43] Graham F. Barlow, 'Vanbrugh's Queen's Theatre in the Haymarket, 1703–9', *Early Music* (November 1989), pp. 515–22, on p. 516.

Fig. 3.2 Theatre ticket design for *The Mock Doctor*, a benefit at the 'Theatre Royal', designed by William Hogarth (?), *c*. 1732. 5 13/16 × 5 11/16 in., Metropolitan Museum of Art, New York, Public Domain.

> A Lay-stall this, Apollo spoke the Word,
> And straight arose a Playhouse from a Turd...
> The Stables have been Cleans'd, the Jakes made Clear,
> Herculean Labours, ne'r will Purge as here.[44]

The Theatre was probably built with the assistance of architect Nicholas Hawksmoor, and it also had a long room for masquerades and assemblies.[45]

---

[44] Daniel Defoe, 'On the Playhouse in the Haymarket', *A Review of the Affairs of France* (3 May 1705), vol. 2, pp. 103–4, as quoted by Barlow, 'Vanbrugh's Queen's Theatre', p. 516.

[45] 'Her Majesty's', *Theatres Trust*, https://database.theatrestrust.org.uk/resources/theatres/show/1993-her-majesty-s-london [Accessed 15 October 2019].

3.3 Mrs Bradshaw: A Seasoned Actress

By 1703, Miss Bradshaw was now styled 'Mrs Bradshaw', moving away from roles of innocent children or ingenues to characterizations of good-hearted women in comedic plays. These roles required rapid repartee, demonstrating her native intelligence and association with positive qualities to enhance her public reputation.

In May 1704, Lucretia performed Weldon's 'From grave lessons and restraint', likely an entr'acte in the farce *Squire Trelooby*, which had been translated from Molière's *Monsieur de Pourceaugnac* by William Congreve, John Vanbrugh, and William Walsh, who performed an act each.[46] The play was described by Congreve as 'a compliment made to the people of quality at their subscription music, without any design to have it acted or printed farther. It made people laugh.'[47]

In March 1705, the play *Love Betray'd* by Burnaby was revived as a benefit night for Lucretia and the actor George Pack, advertised with singing by Mary Hodgson.[48] By 1705, she was playing major roles at the Haymarket, such as Corinna in the new comedy *The Confederacy* (30 October 1705) and Constantia in *The Faithful General* (3 January 1706), characters appropriate to a young actress of sparkling wit.[49] In John Vanbrugh's *Confederacy*, Corinna, the sixteen-year-old daughter of Gripe and Clarissa Gripe, is pursued by Dick Amlet. Despite the fact Amlet poses as a colonel to lie his way into her affections, Corinna still loves him. She sees beyond his masquerade and, appreciating his good qualities, shows herself as a character of patience, humour, and spirit (perhaps the qualities Lucretia herself possessed). Commenting to her maid Flippanta about her father, Corinna states, 'I'm but a girl 'tis true, and a fool too, if you believe him; but let him know, a foolish girl may make a wise man's heart ache; so he had as good be quiet.'[50] In another delightful scene, on receiving Amlet's love letter, Corinna perceptively says the following to Flippanta:

CORINNA: Let me read it, let me read it, let me read it, I say. Um, um, um, *Cupid's*, um, um, um, *darts*, um, um, um, *beauty*, um, *charms*, um, um, um, *angel*, um, *goddess*, um (Kissing the letter), um, um, um, *truest lover*, um, um, *eternal constancy*, um, um, um, *cruel*, um, um, um, *racks*, um, um, um, *tortures*, um, um *fifty daggers*, um, um *bleeding heart*, um, um *dead man.*—Very well, a mighty civil letter I promise you; not one naughty word in it, I'll go lock it up in my work-box.

[46] Baldwin and Wilson, 'The Subscription Musick of 1703–4', p. 42.
[47] Letter from William Congreve to Joseph Kelly, 20 May 1704 in William Congreve, *Works*, ed. D. F. McKenzie and C. Y. Ferdinand, 3 vols (Oxford: Oxford University Press, 2011), vol. 3, p. 160.
[48] Lowerre, *Music and Musicians*, p. 313.   [49] *Biographical Dictionary*, vol. 2, p. 284.
[50] Sir John Vanbrugh, *The Confederacy: A Comedy* (New York and Boston: Wells and Lilly, 1823), p. 39.

FLIPPANTA: Well—but what does he say to you?
CORINNA.   Not a word of news, Flippanta, 'tis all about business.[51]

Unfortunately, despite the brilliance of the performances, the utility of the Queen's Theatre at the Haymarket was sacrificed to the aesthetics of Palladianism. Although Vanbrugh wished to 'create an aesthetically satisfying, unified architectural structure that would in its own terms echo the harmony of the music to be performed within its walls', he did not address practical concerns.[52] The building lacked backstage spaces, and its decoration consisting of large *trompe-l'oeil* paintings on flat surfaces caused severe echoing.[53] The actor Colley Cibber noted about the building,

> every proper Quality and Convenience of a good Theatre had been sacrificed or neglected to shew the Spectator a vast triumphal Piece of Architecture!...For what could their vast Columns, their gilded Cornices, their immoderate high Roofs avail, when scarce one Word in ten could be distinctly heard in it?...At the first opening it, the flat Ceiling that is now over the Orchestra was then a Semi-oval Arch that sprung fifteen Feet higher from above the Cornice; the Ceiling over the Pit, too, was still more raised, being one level Line from the highest back part of the upper Gallery to the Front of the Stage: The Front-boxes were a continued Semicircle to the bare walls of the House on each Side: This extraordinary and superfluous Space occasion'd such an Undulation from the Voice of every Actor, that generally what they said sounded like the Gabbling of so many People in the lofty Isles in a Cathedral....'.[54]

Cibber also added that the location of the theatre 'at that time...had not the Advantage of almost a large City, which has since been built in its Neighbourhood: Those costly Spaces of *Hanover, Grosvenor,* and *Cavendish* Squares, with the many and great adjacent Streets about them, were then all but so many green Fields of Pasture'.[55]

Not only did the theatre have poor acoustics, but Vanbrugh's plans to hold the monopoly on the performance of opera, particularly the new Italian variety, proved financially disastrous. The profitable success of *Arsinöe, Queen of Cyprus* and *Camilla* at Drury Lane in 1705 and 1706 suggested to him and many others

---

[51] Vanbrugh, *The Confederacy*, pp. 41–2.   [52] Barlow, 'Vanbrugh's Queen's Theatre', p. 516.

[53] 'Her Majesty's', *Theatres Trust*, https://database.theatrestrust.org.uk/resources/theatres/show/1993-her-majesty-s-london [Accessed 15 October 2019].

[54] *An Apology for the Life of Mr Colley Cibber, etc.*, vol. i, pp. 321–2, as quoted by 'The Haymarket Opera House', in Francis H. W. Sheppard, ed., *Survey of London: Volumes 29 and 30, St James Westminster, Part 1* (London: London County Council, 1960), pp. 223–50. *British History Online.* http://www.british-history.ac.uk/survey-london/vols29-30/pt1/pp223-250 [Accessed 25 November 2018]. There were later alterations to try to correct the acoustical problems.

[55] *An Apology for the Life of Mr Colley Cibber, etc.*, vol. i, pp. 321–2.

Fig. 3.3 The Italian Opera House at the Haymarket before the fire of 17 June 1789, original drawing by William Capon, etching by Charles John Smith, ca. 1837.
Charles John Smith, *Historical and literary curiosities, consisting of facsimiles of original documents* (London: H.G. Bohn, 1852), p. 294, archive.org, University of California Libraries.

that the genre would be economically sustainable.[56] On 31 December 1708, the Lord Chamberlain ordered all actors working at the Queen's Theatre to return to Drury Lane, with a monopoly on opera given to Vanbrugh at the Haymarket.[57] Although the 'Order of Union' seemingly gave a green light to the production of operas, Vanbrugh quickly realized how ruinously expensive they were to stage.[58] Opera singers were in short supply, and since they were not under contract to any particular theatre, 'they were free to demand any salaries that fancy or rapacity might suggest'; sets and costumes were also costly.[59]

[56] Judith Milhous and Robert D. Hume, eds, *The London Stage 1660–1800, Part 2: 1700–1729: Draft of the Calendar for Volume I, 1700–1711*, p. 373. http://www.personal.psu.edu/hb1/London%20 Stage%202001/lond1707.pdf [Accessed 17 October 2019]. The draft was edited and compiled by Milhous and Hume but has remained unpublished.
[57] 'Theatre historians call Kent's decree the "Order of Union" because it enforced the amalgamation of the two acting companies and returned all actors to Drury Lane'; not to be confused with the 'Act of Union'. See Milhous and Hume, *The London Stage*, p. 373.
[58] Milhous and Hume, *The London Stage*, p. 373. http://www.personal.psu.edu/hb1/London% 20Stage%202001/lond1707.pdf [Accessed 17 October 2019].
[59] Milhous and Hume, *The London Stage*, p. 373. http://www.personal.psu.edu/hb1/London% 20Stage%202001/lond1707.pdf [Accessed 17 October 2019].

Vanburgh anticipated an income of £120 per night for a season of sixty-four nights, with expenses of £6,000 or £94 per night; by April 1708, the total per night of expense was closer to £116, and he did not receive his projected profits.[60] Despite his making alterations to the interior in 1707–8 when boxes were constructed, during the spring of 1708 the Haymarket only offered twenty-nine performances.[61] Not surprisingly, after he was not granted a Royal subsidy Vanbrugh gave up, and in May 1708 he bailed out and left the whole operation to [Owen] Swiney, to end his own financial liability.[62] In July 1708, Vanbrugh wrote to the Earl of Manchester: 'I lost so Much Money by the Opera this Last Winter, that I was glad to get quit of it', then remarked presciently, 'and yet I don't doubt but Operas will Settle and thrive in London'.[63] Indeed, upon the accession of King George I in 1714, the theatre was renamed the King's Theatre and premiered more than twenty-five operas by George Frideric Handel between 1711 and 1739 (see figure 3.3).[64]

It was a notorious fiasco. Elizabeth Barry, Lucretia's mentor, even temporarily retired for a year after her performance as Sophonisba on 17 June 1708, seemingly precipitated by the Order of Union of 1708 and its aftermath.[65] In the ensuing chaos, it is little wonder that Lucretia found her career at the Haymarket untenable.[66]

With the actors Mary Porter, John Verbruggen, and George Park, Lucretia left the Haymarket in November 1707, making terms with Christopher Rich at Drury Lane, who was seducing actors away to reinforce his own company. Unfortunately, actors and actresses were not allowed legally to change playhouses without permission and a formal discharge from the relevant theatre. Bradshaw thus fell under the Lord Chamberlain's displeasure and had to petition him for permission to resume acting.[67] In January 1708, Colonel Henry Brett, a shareholder in the Drury Lane Theatre, wrote a letter to Lord Chamberlain Henry Grey, Earl of Kent on her behalf. Brett was a friend of Addison and Steele, referred to as 'Colonel Ramble' in their series The Tatler, and a member of a literary club which frequented Button's coffee-house. Brett wrote:

[60] Milhous and Hume, Theatrical Papers, pp. 41–2, 71, and 78.

[61] Milhous and Hume, Theatrical Papers, p. 78.

[62] Milhouse and Hume, Theatrical Papers, p. 28.

[63] Bonamy Dobrée and Geoffrey Webb, eds, The Complete Works of Sir John Vanbrugh, 4 vols (London: Nonesuch Press, 1927–8), vol. 4, p. 24.

[64] 'The Architecture of Opera Houses', Victoria & Albert Museum, London. https://www.vam.ac.uk/articles/opera-architecture [Accessed 17 November 2019].

[65] John Harold Wilson, All the King's Ladies: Actresses of the Restoration (Chicago: University of Chicago Press, 1958), p. 115; Martin, 'Prosopographical', p. 65.

[66] Milhous and Hume, The London Stage, p. 402. http://www.personal.psu.edu/hb1/London%20Stage%202001/lond1707.pdf [Accessed 17 October 2019].

[67] MS/LC/5/154, 'To the Managers of the Theatre in Drury Lane' by Lord Chamberlain Kent on 31 December 1707; letterbook copy, f. 159, in Milhous and Hume, The London Stage, p. 373. http://www.personal.psu.edu/hb1/London%20Stage%202001/lond1707.pdf.

Dear Sir:

I hope this will find you in a disposition to suffer Mrs Bradshaw to Play upon the terms she has agreed on with Mr Rich. She has convinc'd me she left Mr Vanbrug for reasons that will very well excuse her, and at least if you shou'd severely think there's no room for favour to her as a Player, I hope you will joine with me in not being able to refuse her any thing as Mrs Bradshaw. If I have been too sollicitous to you in this affair, I hope you'll forgive me for preferring her interest and the interest of ye House to any other consideration.[68]

Brett's missive must have done the trick, because Lucretia was acting with Drury Lane again by January 1708, appearing in *The Taming of the Shrew*, Jonson's *Bartholomew-Fair*, as Ophelia in *Hamlet*, as well as in *The Double Gallant*, a new comedy by Colley Cibber, and old warhorses such as John Crowne's *Sir Courtly Nice*; she averaged nine performances each month and had a benefit on her behalf on 20 April 1708, with her colleagues performing the classic comedy of manners, *The Man of Mode, or, Sir Fopling Flutter* by George Etherege.[69] Her performance as Constantia in John Fletcher's perennially popular play *The Chances*, alongside Robert Wilks as Don John and the veteran actress Anne Oldfield, also received an advance puff in *The Tatler*:

It is a true Picture of Life…wherein Don John and Constantia are acted to the utmost Perfection. There need not be a greater Instance of the Force of Action, than in many Incidents of the Play, where indifferent Passages, and such that conduce only to the Tacking of the Scenes together, are enlivened with such an agreeable Gesture and Behaviour, as apparently shows what a Play might be, though it is not wholly what a Play should be.[70]

As her stature as an actress grew, Lucretia was also trusted to act in important events. Her portrayal of witty lady's maid Luce on 3 June 1708 in Thomas Shadwell's *Bury Fair* was a special performance for the Ambassador of the Czar of Russia; the cost of £29 9/6 for the box used by the Russians for a variety of performances was borne by the British government one year later when the diplomats left without paying.[71] Satirizing his hometown, Shadwell displaces a London spark named Wildish and a French pretender into the provincial world of Bury

---

[68] Milhous and Hume, *Theatrical Papers*, pp. 52–3.

[69] Milhous and Hume, *The London Stage*, p. 429. http://www.personal.psu.edu/hb1/London%20Stage%202001/lond1707.pdf [Accessed 17 October 2019].

[70] Donald F. Bond, ed., *The Tatler*, 3 vols (Oxford: Clarendon Press, 1987), vol. 3, 37 (27–9 June 1710), as quoted in Robert D. Hume and Harold Love, eds, *Plays, Poems, and Miscellaneous Writings Associated with George Villiers, Second Duke of Buckingham*, 2 vols (Oxford: Oxford University Press, 2007), vol. 1, p. 24.

[71] Milhous and Hume, *The London Stage*, p. 439. http://www.personal.psu.edu/hb1/London%20Stage%202001/lond1707.pdf [Accessed 17 October 2019].

St Edmunds; in a satiric self-referent, Shadwell had Wildish remark, 'But one thing I can tell of thy Town, That it can produce a Blockhead'. On 9 October 1708, Lucretia also played Desdemona in *Othello* for the entertainment of the Moroccan ambassador.[72] That year, on order of the Lord Chamberlain, Lucretia was sworn in as a Queen's servant with thirty-four other players at Drury Lane 'for ye better regulation…of the stage', with her name scribbled in the margins of the manuscript with the other reinstated defectors.[73]

One would have hoped that after all the turmoil that Lucretia endured, she would have spent the rest of her days at Drury Lane in some semblance of peace. Indeed, the 1708–9 season was reasonably quiet and orderly, although the theatre lost more than six weeks of performance in the autumn due to the mourning for Prince George. On 12 May 1709, the season also saw the premiere of Susanna Centlivre's *The Busie Body*, which became a stock piece in the repertory until the nineteenth century, its success a great surprise to Georgian theatre goer John Mottley:

> There had been scarce any thing mentioned of it in the Town before it came out, and those who had heard of it, were told it was a silly thing wrote by a Woman, that the Players had no Opinion of it, and on the first Day there was a very poor House, scarce Charges. Under these Circumstances it cannot be supposed the Play appeared to much Advantage, the Audience only came there for want of another Place to go to, but without any Expectation of being much diverted; they were yawning at the Beginning of it, but were agreeably surprized, more and more every Act, till at last the House rung with as much Applause as was possible to be given by so thin an Audience.[74]

In *The Tatler*, Richard Steele had high praise of it: 'The plot and incidents of the play are laid with that subtlety of spirit which is peculiar to females of wit, and is very seldom well performed by those of the other sex, in whom craft in love is an act of invention, and not, as with women, the effect of nature and instinct'.[75] The play certainly has sharp, rapid pace and timing, and it is tempting to think Lucretia, as a leading lady at Drury Lane, had a role, although the identity of the

---

[72] Milhouse and Hume, *The London Stage*, p. 453. http://www.personal.psu.edu/hb1/London%20Stage%202001/lond1708.pdf [Accessed 21 August 2020].

[73] Milhous and Hume, *The London Stage*, pp. 406–7. http://www.personal.psu.edu/hb1/London%20Stage%202001/lond1707.pdf [Accessed 17 October 2019]. See also Arthur H. Scouten and Robert D. Hume, 'Additional Players' Lists in the Lord Chamberlain's Registers, 1708–1710', *Theatre Notebook* 37 (1983), pp. 77–9.

[74] [John Mottley], *A Complete List of All the English Dramatic Poets and of all the Plays ever Printed in the English Language, to the Present Year M, DCC, XLVII*, appended to Thomas Whincop, *Scanderbeg: or, Love or Liberty. A Tragedy* (London: W. Reeve, 1747), pp. 185–6. Like any theatre gossip, this account may not be entirely trustworthy.

[75] *The Tatler* 15 (14 May 1709).

cast was not provided.[76] We do know, however, that Lucretia reprised her role as Ophelia in *Hamlet*, as well as playing Cordelia in *King Lear* for the first time, with Thomas Betterton as Lear and Colley Cibber as the Earl of Gloucester.[77] The role of Cordelia in October 1708 was one she inherited from Elizabeth Barry; Lucretia replaced her during her temporary retirement and played the role until 1714.[78] Lucretia was not performing from Shakespeare's original play, but from Nahum Tate's 1681 tragicomic adaptation: *The True and Ancient History of King Lear and his Three Daughters*, which had a happy ending. In Tate's version, Lear regained his throne and Cordelia married Edgar, a 'tribute to the new practice of having women actors on the Restoration stage'; in the original, the two characters never address each other.[79] Tate adapted the play in 1681, making a parallel between Charles II coming back to his throne and Lear's 'blest Restauration', and his version was popularly performed until the nineteenth century.[80] Lucretia's experience at playing female heroines in love and ingenue daughters would have well prepared her for the role.

Unfortunately, the troupe at Drury Lane was increasingly unhappy about working for Rich, who was taking extra money from actors' benefits; his cheating of leading actress Anne Oldfield by detaining one third of the profits beyond charges led to her formally complaining to the Lord Chamberlain. 'The Lord Chamberlain made no response and took no public action for almost two months, so Rich proceeded to "tax" other performers' benefits in like manner. Kent's seeming lack of concern was a ruse. In actuality, he was in cahoots with the actors'.[81] On 30 April 1709, Kent ordered the Drury Lane management to pay to the respective players who have 'had benefit plays...the full receipts of such plays deducting only from each the sume of £40 for the charges of the house pursuant to the Articles'.[82] The point was clearly to lead Rich into breaching a formal order from the Lord Chamberlain, which he did. The Lord Chamberlain then shut down Drury Lane Theatre on 6 June, imposing an Order of Silence, and on 18 July he gave the Haymarket the right to hire any actors it chose to play comedies and tragedies, breaking the Order of Union. Kent's intent was to put Rich out of

---

[76]   Frederick Peter Lock, 'The Dramatic Art of Susanna Centlivre', PhD Thesis (McMaster University, 1974), chapter IV, *passim* for an analysis of *The Busie Body* and its reception.

[77]   Theatre Calendar in Milhous and Hume, *The London Stage*, p. 444. http://www.personal.psu.edu/hb1/London%20Stage%202001/lond1708.pdf [Accessed 17 October 2019].

[78]   Martin, 'Prosopography', p. 67.

[79]   'Shakespeare in Performance: Stage Production', *Internet Shakespeare Editions*, University of Victoria. https://internetshakespeare.uvic.ca/Theater/production/stage/2779/index.html [Accessed 21 August 2020].

[80]   Sonia Massai, 'Nahum Tate's Revision of Shakespeare's *King Lears*', *Studies in English Literature* 40, 3 (Summer 2000), pp. 435–50, on p. 436.

[81]   Milhous and Hume, *The London Stage*, p. 444. http://www.personal.psu.edu/hb1/London%20Stage%202001/lond1708.pdf [Accessed 17 October 2019].

[82]   TNA LC5/154/437, The National Archives, Kew, as quoted in Judith Milhous and Robert D. Hume, 'The Silencing of Drury Lane in 1709', *Theatre Journal* 32, 4 (December 1980), pp. 427–47, on pp. 435–6.

business. However, not all actors were invited to the Haymarket, and their pay would be curtailed, as two days a week were devoted to Opera; the immediate effect of the Order of Silence was to bar junior actors from making a living during the summer season.[83] The squabbles and their effects on Drury Lane were significant for both the actors and the playwrights. Susanna Centlivre remarked in the *Female Tatler* on the premier of her play, *The Man's Bewitch'd*:

> twas much easier to Write a Play than to get it Represented; that their Factions and Divisions were so great, they seldom continued in the same mind two Hours together; that they treated her…as they do all Authors with Wrangling and Confusion…they had actually cut out the Scene in the Fifth Act. Between the Countryman and the Ghost, which the Audience receiv'd with that wonderful Applause, and 'twas with very great struggling the Author prevail'd to have it in again; one made Faces at his Part, another was Witty upon her's.[84]

Drury Lane eventually opened again for the 1709–10 season under the new management of William Collier, 'a Tory MP with a court interest and an interest in the sleeping [theatre] patent', and his deputy Aaron Hill, an aspiring dramatist with little experience in running a theatre.[85] Hill faced many challenges during his tenure as manager. Queen's had a much stronger company, Drury Lane being led by Barton Booth, 'scarce out of his Minority as an Actor' and George Powell, experienced, but with a reputation for 'heavy drinking and debauchery'.[86]

Despite these challenges, Hill was resilient, writing plays as well, including *Elfrid, or the Fair Inconstant*, staged on 3 January 1710, where Lucretia played the title role.[87] He was also a journalist and entrepreneur, and a founder of the question-and-answer periodical *The British Apollo* (1708–11), an imitation of John Dunton's *Athenian Mercury*. Hill used the paper to puff his roles as manager at Drury Lane and director of the Opera at the Haymarket.[88]

So, although Hill's tenure was short, it was not a complete failure; the crowds flocked to see Lucretia play Arabella Zeal in Charles Shadwell's play *The Fair Quaker of Deal*. Barton Booth played the sea captain Worthy, in love with the Fair Quaker Dorcas Zeal, played by the gifted dancer Hester Santlow. Dorcas loves Worthy in return, but her sister Arabella also loves the sea captain, who rejects her. Arabella's schemes to wreak revenge on Worthy and Dorcas by lies and disguise fail, and the couple are reconciled. In the finale, Worthy asks the pious Dorcas to dance in celebration, and she finally agrees: 'Well rather than spoil your Mirth, I will walk about'. As Santlow was already known for her dancing, it

---

[83] Milhous and Hume, 'Silencing', p. 436.   [84] *The Female Tatler*, 12–14 December 1709.

[85] Christine Gerrard, *Aaron Hill: The Muses' Projector, 1685–1750* (Oxford: Oxford University Press, 2003), p. 25.

[86] Goff, *Hester Santlow*, p. 24.   [87] Gerrard, *Aaron Hill*, p. 27.

[88] Gerrard, *Aaron Hill*, p. 19.

brought the house down; Santlow's line in the epilogue, 'Because with much con-straint I've set my Face. To carry on a Quaker's dull Grimace' also drew applause.[89] Colley Cibber noted of Santlow that the 'gentle Softness of her Voice, the com-posed Innocence of her Aspect, the Modesty of her Dress, the reserv'd Decency of her Gesture, and the Simplicity of the Sentiments, that naturally fell from her, made her seem the amiable Maid she represented'.[90] In the preface to the play, Shadwell not only remarks upon the remarkable performances of Santlow, but also of Lucretia Bradshaw, Mr Pack, and Mr Leigh.

Though the *British Apollo* celebrated Hill as manager of Drury Lane, calling him a 'MIGHTY GENIUS' who 'now sits at thy Helm', Hill could not control the several internal management squabbles between him and a group of actors led by George Powell who 'wanted to handle management their way'.[91] By May 1710, the troupe was openly flouting Hill and refused to act. As theatre historians Milhaus and Hume indicated,

Hill seems to have tried to buy the actors off by scheduling second benefits for five senior actors between 23 May and 6 June, but their open insolence made him 'send an order to Stockdale, not to open ye Doors that night, (last Fryday) [i.e. 2 June] till I sent him a Guard of Constables to keep ye Peace & protect him from being insulted in his Duty'.[92]

There was an ensuing riot, the first and only one in which Lucretia participated, perhaps as an innocent victim, though it is difficult to determine. At 4 p.m., before the play was to start, several actors including Barton Booth, George Powell, Theophilus Keen, and Francis Leigh broke open the doors of the theatre within, Lucretia letting them in 'thro' a private way, from her Lodging'. Lucretia lived next door to the theatre and had direct access to it from her house, though it is not known whether she knew of the actors' intentions.[93]

It was usual for the outer or street doors of the theatre to be opened first, to admit the audience to the hallways that led to the auditorium; the inner doors were, however, closed until an hour before the performances. Unless one had a private box, seats were not reserved, so audience members would often arrive early in hopes of getting a good seat, buying tickets from door keepers. It was often chaotic in the best of times, but the scene quickly turned into a riot-ous melee.[94]

[89] Goff, *Hester Santlow*, pp. 28–9.
[90] Cibber, *An Apology for Colley Cibber*, pp. 347–8; see also Goff, *Hester Santlow*, pp. 27–8.
[91] Gerrard, *Aaron Hill*, p. 27; *Biographical Dictionary*, vol. 2, p. 285.
[92] Milhous and Hume, 'Silencing', p. 444.   [93] *Biographical Dictionary*, vol. 2, p. 285.
[94] Charles Beecher Hogan, *The London Stage 1776–1800: A Critical Introduction* (Carbondale: Southern Illinois University Press, 1968), pp. xx–xxvii, quoted in Wright, *The Georgian Theatre Audience*, p. 24.

A crowd of actors subsequently burst into the manager's office to confront Hill, swords in hands. As Hill related, Powell:

> had shorten'd his sword to stab me in ye Back, and has cut a Gentleman's hand thro', who prevented the Thrust, [the actor] Leigh in ye mean Time, while my brother was held, struck him a dangerous Blow on ye Head, with a stick, from behind. This was done in the open Face of ye day, amongst Numbers of men & women, who came to see ye Play.[95]

The actors also enthusiastically greeted Christopher Rich, who 'happen'd to pass by in ye first Tumult…being huzz'd along ye Passage', indicating he was connected with the event.[96] Although the actors were suspended by the Lord Chamberlain, the melee gave Christopher Rich an opportunity to regain physical possession of Drury Lane and turf Collier and Hill out. It was not until 20 November 1710 that Drury Lane reopened for the season, with an actors' triumvirate of Robert Wilks, Colley Cibber, and Thomas Doggett in charge, and Rich was henceforth kept out of direct management.

By June 1710, Elizabeth Barry had retired, and Lucretia inherited her roles—Lavinia in Otway's *Caius Marius* (1679), considered by *The Spectator* to be her finest performance, 'Lucina in Rochester's *Valentinian* (1684), and Isabella in Southerne's *The Fatal Marriage* (1694)—all of which were parts in tragedies'.[97] She would also later appear in *Julius Caesar*, her costume bill 4s for a hood.[98]

That said, Lucretia only received three of Barry's twenty-one roles still in the repertory, which perhaps indicates either that she was not as valued by the management at this point, or that she may have been in a fraught relationship or in a rivalry with another leading actress.[99] Indeed, in an anonymous ephemeral publication (*A Justification of the Letter to Sir John Stanley*) addressed to the management at Drury Lane written in 1709, the author queried, 'And why is Mrs Bradshaw used with so much insolence, only to gratify the pride of Mrs Oldfield, who cannot bear any one should shine above her?'[100] Oldfield, a renowned actress known for her roles in the plays *Cato* and *Jane Shore*, also wielded considerable financial clout at Drury Lane. In 1711, she became:

---

[95] Milhous and Hume, *Theatrical Papers*, p. 144.   [96] Gerrard, *Aaron Hill*, p. 27.

[97] Martin, 'Prosopography', p. 79.

[98] MS Folger W.b. 111, f. 10a, 23 January 1714. The bill was approved by Wilks and Booth.

[99] Martin, 'Prosopography', p. 80.

[100] *A Justification of the Letter to Sir John Stanley, relating to his Management of the Play-house in Drury Lane*, Music Division, The New York Public Library, Astor, Lenox, and Tilden Foundations, quoted in Judith Milhous and Robert D. Hume, 'Theatrical Politics at Drury Lane, New Light on Letitia Cross, Jane Rogers and Anne Oldfield', *Bulletin of Research in the Humanities* (Winter 1982), pp. 412–29, on p. 420. For the details surrounding the letter, and a full transcription, see Judith Milhous and Robert D. Hume, 'A Letter to Sir John Stanley: A New Theatrical Document of 1712', *Theatre Notebook* 43, 2 (1989), pp. 71–80.

a managing partner and sharer with the triumvirate of actor–managers Cibber, Wilks, and Thomas Doggett. However, Doggett objected to a female sharer, and instead she was offered an assured £200 each year and a charge-free benefit. This rose to 300 guineas a year, and her benefit, according to Cibber's estimate, brought her twice that. This made Oldfield one of the highest paid performers in the company, although the figures are provided by Cibber in his memoirs, and might be suspected of being glossed.[101]

3.4 'Not a More Happy Couple': Mrs Lucretia Folkes

If Lucretia was 'being used with so much insolence' by the Queen Bee, Anne Oldfield, it would not have been the most conducive atmosphere for her to advance in her career. However, Lucretia would have the last laugh. Whereas Oldfield had two settled partnerships, first with Whig MP Arthur Mainwaring and then with Brigadier Charles Churchill, neither man would marry her due to her reputation and social standing; it was suggested in publication that 'she could easily fill the audience at her benefit performance with men who had been intimate with her'.[102] Lucretia, on the other hand, found marriage as her way out of the theatre and into gentility, a contemporary reporting that she was 'taken off the Stage, for her exemplary and prudent Conduct, by Martin Folkes, Esq; a Gentleman of a very considerable Estate, who married her; and such has been her Behaviour to him, that there is not a more happy Couple'.[103]

Lucretia was five to ten years older than Martin, and it was an unusual pairing. The nineteenth-century commentator John Doran, in his work, *Their Majesties' Servants*, wrote about the marriage:

At this period (about 1714) the stage lost a lady who was as dear to it as Queen Anne, namely Mrs. Bradshawe. Her departure, however, was caused by marriage, not by death; and the gentleman who carried her off, instead of being a rollicking gallant or a worthless peer, was a staid, solemn antiquary, Martin Folkes, who rather surprised the town by wedding young Mistress Bradshawe.[104]

[101] J. Milling, 'Oldfield, Anne (1683–1730), actress', *Oxford Dictionary of National Biography* (Oxford: Oxford University Press, 2008). https://doi.org/10.1093/ref:odnb/20677 [Accessed 2 January 2019].
[102] Crouch, 'Public Life', p. 60.
[103] Betterton, *The History of the English Stage*, p. 62. As for Drury Lane, Hester Santlow 'ultimately took over nine roles that had belonged to Lucretia…acquiring most of them during 1714–15'. Goff, *Hester Santlow*, p. 56.
[104] John Doran, *'Their Majesties' Servants', or Annals of the English Stage* (London: Wm. H. Allen and Co, 1865), p. 112.

Their appearances also jarred. Lucretia was reputed to be an exquisitely beautiful woman with a penchant for feathered headdresses; leading actresses were given clothing allowances for professional and personal wear, and 'throughout the period continued to dress at the height of fashion, especially in comedy' to associate themselves with gentility.[105] On the other hand, a contemporary miniature portrait of Martin by Bernard Lens shows a portly and pale figure with a double chin and unibrow, long powdered periwig, and dull brown overcoat, grasping an astronomical quadrant with a telescopic sight in a form familiar to modern surveyors. Not exactly a love token.[106] (see figure 4.9). As Archibald Bower commented, 'Indeed I never looked upon Martin Folkes Esq. as a very fashionable man, or a man who would do any thing merely to comply with the fashion'.[107] Bower was correct in more ways than one.

Both their self-fashioning and their marriage thoroughly subverted the usual sociocultural markers of class and gentility. One of the very first between a member of the gentry and an actress, it was characteristic of the freethinking and larger disregard of social norms that Folkes would evince throughout his life. 'The languages and customs of romantic love were completely transformed as the [eighteenth] century progressed', with the development of 'new cultures of sensibility and romanticism'.[108] The couple was on the cusp of the new celebration of romantic love that occurred in this period; 'loving marriages were taken as key markers of progress, in order to distinguish Enlightened Europeans from uncivilised counterparts'.[109]

The marriage took place on 18 October 1714 in St Helen's Church on Bishopsgate in London, one of the few City churches that survived the Fire of 1666.[110] The church was a burial place for several physicians and natural philosophers such as Sir Thomas Gresham (whose College was the initial meeting place of the Royal Society), Robert Hooke, and physician Jonathan Goddard. In fact, the church had more monuments than any other church in central London,

---

[105] Crouch, 'Public Life', p. 71; Folger W.b. 111, f. 13a, 26 April 1714, Costume Bill, Drury Lane, records 6s for cleaning a feather headpiece for Mrs Bradshaw in King Lear, approved by Wilks, Booth, and Cibber, Folger Shakespeare Library, Washington DC. MS Folger W.b. 110, no. 51 is a bottom of a bill for £1.15s.2d submitted by Thomas Lewis for costumes for Mrs Bradshaw and Ryan, largely approved by Wilks, Booth, and Cibber.

[106] When listed by Farrer, 1908, at Hardwick House, Suffolk, the Lens portrait, National Portrait Gallery 1926, was catalogued as Folkes painted by Lens in 1720. See Edmund Farrer, *Portraits in Suffolk Houses (West)* (London: Bernard Quaritch, 1908), p. 151. However, there is no date on the piece.

[107] Archibald Bower, *Mr Bower's Answer to a Scurrilous Pamphlet* (London: W. Sandby, 1757), p. 95.

[108] Sally Holloway, *The Game of Love in Georgian England: Courtship, Love and Material Culture* (Oxford: Oxford University Press, 2019), p. 1.

[109] Holloway, *The Game of Love*, p. 6.

[110] MS 6832, Diocesan Record Offices, Guildhall Library, London, p. 26: 'Martin Folkes Gentleman. of the parish of/ Hasserten in Yorks Shier and Lucretia/ Bradshaw of the parish of St Andrew/ Holburne Londn were Maried----/ by Mr William Butler October 18th/1714'.

and was known as the 'Westminster Abbey of the City'.[111] Perhaps its antiquarian appeal is why Folkes chose it. They were married by William Butler, LLB, Prebendary of St Paul's. The parish registers reveal that Folkes recorded as his home parish Hasserton, Yorkshire, which was surely false. While the marriage was not a Fleet, or clandestine union, Folkes may have done this to avoid the publicity of the banns and the ensuring social stigma for his family in Norfolk.[112]

The marriage was certainly a surprise to Folkes's mother, Dorothy Hovell Folkes. William Stukeley (who by mistake called Lucretia Mrs Bracegirdle, confusing her with another actress, Anne Bracegirdle) reported that Dorothy 'grievd' at this match 'so much that she threw herself out of a window & broke her arm'.[113] Stukeley was a gossip and had a penchant for spreading hostile rumours about his former friend Folkes, but it was true that Dorothy had reason to worry. Dorothy's sister, Ethelreda Hovell, was married to William Wake, Archbishop of Canterbury (1657–1737); Ethelreda was thus Martin Folkes's maternal aunt. Wake's 'correspondence and engagement with the churches in Europe' and his 'strong commitment to the ideal of protestant union, both at home and with the reformed churches abroad', meant he had an international reputation, and he was respected by many foreign Protestant intellectuals, including Gottfried Wilhelm Leibniz.[114] Folkes's own father sold properties on the behalf of the Corporation of the Sons of the Clergy, incorporated by charter on 1 July 1678 at the instigation of loyalist Anglicans concerned with alleviating 'the lot of needy dependants of Anglican clergy who had suffered for their orthodoxy during the time'.[115] Folkes's marriage to an actress was thus considered a blot on the family reputation.

Nonetheless, to commemorate their marriage, the couple bought, or were given, a richly bound copy of a 'Vinegar Bible' (so called due to a printing error in the headline above Luke XX: the 'parable of the vineyard' becoming the 'parable of the vinegar'). Known for its beauty of 'type, impression, and paper', the Bible featured silver work engraved with the Folkes arms by the master silversmith

---

[111] Henry Benjamin Wheatley and Peter Cunningham, *London Past and Present: Its History, Associations, and Traditions* (Cambridge: Cambridge University Press, 2011), p. 205; W. Bruce Bannerman, *The Registers of St. Helen's Bishopsgate London* (London: Mitchell and Hughes, 1904), p. 176.

[112] John R. Gillis, *For Better, for Worse: British Marriages, 1600 to the Present* (Oxford: Oxford University Press, 1985), p. 90.

[113] William Stukeley, *The Family Memoirs of the Rev. William Stukeley M.D.*, 2 vols, Vol. LXXIII, Publications of the Surtees Society (Durham, London, and Edinburgh, 1882–7), vol. 1, pp. 99–100.

[114] Stephen Taylor, 'Wake, William (1657–1737), Archbishop of Canterbury', *Oxford Dictionary of National Biography*. http://www.oxforddnb.com/view/10.1093/ref:odnb/9780198614128.001.0001/odnb-9780198614128-e-28409 [Accessed 14 Jan. 2019].

[115] A/CSC/1514, London Metropolitan Archives. The people involved were: 1. The Rt Hon. Thomas Lord Wharton; 2. Mary Trevor of St. Giles-in-the-Fields, Co. Middlesex Gentlewoman Dame Mary Kinsey, Widow exix. of Sir Thomas Kinsey Richard Atkyns of Lincoln's Inn, Co. Middlesex, Esq. Mary his Wife Peter Birch of St. Bride's, London, D.D Martha his Wife William Whitchurch of Froome, Co. Som. Esq., James Whitchurch of London, Merchant, Ruth his Wife; 3. Edward Harley of Lincoln's Inn, Esq.; 4. The Hon. Thomas Newport of The Inner Temple, London, Esq., Sir Thomas Rawlinson of London, Kt. Martine Folkes of Greyes Inn Co. Middlesex, Esq. Stowe, Co. N'hants.: Manor of Stowe, etc.

Anthony Nelme, known for the altar candlesticks he wrought for St George's Chapel in Windsor. The Bible also had lavishly dramatic engravings for the vignette head- and tailpieces created by mural painter James Thornhill.[116] Thornhill had connections to the theatre, designing garden sets for Thomas Clayton's opera *Arsinöe, Queen of Cyprus*, performed at Drury Lane in 1705; the Bible would have thus been especially appealing to the newly married couple.[117] Perhaps Folkes's Uncle William hoped it would entice the couple to read it more often.

Despite any public scandal that ensued, the marriage through the 1720s was purportedly quite happy, the antiquary William Cole reporting that Folkes was 'infinitely fond' of Lucretia.[118] Lucretia and Martin had two daughters, Dorothy (b. 1718) and Lucretia (1721–58), as well as a son, Martin (1720–40), and there were lively gatherings at their house on Queen Square. In 1726, the poet John Byrom recorded in his diary visiting Folkes's home with the surgeon William Cheselden where Folkes 'showed us his books and rarities... Mr [George] Graham [the watchmaker] came and stayed till we came away; Mrs Folks complained of having got a great cold; the monkey was very comical, we disputed whether he had reason or no, I said he was a man without reason, which definition Mr Graham said was right; we came away after twelve'.[119] Like many Georgian couples, placing 'Love and Hope at the Helms', the Folkes had sailed through their marriage 'with Safety', complete with their pet monkey.[120]

[116] Harry Carter, *A History of the Oxford University Press, Volume 1: to 1780* (Oxford: Oxford University Press, 1975), p. 171.

[117] '*The Holy Bible, Containing the Old Testament and the New: Newly Translated out of the Original Tongues*, Oxford, John Baskett, 1716', Forum auctions, 20 March 2017. https://www.forumauctions.co.uk/31029/Anthony-Nelme-silver-binding-on-Bible?auction_no=1003&filename=10611-1_2.jpg&view=lot_detail [Accessed 9 January 2020]. See C. H. Collins Baker, 'Sir James Thornhill as Bible Illustrator', *Huntington Library Quarterly* 10, 3 (May, 1947), pp. 323–7.

[118] BL Add MS 5833, f. 158, British Library, London.

[119] John Byrom, *The Private Journal and Literary remains of John Byrom*, 2 parts in 4 (Manchester: Chetham Society, 1854–7), vol. 1, pp. 209–10.

[120] MS 161/102/2, John Lovell to Sarah Harvey, undated (*c.*1757) Wiltshire and Swindon Archives, Chippenham, as quoted in Holloway, *The Game of Love*, p. 5.

4

Folkes and His Social Networks in 1720s London

4.1 Introduction

During the 1720s, Folkes's marriage was followed by a very active period of his building his social reputation and networks by participation in a variety of elite clubs and organizations. Folkes expanded his connections in the Royal Society and became a prominent Freemason. He also became a member of the Society of Antiquaries of London and an Esquire of the Knights Companions at the inaugural ceremony for the Order of the Bath. His membership in these organizations, which often overlapped, could be seen as evidence of his clubbability and affable nature, but in reality Folkes was part of a much larger social movement. In his study of the foundations of the British honours system, Matikkala has shown that in the opening decades of the eighteenth century, aristocratic elite culture and antiquaries were very interested in ideas of revival and institutionalization of organizations with venerable pasts, or at least legends of venerable pasts, in a movement he has termed 'The Chivalric Enlightenment'.[1]

> Just as every period since the Crusades era had produced its own version of 'new knighthood', so too did the age of Enlightenment. The Chivalric Enlightenment was essentially rhetorical, learned, antiquarian and eclectic in drawing from several parallel and overlapping currents. Owing to the contested nature of the Enlightenment itself, the so-called 'Enlightened historians' often scorned antiquarian pursuits, but besides this anti-medievalism and antiquarianism, there was also 'enlightened interest in the past'.

The Society of Antiquaries was revived in 1717, as was the Grand Lodge of the Freemasons, formed from the unification of four existing Lodges in London.[2] There were many common members in all these organizations. William Stukeley, who served as the first secretary of the Society of Antiquaries and who was initiated as a Freemason in January 1721 (he was master of the Lodge meeting at

---

[1] Antti Matikkala, *Orders of Knighthood and the Formation of the British Honours System* (Cambridge: Cambridge University Press, 2008), p. 43.
[2] Matikkala, *Orders of Knighthood*, p. 48.

Martin Folkes (1690–1754): Newtonian, Antiquary, Connoisseur. Anna Marie Roos, Oxford University Press (2021).
© Anna Marie Roos. DOI: 10.1093/oso/9780198830061.003.0004

Fountain Tavern on London's Strand), approached Freemasonry from an anti-quarian point of view, believing it represented the 'remains of the mysterys of the antients'.[3] The 'friendly relation between...Antiquaries and Freemasons' was 'said to have been initiated by Lord Coleraine, Vice President of the Antiquaries since 1727 and Grand Master of the Grand Lodge in 1727–8'.[4] The Royal Society was also part of this shared brotherhood. It would display, like the other organiza-tions, a growing antiquarian interest in its own material and cultural heritage, particularly under the vice presidency of Folkes from 1723 until Newton's death in 1727.

This chapter will begin by demonstrating that Folkes's participation in these societies was a reflection of his enlightened interest in the past, and was done to cement his social connections and promote an antiquarian programme in the Royal Society. Ultimately, through his experimental work and social ties Folkes hoped to build a group of supporters in the Royal Society to continue antiquarian pursuits as well as the Newtonian programme that emphasized iatromechanics (medical application of physics), astronomy, and the new mathematics. Sir Hans Sloane, on the other hand, as secretary of the Royal Society, promoted research in antiquarianism, but primarily natural history, combining a providential and com-mercial curiosity about flora and fauna.[5] After Newton's death, Folkes had good reason to believe that with his social connections he could make a credible bid for the presidency against Sloane in a Royal Society that was divided not only by intellectual interests but also by political ones. The legacy of the Jacobite revolt of 1715 and differing views among the Fellows about the role of foreign natural philosophers in the Royal Society also set members apart from each other. These were divisions that Folkes ultimately could not navigate successfully, and Sloane's personal and institutional networks were also formidable. Thus Folkes did not become president of the Society until after Sloane had resigned in 1741.

Defeated in his bid for the presidency, Folkes could, however, carry out the Newtonian legacy, or at least his version of it, his antiquarian interests shaping his co-editing with Thomas Pellett of Newton's *Chronology of Ancient Kingdoms Amended* (1728) and *Observations Upon the Prophecies of Daniel and the Apocalypse of St John* (1733). Although a Mason, Folkes was a sceptic throughout his life about organized Christian religion, treating the Bible as a purely historical source. His editorship of Newton's works was thus for him more a historical

---

[3] Bodl. MS Eng. misc. e. 666, William Stukeley, Abstract of my Life 1750, Bodleian Library, Oxford, as transcribed in William Stukeley, *The Family Memoirs of the Rev. William Stukeley, M.D.*, 2 vols, Vol. LXXIII, Publications of the Surtees Society (London: Surtees Society, 1882–7), vol. 1, p. 51, and quoted in Stuart Piggott, *William Stukeley: An Eighteenth-Century Antiquary* (New York: Thames and Hudson, 1995), p. 76.

[4] Joan Evans, *A History of the Society of Antiquaries* (Oxford: Oxford University Press, 1956), pp. 54–5, as quoted in Matikkala, *Orders of Knighthood*, p. 48, footnote 139.

[5] James Delbourgo, *Collecting the World: The Life and Curiosity of Hans Sloane* (Cambridge, MA: Harvard University Press, 2018), p. 160.

project and a reflection of his devotion to Newton, rather than one bound with devotion to the Christian creed.

4.2 Folkes and Freemasonry

Folkes served as Deputy Grand Master of the Premier Grand Lodge in London from 1724 to 1725, his influence contributing to an extant surge of popularity and expansion of its practice in the first half of the eighteenth century. Early modern Freemasonry did owe its origins, at least in part, to previous traditions and associations of English medieval stonemasons who specialized in the cutting of freestone, finely grained sandstone or limestone used for tracery, gargoyles, or capitals and cornices.[6] In the seventeenth and eighteenth centuries, antiquarian interest in the material culture of the past contributed to Freemasonry's revival and growing popularity, seeing its gradual transition from an operative Masonic guild to what is termed 'gentlemen's freemasonry', 'speculative', 'symbolic', or non-operative Masonry. Part of this transition from an 'operative' to 'non-operative' Freemasonry was due to the 1717 formation of Grand Lodge from four previous Masonic Lodges who assembled at the Goose and Gridiron alehouse in London, 'to which each lodge would belong and send representatives'.[7] To control the organization and prevent 'unauthorised lodges' from proliferating, the Grand Lodge published an approved list of Lodges in 1723, and extended its authority into the English counties.[8] With this firm administrative base, the Grand Lodge and non-operative Freemasonry thrived.

Concomitantly, the 1720s saw attempts to ground the history of Masonry, and the Grand Lodge itself, in a venerable past. Legendary histories of the craft of stonemasonry compiled in Britain, copied from early fifteenth-century manuscripts by early modern antiquaries, were known as the 'Old Charges', and were perceived as a link between medieval stonemasons and Enlightenment Freemasonry. Part of these 'Old Charges' consisted of a fifteenth-century manuscript (Add MS 23198 in the British Library) known as the 'Cooke' manuscript.[9] The manuscript began with a paean to God, who gave knowledge of the crafts, including geometry. The scribe considered geometry the root of the seven liberal

---

[6] Andrew Prescott, 'The Old Charges', in Henrick Bogdan and Jan A. M. Snoek, eds, *Handbook of Freemasonry* (Leiden: Brill, 2014), pp. 33–49, on p. 33.

[7] Jessica Harland-Jacobs, *Builders of Empire: Freemasons and British Imperialism, 1717–1927* (Chapel Hill: University of North Carolina Press, 2007), p. 23; Although the 1717 founding is the received view, work by Professor Andrew Prescott, University of Glasgow, suggests that the Grand Lodge was founded later, in 1721. See 'It all started when? Four historians ask whether the Grant Lodge was formed in 1717', *Freemasonry Today*, 12 June 2018. https://www.freemasonrytoday.com/more-news/lodges-chapters-a-individuals/it-all-started-when-four-historians-ask-whether-the-first-grand-lodge-was-formed-in-1717 [Accessed 24 August 2020].

[8] Harland-Jacobs, *Builders of Empire*, p. 24.   [9] Prescott, 'The Old Charges', p. 38.

arts and traced its origins to the son of Lamach mentioned in Genesis before moving onto the influences of Pythagoras, and Ham, Noah's son upon the revival of Masonic practice. The manuscript scribe then moved onto Euclid, before describing how Masonry came to England via the Anglo-Saxon King Aethelstan, who promulgated a series of ordinances regulating the craft, its practices, and pay.[10] As Bogdan and Snoek have noted,

> Much of the symbolic language used by the authors of the Old Charges was, understandably, influenced by the Bible, and this continued as modern Freemasonry developed in the first decades of the eighteenth century. The combination of speculation on the art and craft of stonemasonry with Christian symbolism naturally made the Temple of Solomon and its architect Hiram Abiff ideal symbols for Freemasonry, as evidenced by the legend of the third degree of Freemasonry [the Master Mason] which was adopted in the first half of the 1720s. The Master mason degree and the Hiramic legend spread throughout the masonic world....[11]

The Cooke manuscript was displayed on 24 June 1721 at a meeting of the Grand Lodge by George Payne, a former Grand Master, and calligraphic copies were made of it for reference for Dr James Anderson's *Book of Constitutions* (1723), which became the basis of modern English (and American) Freemasonry.[12] Anderson, who was Deputy Grand Master of the Grand Lodge, had been directed to amalgamate the apocryphal records and the 'Old Charges' into a document which he published as the *Constitutions*, so as to give Freemasonry a distinguished genealogy and a codified set of ritual practice.

Early modern Freemasonry was overall characterized by: rituals of initiation (those for the three degrees of Entered Apprentice, Fellow Craft, and Master Mason); meetings at local taverns (the Lodges were named after the public houses in which they met); symbols such as the Temple of Solomon or the Eye in the Triangle as an icon for God; tools and objects adopted from medieval Freemasonry, such as the square, compass, the white apron, gloves, and tracing boards; a hierarchical structure; activities concerning the individual quest for improvement, reflected in educational lectures in the Lodges (often in natural philosophy); some emphasis on charitable works; an emphasis on a fraternal brotherhood with convivial feasts; and the practice of secrecy.[13] The only stated

---

[10]   Prescott, 'The Old Charges', p. 39.

[11]   Henrick Bogdan and Jan Snoek, 'Introduction', in *Handbook of Freemasonry*, pp. 1–10, on pp. 3–4.

[12]   [James Anderson], *The Constitutions of the Free Masons. Containing the History, Charges, Regulations, &c. of that most Ancient and Right Worshipful Fraternity. For the Use of the Lodges* (London: William Hunter, 1723).

[13]   This list of characteristics was adapted from Bogdan and Snoek, 'Introduction', in *Handbook of Freemasonry*, p. 2.

requirement for membership was belief in a supreme being, the Great Architect of the Universe.[14]

Elliott and Daniels have estimated that in the 1720s some forty-five per cent of the Fellows of the Royal Society were Freemasons, though they admit 'further analysis is required before its precise nature can be ascertained.'[15] A more verifiable means of identification can be accomplished by comparing the names of the 1723, 1725, and 1730 lists of Masons in the first meeting Minute Book of Grand Lodge, and the list of Fellows of the Royal Society.[16] Using this methodology, 'about one in five of the Society's members were Freemasons...eighty-nine of them have been identified as Masons.'[17] On the other hand, there are persons, including Grand Officers, who are mentioned in the Grand Lodge Minute Book but do not appear on the membership lists, so exact calculations are difficult.[18] Nonetheless, it is safe to say that there was exceptional interest in Freemasonry among early eighteenth-century Fellows of the Royal Society. As vicar and satiric English poet James Branston ironically remarked, 'Next Lodge I'll be Freemason, nothing less, Unless I happen to be FRS.'[19]

What was the appeal of Freemasonry to natural philosophers? Some of it was polite sociability and free discourse within the bounds of a club or association. By the eighteenth century, clubs and associations were part and parcel of London's civil society.[20] In 1711, the 3rd Earl of Shaftesbury argued that sociability was as important to man's survival as was his need to eat and drink.[21] He wrote in 'defence only of the Liberty of the Club, and of that sort of Freedom which is

[14] Harland-Jacobs, *Builders of Empire*, p. 5.

[15] Paul Elliott and Stephen Daniels, 'The "School of True, Useful and Universal Science"?: Freemasonry, Natural Philosophy, and Scientific Culture in Eighteenth-Century England', *The British Journal for the History of Science* 39 (2006), pp. 207–29, on p. 213; H. Lyons, *The Royal Society, 1660–1940: A History of its Administration under its Charters* (Cambridge: Cambridge University Press, 1944); J. R. Clarke, 'The Royal Society and Early Grand Lodge Freemasonry', *Ars Quatuor Coronati* 80 (1967), pp. 110–19; D. Clements and B. Hogg, *Freemasons and the Royal Society: An Alphabetical List of Fellows of the Royal Society who were Freemasons* (London: Library and Museum of Freemasonry, 2012).

[16] Fiona E. Pollard, 'An Analysis of the Emergence of Early Masonic Symbolism', M. Phil (University of London, SOAS, 1997), p. 85. The Minute Books are in the London Library and Museum of Freemasonry. A print edition is: W. J. Songhurst, ed., *The Minutes of the Grand Lodge of Freemasons of England 1723–39*, Masonic Reprints Volume X (London: Quatuor Coronati Antigrapha, 1913).

[17] Pollard, 'Emergence of Early Masonic Symbolism', p. 85.

[18] 'Admissions and Lodge Meetings', in Róbert Péter, Jan A. M. Snoek, and Cécile Révauger, eds, *British Freemasonry, 1717–1813*, 5 vols (New York and Oxford: Routledge, 2016), vol. 5, e-book.

[19] As noted by Ric Berman, *The Foundations of Modern Freemasonry* (Brighton: Sussex University Press, 2012), p. 98.

[20] India Aurora Mandelkern, 'The Politics of the Palate: Taste and Knowledge in Early Modern England', PhD Dissertation (Berkeley: University of California, Spring 2015), p. 59; See also Peter Clark, *British Clubs and Societies 1580–1800: The Origins of an Associational World* (Oxford: Oxford University Press, 2000).

[21] See Anthony Ashley Cooper, Earl of Shaftesbury, 'Sensus Communis, or an essay on the Freedom of Wit and Humour', in *Characteristics of Men, Manners, Opinions, Times, vol. 1* (London: s.n., 1711), as quoted in Mandelkern, 'The Politics of the Palate', p. 59.

taken amongst Gentlemen and Friends, who know one another perfectly well'.[22] Such conversations liberated discourse from the 'formality of business and the Tutorage and dogmaticalness of the schools'.[23] Shaftesbury stated that in such polite discourse, 'We polish one another, and rub off our Corners and rough Sides by a sort of amicable Collision. To restrain this, is inevitably to bring a Rust upon Mens Understanding. 'tis a destroying of Civility, good Breeding, and even Charity it-self'.[24] Nearly forty years later, the 4th Earl of Chesterfield also stressed the importance of club-like activity to a gentleman's education: 'It is by conversations, dinners, suppers, [and] entertainments in the best companies', he wrote in 1750, 'that you must be formed for the world'.[25]

In her discussion of early eighteenth-century arguments between materialists and anti-materialists, and the formal foundation of modern Freemasonry with the establishment of the Grand Lodge in London in 1717, Margaret Jacob has also suggested that 'tolerant-minded Newtonians…invented a new form of ritual to worship the Grand Architect of the Universe'; Jacobs argued, in short, that Freemasonry was an innovation sprung 'from Newtonian inspiration'.[26] Garry Trompf agreed, writing 'The more one ponders Newton's axial principles…and then one relates this covert, Talmudically-inspired unorthodoxy to his fascination for the mysterious proportions of the Solomonic temple, the more one can sense the milieu of early Freemasonry'.[27]

Folkes, as a sociable natural philosopher interested in mathematics and architecture, became a fervent Mason. Between 1719 and 1742, Folkes proposed eleven fellow Masons of the Bedford Head in Covent Garden and the Maid's Head Lodge in Norfolk, near his seat of Hillington Hall, as Fellows of the Royal Society. Although Folkes's membership in the Bedford Head may have been due to the tavern's reputation for food and gaming, it was also a 'location for scientific lectures given by Desaguliers and James Stirling, among others', combining merry-making, natural philosophy, and sociability with esoteric philosophy and hermeticism.[28] In a retrospective article, the August 1798 edition of the *Monthly Magazine* noted (perhaps hagiographically) that 'When Dr Desaguliers and

---

[22] Earl of Shaftesbury, *Sensus Communis*, p. 51.
[23] Earl of Shaftesbury, *Sensus Communis*, p. 51.
[24] Earl of Shaftesbury, *Sensus Communis*, p. 44.
[25] See Letter 118, dated 6 August 1750, in Philip Dormer Stanhope, *Letters Written by the Late Right Honourable Philip Dormer Stanhope, Earl of Chesterfield, to his Son, Philip Stanhope, Esq.* (London: E. Lynch, 1774), p. 38, as quoted in Mandelkern, 'The Politics of the Palate', p. 59.
[26] Betty Jo Teeter Dobbs and Margaret C. Jacob, *Newton and the Culture of Newtonianism* (Amherst: Humanity Books, 1995), p. 102; Elliott and Daniels, 'School of True, Useful and Universal Science', p. 209.
[27] Garry W. Trompf, 'On Newtonian History', in Stephen Gaukroger, ed., *The Uses of Antiquity: The Scientific Revolution and the Classical Tradition* (New York: Springer, 2013), pp. 213–49, on p. 234–5.
[28] Berman, *The Foundations of Modern Freemasonry*, p. 102.

Martin Folkes presided, science and decorum were strictly attended to, and philosophical lectures were given in the principal lodges in London'.[29]

Among the FRS at The Bedford Head were mathematician Brook Taylor (1685–1731), who helped adjudicate the Newton–Leibniz calculus dispute; Thomas Pellett (1671–44), who assisted Folkes with editing Newton's *Chronology of Ancient Kingdoms Amended*; astronomer John Machin (1686–1751); John Arbuthnot (1667–1735); and William Rutty (1687–1730), elected the Royal Society secretary in 1727.[30] The 9th Earl of Pembroke, Henry Herbert, was also a member of Bedford Head and friend of Lennox, and shared numismatic interests with Folkes. The Bedford Head Lodge included Huguenot financiers such as Messrs Cantillon, Varenne, Desbrostes, and Botelcy,[31] demonstrating again Folkes's close ties with the elite Huguenot community.

The Lodges of the early eighteenth century were 'ruled by grand masters drawn from the peerage, strictly hierarchical in structure yet curiously egalitarian at their meetings and banquets' as well as in admissions, and this was particularly true of the Grand Lodge.[32] As Clark has noted,

> This picture of enhanced respectability, but not social exclusivity, may well reflect conflicting institutional and ideological pressures within freemasonry. Although masonic commentators throughout the period stressed the role of freemasonry as a unifying force in society, the…Grand Lodge seems from early on to have sought to advance the order's fame and fashionability by raising the social threshold of membership.[33]

Indeed, as Berman has indicated, 'Folkes [was] instrumental in encouraging a succession of aristocrats to join Freemasonry. Four of the first five noble Grand Masters were FRS and were friends of Folkes: Montagu, appointed in 1721 (elected FRS in 1718); [Francis Scott, Earl of] Dalkeith, appointed 1723 (elected FRS in 1724); Lennox, appointed 1724 (elected FRS in 1724) and Paisley, appointed in 1725 (elected in 1715)'.[34] Although the second edition of the *Masonic Constitutions* (1738) boasted 'many Noblemen and Gentlemen of the first Rank desir'd to be admitted into the *Fraternity*', the claim was not an idle one.[35]

Folkes's connections with the Grand Masters meant that he was elected Deputy Grand Master in 1724, serving his friend the affable Lennox. The two often dined

[29] *Monthly magazine, or British register* 6 (August 1798), pp. 91–3.
[30] Berman, *The Foundations of Modern Freemasonry*, p. 107.
[31] Berman, *The Foundations of Modern Freemasonry*, p. 101.
[32] Margaret Jacob, *The Radical Enlightenment: Pantheists, Freemasons and Republicans* (London: George Allen and Unwin, 1981), pp. 116 and 119.
[33] Clark, *British Clubs and Societies*, pp. 323–4.
[34] Berman, *The Foundations of Modern Freemasonry*, p. 104.
[35] James Anderson, *The New Book of Constitutions of the Antient and Honourable Fraternity of Free and Accepted Masons* (London: Caesar Ward and Richard Chandler, 1738), p. 115.

together; the diary of poet, Mancunian, and Jacobite John Byrom (1692–1763) recorded a meal he had at the Pontac's Head Tavern with Lennox and Folkes on 11 March 1725, after which he came by carriage to the Royal Society with them and 'we talked about masonry and shorthand'.[36] (Byrom had just invented his own system of shorthand, and taught Folkes's son his method). Shortly afterwards, Lennox wrote to Folkes on 27 June 1725:

> I have been guilty of such an omission that nobody less than the Deputy Grand Master of Masonry can make up for me. St. John's day, being the *great & important day*, was entirely out of my head, so much that I have never once cast an eye upon the report of the Committee upon Charity; which I ought to have return'd a week ago; therefore I beg you would make my excuses to Brother Sorel to whom I will return it in a post or two, with a few remarks of my own…I desire you would present my humble service to Mrs. Folkes, I hope she was entertained at the Instalment.[37]

Lennox's letter was a reference to the anniversary festival on St John the Baptist's Day held on 24 June at Merchant Taylors' Hall, which the *Evening Post* reported as a 'very handsome Appearance and all Things…transacted with the utmost Order and Unanimity'.[38] Lennox and Folkes were accompanied by Francis Sorrell and George Payne as Grand Wardens. Lennox also referred in his letter to Folkes's recent nomination to the 'politically important and highly visible committee for managing the Bank of Charity', a mutual aid society characteristic of lay religious initiatives of the period.[39] The proposal read that 'in order to promote and extend the old charitable Disposition of Masons, that a common Stock be formed, and the Money then arising be put into the Hands of a Treasurer, a Brother of known Worth and Integrity, at every Communication, for the Help and Relief of distressed Brethren throughout the World'.[40] The total relief available was three pounds. Other than Folkes and Lennox, Payne, Sorrell, and Dalkeith were placed on the committee.

Folkes was also active in another Lodge that met at the St Paul's Churchyard tavern. In 1725, as Byrom related,

> Tom Bentley was there, but would not go with us to Paul's Churchyard, where Mr Leycester and I went, Mr Graham, Foulkes, Sloan, Glover, Montagu…There

---

[36] John Byrom, *The Private Journal and Literary Remains of John Byrom* (Manchester: Chetham Society, 1854), vol. 1, part 1, p. 92.

[37] RS/MS/865/1. Charles Lennox to Martin Folkes, 27 June 1725, Royal Society Library, London.

[38] *The Evening Post*, Tuesday, 23 June to Thursday, 25 June 1724, No. 2327, p. 1. This event was also reported in the *Weekly Journal or British Gazetteer* of 27 June 1724.

[39] Berman, *The Foundations of Modern Freemasonry*, p. 99.

[40] *A short Account of the Rise and Establishment of the General Fund of Charity for the Relief of Distressed Masons* (London: J. Scott, 1754), p. 1.

was a lodge of the Freemasons in the room over us, where Mr Foulkes, who is
deputy grand master, was till he came to us. Mr Sloan was for taking me up stairs
if I would go: I said I would, and come back if there was anything I did not like,
and then he bid me sit down.[41]

'Mr Sloan' was Hans Sloane's nephew, William Sloane FRS, and member of the
Lodge that met at the Dolphin in Tower Street; 'Glover' was Philips Glover FRS,
the High Sheriff of Lincolnshire; 'Mr Leycester' was Ralph Leycester FRS, known
to Byrom as they both went to Trinity College, Cambridge; 'Montagu', John, 2nd
Duke of Montagu, and 'Mr Graham' or George Graham, the watchmaker, FRS
and friend of Folkes. According to Byrom, there was another meeting at the tavern on 20 April 1725, and the St Paul Churchyard's Lodge was featured in a copy
of the earliest printed pocket list of Freemasons' Lodges and meetings, the existence of which had long been suspected but remained unknown until the last decade; a copy was tucked into a collection of papers containing letters from Lennox
to Folkes.[42] The list was created when Philip Wharton, 1st Duke of Wharton, was
Grand Master in 1722, and it served as the 'official record of established lodges'[43]
(see figure 4.1).

Folkes was not only a 'prestigious figure in the scientific and antiquarian com-
munities' in Britain and the Continent, but also a visible ambassador of
Freemasonry's ideals of free thought and sociability and of Newtonian scientific
principles. It was rumoured that Folkes was even the author of the scurrilous
Relation Apologique (1738) for Freemasonry, which applied Newtonian principles
to government and communicated Masonry as a primarily scientific institution.[44]
The *Relation Apologique* stated that the Freemasons:

seem to be able to send their thoughts and attention into the universe and dili-
gently observe all its blazing bodies and wonders in every sphere, course and
movement. On the other hand, their contemplations stretch down to the
harbour and interior of the Earth or almost to its centre, to investigate the prod-
uct of the underworld...mineral bodies and precious gems. From there...they
lift their thoughts up to the Earth, in order to investigate the shape and inherent
utility of trees, crops and all sorts of herbs...These diligent free masons also

[41] Berman, *The Foundations of Modern Freemasonry*, p. 108; Byrom, *Private Journal*, 6 April 1725,
vol. 1, part 1, p. 109.
[42] Presented by the firm of Campbell Hooper, solicitors late of Old Queen Street, Westminster, 20
May 2010 to the Royal Society Library, London. RS/MS/865/2, Royal Society Library, London. My
thanks to Keith Moore for this information. See also Péter, Snoek, and Révauger, eds, *British
Freemasonry*, vol. 5, pp. 1–4.
[43] Péter, Snoek, and Révauger, eds, *British Freemasonry*, vol. 5, p. 1.
[44] William Eisler, 'The Construction of the Image of Martin Folkes (1690–1745), Part I', *The Medal*
58 (2011), pp. 1–29, on p. 5; see also Andreas Önnerfos, 'The earliest account of Swedish Freemasonry?
Relation apologique (1738) revisited', *Ars Quatuor Coronatorum* 127 (2014), pp. 1–34.

Fig. 4.1 List of Freemasons' lodges and meetings, engraved by John Pine, © The Royal Society, London.

meditate carefully and investigate the changing notion of weather and the movements of Oceans and the wild sea within its limitations.[45]

While Newtonianism and Freemasonry had definite alignments, this is not, however, to say that the argument advanced by Jacob about Freemasonry and its connections to *political* beliefs is entirely correct. Jacob argued that the Jacobites, led by James II and his eldest son 'The Old Pretender', were 'viewed as impediments to the progress promulgated by Newtonians, Protestant Whigs, and Freemasons', represented by individuals such as Desaguliers and Folkes.[46] Jacob then noted that 'The Jacobite cause was totally antithetical to the principles of Desaguliers and his friends [such as Folkes], and indeed one attempt by the Jacobite Philip, 1st Duke of Wharton to take control of the [Masonic] movement in the early 1720s was firmly

---

[45] J.G.D.M.F.M., *Relation Apologique et Historique de la Societe des Franc-Maçons* (Dublin, False imprint, 1738), pp. 47–9. See also Andreas Önnerfos, 'Secret Savants, Savant Secrets: The Concept of Science in the Imagination of European Freemasonry', in *Scholars in Action: The Practice of Knowledge and the Figure of the Savant in the 18th Century*, 2 vols (Leiden: Brill, 2013), pp. 433–57.

[46] Eisler, 'The Construction of the Image of Martin Folkes', p. 8.

thwarted'.[47] Jacob's conclusions that Newtonians were always Whigs, however, are not borne out by the work of Guerrini, who has demonstrated that early eighteenth-century Scottish Newtonians like Archibald Pitcairne and David Gregory were firm Tories inclined towards Jacobitism, forming collegial relationships with English Jacobites like physician John Friend, FRS.[48] It seems that Lodges instead 'contained within their copious embrace royalist, though Protestant, elements, and even a Jacobite strain—in short, radicals from both ends of the spectrum'.[49]

Furthermore, while Jacob was correct in stating that there was a schism in Freemasonry, the delineations were more nuanced, and it was not just one of Whig/Newtonian versus Tory/Jacobite. The schism in the Masons was partially occasioned by the election of the Whiggish John, 2nd Duke of Montagu, as Grand Master in 1721, giving 'rise to the concern that the society had fallen into the pocket of the government. Even the Whig Prime Minister and Freemason, Robert Walpole, had no compunction in using the network of lodges as a system of espionage'.[50] However, the schism was also occasioned by the opposition of the conservative faction, led by Philip, 1st Duke of Wharton (who had Jacobite sympathies), to the Whig-led innovations exemplified by Dr James Anderson's *Book of Constitutions* (1723). As mentioned, Anderson had been charged by Montagu to write and publish the *Constitutions*. The history, rich symbolism, and allegory of Freemasonry resonated with Montagu, who was dedicated to heraldic and antiquarian studies. Montagu's 'experience of overseeing the employment of historical texts to underwrite an ancient provenance was to come into play again in his self-appointed task of reviving the Order of the Bath', in which Folkes also took part (see section 4.3).[51]

In his work, Anderson argued that, with the founding of the United Lodge of England in 1717, there was a continuity between ancient operative freemasonry as a medieval craft guild, and the speculative or symbolic Freemasonry which the London Lodges practised. In 'tracing speculative Freemasonry back to the antediluvian membership of the Biblical patriarchs, Enoch and Adam, then to Grand Master Moses and beyond, Anderson had effectively reconstructed the history of Western Civilisation and Christendom as one great lodge'.[52] So, the divisions in the Masons were thus not merely political, but represented a conflict as to how the Masons were presented *historically*. Anderson and his followers saw an

---

[47] Jacob, *The Radical Enlightenment*, pp. 127–8.
[48] Eisler, 'The Construction of the Image of Martin Folkes', p. 9; Anita Guerrini, 'The Tory Newtonians: Gregory, Pitcairne and Their Circle', *Journal of British Studies* xxv (1986), pp. 288–311.
[49] Ronald Paulson, *Hogarth: The Modern Moral Subject, 1697–1732*, 3 vols (Cambridge: The Lutterworth Press, 1991), vol. 1, pp. 114–15.
[50] Marie Mulvey-Roberts, 'Hogarth on the Square: Framing the Freemasons', *British Journal for Eighteenth-century Studies* 26 (2003), pp. 251–70, on p. 255.
[51] Andrew Hanham, 'The Politics of Chivalry: Sir Robert Walpole, the Duke of Montagu and the Order of the Bath', *Parliamentary History* 35, 3 (2016), pp. 262–97, on p. 266.
[52] Mulvey-Roberts, 'Hogarth on the Square', p. 255.

ancient past as a means to support the legitimacy of speculative Freemasonry; Wharton and his group thought these innovations discredited the order.

The turmoil continued, as Montagu was defeated in the next election in 1722/3 by Wharton, and as a compromise, Desaguliers was elected Deputy Grand Master and Anderson the warden, and Anderson's *Book of Constitutions* was approved.[53] Wharton only lasted one more year in office, being replaced in 1724 by Dalkeith by a single vote; Wharton was an alcoholic and eventually went into severe debt. Facing a charge of treason for serving as a lieutenant colonel in the Jacobite forces in the Spanish army fighting England, Wharton eventually fled to France, and by 1729 he was in dire poverty. As Charles Lennox reported to Folkes,

> this is all the news that Paris affords, except only a thing that I had almost forgot to tell you, which is that the Duke of Bedford is here, & has had two conferences with the Duke of Wharton, up two pair of stairs, at the English Coffee house, over a bowl of Punch. & those that have seen the latter, tell me that no Theatre discarded Poet, was ever half so shabby. & that none of Shykespears stroling knights of the Garter, had ever so dirty a Star, & Ribbon.[54]

As for Folkes, he was an adherent of Anderson's faction, whose attempts to historicize Masonry appealed to his antiquarian interests.[55] Folkes provided Lennox with manuscripts supporting Masonry's long history and discussed the several editions of Anderson's work that were published in the 1720s and 1730s. In 1725, Lennox noted in correspondence, 'I thanke you for the Old Record you sent me, it is really very curious, & a certain proof at least, of our antiquity, to the unbelievers.'[56] In 1737, however, Folkes gently disabused Lennox of the idea that there was a connection between his own ancestral house and Masonic history:

---

[53]   Paulson, *Hogarth: The Modern Moral Subject, 1697–1732*, vol. 1, p. 115.

[54]   RS/MS/865/5, Charles Lennox, 2nd Duke of Richmond and Duke of Aubigny, to Martin Folkes, Paris Wednesday, 4 May N.S. 1729, Royal Society Library, London.

[55]   Eisler has claimed Folkes was a Jacobite based upon his interpretation of the iconography of his personal medals, and indeed Folkes's numismatic collection contained some Jacobite touch-pieces and medals. When Folkes's collection was auctioned after his death in 1756, the auctioneers were careful to list the pieces of the Stuarts succeeding the Old Pretender by neutral names, such as 'le Chevalier'. However, as Neil Guthrie noted, it was difficult to tell if the 'buying public regarded these things of at least some danger or as just antiquarian curiosities?' (p. 136). For Folkes, it seemed to be the latter. While there is no denying Folkes knew Jacobites professionally and personally, such as John Byrom, and that he was an adherent of free discourse and religiously unorthodox, there is no direct evidence to suggest he was a Jacobite himself. See Eisler, 'The Construction of the Image of Martin Folkes', pp. 1–29, and 'Part II', *The Medal* 59 (Autumn 2011), pp. 4–16; Neil Guthrie, *The Material Culture of the Jacobites* (Cambridge: Cambridge University Press, 2013), pp. 135–6.

[56]   RS/MS/865/1, Lennox to Folkes, 27 June 1725, Royal Society Library, London. There was frequent exchange of manuscripts and artwork between Lennox and Folkes. The posthumous auction catalogue of manuscripts in Folkes's collection indicates a 'Historical Account of Two Pictures, relating to the Death of King Henry Darnley of Scotland, one in the Possession of the Earl of Pomfret, and the other belonging to the Duke of Richmond'. *A Catalogue of the Entire and Valuable Library of Martin Folkes, Esq.* (London: Samuel Baker, 1756), p. 155.

Sir Th[omas] Pendergast took notice that in Andersons copy there were the words *uxoris ejus* which being no part our inscription made him think with a greate deal of reason that Inscription could not have been just the same as that on your Graces; but upon examining it those words are not to be made out to that. Anderson was certainly mistaken about them and they were never there.[57]

Folkes's allegiance to Anderson's cause was possibly noted by William Hogarth (1697–1764), who, in one of his prints, *The Mystery of Masonry Brought to Light by the Gormogons*, satirized these divisions in the Masonic order (see figure 4.2). Hogarth's work was published in 1724, the same year that Folkes was elected Deputy Grand Master of Grand Lodge. Hogarth was himself a Freemason, a member of the Lodge meeting at the Hand and Apple Tree in 1725 and, five years later, of the newly constituted Corner Stone Lodge, as well as the designer of the first Steward's Jewel.[58]

At the time of Wharton's loss of the Grand Master's election with 'rumors of the Jacobite faction's withdrawal…someone of the Desaguliers–Anderson faction

Fig. 4.2 William Hogarth, *The Mystery of Masonry Brought to Light by Ye Gormagons* [sic], Dec. 1724, 8.5 × 13.5 in., Metropolitan Museum of Art, New York, Public Domain.

[57] Goodwood MS 110, Folkes to Lennox, 23 July 1737, West Sussex Record Office.
[58] Mulvey-Roberts, 'Hogarth on the Square', p. 251.

invented the fiction of the Gormogons as a splinter Freemasonry being carried off by Wharton and his friends'.[59] A mock notice of this event appeared in the *Daily Post* of 3 December 1724, indicating that the Gormogons were supposedly formed by Chin-Quaw Ky-Po, the 1st Emperor of China.[60] Hogarth used this newspaper notice as his inspiration, directing his humour at both Wharton and the Folkes–Anderson–Desaguliers camp, portraying them in a satiric Masonic procession; regular Masonic processions began in the 1720s for the Annual Feast of London.[61]

The satiric procession is shown accompanied by a monkey in a Freemason's apron and gloves, symbolic of the Gormogons aping the true Freemason's craft. Wharton was caricatured as a knighted Don Quixote with an enormously plumed helmet, pointing at the Gormogons leaving the tavern where the Lodge meeting was being held.[62] In the procession, Paulson and Mulvey-Roberts have identified James Anderson, 'a reputed sycophant', as being in the 'humiliating posture of being trapped in a ladder with his nose pointing towards the buttocks of an old woman', a caricature of Desaguliers; Anderson is portrayed about to give Desaguliers what was popularly known as the 'Mason kiss'.[63] A version of Samuel Butler's poem *Hudibras* (1684), published as *The Freemasons; an Hudibrastic Poem in London* by A. Moore in 1723, refers to the 'Mason kiss', implying that during the initiation ceremony 'Masons had their backsides flogged and their right buttock branded and that they kissed each other's bare bottoms for identification'.[64]

In the *Gormogon* print, the Folkes–Anderson–Desaguliers adherents are led by the Chinese Emperor (labelled as 'A') referred to in the article in the *Daily Post*. The Chinese emperor has a protruding lower lip, bumped nose, and portly girth, with the added fillip of a Mandarin's thin moustache. Although with the proviso that likeness is not proof, Folkes also had very similar and distinctive physiognomy (sans the moustache, see figures 4.4a and 4.4b). In 1720, a drawing attributed to Hogarth was done of a character identified as Folkes sitting in a coffee-house, so the artist was probably familiar with his features (see figure 4.3). In a firmer attribution, Hogarth painted an oil portrait of Folkes *c.*1740 with the same facial and bodily characteristics. This portrait is now in the Royal Society Collections, presented by Folkes himself to the Society in 1742 (see cover image).[65]

In the *Gormogon* print, the cartoon Desaguliers looks in wonderment at the Emperor/Folkes character at the head of the procession. This adoring gaze may

---

[59] Paulson, *Hogarth: The Modern Moral Subject, 1697–1732*, vol. 1, p. 115.

[60] Paulson, *Hogarth: The Modern Moral Subject, 1697–1732*, vol. 1, pp. 115–16.

[61] Bogdan and Snoek, 'Processions', in *Handbook of Freemasonry*, e-book.

[62] Paulson, *Hogarth: The Modern Moral Subject, 1697–1732*, vol. 1, p. 118.

[63] Mulvey-Roberts, 'Hogarth on the Square', p. 257.

[64] Yasha Beresiner, 'Masonic Caricatures: For Fun or Malice: 300 Years of English Satirical Prints', *Ars Quatuor Coronatorum* 129 (September 2016), pp. 1–38, on p. 3. See also S. B. Morris, 'New Light on the Gormogons and Other Imitative Societies', *Ars Quatuor Coronatorum* 126 (2013), pp. 15–70.

[65] William Hogarth, *Portrait of Martin Folkes*, *c.*1740, oil on canvas, 735 mm x 620 mm, The Royal Society, London.

Fig. 4.3 William Hogarth (attributed to), *Examining a watch; two men seated at a table, the older (Martin Folkes) looking through his eyeglasses at a watch, a paper headed 'Votes of the Commons' (?) on the table.* Pen and brown (?) ink and wash, over graphite, *c.* 1720, 125 × 185 mm, Image 1861,0413.508, © The Trustees of the British Museum.

have been a reflection of their hierarchical relationship in the Royal Society, as Vice President Folkes often sat in the Presidential Chair covering for an aged Newton (Newton died in 1727), and at the time he was Newton's heir apparent for the presidency; Desaguliers served as the Society's experimental demonstrator. The Emperor/Folkes's head is, however, not crowned with the laurel leaves of wisdom, but with an exotic wreath of flowers and possibly bacchanalian grapes. This bunch of grapes was also portrayed on the inn sign, and suggests probable allusions to his and the Masons' overindulgences in their Masonic tavern meetings. (In 1718, Folkes had himself portrayed by Richardson in an Epicurean 'kit-cat' style portrait (see figure 4.6)) Emperor/Folkes carries a globe, indicative of his status as an astronomer, and on the globe is a constellation of a lamb, perhaps an allusion to the heraldic device of the golden fleece he used for his bookbindings. Folkes copied the device from the playwright Hilaire-Bernard de Requeleyne, baron de Longepierre (1659–1721), who decorated all his books with the device in celebration of the success of his *Médée* (1694) (see figure 4.5).[66]

    The Emperor/Folkes character's head is also backlit, the Sun rays visually mimicking the Masonic Lodge symbol of the Sun which the cartoon of Sage Confucius

---

[66] William Younger Fletcher, *Bookbinding in England and France* (London: Seeley and Company, 1897), p. 56.

Fig. 4.4a Ottone Hamerani: *Martin Folkes*, 1742, bronze, 37 mm., reversed, collection and photo of the author.

Fig. 4.4b Enlarged view of Folkes in the Gormogon print, Metropolitan Museum of Art, New York, Public Domain.

Fig. 4.5 1751 English armorial binding for Martin Folkes (1690–1754) with his crest of the golden fleece. Horace, [*Epistulae*. Liber 2. 1. English & Latin] *Q. Horatii Flacci epistola ad Augustum. With an English commentary and notes*. (London: W. Thurlbourn, 1751), PA6393 E75 1751 cage, Folger Shakespeare Library.

has around his neck, a mock embodiment of ancient enlightened Masonic wisdom that the Folkes–Desaguliers– Anderson faction advocated. Folkes's head may also have been 'eclipsing' the light. On 9 May 1724, a few months before Hogarth made the Gormogon print, Folkes, Desaguliers, and other Fellows of the Royal Society were making observations of a total solar eclipse (see section 4.4). Desaguliers had made a journey to Bath to give a lecture on the eclipse at a Masonic Lodge meeting at the Queen's Head, where the Whig politician John, Lord Hervey (1696–1743) was inducted into the craft.

> Dr Desaguliers, from Five this afternoon to the Time of the most Eclipse, read a lecture on this occasion…the Gentlemen, between 30 and 40, giving him three Guineas each to hear him, and he gave those ingenious and learned gentlemen great satisfaction for their money. This night at the Queen's Head Dr Desaguliers is to admit into the Society of free and accepted Freemasons several fresh members, among them are Lord Cobham, Lord Harvey, Mr Nash and Mr Mee, with many others. The Duke of St Albans and Lord Salisbury are here and about 10 other Lords English and Irish.[67]

[67] Letter from Bath dated 11 May 1724 published in *Parker's London News or the Impartial Intelligencer*, 18 May 1724, as quoted in Ric Berman, 'The Architects of Eighteenth-Century English Freemasonry, 1720–1740', PhD Dissertation (University of Sussex, 2010), p. 265.

Perhaps in his astronomical allusions, with the bulky figure of Folkes eclipsing the Sun, Hogarth was portraying the wisdom of Freemasonry 'eclipsed' by the squabbles of its members, casting shadows on true 'Enlightenment'.

4.3 The Order of the Bath

Freemasonry was not the only order predominant in the 1720s. The 'Order of the Bath', created in 1725, was represented as a revival of the British Chivalric Enlightenment within an antiquarian context. The idea originated with the antiquarian research of John Anstis, Garter King at Arms, 'who took the name from the medieval practice of ritual bathing on special occasions when new knights were created (the more usual practice being that of dubbing with a sword)'.[68] The prominent Whig John, 2nd Duke of Montagu soon took an active interest in restoring the old knighthood, as it validated 'the position of the Hanoverian royal family as Britain's rulers through the memorialisation of their ancestral connections with English monarchs of the distant past'.[69] King George I could be made to seem more 'English' and be identified as a 'modern upholder of chivalric values'.[70] With the encouragement of Robert Walpole, Montagu invited the 'cream of the Whig establishment' to be members, including Lennox, who was assured that receiving this honour would not preclude him from receiving the Order of the Garter.[71]

The order was created on 18 May, the statutes were given on 23 May, and the inauguration was on 17 June 1725 at Westminster Abbey.[72] Folkes attended as one of the 111 esquires, of whom each knight had been required to nominate three; the esquires were 'required to be armigerous relatives, friends or retainers of the knights'.[73] Montagu, no longer Grand Master of the Masons, became the Great Master, in the interim joining the Society of Antiquaries on 28 April 1725, as did Grey Longueville, Bath King of Arms, three months later on 28 July.[74]

Montagu was fascinated with heraldry, antiquity, and genealogy, his library at Boughton House crowned by a frieze of the original Knights of the Garter; as one visitor in 1724 recorded, 'the coats of arms and pedigree in wood work over the salloon chimney was remarkable'.[75] Just as the Masons with their *Constitution*

---

[68] 'Accession of George I', *College of Arms*. https://www.college-of-arms.gov.uk/news-grants/news/item/105-accession-of-george-i [Accessed 26 August 2019].

[69] Andrew Hanham, 'The Politics of Chivalry: Sir Robert Walpole, the Duke of Montagu and the Order of the Bath', *Parliamentary History* 35, 3 (2016), pp. 262–97, on p. 296.

[70] Hanham, 'The Politics of Chivalry', p. 275.   [71] Hanham, 'The Politics of Chivalry', p. 287.

[72] John Pine, *The Procession and Ceremonies Observed at the Time of the Installation of the Knights Companions of the Most Honourable Military Order of the Bath* (London: S. Palmer and J. Huggonson, 1730), p. 1.

[73] Hanham, 'The Politics of Chivalry', p. 292, footnote 141.

[74] Matikkala, *Orders of Knighthood*, p. 49.

[75] John Montagu Douglas Scot, ed., *Boughton: The House, Its People and Its Collections* (Hawick: Caique Publishing, Ltd, 2016), p. 152.

tried to claim history to bolster the legitimacy of speculative Freemasonry, the Order of the Bath was one means to 'claim the use of history for the government'; the other was the institution of the Regius Chairs of Modern History in Oxford and Cambridge in 1724.[76] The structural similarities between the Order of the Bath and the Freemasons were striking: formal processions, quasi-religious ceremony, drinking, and feasting on a grand scale, with a precisely delineated hierarchical order. As Matikkala has perceptively noted, 'for the social elite and their circle, the "enlightened" curiosity and ideas were not incompatible with interest in the Middle Ages, ceremonial splendour, or the orders of knighthood, which inevitably created hierarchical social distinctions'.[77]

These hierarchical ceremonies were precisely recorded by John Pine (1690–1756), a Mason who was a member of the Horn Tavern, one of the four Lodges which formed the Grand Lodge in 1717.[78] Pine engraved the frontispiece of the 1723 *Constitutions of the Free-Masons* designed by Sir James Thornhill, and became the principal engraver for the Grand Lodge, producing the annual lists of Lodges from 1725 until 1741, which gave details of their time and meeting.[79] Pine's 'output was wide-ranging, comprising not only book illustration' (such as that for Daniel Defoe's *Robinson Crusoe*) but also 'heraldry, maps and facsimiles of historical documents'.[80] In 1747, Pine also produced the twenty-four sheets of John Rocque's first comprehensive street plan of London, which received support from the City Corporation and the Royal Society and which was dedicated to Martin Folkes, who was then sitting President.[81] As Prescott has noted, 'Freemasonry apparently brought him in contact with the antiquary and scientist William Stukeley', and Pine engraved some of Stukeley's drawings to illustrate his historical compilation *Itinerarium Curiosum*. Pine's interest in wider philosophical issues is evident in his illustrations of Henry Pemberton's 1728 *View of Newton's Philosophy*, a popular account of Newton's theories, again showing the close relationship between Masonry, natural philosophy, and antiquarianism.[82] A close friend of Hogarth, Pine was commissioned to engrave the procession and ceremonies for a lavish folio, *The Procession and Ceremonies Observed at the Time of the Installation of the Knights Companions of the Most Honourable Military Order of the Bath*, printed by Samuel Palmer and John Huggonson in 1730 and considered his first major work.

---

[76] Matikkala, *Orders of Knighthood*, p. 49.   [77] Matikkala, *Orders of Knighthood*, p. 49.

[78] Andrew Prescott, 'John Pine: A Sociable Craftsman', *MQ Magazine* 10 (July 2004), p. 8. http://www.mqmagazine.co.uk/issue-10/p-08.php [Accessed 15 August 2019].

[79] Andrew Prescott, 'John Pine: A remarkable 17th Century Engraver and Freemason', *Freemasonry Matters*. https://freemasonrymatters.co.uk/latest-news-freemasonry/john-pine-a-remarkable-17th-century-engraver-and-freemason/ [Accessed 12 January 2020].

[80] Prescott, 'John Pine: A Sociable Craftsman', p. 7.

[81] John Rocque, *Plan of the cities of London and Westminster and borough of Southwark ... with the contiguous buildings is humbly inscribed* (London, 1749).

[82] Prescott, 'John Pine: A Sociable Craftsman', p. 9.

Fig. 4.6 Jonathan Richardson the Elder, Martin Folkes, oil on canvas, 1718, H 76 × W 63 cm, LDSAL 1316; Scharf Add. CVIII, by kind permission of the Society of Antiquaries of London.

Joseph Highmore (1692–1780), a member of the Masonic Lodge at the Swan at Greenwich, painted the Knights of the Bath in their regalia. Using the work of Highmore, Pine engraved every knight and attendant, each identified by his coat of arms, 'this meticulous record no doubt commending him to the College of Arms and influencing his appointment as Bluemantle Pursuivant in 1743'.[83] Grey

---

[83] Susan Sloman, 'Pine, John (1690–1756), Engraver', *Oxford Dictionary of National Biography*. https://www-oxforddnb-com [Accessed 6 August 2019]. Bluemantle Pursuivant is a junior officer of arms in the London College of Arms, said to date from the fifteenth-century Order of the Garter.

Longueville himself provided an endorsement of the volume, having 'inspected the Designs or Drawings' and certified the 'Truth of them'. This verification of empirical details is reminiscent of the work of Steven Shapin, which demonstrated that gentlemen-members of the Royal Society attested to the truth of observations in natural philosophy made by those below the gentry; 'gentlemanly conversation was the model for the new experimental science and…gentlemen were regarded as uniquely reliable truth-tellers'.[84] It seems that this proviso applied to artistic rendering of heraldic processions, showing a similar sensibility.

Pine's volume was presented to King George I, the funds raised by subscription. In the subscription list were a substantial number of attendees that were Masons, including the Duke of Montagu (the Grand Master of the Order), Lennox, the Earl of Deloraine and the Earl of Inchiquin, and of course Martin Folkes. Folkes, along with fellow Mason Thomas Hill, attended Lennox, who was installed by proxy due to illness from smallpox. The description of the ceremony was almost microscopic in detail, both in text and image, including illustrations of the knights' and squires' caps, mantle, spurs, collars, and swords and stars that would not have been out of place in a *Philosophical Transactions* article describing the plumage of exotic beasts or the petals of flora (see figures 4.7 and 4.8).

A symmetric diagram of the orders of the stalls and the ceremony was accompanied by engravings of the dining table at the installation ceremony dinner, the knights' table, which was described as '96 ft Long And 3 Foot. 10 Inch Broad' with 218 courses featuring 218 plates of food, including exotic dishes such as Italian Lobsters, with a good quantity of venison, salmon and other game: all luxury foods. The esquires contented themselves with three, '50 foot long, 3 foot six broad' places to dine and forty-nine plates of food: the hierarchy was indicated quantitatively. Pine included an engraving of the meal itself, portraying the court ladies in a stall who watched the meal. The ensuing 'plan of the entertainment for the supper' followed, the dishes in geometric arrangements resembling gustatory flowers.

As Lennox could not attend, on 19 June 1725 Folkes wrote him a letter addressed to his country estate at Goodwood describing the ceremony. He first thanked him for securing tickets for the ceremony for his wife, due to the 'great sollicitations that have been made every where for them…and indeed the Ceremony and ball was as magnificent as can possibly be conceived…the Knights with a great number of Nobility and persons of distinction all well dressed made a very splendid appearance'.[85] Folkes also mentioned 'the entertainment at Westminster I think was hardly equall to the rest of the solemnity, tho very great:

---

[84] Keith Thomas, 'Gentle Boyle', review of Steven Shapin's *Social History of Truth: Civility and Science in 17th-Century England*, London Review of Books 16, 18 (22 September 1994), pp. 14–16. https://www.lrb.co.uk/v16/n18/keith-thomas/gentle-boyle [Accessed 10 March 2019].

[85] Goodwood MS 110, Letter 24, Folkes to Lennox, 19 June 1725, West Sussex Record Office.

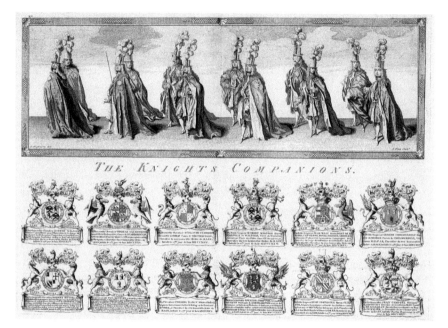

Fig. 4.7   Knights of the Order of the Bath, with elaborate plumage. John Pine, *The Procession and Ceremonies Observed at the Time of the Installation of the Knights Companions of the Most Honourable Military Order of the Bath* (London: S. Palmer and J. Huggonson, 1730), GR FOL-353, Bibliothèque nationale de France, département de l'Arsenal, Public Domain.

but Heidegger performd his part to admiration'. Johann Jacob Heidegger was the leading impresario of masquerades in London from 1715 until the 1730s, and was employed to organize public celebrations such as the illumination of Westminster Hall in October 1727 for the coronation of George II. He later became joint manager of the Haymarket Theatre to produce opera with Handel. Not surprisingly, Heidegger also ran the annual feast of the Grand Masonic Lodge.[86] In the conclusion to Folkes's letter to Lennox, he emphasized that he would get the Duke a promised magic lantern and 'spare sliders for the microscope', as Lennox was also a devotee of natural philosophy, and praised the 'cold supper' at the Ceremony as the 'most elegant and best orderd of any thing I ever saw of the kind'.[87]

When he became President of the Royal Society, Folkes would take the same emphasis on lavish feasting to the Thursday's Club, which later became the Royal Society Dining Club. As Mandelkern indicated:

---

[86]   Peter Robert, 'Theatre', in *British Freemasonry*, vol. 5, p. 374, footnote 13.
[87]   Goodwood MS 110, Letter 24, Folkes to Lennox, 19 June 1725, West Sussex Record Office.

Fig. 4.8 Portrait of Folkes, Thomas Hill, and Matthew Snow, the 'Knights Companions' for Charles Lennox, the Duke of Richmond for the Installation Ceremony for the Order of the Bath. John Pine, *The Procession and Ceremonies Observed at the Time of the Installation of the Knights Companions of the Most Honourable Military Order of the Bath* (London: S. Palmer and J. Huggonson, 1730), GR FOL-353, Bibliothèque nationale de France, département de l'Arsenal, Public Domain. Folkes is portrayed gesturing to the right towards the procession.

The Thursday's Club was unusual among gentleman's clubs for entertaining high numbers of guests. Guests comprised on average about one fifth of each meal's attendance, for which the Royal Society's then president, the antiquarian and mathematician Martin Folkes, was undoubtedly responsible. Ever since his election to the presidency in 1741, Folkes had sought to spread the Royal Society's cultural influence across the Continent, and he saw the Thursday's Club as an untapped public relations tool. Folkes's involvement most likely catalysed the club's formalization, as Colebrooke began keeping detailed records almost immediately after Folkes's election to the Thursday's Club in 1747.[88]

The Thursday's Club's formalization occurred at the same time as the exchange of the *Mémoires* and *Philosophical Transactions* between the Royal Society and the Académie des Sciences.[89] During the 1740s, due to Folkes's influence, nearly fifty per cent of the Royal Society's membership was foreign, helping to internationalize Newtonian natural philosophy and build further ties with European scientific institutions.[90]

4.4 The Royal Society in the 1720s

From his time in the Masons, and as an attendant in the Order of the Bath, Folkes understood well the social ritual of clubs and feasting, and how it could engender conviviality, cement relationships, and reinforce social hierarchies. Within the Royal Society itself, Folkes spent much of the 1720s building connections and positioning himself to inherit the Royal Society presidency upon Newton's death. He performed a number of tasks to make himself indispensable to the Society and built a group of supporters who wished to continue the Newtonian programme of mixed mathematics and astronomy. For instance, in March 1721, he accompanied James Jurin to compare the resolution and magnifying power of the 123-foot-long 'great telescope at Wanstead' with John Hadley's new 'catadioptrick' telescope which had a parabolic mirror to improve the Newtonian model by eliminating chromatic aberration;[91] Folkes and Jurin confirmed that Hadley clearly saw the shadows of the first and third of Jupiter's moons and, in May 1722, the fine gradations in the rings of Saturn.[92] In 1729, Hadley became one of the

---

[88] Mandelkern, 'The Politics of the Palate', p. 64; James McClellan, *Science Reorganized: Scientific Societies in the Eighteenth Century* (New York: Columbia University Press, 1985). See also Sir Archibald Geikie, *Annals of the Royal Society Club; the record of a London dining-club in the eighteenth and nineteenth centuries* (London: Macmillan and co., 1917), particularly chapter two; Thomas E. Allibone, *The Royal Society and Its Dining Clubs* (Oxford: Pergamon Press, 1976).

[89] Mandelkern, 'The Politics of the Palate', p. 64.

[90] See Roderick W. Home, 'The Royal Society and the Empire: the Colonial and Commonwealth Fellowship Part 1: 1731–1847', in *Notes and Records of the Royal Society* 56, 3 (2002), pp. 307–32.

[91] It was called a 'catadioptrick' telescope instead of a reflecting telescope because it combined mirrors (the concave objective and the plane secondary) with a lens (the eyepiece). Newton had purchased the 123-foot lens made by Christiaan Huygens and had a maypole erected there so James Pound, the vicar of Wanstead and a client of Newton's, could use the lens in an aerial telescope. See Jim A. Bennett, 'Catadioptrics and Commerce in Eighteenth-century London', *History of Science* 44 (2006), pp. 247–77, on p. 255.

[92] RS/JBC/12, 2 March 1720/1, and RS/Cl.P/8i/67, Royal Society Library, London. The account of the telescope was published in John Hadley, 'An Account of a Catadioptrick Telescope, Made by John Hadley, Esq; F. R. S. With the Description of a Machine Contriv'd by Him for the Applying It to Use', *Philosophical Transactions* 32, 376 (1722–3), pp. 303–12.

Vice Presidents of the Society; the next year he invented the reflecting octant (often confusingly referred to as Hadley's quadrant), which was capable of the necessary accuracy of ± 1 minute of arc when determining longitude (the most important problem in eighteenth-century navigation) by the lunar distance method, or the distance between the Moon and a known star.[93] Folkes later would come to own a 'Hadley's quadrant 2 feet radius in a box, by Jackson'.[94] Bernard Lens also painted a miniature of the young Folkes holding an astronomical quadrant with a diagonal or transverse scale and a telescopic sight on the index arm (see figure 4.9).[95] This portrait began Folkes's practice, continued for the next two decades, of using portraiture and portrait busts with astronomical or Newtonian iconography to signal his prowess in natural philosophy (see chapter five).

Internal Royal Society correspondence for the 1720s is scantier about Folkes's astronomical activities, probably because he was present at most meetings. However, Folkes's work in astronomy can be well elucidated by the correspondence he maintained during this time with Charles Morgan, Master of Clare Hall. Folkes served as intermediary so that Morgan could keep in touch with the London scientific world, sending him information about recent experiments at the Royal Society and news of the 'activities of Newton and of Morgan's Fellow colleague, Whiston'.[96] Folkes, for instance, passed along Newton's table of atmospheric refraction, which was not made public until Halley appended it to a 1721 article in the *Philosophical Transactions*.[97] When, in 1727, James Bradley, the Savilian Chair of Astronomy at Oxford, discovered the aberration of light, conclusive evidence for the movement of the Earth, Folkes related the discovery to Morgan, also reporting he had just taken leave of William Whiston, who was

---

[93] See John Hadley, 'The Description of a new Instrument for taking Angles', *Philosophical Transactions* 37, 420 (August–September 1731), pp. 147–57.

[94] See *A Catalogue of the Genuine and Curious Collection of Mathematical Instruments…of Martin Folkes, Esq* (London: Samuel Langford, 1755), p. 4.

[95] *Martin Folkes*, by Bernard Lens, National Portrait Gallery, London. https://www.npg.org.uk/collections/search/portraitExtended/mw02269/Martin-Folkes? [Accessed 12 January 2020]. Firstly, the instrument in the miniature cannot be a Hadley quadrant; the instrument Folkes holds must be supported on a stand, otherwise it would not be possible for him to hold it in the manner portrayed. Secondly, the telescopic sight on a Hadley octant is fixed to the frame, not on the index arm as Lens portrayed. As Hadley did not invent his quadrant until 1730, this also dates the Lens portrait more accurately to the 1720s, when Folkes was establishing his scientific reputation in the Royal Society. My deep thanks to Jim Bennett for this information.

[96] John Gascoigne, *Cambridge in the Age of the Enlightenment: Science, Religion and Politics from the Restoration to the French Revolution* (Cambridge: Cambridge University Press, 1989), p. 158.

[97] Charles Morgan papers [uncatalogued and not foliated] Clare College Archives, Cambridge. NRA 33326; Derek T. Whiteside, 'Kepler, Newton and Flamsteed on Refraction Through a "Regular Aire"', *Centaurus* 24 (1980), pp. 288–315, on p. 293. As Whiteside stated, Newton's tables assume that there is a field of 'atmospheric optical forces centrally around the Earth, and a corpuscle of light entering—or otherwise passing through—the air will in effect be "attracted" towards the Earth's centre as though it were a gravitating particle. And the problem of atmospheric refraction is, in this model, reduced to determining the variation of the refractive force with the distance away from that centre which best fits the observed empirical reality'; Edmond Halley, 'Some Remarks on the Allowances to be made in Astronomical Observations for the Refraction of the Air', *Philosophical Transactions* 31, 368 (May–August 1721), pp. 169–72.

Fig. 4.9   Bernard Lens (III), miniature of *Martin Folkes*, watercolour and bodycolour on ivory, ca. 1720, 83 mm × 64 mm, © National Portrait Gallery, London.

'very cocksure' and 'sanguine' of his proposal to find longitude at sea through observing the eclipses of Jupiter's satellites with an improvement to Hadley's telescope.[98] As Barrett has indicated about Whiston's method,

> Its two main inconveniences were the fact that Jupiter's satellites were not visible for about 2 months per year as their orbits drew close to Jupiter, and that the accurate observations necessary to use them were extremely difficult on a moving vessel. Whiston combatted the second in this proposal with his newly invented telescope. He proposed combining seven eye glasses such that the observer could keep Jupiter and the satellites in view in at least one of the glasses despite standing on a rocking deck. He likewise argued that this telescope could be used to observe lunar appulses during the months when the eclipses couldn't be seen, even though this was more complicated.[99]

---

[98]   Folkes to Morgan, 3 December 1728, College Letterbook, Clare College Archives, Letter 89b.

[99]   Katy Barrett, 'Mr Whiston's Project for Finding the Longitude', Board of Longitude. https://cudl. lib.cam.ac.uk/view/MS-MSS-00079-00130-00002/2 [Accessed 18 June 2019]. An appulse is when two or more astronomical objects appear close to each other in the sky, in this case the Moon passing by another body.

By this point, Whiston was a figure of fun for his several longitude schemes, including the use of bomb vessels, magnetic variation, and measures of solar eclipses. Folkes shared his scepticism with Morgan about Whiston's 'spirit levels' and 'telescopes and instruments to a radius of four feet'.[100]

Folkes also sent to Morgan astronomical charts of solar and lunar eclipses compiled by Edmond Halley, likely in preparation for Halley's project to gather crowdsourced data for the total solar eclipse of 1715, the first total solar eclipse London had seen for more than 800 years.[101] Halley hoped to be able to measure the eclipse, which would enable him to predict the trajectory and timing of future ones more accurately, and he did so in an early form of citizen science, soliciting data from the public. Halley 'caused a small Map of England, describing the Tract and Bounds' of the eclipse 'to be dispersed all over the Kingdom', with a 'Request to the Curious to observe what they could about it, but more especially to note the Time of Continuance of total Darkness', which, Halley believed, 'required no other Instrument than a Pendulum Clock with which most Persons are furnish'd'.[102] Folkes was present at the Royal Society meeting when Halley proposed writing a letter to Flamsteed to send observations from Greenwich.[103]

Another series of manuscripts in the Morgan papers were in Folkes's hand, concerning observations of the total solar eclipses of 12 May 1706 and 22 May 1724, with calculations of predicted length and extent of the penumbra and actual observations which had to be adjusted for parallax. These papers seem to be the material enclosed in a letter (now separated) that Folkes wrote to Morgan of 23 June 1724.[104] Folkes wrote:

I have put in the box a little piece of Archimedes I got for you some time since which I think you wanted and Dr Halleys and Will Whistons scheme of the late Eclipse by which you will see how surprisingly over the mark the former has been. Mr Whistons would have been no ill thing had not he gone after the other, in which he acted in my opinion very imprudently as Dr Halley had desir'd to go upon from a great number of observations of [his] own [which] the other could possibly know nothing of.

[100] Folkes to Morgan, 3 December 1728, College Letterbook, Clare College Archives, Letter 89b.

[101] Charles Morgan papers [uncatalogued and not foliated] Clare College Archives, Cambridge. NRA 33326. There had been three eclipses visible elsewhere in the area now known as the United Kingdom during the seventeenth century in 1652, 1654, and 1699.

[102] Edmond Halley, 'Observations of the late total eclipse of the Sun on the 22d of April last past, made before the Royal Society at the house in Crane Court in Fleet-Street London', Philosophical Transactions 29, 343 (1714), pp. 245–62, on pp. 245–6.

[103] Charles Morgan papers [uncatalogued and not foliated] Clare College Archives, Cambridge. NRA 33326.

[104] Martin Folkes to Reverend Charles Morgan, 23 June 1724, Osborn Collection, Box 19572, Beinecke Library, Yale University, New Haven, CT.

Folkes was referring to the fact that Halley made predictions of the eclipse show-ing its path that would pass 'just to the west of London, with the eclipse showing totality at Reading and Windsor'.[105] Actually, Halley was off by some twenty-five miles, and well 'over the mark'; his inaccuracy frustrated vast numbers of would-be observers 'who travelled west of London only to be greeted by a view of a par-tial eclipse'.[106] Halley's 1715 and 1724 eclipse maps did not appear to have been made for his own financial gain.[107] On the other hand, Whiston had imitated Halley's successful eclipse map with a different predicted path, including promin-ent advertisements for other astronomical products, in an 'obvious and rather crude attempt to copy Halley's successful formula', including a totality oval as an extra feature.[108] It was not Whiston's lack of accuracy that Folkes found objec-tionable (the actual eclipse path was actually between the two astronomers' estimations), but rather Whiston's commercial and uncollaborative behaviour.[109] As for Folkes, he admitted to Morgan, 'I had rather worse luck my self than at the last Eclipse, I was upon Salisbury Plain but for which thick clouds saw nothing at all, only that it was very dark. The same fate attended almost every body in this Island, but abroad they were better'.[110] Folkes's experience was shared by William Stukeley who sketched an almost apocalyptic scene of the eclipse from the same location (see figure 4.10).

In addition, Folkes sent to Morgan a star chart of his telescopic observations of the V-shaped Hyades cluster, comparing his naked-eye observations to those he made when he used a telescope; double stars abound in the cluster, so being able to differentiate with the naked eye between, for instance, the Sigma Tauri stars, is a sign of prowess.[111] Folkes may have been indulging in a bit of showing off to the master of his college, whom he urged to visit him at Hillington.

As we have seen, Folkes sent mathematical papers by Roger Cotes to Morgan, but he also included Cotes's designs for astronomical instruments. In a letter to his uncle John Smith of 10 February 1707/8, Cotes indicated he had 'lately hit

---

[105] Alice N. Waters, 'Ephemeral Events: English broadsides of Early Eighteenth-Century Solar Eclipses', *History of Science* 37 (1999), pp. 1–43, on p. 20.

[106] Waters, 'Ephemeral Events', p. 21.   [107] Waters, 'Ephemeral Events', p. 21.

[108] Geoff Armitage, *The Shadow of the Moon: British Solar Eclipse Mapping in the Eighteenth Century* (Tring: Map Collector Publications Ltd, 1997), quoted in Jay M. Pasachoff, 'Halley as an Eclipse Pioneer: His Maps and Observations of the Total Solar Eclipses of 1715 and 1724', *Journal of Astronomical History and Heritage* 2, 1 (1999), pp. 39–54, on p. 51.

[109] For an analysis of the two maps, see Owen Gingerich, 'Eighteenth-century Eclipse Paths', in Owen Gingerich, ed., *The Great Copernicus Chase and Other Adventures in Astronomical History* (Cambridge: Cambridge University Press, 1992), pp. 152–9; See also John Westfall and William Sheehan, *Celestial Shadows: Eclipses, Transits, and Occultations* (New York: Springer, 2014), p. 115.

[110] Martin Folkes to Reverend Charles Morgan, 23 June 1724, Osborn Collection, Box 19572, Beinecke Library, Yale University, New Haven, CT.

[111] Charles Morgan papers [uncatalogued and not foliated] Clare College Archives, Cambridge. NRA 33326. Sigma is probably not a true double star, as the two stars are nine light-years apart. See Bob King, 'Happy Nights with the Hyades', *Sky and Telescope* (30 January 2019). https://www.skyand telescope.com/observing/happy-nights-with-the-hyades [Accessed 14 December 2019].

The appearance of the Total Solar eclipse *from Haradon hill May 11, 1724.*

Fig. 4.10 *The appearance of the total solar eclipse from Haradon Hill near Salisbury, May 11 1724,* by Elisha Kirkall after William Stukeley, 1984–453, © The Board of Trustees of the Science Museum.

upon a device which I believe will be of very good use for observing eclipses'.[112] The device was an equatorially mounted telescope 'fixed parallel to the earth's axis of rotation, and rotating at the same speed of the earth'.[113] By adjusting a prism or mirror, it was possible to keep an object in view for a required length of time; Cotes described the telescope in an unpublished paper for Clare Hall, which was communicated by Folkes to Morgan. Where Folkes had acquired the manuscript, it is not certain, but he may have sent the account to Morgan, as such a telescope would have been useful for Halley's eclipse observations; in the words of Cotes, such a telescope 'may be directed to the Pole of the world; and thereby Eclipses and any other such appearances be observed more easily than is usual'.[114]

4.5 Antiquarianism and the Royal Society in the 1720s

The papers Folkes sent to Morgan also show Folkes's developing interest in antiquarian topics—particularly those concerning Arabia and the Middle East—which he promoted in the Royal Society. In his appraisal of England's political,

[112] Ronald Gowing, *Roger Cotes: Natural Philosopher* (Cambridge: Cambridge University Press, 1983), p. 87.
[113] Gowing, *Roger Cotes*, p. 87.
[114] Charles Morgan papers [uncatalogued and not foliated] Clare College Archives, Cambridge. NRA 33326.

religious, and cultural scene following the Restoration, Gilbert Burnet singled out the esteemed state of learning at the universities of Cambridge and Oxford, and 'chiefly the study of the oriental tongues', for there were professorships of Arabic established at Oxford and Cambridge in the 1630s. However, the reality was that in early modern England 'private instruction and solitary study remained central for the acquisition' of Arabic, although Richard Busby bestowed a considerable collection of Oriental manuscripts to Trinity College, Cambridge in 1697.[115] In 1706, Halley published Apollonius's *De Sectione Rationis*. In 1710, he published *Conics*, Books V–VII by the same author, this time from an Arabic manuscript. Using a few pages already translated by Edward Bernard, he thus attempted to teach himself the language. Folkes recalled that:

> this he did with such success, though his being so great a Master of the Subject, that I remember the Learned Dr Sykes, (our Hebrew Professor at Cambridge, and the greatest Orientalist of his time, when I was at that University,) told me, that Mr Halley talking with him upon the subject, shew'd him two or 3 passages which wanted Emmendation, telling him what the Author said, and what he shou'd have said, and which Dr Sykes found he might with great ease be made to say, by small corrections, he was by this means enabled to make in the Text. Thus, I remember, Dr Sykes expresst himself, Mr Halley made Emendations to the Text of an Author, he could not so much as read the language of.[116]

Individual Arabic manuscripts were thus worthy of note. Folkes sent Morgan a copy of a letter sent to him from Mohammed Ben Ali Abgali, the Moroccan Ambassador to Britain, in French translated from the Arabic; the letter thanked Folkes for his friendship. Abgali spent eighteen months in England from August 1725 to February 1727 as ambassador to King George I. During his stay, he was elected as a Fellow of the Royal Society and was hosted at Goodwood by Lennox.[117] As John Byrom recalled in his diary, Lennox 'proposed the Morocco ambassador to be a member, so we all balloted and chose him immediately, Sir Isaac [Newton] in the chair, the room was pretty full of people'[118] (see figure 4.11).

---

[115] Mordechai Feingold, 'Learning Arabic in Early Modern England', in Jan Loop, Alastair Hamilton, and Charles Burnett, eds, *The Teaching and Learning of Arabic in Early Modern Europe* (Boston and Leiden: Brill, 2017), pp. 33–56, on pp. 38–9 and p. 49.

[116] Martin Folkes, 'Memoir of Dr. Edmond Halley', in Eugene Fairfield MacPike, ed., *Correspondence and Papers of Edmond Halley* (London: Taylor and Francis, Ltd, 1937), pp. 1–13, on p. 10.

[117] The copy in French is in the College Letterbook, Clare College Archives, Letter 89. The original in Arabic is in RS/MS 790/3 Royal Society Library, London, Abgali to Folkes, dated 1726: 'As for what you said concerning the obligations you have towards the two illustrious men, the Dukes of Richmond and of Montagu, for they have been the cause why we met you, you are telling the truth, but in fact we ourselves owe them and you the kind deed, they were the ones who made us meet you, so that if you consider things carefully, you will see that the obligation is entirely on our part, towards you, and towards your above-mentioned friends'.

[118] Byrom, *Private Journal*, 24 March 1726, vol. 1, part 1, p. 228.

Fig. 4.11   Enoch Seeman, *Mohammed Ben Ali Abgali,* oil on canvas, ca. 1725, 60 × 40 inches (152.5 × 101.5 cm), Private Collection, courtesy of Ben Elwes Fine Art.

Abgali's letter was translated into French by Solomon Negri (1665–1727) (Sulaiman ibn Ya'kub Al-Shami Al-Salihani). Negri was a Greek Orthodox Christian immigrant from Damascus, who, after living in Rome and Holland, came to London, where he taught Arabic to students at St Paul's under the Mastership of Orientalist John Postlethwaite.[119] Negri also did translation work on behalf of the monarchy for foreign ambassadors, was involved in the production of an Arabic Psalter and New Testament through the Society for Promoting Christian Knowledge, and assisted Humphrey Wanley with identifying Oriental manuscripts, as well as Dr John Friend and Dr Richard Mead separately in the translation of Arabic texts on medicine.[120] Folkes met Negri via his friendships with the Duke of Montagu and Lennox.

[119]   John-Paul A. Ghobrial, 'The Life and Hard Times of Solomon Negri: An Arabic Teacher in Early Modern Europe', in Jan Loop, Alastair Hamilton, and Charles Burnett, eds, *The Teaching and Learning of Arabic in Early Modern Europe* (Leiden: Brill, 2017), pp. 310–31, on p. 319; Feingold, 'Learning Arabic in Early Modern England', p. 53.

[120]   TNA PROB 31/50/413, The National Archives, Kew; Admission to Bodleian Library, Admissions register, 1683–1833, Bodl. MS e. 534. f. 32r., Bodleian Library, Oxford, as noted in Ghobrial, 'The Life and Hard times of Solomon Negri', p. 328, footnote 76.

Also included in Folkes's correspondence to Morgan was an account of a 'petrify'd city' called Ouguela or Rassem which held 'men and women and children, some in the streets, some in the shops and houses, and in as great variety as if alive and following their several occupations', much like is seen in Pompeii.[121] In August 1728, an official envoy of the Regency of Tripoli, Cassem Algaida Aga, came to London on a diplomatic mission to George II; this was important, as Tripoli was being threatened by the French. Jezreel Jones, appointed as the Royal Society clerk succeeding Halley, served as translator while Cassem Aga made a declaration to the King about a number of topics, including their practice of smallpox inoculation and the apocryphal city of Rassem.[122]

The story quickly reached scholarly and political circles. In his letter to Morgan, Folkes wrote, 'the accounts he gives being somewhat astonishing…what foundation there is God only knows. But I have my self found that the Tradition was as long ago in the Levant as Baumgarten's travels into these parts who travelled almost 220 years ago'.[123] Folkes then went on to list the reports of the town he had collected from Mohammed Ben Ali Abgali and Sir Clement Cottrell, the King's Master of Ceremonies, who communicated it to the Secretary of State. Folkes also discussed Rassem with Lennox, who in his rambling manner, wrote to Folkes with some scepticism:

> I find Ougela [Ouguela], which I always took to be the name of the place, is a City only in the neighbourhood of it, but that the true name is Rassem; & that it is not a town but a whole Country that is petrified, in which might be some villages. but I have since heard that Sr. Hans Sloane has, in a letter he lately writ to some of the academy here, say'd that there has been some discoverys in England, which makes every body believe the whole account, & every one that has been given hitherto, of it, false, & a cheat. therefore if there has been really any such thing, I beg you would let me know it.[124]

By the 1740s, Folkes passed on his collection of sources about the city to Thomas Shaw, Chaplain to the English Factory in Algiers, who gave a lecture on them at the Royal Society in 1745 when Folkes was President; interest was probably

---

[121] College Letterbook, Clare College Archives, Letter 89. This was a copy in Morgan's hand of a letter Folkes sent to him on 3 December 1728.

[122] See Cassem Aga, *An account of the success of inoculating the small-pox in Great Britain, for the years 1727 and 1728. With a comparison between the mortality of the natural small-pox, and the miscarriages in that practice; as also some general remarks on its progress and success, since its first introduction. To which are subjoined, I. An account of the success of inoculation in foreign parts. II. A relation of the like method of giving the small-pox, as it is practised in the kingdoms of Tunis, Tripoli, and Algier. Written in Arabic by his excellency Cassem Aga, ambassador from Tripoli. Done into English from the French of M. Dadichi, His Majesty's interpreter for the eastern languages* (London: J. Peele, 1729).

[123] Charles Morgan papers [uncatalogued and not foliated] Clare College Archives, Cambridge.

[124] RS/MS/865/6, Charles Lennox, 2nd Duke of Richmond and Duke of Aubigny, to Martin Folkes, Chantilly 23 August N.S. 1729, Royal Society Library, London.

stimulated by the excavations at Pompeii and Herculaneum, reports of which were published in the *Philosophical Transactions* of 1740.[125] Shaw's method of verification was not a field expedition, but the collecting of incidences in the Baconian tradition combined with 'praeter-philological methods of exact analysis, checking, comparing of sources and calculation of general probability' typical to the philological tradition of antiquarianism, in which linguistic and verbal remains were used to understand the historical record.[126] Folkes, however, went a step further, using direct empirical evidence drawn from natural history techniques in looking at fossils to examine the possibilities of spontaneous petrification. Following Shaw's lecture, he presented his own report of a body of a petrified man he had seen in 1734 at the Villa Ludovisi at Rome, known for its collection of Roman antiquities and statuary, including *The Dying Gladiator*. These remains, purportedly those of a man discovered frozen in the Alps, were famous; John Evelyn, in his diary entry for 10 November 1644, remarked:

> In the villa-house is a man's body flesh and all, petrified, and even converted to marble, as it was found in the Alps, and sent by the Emperor to one of the Popes; it lay in a chest, or coffin, lined with black velvet, and one of the arms being broken, you may see the perfect bone from the flesh which remains entire.[127]

As for Folkes, he cited travel accounts by Richard Lassells and Athanasius Kircher, but discounted them, noting that what he saw was a heap of bones surrounded by a white crust, and provided a detailed description and scale drawing for an article in the *Philosophical Transactions*[128] (see figure 4.12). The mummy had obviously disintegrated by that point. Soon after Folkes's elucidation of the matter, ideas of spontaneous petrification would be abandoned by the Royal Society.

One of Folkes's firmest supporters in the Royal Society of his growing interests in antiquarianism and history was James Jurin (1684–1750), who Folkes had probably met at Cambridge, as Jurin was a Fellow at Trinity when Folkes was an undergraduate. Not only did Jurin promote Newtonianism in the Society, but he also worked with Folkes to preserve the Royal Society's heritage and material culture. Jurin was a physician of Huguenot descent who, after a period of being a schoolmaster and giving public lectures in natural philosophy in Newcastle, came to London in 1716 to establish a medical practice. Jurin attended his first Royal

[125] Cornel Zwierlein, *Imperial Unknowns: French and British in the Mediterranean, 1650–1750* (Cambridge: Cambridge University Press, 2016), p. 289.

[126] Zwierlein, *Imperial Unknowns*, p. 289.

[127] T. Corey Brennan, 'The 1644 Visit of the Englishman John Evelyn to the Villa Ludovisi', *Archivio Digitale Boncompagni Ludovisi*. https://villaludovisi.org/2012/12/03/1644-english-diarist-john-evelyn-visits-the-villa-ludovisi [Accessed 25 June 2019].

[128] Martin Folkes, 'An Account of Some Human Bones Incrusted with Stone, now in the Villa Ludovisia at Rome: Communicated to the Royal Society by the President, with a Drawing of the Same', *Philosophical Transactions* 43, 477 (1744), pp. 557–60.

Fig. 4.12 Figure of 'petrified man', from 'An Account of some human Bones incrusted with Stone, now in the Villa Ludovisia at Rome: communicated to the Royal Society by the President, with a Drawing of the same', *Philosophical Transactions* 43, 477 (1744), p. 558, Image from the Biodiversity Heritage Library. Contributed by Natural History Museum Library, London. www.biodiversitylibrary.org.

Society meeting on 25 October 1716 accompanied by Robert Smith, who had recently become Plumian Professor of Astronomy at Cambridge on the death of Roger Cotes. A year later, Folkes nominated Jurin to the Royal Society, and in 1718 and 1719 Jurin made nineteen presentations to the Royal Society on iatro-mechanism, capillary phenomena, and translations of Leeuwenhoek's letters.[129] In 1721, Jurin became secretary of the Royal Society.

Two years later, Francis Hauksbee, the instrumental demonstrator, had just been elected keeper of the Library and Repository, and on 9 May 1723 it was 'Ordered that a Committee be appointed to inspect the State of the Library's and Repository'.[130] Folkes and Jurin headed the committee and Folkes wrote the committee report as part of a more general effort to improve these facilities and preserve the early heritage of the Royal Society. The committee minutes indicated, for example, that 'fresh Notice be given to the Cheese Monger, who is in possession of the Cellar under the Repository, to Remove at Michaelmas next, or as

[129] Andrea Rusnock, 'Introduction', *The Correspondence of James Jurin (1684–1750): Physician and Secretary of the Royal Society* (Amsterdam and Atlanta, GA: Rodopi, 1996), p. 12.
[130] RS/CMO/2/311, Minutes of a meeting of the Council of the Royal Society, 9 May 1723, Royal Society Library, London.

soon as his Lease is expired.'[131] The meeting minutes also revealed that the Library had its own odour of decay:

> Many of the Manuscripts as well as other Valuable Books were much out of Repair and in Danger of being Spoiled through want of new Binding...The Council were informed that two large Manuscripts which were then upon the Table had been offered to Dr [Richard] Mead for Sale by an unknown Person, but he finding they belonged to the Arundel Library had Stoped and Secured them for the Society.[132]

From the report of missing manuscripts (the *Prophecies of Merlin* and *Biblia hieronimi*) in the Arundel Collection, it was clear that the Library rules made in the seventeenth century had been contravened, particularly: '4. That for securing the bookes and to hinder their being imbezilled noe booke shall be Lent out of the Library to any person whatever.'[133] To be fair, Folkes himself violated this rule, the Committee Report showing he borrowed and returned 'Zimmerman Cometoscopia, quarto, 1682'.[134]

The Arundel Library of *c.*3,000 books and *c.*500 manuscripts was a donation from Henry Howard, 6th Duke of Norfolk and 22nd Earl of Arundel, given to the Royal Society at the urging of John Evelyn. The regulations for the donation specified that the Norfolk Library would be kept separately from the rest of the Royal Society's collections, and that there would be 'an exact catalogue of all the books' of the collection.[135] The only catalogue for the Library was the alphabetical list *Bibliotheca Norfolciana* (1681), compiled largely by William Perry FRS (*c.*1650–96), first librarian of the Society; Fellows and librarians annotated their own copies, adding entries or crossing them out. Folkes reported that in the Arundel Library itself the books were not classed or in alphabetical order, 'but number'd throughout, viz the Printed books from 1 to 3285 omitting the numbers between 900 and 1000 as also between 1300 and 2000...We have lookt over these number and added the account of such as we found wanting, but cannot tell

---

[131] RS/CMO/2/311, Royal Society Library, London.

[132] RS/CMO/2/312, Minutes of a meeting of the Council of the Royal Society, 27 June 1723, Royal Society Library, London.

[133] RS/CMO/2/312, p. 3v. The Report of the Committee Appointed to Inspect the State of the Library's and Repository, 1723; RS/CMO/1/247, Minutes of meeting, 26 December 1678, Royal Society Library, London.

[134] RS/CMO/2/312, p. 2v. The Report of the Committee Appointed to Inspect the State of the Library's and Repository, 1723, Royal Society Library, London. This was Johann Jacob Zimmermann (1642–93) who in his *Cometoscopia* (1681, 2 parts) mapped trajectories of these comets across the starry skies and provided 'astrotheological' prophecies and interpretations of their meaning. One wonders if this had anything to do with Folkes's later editing of Newton's *Chronology*.

[135] Linda Peck, 'Uncovering the Arundel Library at the Royal Society: Changing Meanings of Science and the Fate of the Norfolk Donation', *Notes and Records of the Royal Society of London* 52, 1 (1998), pp. 3–24, on p. 4 and p. 8.

always what books they belongd to, for want of a Numerical Catalogue'.[136] There were several books missing, including a copy of Galileo's *Sidereus Nuncius* lent to Edmond Halley (and then returned) as well as printed books and manuscripts not catalogued that were given to the Society as gifts.

There was also an 'interleavd Catalogue of the Bodley' in Oxford, into which other of the Royal Society's books donated by fellows such as George Ent and Francis Aston had 'been alphabetically inserted'.[137] This particular part of the Library seemed to be more complete and better organized, 'with shelves in the respective Classes, figured according to the order of the Alphabet from A to T' and the catalogue and the number and placement of the books in agreement, 'excepting the small amount we have taken notice of'.[138] We can glimpse the connoisseur's pleasure Folkes took in examining the Library, characterizing the woodcuts in Albrecht Dürer's *Historia Mariae et Parthenices* (Life of the Virgin) as 'figuras pulcherrimus' (very beautiful figures).[139]

As to the Repository, all Folkes could report is 'as the Curiositys there are not numberd, and we find no Catalogue, we are not able to give any particular account of them'.[140] Unlike Sloane, Folkes was not a natural historian, so while the instruments and art objects would have been of interest to him, the greatly decayed taxonomic specimens were less compelling (see chapter seven).

In fact, it would not be until 1730 that the Library and Repository Committee was revived. The Committee, headed by Royal Society secretary Cromwell Mortimer, worked from 15 January 1729/30 until 30 October 1733 and went systematically through the collection, checking the objects against Nehemiah Grew's 1681 museum catalogue, making observations such as:

Ordered that Mr Phillips do prepare and make a Cabinet to contain the Human curiosities, after the same model as those he formerly made, and place it in the Gallery between the Chimney and window. The Committee proceeded to examin the Quadrupeds and numbered them to 54. And [Cromwell] Mortimer was desired to proceed in making out a catalogue of the Serpents.[141]

[136] RS/CMO/2/312, p. 3r., Royal Society Library, London.
[137] RS/CMO/2/312, p. 2r., Royal Society Library, London.
[138] RS/CMO/2/312, f. 2r., Royal Society Library, London. Folkes reported 'The running Catalogue before mention'd extends to the IVth shelf of the Class T omitting 2 small Classes N and Z, in the former of which are 95 volumes in 12°, Duplicates in this Library, and in the latter are some new books not yet enterd in the Catalogues. On the vth and VIth Shelves of T are some more new Books, on the VIIth several unsorted pamphlets and loose papers, and on the lowest shelf of the 3 last Classes R.S & T are 54 volumes of Duplicates as well Folios or Quartos & Octavos, amongst which are som[e] odd Volumes of Rhymers Foedera.'
[139] RS/CMO/2/312, f. 3r., Royal Society Library, London. Albrecht Dürer, *Divae Parthenices Mariae Historiam ab Alberto Durero* [Life of the Virgin] (Nürnberg: Hieron. Hölzel, 1511).
[140] RS/CMO/2/213/1, f. 5r., Royal Society Library, London.
[141] RS/CMB/63/50, Minutes of a meeting of the Repository Committee of the Royal Society, 2 April 1733, Royal Society Library, London.

Of the 5,217 objects that the Repository should have contained, just a fraction of these (1,775) were found, slightly fewer than Grew's catalogue.[142] There is no doubt the collection was not taken care of properly. The Committee acknowledged that the Repository was 'wretched' due to improper storage and damp and promised that there would be 'proper care taken hereafter to preserve' it 'in good condition', for pragmatic reasons.[143] The Committee noted:

> The Collection of the Royal Society hath formerly made a considerable figure and though at present it is so much reduced, they hope, by their care in recovering and preserving what is left, to incite the curious part of the world to be as generous to it as they have formerly been. There are many ingenious Anatomists Fellows of the Royal Society, who, the Committee hope, will be by these means encouraged to enlarge the Cabinet of human Curiosities with some of their excellent preparations. Some Gentlemen curious in their enquiries into Nature may be engaged to supply such Curiosities, which formerly had place in the Repository, but are now entirely perished. Others may be induced to deposit their Collections here, as a sure means of rendering them usefull to the publick: and will have the Satisfaction to know that what they have collected with so much industry and expence, will here remain safe and entire.[144]

The 'wretched' state of the Repository during this period was taken by scholars as a symbol of 'the Society's decline after the intellectual and experimental dynamism of its founding years', but this may be too simplistic an interpretation.[145] It might be fairer to say that there were several attempts to restore the collection, and a number of specimens were given to the Society from the museum's beginnings in an incomplete state because they were rare or difficult to replace, such as a giant stag beetle missing its head. What we might see as piles of rubbish were specimens to be observed, or used or reused in experimentation. As Werrett has demonstrated, the idea that 'second-hand' or 'repaired' was inferior was only a nineteenth-century construct, even in natural philosophy. 'An acceptance or approval of fragments and broken things, or what art historians call *Ruinenlust* (an appetite for ruins), was also apparent in the sciences'.[146]

The same attitude could be said to have been evinced for instruments left to the Society, and here Folkes took a more definite interest. In 1723, Folkes wrote a

[142] Jennifer M. Thomas, 'A "Philosophical Storehouse": The Life and Afterlife of the Royal Society's Repository', PhD Dissertation (Queen Mary, University of London, 2009), p. 35.

[143] RS/CMO/3/55, 18 February 1733/4, Royal Society Library, London; Thomas, 'Philosophical Storehouse', p. 35.

[144] RS/CMO/3/55, 18 February 1733/4, Royal Society Library, London.

[145] Alice Marples, 'Scientific Administration in the Early Eighteenth Century: Reinterpreting the Royal Society's Repository', *Historical Research* 92, 255 (February 2019), pp. 183–214, on p. 184.

[146] Simon Werrett, *Thrifty Science: Making the Most of Materials in the History of Experiment* (Chicago: University of Chicago Press, 2019), p. 112.

detailed article about the provenance and decayed condition of a set of twenty-six microscopes bequeathed by Maria van Leeuwenhoek, the daughter of the famous natural philosopher Antonie van Leeuwenhoek (1632–1723). The year before, Jurin had renewed correspondence with the Dutch microscopist, sending along a volume of the *Philosophical Transactions* as a measure of the Society's esteem, translating some of his previous letters to the Society in Dutch and having them published.[147] Jurin, for instance, proposed to Leeuwenhoek to use pieces of silver wire to measure microscopic objects 'instead of hair or grains of Sand' and asked him if he would see if there were animalcules in smallpox or skin pustules.[148]

Jurin's efforts bore fruit in the bequest Leeuwenhoek left to the Society, presented on 7 November 1723 by Abraham Eden. Folkes reported that 'the legacy consists of a Small Indian Cabinet, in the drawers are which are 13 little boxes or cases, each containing Two Microscopes, handsomely fitted up in Silver, all which, not only the glasses, but also the Apparatus for managing of them, were made with the late Mr Lewenhoecks own hand'.[149] Leeuwenhoek had provided each instrument with a permanently mounted object, such as lime tree wood, part of a whale's eye or elephant's tooth, 'being desirous the Gentlemen of the Society, should without trouble be enabled to examin many of these objects, on which he had made his most considerable discoverys'.[150]

In the next two decades, the microscope maker John Cuff (1708–72) busily sold his instruments with improved micrometers and optics to several of the Royal Society Fellows, the Swiss naturalist Abraham Trembley (1710–84) made discoveries about generation with hydra using innovative microscopic techniques, and Henry Baker received the Copley Medal (1744) from the Society for his microscopic observations of saline crystals (see chapter seven). The afterlives of Leeuwenhoek's donations for research purposes would thus have been especially relevant.[151] Folkes indeed remarked that he hoped 'some of the Society would pursue those Enquiries, that late Possessor of these Microscopes was so deservedly famous for', and use the instruments in current research; the microscopes' importance to the heritage of the Royal Society did not mean they were only totemic. Folkes's hope was not a vain one; in his 1721 account of the Royal Society, Luigi Ferdinando Marsigli noted, 'everyone studies and carries out tests at home and if they happen to need certain instruments…they go to Mr Desaguliers…When they have

[147] Wellcome MS 6143, Letters from Jurin to Antonie van Leeuwenhoek, 1722–4, Letter of 22 February 1721/2, in Rusnock, *The Correspondence of James Jurin*, pp. 86–8.

[148] Wellcome MS 6143, letter of 24 December 1722 in Rusnock, *The Correspondence of James Jurin*, pp. 120–2, and 'Introduction' in *The Correspondence of James Jurin*, p. 18.

[149] RS/Cl.P/2/17, f. 1r., 'Some Account of Mr Leeuwenhoek's curious microscopes lately presented to the Royal Society' by Martin Folkes, 1723, Royal Society Library, London.

[150] RS/Cl.P/2/17, f. 1v., 'Some Account of Mr Leeuwenhoek's curious microscopes lately presented to the Royal Society' by Martin Folkes, 1723, Royal Society Library, London.

[151] See Mark J. Ratcliff, *The Quest for the Invisible—Microscopy in the Enlightenment* (Aldershot: Ashgate, 2009), pp. 27–8.

done, they set it down on paper to take it to the Society where the experiment is performed in the demonstration room, after which they take everything back home so no apparatus ever remains there'.[152] In June 1728, the Leeuwenhoek microscopes were moved from the Repository along with a telescope of Christiaan Huygens (which had its optics retrofitted by George Graham) to a lockable closet for security, but also so they were more available for experimentation.[153] This may well be why the microscopes eventually disappeared from the Repository, as well as the fact that while 'several of the objects yet remain before the Microscopes ... the greater number' were 'broken off'.[154] Folkes also remarked that he could not immediately verify Leeuwenhoek's observations with the instruments, as 'we are under very great disadvantages for want of the experience he had', but as so many others had, 'perfectly confirm'd' his observations, 'there can surely be no reason to distrust his accuracy'.[155] As his microscopes were quite difficult to use, Leeuwenhoek had faced much scepticism about the veracity of his observations, and with his comments about their difficulty of use Folkes preserved the scientific reputation of the Dutch natural philosopher.[156]

During his time in the Royal Society in the 1720s, Folkes had shown himself to be an actively contributing member, demonstrating his burgeoning antiquarian interests, whether in the apocryphal ancient city of Rassem or preserving the history and heritage of the Royal Society itself via the reform of its Library, its Repository, and its instruments. Folkes also promoted a Newtonian programme via his astronomical research. He even helped spread one of the most famous stories of the Newtonian legacy. After leaving Cambridge, he remained in touch with both Charles Morgan at Clare and Robert Greene. In 1727, the year of Newton's death, Greene published his book (*Principles of the Philosophy of the Expansive and Contractive Forces or an Inquiry into the Principles of the Modern Philosophy*), a work described by Augustus de Morgan as a 'volume ... of 981 good folio pages [which] treats of all things, mental and material. The author is not at all mad, only

[152] A. McConnell, 'L. F. Marsigli's Visit to London in 1721, and his Report on the Royal Society', *Notes and Records of the Royal Society of London* 47 (1993), pp. 179–204, on p. 191.

[153] RS CMO/3/7, Royal Society Library, London. 'Ordered That Mr Hugens's three large Telescope Glasses with their Furniture and also Mr Lewenhoeks Microscopes etc be reposited under a new Lock in the Closet in the Council Room', as mentioned in Thomas, 'Philosophical Storehouse', pp. 91–2. James Bradley offered the 'glass and old furniture of Mr Huygen's large telescope', asking the Society 'to accept of such new additions and improvements which his Uncle (Mr Pound) had made to the Furniture & apparatus viz curious micrometer, Made by Mr Graham, a new eye glass, a new directr to the sight & a new tin tube to carry the object glasse'. RS/JBO/14/235.

[154] RS/Cl.P/2/17, f. 1v., Royal Society Library, London. 'Some Account of Mr Leeuwenhoek's curious microscopes lately presented to the Royal Society' by Martin Folkes, 1723. Some of the damage could have occurred due to shipping the instruments as well.

[155] RS Cl.P/2/17, f. 3v., Royal Society Library, London. 'Some Account of Mr Leeuwenhoek's curious microscopes lately presented to the Royal Society' by Martin Folkes, 1723. See also Ratcliff, *The Quest for the Invisible*, p. 27.

[156] For the reception of Leeuwenhoek's work, see Edward Ruestow, *The Microscope in the Dutch Republic: The Shaping of Discovery* (Cambridge: Cambridge University Press, 2004), pp. 28–30.

wrong on many points'.[157] However, in the midst of his errors, Greene mentioned the following:

> This was written by me when I reflected that Newton's theory of gravity is the beginning of everything...which famous proposition, all things considered, originates, as disclosed to our knowledge, from an apple. This I heard from a most ingenious and most learned man, and also the finest, and friendliest to me, Martin Folkes Esquire, truly meritorious Fellow of the Royal Society.[158]

This was one of the first published accounts of the apple story, also related in print by Voltaire in his *Essay on the Civil Wars in France* (1727) and *Letters concerning the English Nation/Lettres philosophiques* and by William Stukeley in his *Memoirs of Sir Isaac Newton's Life* (1752).[159]

4.6 Bidding for the Royal Society Presidency

After Newton died on 20 March 1727, Sir Hans Sloane became President, with Vice Presidents including Newton's physician and collector Dr Richard Mead, Folkes, and antiquary Roger Gale.[160] He was President at a time when rivalry in the Royal Society was deep seated. Several natural philosophers, particularly the astronomers and mathematicians, felt that when Sloane was secretary he had promoted cronyism and 'brought down the reputation of the Royal Society by skewing publication towards his own natural historical interests and correspondents'.[161] John Flamsteed wrote in a letter to his friend Abraham Sharp in October 1704: 'Our Society decays and produces nothing remarkable, nor is like to do it, I fear, whilst 'tis governed by persons that either value nothing but their own interests, or understand little but vegetables, and how, by making a bouncing noise, do

---

[157] Augustus de Morgan, *A Budget of Paradoxes* (London: Longmans, Green, and Co., 1872), pp. 80–1.

[158] Robert Greene, *The Principles of the Philosophy of the Expansive and Contractive Force, or an Enquiry into the principles of the Modern Philosophy, that is into the General Chief rational Sciences* (Cambridge: Cornelius Crawfield, 1727), p. 972. [Haec a me Scripta fuerunt, cum Newtoni Gravitationem reputarem esse omnium Rerum Principium, neque quicquam fere jam etiam Corrigo, nisi quod Asseram Pressionem non forsan esse Aequabilem per Totam Rerum Compagem, nec Materiam esse Similarem, vel Proportionalem ipsius Ponderi; Quae Sententia Celeberrima, Originem ducit, uti omnis, ut fertur, Cognitio nostra, a Pomo; id quod Accepi ab Ingeniosissimo & Doctissimo Viro, pariter ac Optimo, mihi autem Amicissimo, *Martino Folkes* Armigero, Regiae vero Societatis Socio Meritissimo; Quem hic Honoris Causa Nomino].

[159] Of course, John Conduitt recorded the apple story in manuscript (Cambridge University Library, Keynes MS 130.4, pp. 10–12), as did Abraham De Moivre (Joseph Halle Schaffner Collection, University of Chicago Library, MS 1075.7). For the relationship between Folkes and Voltaire, see chapter nine.

[160] RS/CMO/2/332, Minutes of a meeting of the Council of the Royal Society, 27 March 1727, Royal Society Library, London.

[161] Marples, 'Scientific Administration', p. 192.

cover their own ignorance'.[162] Jurin noted to Folkes that Newton himself had fre-
quently said, 'That Natural history might indeed furnish materials for Natural
Philosophy; but, however, Natural History was not Natural Philosophy'.[163] Some
members may also have recalled William King's lampoon of the *Philosophical
Transactions* in his *Transactioneer* (1700) where Sloane's curiosity-mongering was
viciously satirized. The following year, Edward Young lampooned Sloane in his
Love of Fame: The Universal Passion, calling him 'the foremost Toyman of his
Time', who collected and purveyed curiosities of little value:

> His nice Ambition lies in curious Fancies,
> His Daughter's Portion a rich shell inhances,
> And Ashmole's Baby-house[164] is, in his View,
> Britannia's golden Mine, a rich Peru!
> How his eyes Languish? how his thoughts adore
> That painted coat which Joseph never wore?
> He shews on holidays a sacred pin,
> That toucht the ruff, that toucht Queen Bess's chin.[165]

Although Sloane's election faced such studied opposition, he defended himself
with 'concentrated indifference':

> I have formerly done a great deal for the service of the Society, which I did by
> inclination very willingly, and even have continued it to this time, when my for-
> mer services had render'd them less necessary…When I had the honour to be
> made President, I did what I could to make the Society and that design of their
> institution for great purposes known, and was going onto think of hindering the
> admission of members who did not deserve that respect, to employ members fit
> to prosecute their affairs with vigour, and to encourage foreigners or absent fit
> persons \to help us/, that are here, with their observations, and did endeavour to
> show them by example that the officers of the society should every one in their
> place exert themselves, as I intended to do.[166]

---

[162] John Flamsteed to Abraham Sharp, 21 October 1704, quoted in *An Account of the Reverend John
Flamsteed, the First Astronomer Royal*, ed. F. Baily (London, 1835), p. 218, as quoted in Marples,
'Scientific Administration', p. 192.

[163] James Jurin to Martin Folkes, in Nichols, *Literary Anecdotes*, vol. 3, p. 320.

[164] A reference to the Ashmolean Museum.

[165] Edward Young, *Love of Fame, the Universal Passion. In Seven Characteristical Satires*, 2nd ed.
(Dublin: Sarah Powell, for George Ewing, 1728); Satire IV, ll, pp. 113–22, as quoted by Barbara
Benedict, 'Collecting Trouble: Sir Hans Sloane's Literary Reputation in Eighteenth-Century Britain',
Eighteenth-Century Life 36, 2 (Spring 2012), pp. 111–42, on p. 128.

[166] BL MS Sloane 4026, f. 338r., 20 November 1727, British Library, London, as quoted in Marples,
'Scientific Administration', p. 193.

That said, Sloane tried at first to be conciliatory, moving away from the rather vigorously imperious way Newton chaired meetings; the Council minutes noted that he:

> took occasion to mention an Alteration he Judged ought to be made in a matter of form hitherto used in the Ordinary meetings, which is to make a difference of solemnity in laying the Mace when the President is in the Chair, and not laying it when the Chair is Supplyed by a Vice President. He observed that there was no foundation for making any Such difference, and that as the Vice President is invested with the powers of a President in every respect in his absence, he ought also to be Attended with the same Solemnity in the Ordinary meetings.[167]

Despite Sloane's efforts, a contested election was held between Folkes and Sloane on Anniversary Day in November 1727. To position himself for victory, Folkes began nominating friends such as Benjamin Hoadly and relatives such as his brother William as Fellows, to ensure votes in his favour.[168]

The most detailed account of the division in the Society, Folkes's bid for the presidency, and his ultimate defeat were provided by the diary of John Byrom. As mentioned, Byrom was a Jacobite poet and inventor of a shorthand system, the *New Universal Shorthand*, officially taught at Oxford and Cambridge and used by the clerk of the House of Lords.[169] The entry in Abraham Rees's *Cyclopaedia* remarked. 'Mr. Byrom has completely succeeded in the invention and establishment of his System…founded on rational and philosophical principles'.[170] Adherents of his system included 'clergy, Heads of Cambridge Colleges, Members of Parliament, and aristocrats: among others, Horace Walpole, Charles and John Wesley, the Scottish astronomer James Douglas, 14th Earl of Morton' and Folkes.[171] Byrom met Folkes due to D'Anteney, a virtuoso interested in geometry and cryptography, who advised him to secure Folkes's patronage by teaching him shorthand 'for nothing but his approbation only, because, he said, Mr. Folkes was a clever man and his approbation would be of service'. To this idea Byrom

---

[167] RS/CMO/2/332, Minutes of a meeting of the Council of the Royal Society, 27 March 1727, Royal Society Library, London. The literature on Newton as President of the Society and his imperious management style is vast, from Henry Lyons, *The Royal Society, 1660–1940* (Cambridge: Cambridge University Press, 1949), to Richard Westfall's *Never at Rest: A Biography of Isaac Newton* (Cambridge: Cambridge University Press, 1983), p. 682, which characterized his presidency as 'imperious as well as imperial'.

[168] RS/JBO/14/71, Royal Society Library, London.

[169] There has only been one recent book-length study of Byrom: Joy Hancox's *The Queen's Chameleon: The Life of John Byrom. A Study of Conflicting Loyalties* (London: Jonathan Cape, 1994), which, from its reviews, seems to have serious flaws. See, for instance, Paul Korshin's review *in Journal of British Studies* [Albion] 27, 4 (Winter 1995), pp. 674–5.

[170] Abraham Rees, 'Stenography' in *The Cyclopaedia: Or Universal Dictionary of Arts, Sciences, and Literature*, 41 vols (London: Longman, Hurst, Rees, Orme, and Brown, 1802–19), vol. 34. For the division in the Society, see John Heilbron, *Physics at the Royal Society during Newton's Presidency* (Los Angeles: William Andrews Clark Memorial Library, 1983), p. 37.

[171] 'Feature of the Month: Byrom's Shortland', University of London, Senate House Library. https://archive.senatehouselibrary.ac.uk/blog/feature-month-byroms-shorthand [Accessed 8 August 2019].

responded: 'I told him I was of his mind in that, and would show it to Mr. Folkes when he pleased; he thought that Mr. Folkes at the club, when he commended it, looked as if he was eager to know it'.[172] Byrom's pupils typically paid five guineas and took an oath of secrecy.[173]

Byrom first mentioned Folkes in a letter to his wife of 18 July 1723,

> went to the Sun in Paul's church yard into the club, into which Mr Foulks introduced me; there was Mr Graham [George Graham FRS], Hawksbee [Francis Hauksbee, clerk of the Royal Society], Brown [this could be Sir William Browne FRS, whose daughter Mary had married Folkes's brother William], Heathcote [possibly Gilbert Heathcote who was Hans Sloane's factor], Pemberton [Henry Pemberton, who became Gresham Professor of Physic], Snead.[174]

The club was an association of those who repaired to the tavern weekly after Royal Society meetings, and Byrom spent several evenings over ensuing years in discussions with Folkes and his coterie from the Royal Society, where, for example, 'they fell a talking about morality by the fireside till Mr. Hauksbee and Hoadly came in; then Mr. Graham desired us, privately, to let that discourse alone'.[175] A few days later, they spoke about 'the art of memory, Mr Folkes got thirty words', and other times they met at Folkes's house or Pontac's Head to discuss a dizzying array of books and ideas: 'Fran. Hoffman's strange book of strange miscellanies', Chambers' *Cyclopaedia*, Armand's *Grammaire raisonné*, and magnetic needles in compasses.[176] It was evident that Byrom sought Folkes's patronage and friendship, something that would serve him well, as he was elected to the Royal Society on 17 March 1724.[177]

On 14 February 1726, Byrom went to the home of William Stukeley for 'he had a great many stones, shells, and other curiosities in his room; Mr Folkes going by Bridge's auction[178] we called on him out of the window, and we went there after

---

[172] Byrom, *Private Journal*, 21 December 1725, vol. 1, part 1, p. 186.

[173] 'John Byrom', *Dictionary of National Biography*, 60 vols (Oxford: Oxford University Press, 1885–1900), vol. 8, p. 130. Byrom did indeed teach Folkes shorthand, or at least joked about doing so. He records in a letter to his wife of 3 June 1727, 'I should have answered yours last post, but I was so late in company with some of our Society, at my scholar Mr Folkes's chambers, that I had not time'. Byrom, *Private Journal*, vol. 1, part 1, p. 263.

[174] Byrom, *Private Journal*, 18 July 1723, vol. 1, part 1, p. 51.

[175] Byrom, *Private Journal*, 21 December 1725, vol. 1, part 1, p. 185. Possibly Dr Benjamin Hoadly (1706–57).

[176] Byrom, *Private Journal*, 4 January 1725/6 to 13 January 1726, vol. 1, part 1, pp. 188–190, p. 195, p. 200.

[177] By 1741, Byrom was in arrears in his membership dues, and on 16 November 1741 it 'was ordered, that Mr Byrom have liberty to take up his Bond upon making up his past Payments to the Sum of twenty five Pounds against Christmas next'. This was commuted to him paying ten guineas on 19 January 1741/2. CMO/3/90, CMO/3/93, CMO/3/96, Royal Society Library, London. Byrom later endorsed a Manchester surgeon, Charles White, as FRS in 1761, and Thomas Percival, a writer from Lancashire in 1756. See EC/1756/13 and EC/1761/13, Royal Society Library, London.

[178] This was the sale of the library of John Bridges, Esq., one of the most extensive book sales in the first quarter of the eighteenth century, with 4,313 items and a price realized of £4,001. Byrom and Folkes were regular attendants. See Byrom, *Private Journal*, 13 January 1726, vol. 1, part 1, p. 204.

him; I told Dr Stukeley by the way how the Romans and Greeks practiced short-hand, which he said he did not think at before, and said I should do well to write a dissertation upon it'.[179] Byrom complied, and delivered a paper to that effect to the Royal Society in April 1727 in which he also argued (tongue in cheek) that Egyptian hieroglyphics were a sort of shorthand.[180] An enthusiastic collector of books and a buying agent for the Chetham's Library in Manchester, Byrom also performed favours for Folkes, buying a work of Macrobius for 6s. 6d. at Pierre de Varennes's book auction held at his famous shop at 'Seneca's Head' in the Strand.[181] This type of circulation of objects through the auction of natural philo-sophical books, instruments, or natural history objects came to its peak in the eighteenth century, when enterprising auctioneers staged events where 'thrifty science met public science' and elite scientific material culture was redistributed to other natural philosophers as well as middling practitioners. Appealing to their audience's gentility, tradition, and erudition, the auction site became a public space for polite education as well as a commercialization of learned leisure.[182]

Ultimately, however, Byrom's friendship with Folkes and his Jacobite procliv-ities would prove to be to Folkes's detriment. On 27 April 1727, there had been an embarrassing discussion in the Council about making King George I the patron of the Royal Society, as previous monarchs had been, as 'the said design' previ-ously had 'proved of no Effect'.[183] The Royal Coat of Arms had even been painted in readiness in the Charter Book, but as Newton had a distrust of Hanoverian rulers because of the Saxon origins of Gottfried Wilhelm Leibniz, his rival in the calculus dispute, this distrust 'carried over to his followers'.[184] According to Heilbron, Newton worried that the new King 'might endeavour to moderate (The Royal Society's and Newton's) treatment of Leibniz, and perhaps even to find a place for his philosopher in the direction of the Society's affairs'.[185] George I was also frankly indifferent to the Society and its aims.[186]

Now, however, that Newton was dead, Sloane, as acting President, wished to approach the monarch again to ask for his patronage of the Society. Lennox,

---

[179]   Byrom, *Private Journal*, 13 January 1726, vol. 1, part 1, p. 203.

[180]   Byrom, *Private Journal*, vol. 1, part 1, pp. 237–48.

[181]   'John Byrom Collection', Chetham's Library, Manchester. https://library.chethams.com/collections/printed-books-ephemera/john-byrom-collection/ [Accessed 10 August 2019]; Varennes was a Huguenot refugee who provided items for British collectors and booksellers. For Pierre De Varennes, see Julian Roberts, 'The Latin Trade' in John Barnard, D.F. McKenzie and Maureen Bell, eds, *The Cambridge History of the Book in Britain, volume IV 1557–1695* (Cambridge: Cambridge University Press, 2008), pp. 141–73, on p. 168.

[182]   Werrett, *Thrifty Science*, p. 149.

[183]   RS/JBO/14/76. See also RS/CMO/2/306, Royal Society Library, London.

[184]   Rusnock, *Correspondence of James Jurin*, p. 33.

[185]   John Heilbron, *Physics at the Royal Society* (Los Angeles: University of California Press, 1983), p. 36. See also Sorrenson, 'Towards a History of the Royal Society', p. 44.

[186]   Richard Sorrenson, 'Towards a History of the Royal Society in the Eighteenth Century', *Notes and Records of the Royal Society of London* 50, 1 (1996), p. 33.

known for his pro-Hanoverian views, was asked to serve as a go-between. Even though Lennox was a grandson of Charles II, he still had a close relationship with George I, serving as his *aide-de-camp*. There also was a discussion about the manner of the monarch's address for a 'paper Containing a form of words to be Spoken to his Majesty upon this occasion'. The draft paper was read twice, and a question was put to ballot: 'Whether the words to be Spoken to his Majestie upon this occasion Shall be in the form exprest in the paper read before the Society or not'.[187]

Byrom, a Jacobite, was known for his outspokenness at meetings about Hanoverian rule, and opposed an address to the King and to the Prince of Wales. Due to Byrom's comments and those of other members, the motion was tabled until the next meeting, though it was decided to invite King George I to the Royal Society as a patron. Sloane then reported, in a subsequent meeting of 11 May 1727, that the monarch had been 'pleased to signe his Name as Patron in the Charter', and then stated, 'he had good reason to believe that his Royal Highness would not refuse to do the Society the honour of Subscribing by his name in the book, if he were addressed in a Proper Manner'.[188] Sloane also asked the Society to consider in what manner to ask the Prince of Wales for his patronage, drew up another paper about the matter of address, and proposed another ballot about the same issue. The minutes recorded that 'some debates' subsequently arose about the manner of proceedings; with 'adverse objections being made against putting Question of this Nature to a Ballot, it was agreed *Nemine Contradiente*' [without dissent]'.[189]

While Byrom had expressed his dissent over the manner of address to King George, Cyriacus Ahlers, Surgeon to George I of England, made a great show of writing Byrom's name down and asking for its precise spelling. Ahlers was a regular attendant at Royal Society meetings, having been previously sent by George I to Guildford to investigate the notorious Mary Toft, 'the rabbit-breeder', who claimed she was giving birth to a litter of the rodents. Ahlers made a report of his findings on 8 December 1726, showing the parts of dismembered rabbits that she 'delivered' had been cut previously cut with a knife. 'He also shewed a small grain or pellet taken out of the Guts which with a magnifying Glass appeared to be the Excrement of a grass food', clearly showing Toft's fraud. For his discovery, Folkes subsequently proposed Ahlers as a Fellow. Ahlers presented his book on the Toft case in the next meeting, and he was elected FRS on 16 March 1726/7.[190]

---

[187] RS/JBO/14/77, Royal Society Library, London.

[188] RS/JBO/14/86, Royal Society Library, London.

[189] RS/JBO/14/86, Royal Society Library, London.

[190] RS/JBO/14, p. 23, p. 25, p. 59; see Cyriacus Ahlers, *Some Observations Concerning the Woman of Godlyman in Surrey: Made at Guilford on Sunday, Nov. 20. 1726: Tending to Prove Her Extraordinary Deliveries to be a Cheat and an Imposture* (London: J. Roberts, 1726).

Byrom then recorded that:

Sir Hans [Sloane] being in the chair, and after reading a paper or two, he bid Dr Jurin read the statutes…I thought there might possibly be some design in this, and that Sir Hans might perhaps make some remarks, and particularly upon my late conduct, so I took out my pen, and as Jurin read I wrote; at last Sir Hans having taken notice of that part of the statue which is particularly about the admission of strangers and saying that it was because things might be reported out of the Society.[191]

Byrom subsequently asked Philips Glover to request of Folkes if he could speak, and when assent was given Byrom noted that:

since we were upon this topic of the statutes, which was certainly very proper, and since Sir Hans had mentioned the reason for that about strangers, and the danger there might be from them to the Society, I desired to know something in relation to that danger that might arise from the behaviour of the members one to another, and, in particular, should be glad to know what the Society thought of this behaviour, that when a member had upon occasion spoken his mind with freedom, without any design of giving offence, one of the members should enquire his name, and desire to know the particular way of spelling, and take it with all this care and exactness in writing,—whether this might not reasonably give occasion to suspect some other design than a gentleman's bare curiosity.[192]

At that moment, Folkes intervened and made a long speech,

importing that every member of this Society had a right to speak his mind upon occasion ; that there had been certain debates, indeed, about addressing his Majesty, not that anybody had ever opposed the thing, but only the form and manner of it; that it had had its course, and been voted and carried in the manner it was ; but that it was very hard that any member who might think differently should be taken notice on in a manner that he must needs say was enough to give suspicion that something besides bare curiosity was at the bottom of so particular an enquiry, and therefore he thought it was very reasonable for the Society to take so much notice of it as to desire Mr. Ahlers to explain himself.[193]

[191] Byrom, *Private Journal*, vol. 1, part 1, p. 253.
[192] Byrom, *Private Journal*, vol. 1, part 1, p. 253. A summary of this conversation was also transcribed in the Royal Society Journal Book, RS/JBO/14, pp. 90–2, Royal Society Library, London.
[193] Byrom, *Private Journal*, vol. 1, part 1, pp. 253–4.

Two other Fellows of the Society agreed: John White (1699–1769), who would become a Whig MP, and Robert Ord, a lawyer educated at Lincoln's Inn. Ord added that Ahlers's notetaking 'looked like a design to intimidate a member, and break in upon that liberty of speech in which the very essence of the Society consisted'.[194] Mr Ord took notice of the

> representations which had been made of the late debates about the form of addressing for the honour of our kings, that it had been said there were Jacobites in the Society. Dr. [Robert] Nesbit spoke, and said it had been so talked in the public coffeehouses, and Mr. Byrom's name and his own had been particularly mentioned under the notion of their being enemies to the government ; that for his part he had always thought it a very great honour to have his Majesty's name upon the books of the Society, but that strange interpretations had been made of the opposition to the particular manner of obtaining it....[195]

Many times in the conversation, Byrom recorded Sloane as wishing to drop the matter, saying 'it was best to dismiss it, and return to the business of the Society'. But, Folkes persisted, 'begging leave to differ with him in that, since he had a power; and Sir Hans denying it, he showed him the statute concerning the behaviour of members to one another'.[196] Finally, Ahlers was questioned as to his intentions, and he replied 'curiosity', denying 'any evill design'.[197] Ord mentioned his writing down names in his pocket-book, and 'Mr. Ahlers took him up and said, Sir, I have no pocket-book. Mr. Folkes said, Perhaps it was an almanac—to which he said nothing'.[198] Sloane then remarked, 'Well but, Mr. Folkes, now all is over, don't you really think that it was very wrong, when his Majesty's name was thought proper, or any member to desire it to be deferred to consider of it for a week'?[199] Folkes then said, 'Sir, as that is over, it is certainly improper to bring it on again; that is not the dispute at present, but whether this unusual observation upon a member's speaking, and thus noting down his name, was not an infringement upon a member's liberty'.[200]

Apart from his rivalry with Sloane, Folkes feared that the polite sociability and liberty of free discourse within the bounds of a club or association were being threatened. Certainly, the Royal Society was still bound by social hierarchies, with those of 'high social status joining easily', and those of middling status less so.[201] However, eighteenth-century societies 'provided a setting in which private friendship and formal organization were judiciously balanced. The inhibitions imposed

---

[194] Byrom, *Private Journal*, vol. 1, part 1, p. 254.
[195] Byrom, *Private Journal*, vol. 1, part 1, p. 255.
[196] Byrom, *Private Journal*, vol. 1, part 1, p. 256.   [197] RS/JBO/14, p. 86.
[198] Byrom, *Private Journal*, vol. 1, part 1, p. 256.
[199] Byrom, *Private Journal*, vol. 1, part 1, pp. 256–7.
[200] Byrom, *Private Journal*, vol. 1, part 1, p. 257.
[201] Sorrenson, 'Towards a History of the Royal Society', p. 33.

by the hierarchical bonds of traditional society [were] diminished'.[202] There was a privilege and liberty of speech that was expected among the Fellows within 'a Society of putative equals'.[203]

Sloane then, in a rather imperious manner, had it noted in the minutes that 'he and whatever members he thought fit to join with him should be authorized in the name of the Society to addresse his Royal Highness in such a manner as should be deemed most proper'.[204] In a letter to Byrom, his friend Dr Deacon (a physician and nonjuring clergyman) subsequently remarked 'You are a strange fellow, to affront the Royal Society President; your philosophical alphabet [shorthand] will not pass now'.[205]

After the heated debate, Byrom went to Tom's Coffee-House on the Strand[206] with Folkes, where another of Folkes's supporters, John Machin, Gresham Professor of Astronomy and mathematician, and secretary of the Royal Society, stated he 'thought one step further should have been taken, and Mr Ahlers obliged to ask pardon of the Society'. After they left, Folkes took hold of Byrom's arm, and told him to call on Mr D'Anteney and inform him what transpired. D'Anteney was rumoured to be close to the Hanoverian regime, and perhaps an informant. A few days later, Byrom recorded,

> I had only time to ask Mr. D'Anteney as we came out whether he had heard anything of our hurry at the Society? Yes, he said, that they had told him that I had made speeches, he supposed for diversion; I just hinted to him that we had had more disputes about my name being taken down, and desired him not to receive any impressions to my disadvantage, but to stay his judgment till he heard the matter from Mr. Folkes or Ord; he said I might be sure he would not.[207]

Byrom subsequently joked to his wife, 'the shorthand men I'm told are resolved to stand by their Grand Master [Folkes]; and if Liberty leave the Royal Society, 'tis confidently asserted that she will retire to the Shorthand one, where a seat is prepared for her reception'.[208]

Unfortunately for Folkes, this joke was not made manifest in the outcome of the 30 November 1727 Anniversary Day Elections for the Royal Society presidency and members of Council. Until the election, he remained active in the Society, serving

---

[202] John Money, *Experience and Identity: Birmingham and the West Midlands 1760–1800* (Manchester: Manchester University Press, 1977), p. 98, as quoted by Sorrenson, 'Towards a History of the Royal Society', p. 33.

[203] Sorrenson, 'Towards a History of the Royal Society', p. 35.   [204] RS/JBO/14, p. 87.

[205] Byrom, vol. 1, part 1, 14 May 1727, p. 251.

[206] The coffee-house was behind 216 Strand, where Twinings Tea was founded; the proprietor Thomas Twining wished to expand into tea-selling, and built the tea-house next to his original coffee-house. https://exploring-london.com/tag/toms-coffee-house/ [Accessed 29 December 2019].

[207] Byrom, *Private Journal*, 20 May 1727, vol. 1, part 1, p. 258.

[208] Byrom, *Private Journal*, 20 May 1727, vol. 1, part 1, p. 260.

as an auditor of the Treasurer's accounts and presenting a paper discussing hygrometers and their imperfections on 15 June.[209] But the stars simply did not align. Folkes had just suffered several misfortunes: burglars invaded his house on Queen Square, robbing him of some family silver earlier in the month, and it was known in advance that Folkes did not stand much of a chance in the election as the Earl of Oxford was firmly in support of Sloane. Folkes at the least had some small comfort that Newtonian Richard Mead supported him, but it seems his remarks about liberty of discussion in the Society and his friendship with the Jacobite Byrom had tainted his reputation.[210]

Byrom wrote to Richard Hassell FRS on 14 November 1727 to encourage Folkes's adherents to vote in their favour:

> I was last night at a committee of the Royal Society for carrying on the grand business of the election, where it was agreed, upon casting up the lists on both sides, that we were like to have a very hard struggle; I will not say, for omen's sake, that we may possibly lose it, but it was thought necessary that somebody should write to everybody that was not in town, and press them not to fail at the day ; and I was desired to write to you and Sir Peter [Leycester], not that we had any mistrust of you, but to put you in mind to use all your rhetoric with Sir Peter, and to bring him to town will-he nill-he, though he should be forced to go out of town the next morning.[211]

When Byrom came to London to vote, he told his wife of the coffee-house gossip that there was a 'vast interest made to keep in Sir Hans; Lord Cadogan has been with all the noblemen, et. They say he has seventeen near relations in the Society…He is to get it by two votes, or by other accounts more. Our hopes are in the freedom of balloting, nor do we yet despair of victory. They are resolved to have Jurin out of the secretaryship if they can'.[212] The next day, the news continued to be bad. When at his house in Queen Square, Folkes read a dispiriting letter from the Archbishop of Canterbury vouching for Sloane to a gathering including George Graham, James Jurin, Byrom, Philips Glover, and John White, all of whom had supported Folkes in the May debates about free discussion. After a desultory Royal

---

[209] RS/RBO/13/5, 'An account of a Tract de hyrometris et corum defectibus' by Martin Folkes, Royal Society Library, London.

[210] *Daily Post*, 8 November 1727, Issue 2356: 'Lost from the House of Martin Folkes Esq in Queen's Square by Ormond Street, the Cover of a Silver Caudle Cup, old-fashion'd, with three Knobs like Feet, as also an old-fashion'd Silver Hand Candlestick; they are both mark'd with a Coat of Arms, viz either a single Flower-de-lis along, or impaled with a single Half-Moon; there was also taken with them a small Silver Strainer: If offered to be sold or pawn'd pray stop them they being stolen, and give Notice at the said Place, where you shall be reasonably rewarded for your trouble'.

[211] Byrom, *Private Journal*, 14 November 1727, vol. 1, part 1, p. 270.

[212] Byrom, *Private Journal*, 25 November 1727, vol. 1, part 1, pp. 272–3.

Society meeting, Folkes, White, Glover, and Byrom walked home, diverting themselves by reciting (no doubt drunkenly) the following Bellman's verse:

> If that we do believe a future state
> Let us repent before it is too late;
> Although we now may be in health and strength,
> The life of man is but a span's length:
> Let's make our calling and election sure.
> *Past one o'clock!*[213]

(Bellman's poetry was a feature of the Christmas season, printed sheets of doggerel verse purporting to be the good wishes of the parish bellman or night watchman to the individual reader; they were left at houses free of charge in hopes of a Christmas token of appreciation).[214]

As predicted, Folkes lost by a large margin. From Byrom's account, the meeting was contentious, Jurin proposing that foreign members should be allowed to vote and the question was 'put to ballot whether they had a right', but that motion was refused, and 'Mr Folkes protested'.[215] Folkes's and Jurin's relationships with Italian and French natural philosophers would have been to their electoral advantage.

When the votes were tallied, Byrom, Folkes, and Jurin were thrown out of the Council, Jurin having only 65 votes, and 'Mr. Folkes 68 votes, to the surprise of most or all people there; after that there was no further contest'. Byrom's Jacobite proclivities, and the discussion about letting foreign Fellows vote, apparently lost more supporters in Folkes's party than anticipated. There had been a heavy increase in foreign members during the first two decades of the eighteenth century (twenty-four between 1700 and 1710 and forty-eight between 1710 and 1720), and there may have been fears that they would have undue influence over Society business.[216] Folkes retreated with his friends in a separate room to dinner at Pontac's Head, the usual destination after the St Andrew's Day elections.[217] Pontac's was known for the best Bordeaux wine in London, as the owners possessed the Haut-Brion vineyard. It must have been quite a consolatory meal.

---

[213] Byrom, *Private Journal*, 26 November 1727, vol. 1, part 1, p. 273.

[214] Michael Twyman, ed., 'Bellman's Verses', in *The Encyclopedia of Ephemera: A Guide to the Fragmentary Documents of Everyday Life for the Collector, Curator, and Historian* (New York: Routledge, 2018).

[215] Byrom, *Private Journal*, 30 November 1727, vol. 1, part 1, p. 274.

[216] Maurice Crosland, 'Explicit Qualifications as a Criterion for Membership of the Royal Society: A Historical Review', *Notes and Records of the Royal Society of London* 37, 2 (1983), pp. 167–87, on p. 176.

[217] RS/MM/20, Miscellaneous Manuscripts, 1739 newspaper clipping noting 'To-morrow being St. Andrew's Day, the Royal Society will hold their Annual Meeting…for the Election of a President and other Officers…after they will proceed to Pontack's where a splendid Entertainment will be provided for them'.

As Jurin lost his secretaryship, being replaced by William Rutty, at a meeting on 7 December he handed in the papers in his possession to the Society. There was still bad blood. When Folkes proposed Jurin should have the thanks for the Society for his services, Sir Hans's nephew William protested, saying that Jurin had defamed his uncle in the coffee-houses. Jurin responded:

I second Mr. Sloane's motion, and desire it may be enquired into, and that Mr. Sloane would name particulars but Sir Hans would hear nothing of either side further about such matters; he said we had for five months neglected the business of the Society for disputes, which by the grace of God he would put an end to. I spoke for Jurin having thanks, as did others, and at last it was minuted down.[218]

Jurin would have his revenge, dedicating the thirty-fourth volume of *Philosophical Transactions* to Folkes, his motive for doing so:

the same which induc'd the greatest Man that ever liv'd, to single you out to fill the Chair, and to preside in the Assemblies of the Royal Society, when the frequent Returns of his Indisposition would no longer permit him to attend them with his usual Assiduity…But I shall not offer here to draw your Character: it is sufficient to say, that Mr Folkes was Sir Isaac Newton's Friend…That Great man was sensible, that something more than knowing the Name, the Shapes and obvious Qualities of an Insect, a Pebble, a Plant, or a Shell, was requisite to form a Philosopher.[219]

Jurin continued, claiming that although Newton did not despise 'so useful a Branch of Learning as Natural History…he judg'd that this humble Handmade to Philosophy…must very much forget her self, and the Meanness of her station, if ever she should presume to claim the Throne, and arrogate to her self the Title of the Queen of Sciences'. Jurin did not appear in the records of the Royal Society again until 1741, when Folkes was elected President.[220]

Ever the gossip, antiquary Thomas Hearne reported 'the great struggle…carried for Sir Hans by a great majority, and Folkes at the same time was put out from being of the Council. This Folkes is an ingenious Man, but forward & conceited and 'twas foolish to oppose Sir Hans'.[221] The Oxonian and spiteful Hearne

[218]   Byrom, *Private Journal*, 30 November 1727, vol. 1, part 1, p. 278.
[219]   James Jurin, 'Dedication to Martin Folkes, Esq; Vice-President of the Royal Society', *Philosophical Transactions* 34, 392 (1727). See also Rusnock, 'Introduction', *The Correspondence of James Jurin*, pp. 34–5.
[220]   Rusnock, 'Introduction', *The Correspondence of James Jurin*, p. 35.
[221]   Charles E. Doble, ed., *Remarks and Collections of Thomas Hearne*, 11 vols (Oxford: Clarendon Press, 1885–1921), Entry for 8 December 1727, vol. 9, p. 379. My thanks to Will Poole for giving me access to the complete diary.

took great delight in recording that Folkes was 'a Cambridge man by Education'. The news of Folkes's defeat reached the London newspapers, and the *Evening Journal* on 4 December reported: 'Martin Folkes, Esq. who was a Candidate for President of the Royal Society, is now left out the list of the Council of that honourable Society. *Tantaene animis caelestibus irae?* [Do the heavenly minds have such great anger? *Aeneid* 1.8–11]'. As for Folkes, he retreated, licked his wounds, and began formulating a strategy to be an ambassador for Newtonianism at home and abroad. Part of this strategy would be his editing of Newton's posthumous works concerning chronology.

4.7 Folkes's Religious Beliefs and his Editions of Newton's *Chronology* and *Observations*

Folkes's loss of the Royal Society presidency taught him an important lesson about the value of public reputation and ambitions for office. It was a lesson that ultimately shaped his co-editorship with Thomas Pellett of Newton's *Chronology of Ancient Kingdoms Amended* (1728) and *Observations Upon the Prophecies of Daniel and the Apocalypse of St John* (1733). Outwardly, Folkes's participation in Freemasonry and in the Order of the Bath marked him as a member of the established order and a close associate of powerful noble patrons. Folkes's private religious beliefs, however, were more controversial, even going beyond Newton's private denial of the Holy Trinity.

Although, according to the *Constitutions*, a Freemason could not be a 'stupid Atheist, nor an irreligious Libertine', Wharton, Grand Master in 1722–3, was the founder of the first hell-fire club in London. Folkes followed suit. Stukeley noted that 'When I lived in Ormond street in 1720…[Folkes] set up an infidel Club at his house on Sunday evenings where…others of the heathen stamp, assembled'.[222] Stukeley then went on to claim that Folkes thought 'there are no differences between us and animals; but what is owing to the different structure of our brain, as between man and man' and that he called himself 'the Godfather of all Monkeys'. Folkes's infidel club included Lennox and the Duke of Montagu, the first aristocratic Grand Master.[223] Two years earlier, Folkes had Jonathan Richardson paint him in 'undress' that men wore at home in informal surroundings—a louche blue velvet (see figure 4.6).

Just as Folkes was informal in private dress, he was informal in his private religious beliefs throughout his adult life; although we do not have enough archival evidence to track subtle stages of his religious proclivities, it seems from evidence

---

[222] Stukeley, *Family Memoirs*, vol. 1, pp. 99–100.
[223] Robert Peter, 'Admissions and Lodge Meetings', *British Freemasonry*, vol. 5, p. 2; *Constitutions* (1723), p. 50.

that he was consistently a religious sceptic throughout his life, while outwardly avoiding outright public heresy to preserve his social status. Folkes simply did not identify with the vindication of religion, natural and revealed, admitting to his friend John Byrom that he was 'a heretic about the book of Daniel'.[224] Folkes remarked to Byrom that his heretical beliefs stemmed from reading the radical deist Anthony Collins's *Discourse of the Grounds and Reasons of the Christian Religion* (1724). Nor was Folkes a believer in natural theology, also admitting to Byrom that he thought William Derham's *Astro-Theology*, a corollary to the Boyle Lectures that used natural history and teleology to promote and prove natural theology, was 'a silly book'.[225] Folkes also had little time for 'popery': pride of place in his art collection was a mezzotint of Sir Godfrey Kneller's 'Spanish Fryar', the subject from Dryden's play of the same name featuring Anthony Leigh in the role.[226]

> In Act II, scene iii, Father Dominic, confessor of Elvira, wife of Gomez, is bribed by Lorenzo that he may obtain access to her…[Dramatist Colley] Cibber recalling that: 'In the canting, grave Hypocrisy [of Leigh's portrayal] he stretcht the Veil of Piety so thinly over him, that in every Look, Word, and Motion you saw a palpable, wicked Slyness shine through it…'.[227]

To be fair, Stukeley's caustic comments may have been raised because, when younger, he had been exposed to heterodox notions from Folkes, who was then his close friend. As stated above, Stukeley was an active Freemason when it was newly fashionable after the appointment of his patron the Duke of Montagu as Grand Master in 1721. Stukeley became a Mason of the Salutation Tavern that year; by 1729, Stukeley admitted to Samuel Dale that he was 'in a manner voyd of religion'.[228] Stukeley's redemption and subsequent taking of holy orders as a priest in the Church of England in 1729 may have made him uncomfortable about his previous indiscretions.[229]

[224] Byrom, *Private Journal*, 30 March 1736, vol. 2, part 1, p. 27.
[225] Byrom, *Private Journal*, 14 December 1725, vol. 1, part 1, p. 181.
[226] Lot 28, Group of five prints of actors in character, Forum Action, Online Sale: Music, Opera, Ballet and Theatre, 20 November 2017. Folkes's collector's mark is on the bottom recto of the print. https://www.forumauctions.co.uk/index.php?option=com_timed_auction&auction_no=2031&lang=en&lot_id=38896&view=lot_detail [Accessed 3 August 2020].
[227] Sir Godfrey Kneller, Anthony Leigh, 1689, National Portrait Gallery, London. https://www.npg.org.uk/collections/search/portraitExtended/mw03857/Anthony-Leigh [Accessed 3 August 2020].
[228] Bodl. MS Eng. misc. e. 126 Stukeley, f. 83. Bodleian Library, Oxford. David Haycock, *William Stukeley: Science, Religion and Archaeology in Eighteenth-Century England* (Woodbridge: The Boydell Press, 2002), p. 51.
[229] George Kolbe, 'Godfather to all Monkeys: Martin Folkes and his 1756 Library Sale', *Asylum* (April–June 2014), pp. 38–92, on p. 41. Stukeley may also have been envious of the social position of Folkes, his man and neighbour, who indirectly provided him with a living in London as vicar of St George's in Queen Square via his uncle William Wake, an antiquary, a numismatist and, as mentioned, the Archbishop of Canterbury. Folkes was also friends with the Duke of Montagu, who was instrumental in giving the position to Stukeley.

Although the precise line that divided deism, natural religion, and the revealed religion of the orthodox Christian was subtle in this period, Folkes's travels to Italy in 1733 to 1735 (see chapter five) confirmed his freethinking beliefs. He noted in his diary that he considered Catholicism and Protestantism 'as only a tool of the state to keep the vulgar in awe', noting that Venetian intellectuals were 'fond of any sort of freethinking book, and with what eagerness they speak on the subject and enquire after the ways of thinking of a nation more used to liberty of thought than themselves'.[230] That said, Folkes did go on to say 'The bible is in no ways known and I have met with stupendous instances of mistakes about the things contained in it...and it has been with the greatest surprise several have heard me speak of the measure of learning of all sorts that very antient...book contains'.[231] Folkes repeated these assertions in correspondence to Lennox during his Grand Tour, writing that:

> Another thing very new is the general face of external devotion through all the Popish countrys, and the ignorance that accompanys it is somewhat stupendous...But I think Bavaria is the most superstitious of all the places I have seen; there is no where else but I have behavd in the Churches sufficiently well to give no offence; at Munich I could not, No; I have ever kneeld to the best of my knowledg, as making no difficulty to bow my self in the house of Rimmon.[232]

On his return from his travels, Folkes also found his personal reputation threatened due to remarks he made during his visit to the Vatican to see the *Codex Vaticanus*;[233] the fourth-century AD Uncial *Codex Vaticanus* is one of the most famous authoritative biblical texts, one of two of the oldest Biblical manuscripts in parchment that are still extant, and a cornerstone of the Septuagint and the New Testament.[234] Although a few scholars had gained access to the *Codex*, 'it was not consulted properly until the end of the nineteenth century...The Vatican seemed reluctant to allow the readings of a manuscript whose text differs so much from the *textus receptus* to be freely available...', particularly its omission

---

[230] Bodl. MS Eng. misc. c. 444, Martin Folkes, 'Journey from Venice to Rome', f. 24r., Bodleian Library, Oxford.

[231] Folkes, 'Journey from Venice to Rome', f. 24r.

[232] Goodwood MS 110, Folkes to Lennox, Venice, 8 July 1733, West Sussex Record Office.

[233] Bibl.Vat., Vat gr. 1209, no. B or 03 Gregory-Aland, δ 1 von Soden.

[234] RS/MS/750/47 and 48, Draft letter and letter of Folkes to Philips Glover, 4 May 1737, Royal Society Library, London; J. Keith Elliot, 'The Text of the New Testament', in Allan J. Houser and Duane F. Watson, *A History of Biblical Interpretation, Vol. 2: The Medieval Through the Reformation Periods* (Grand Rapids, Michigan and Cambridge: William B. Eerdmans Publishing, 2003), vol. 2, p. 244; Bruce Metzger and Bart D. Ehrman, *The Text of the New Testament: Its Transmission, Corruption, and Restoration*, 4th ed. (New York and Oxford: Oxford University Press, 2005), p. 15. The *Codex* is one of four uncial codices being completed in the capital letters of the majuscule script; the others are the *Codex Alexandrinus*, the *Codex Sinaiticus*, and the *Codex Ephraemi Rescriptus*, the latter two only known in the nineteenth century. See Rob Iliffe, *Priest of Nature: The Religious Worlds of Isaac Newton* (Oxford: Oxford University Press, 2017), p. 464, chapter eleven, footnote 33 (ebook edition).

of 1 John 5:7–8, where the Holy Trinity is averred.[235] On 18 June 1521, when compiling his third edition of the Greek New Testament, the humanist Desiderius Erasmus was sent extracts from *Codex Vaticanus* in a letter to Paulus Bombasius, the prefect of the Vatican Library.[236] Erasmus realized the Trinitarian phrase, called the *Comma Johanneum* (as comma means a short clause of a sentence), appeared in no Greek text, whether manuscript, patristic, or in translation; this omission included the *Codex Vaticanus*. As Metzger and Ehrman have explained,

> In an unguarded moment, Erasmus may have promised that he would insert the Comma Johanneum…in future editions if a single Greek manuscript could be found that contained the passage. At length, such a copy was found—or was made to order! As it now appears, the Greek manuscript had probably been written in Oxford about 1520 by a Franciscan friar named Froy (or Roy), who took the disputed words from the Latin Vulgate.[237]

De Jonge has countered that Erasmus ultimately inserted the *Comma Johanneum* into his third edition of the Greek New Testament, due not so much to any promise that may have been made or to the existence of the Oxford manuscript, but because of his scholarly thoroughness and to preserve his reputation.[238] It can be confidently asserted, however, that it was with Erasmus 'that the *Codex Vaticanus* began to play a role in the textual criticism of the New Testament'.[239]

As a result, many early modern scholars 'travelling through Europe sought out ancient manuscripts to check whether the passage was present', and Folkes was no different.[240] However, in a letter to his friend Philips Glover, Folkes noted that during his visit to Rome he had remarked in private conversation of the absence of the Trinitarian formula in the *Codex Vaticanus*, 'and as I understood my name was mentioned, and somewhat I have said now construed to imply a reflection I am sure I never meant', namely anti-Trinitarian heresy; Folkes may have even

---

[235] Elliot, 'The Text of the New Testament', vol. 2, p. 244. In the orthodox *textus receptus*, 1 John 5:7–8 reads: 'in heaven, the Father, the Word, and the Holy Spirit, and these three are one. 5:8 And there are three that testify on earth' [ἐν τῷ οὐρανῷ, ὁ πατήρ, ὁ λόγος, καὶ τὸ ἅγιον πνεῦμα, καὶ οὗτοι οἱ τρεῖς ἕν εἰσι. 5·8 καὶ τρεῖς εἰσιν οἱ μαρτυροῦντες ἐν τῇ γῇ]. Metzger and Ehrman also noted that the authorities of the Vatican Library 'put continual obstacles in the way of scholars who wished to study it in detail…for some reason that has never been fully explained' (Metzger and Ehrman, *The Text of the New Testament*, p. 68).

[236] Henk Jan de Jonge, 'Erasmus and the Comma Johanneum', *Ephemerides Theologicae Lovanienses* LXVI (1980), pp. 381–9, on p. 389.

[237] Metzger and Ehrman, *The Text of the New Testament*, p. 146. The outline of the story is repeated in L. D. Reynolds and N.G. Wilson, *Scribes and Scholars: A Guide to the Transmission of Greek and Latin Literature* (Oxford: Oxford University Press, 2013), p. 162.

[238] Metzger and Ehrman, *The Text of the New Testament*, p. 147; De Jonge, 'Erasmus and the Comma Johanneum', p. 381. De Jonge does not believe Erasmus ever made such a rash promise.

[239] De Jonge, 'Erasmus and the Comma Johanneum', p. 389.

[240] Iliffe, *Priest of Nature*, p. 372.

spoken about deism.[241] Philips Glover FRS (d. 1745) was a dissenter who later became High Sheriff of Lincolnshire in 1727–8. Glover wrote on controversialist topics; his *Letter to the Reverend Dr Waterland Concerning the Nature and Value of Sincerity* (1733) vehemently criticized Daniel Waterland's (1683–1740) stringent view of orthodoxy. Waterland was the 'most indefatigable upholder of Trinitarian orthodoxy of the time' and a fervent anti-Newtonian and critic of Samuel Clarke, so Glover could be thought of as generally sympathetic to Folkes's views.[242] But it seems from the tone of Folkes's letter that Glover would not countenance public anti-Trinitarianism.

As is well known, Newton himself privately held anti-Trinitarian/Arian beliefs, and in November 1690 he had completed a study of 1 John 5:7–8 which he sent to John Locke, telling him 'I have done it more freely because to you who understand the many abuses wch they of ye Roman Church have put upon ye world, it will scarce by ungratefull to be convinced of one more yn is commonly believed'.[243] Newton told Locke he thought Trinitarianism was the result of the corruption of the biblical text, as 'some of ye Latines interpreted the spirit, water and blood of the Father, Son and Holy Ghost to prove them one'.[244] St Jerome, Newton thought, then adopted this false interpretation in his Vulgate. Newton kept secret his belief that the Athanasian Creed, averring the Trinity, was one of the 'pious frauds of ye Roman Church'. Newton's reticence was necessary, as in 1710 William Whiston lost the Lucasian Chair of Mathematics at Cambridge (Newton's former post) for similar views. Whiston then ferociously 'charged Newton, in several works, to have shared his public opinions', leading divines such as Waterland to have 'private concerns regarding Newton's heterodoxy'.[245] Whiston subsequently omitted the spurious part of 1 John 5:7 in his own *Primitive New Testament* (1745), using the best three Ancient Greek manuscripts he knew of at the time.[246]

Folkes, as a known freethinker and disciple of Newton, editor of his *Chronology of Ancient Kingdoms Amended* (1728) and *Observations Upon the Prophecies of Daniel and the Apocalypse of St John* (1733) (see below, p. 83), clearly had to do some damage control with Glover. From the letter, it seems gossip that Folkes was an anti-Trinitarian had also spread to Jonathan Richardson the Younger, the son of his friend, the artist Richardson the Elder, who also lived in Queen Square.

[241] RS/MS/790/47, Folkes to Philips Glover, 4 May 1737, Royal Society Library, London.

[242] Scott Mandelbrote, 'Eighteenth-Century Reactions to Newton's Anti-Trinitarianism', in James E. Force and Sarah Hutton, eds, *Newton and Newtonianism. New Studies* (Dordrecht and Boston: Kluwer, 2004), pp. 93–111, on p. 98.

[243] Newton to Locke, 14 November 1690, in *The Correspondence of Isaac Newton*, vol. 3, p. 83 and p. 108. See also Iliffe, 'The Comma', in *Priest of Nature*, pp. 372–5.

[244] Newton to Locke, 14 November 1690, in *The Correspondence of Isaac Newton*, vol. 3, p. 83.

[245] Mordechai Feingold, 'Isaac Newton, Heretic? Some Eighteenth-Century Perceptions', in Elizabethanne Boran and Mordechai Feingold, eds, *Reading Newton in Early Modern Europe* (Leiden: Brill, 2017), pp. 328–45, on p. 344.

[246] See James E. Force, *William Whiston: Honest Newtonian* (Cambridge: Cambridge University Press, 1985).

Folkes recounted to Glover that he went twice to see the document in the Vatican with friends, comparing its phraseology with 'a common Greek Testament'. Folkes continued,

> I was accordingly askt by several persons and several times about it after my return to all which I gave the answer that appeared to me to be true, that the [phrase] *there are three that have been receiv'd in Heaven, etc is not in that MS* and with regard to what has been said of your self on those occasions, I have said that either you saw some other Manuscript as the famous one in which it might possibly be, or that you was fallen into some mistake about it.

In his letter to Glover, Folkes also denied assertions by divines such as Samuel Clarke that the text had been altered since 'Dr Burnets time of it', as he had 'examind the leaf which...bore all the marks of antiquity and gave no ground to suspect any erasure in it'. Here Folkes referred to the Latitudinarian divine, Gilbert Burnet, who analysed 1 John 5:7–8 in his *Some Letters Containing an Account of What Seemed Most Remarkable in Switzerland, Italy, etc.* (1686/7). Because the passage had been rejected by anti-Trinitarians as a later addition,

> Burnet was greatly interested in the controversy, and took care in his travels to examine all the Antient Manuscripts of the New Testament, concerning that doubted passage of St. John's Epistle. In the subsequent account of his findings he described those manuscripts in which the passage was to be found, those in which it was not, and those in which it had been added in the margin or at the foot of the page. Although Burnet made little attempt to analyse the results of his researches, when he did comment it was to defend the authenticity of the Trinitarian text.[247]

Despite his defence of the text, Burnet's work sparked suspicions of his orthodoxy, and he became part of a larger pamphlet war on the doctrine of the Trinity in the 1690s that involved the French Roman Catholic theologian Jacques Bossuet, and his own High Church Anglican clergy.[248] As Folkes wished to avoid the same fate as Burnet, he indicated to Glover it were all 'better dropt as a mistake', and hoped his account would be 'satisfactory both to yourself and to your father...and this affair may give you no further uneasiness...and that both Mr. R[ichardson] and yourself will think in the most favourable manner of me'.[249] Folkes both drafted the letter, and had a revised copy made in secretary hand.

---

[247] Martin Greig, 'The Reasonableness of Christianity? Gilbert Burnet and the Trinitarian Controversy of the 1690s', *Journal of Ecclesiastical History* 44, 4 (October 1993), pp. 631–51, on pp. 641–2.
[248] Grieg, 'The Reasonableness of Christianity?' p. 643.
[249] RS/MS/790/47, Folkes to Philips Glover, 4 May 1737, Royal Society Library, London.

Richardson the Younger also wrote to Folkes and expressed his satisfaction at the outcome of the affair.[250]

Despite his scepticism about the Trinity, Folkes considered the Bible an authoritative *historical* source (one among many). He also considered scripture to be a source for understanding the origins of Freemasonry and was particularly interested in the application of mathematics, astronomy, and metrology to biblical studies. As such, he was probably influenced by the 'circle of divines' which orbited Sir Isaac Newton in Cambridge and included among its members the philosopher and theologian Samuel Clarke, the mathematician and scholar William Whiston, and the controversialist John Jackson'.[251] After Newton's death, Folkes thus helped Thomas Pellett edit a manuscript for a posthumous publication of Newton's *Chronology of Ancient Kingdoms Amended* (1728), 'a truncated version of Newton's more radical manuscript'.[252] Newton's chronology ordered events from Greek, Persian, Jewish, and Assyrian mythology and history on a timeline from civilization's beginnings (*c.* 1125 BCE) to the conquests of Alexander the Great. Compared to more traditional chronologies, Newton's idiosyncratic timeline shortened the history of civilization by five to six centuries, as he attempted to synchronize pagan myths and scripture. Newton also used astronomical methods to get precise dates for events described in Greek myth. Buchwald and Feingold have demonstrated the efforts Newton made to extract an astronomical description from the *Commentary on Aratus's Phaenomena* by Hipparchus, so he could use it to calculate the date when the observations were made based on the precession of the equinoxes. Newton thought Hipparchus had described observations made by Chiron the centaur, who nurtured Jason the Argonaut to prepare for their expedition. As Jonathan Rée indicated,

> Newton referred to Clement of Alexandria, who wrote in the second century about the 'asterisms' or star patterns associated with someone called Chiron— presumed to be Chiron the Centaur, who nurtured Jason the Argonaut. 'Now Chiron delineated the asterisms', Newton wrote, 'as … Clemens Alexandrinus informs us: for Chiron was a practical astronomer'. Newton concluded that the star map passed down by the classical Greeks was originally devised by Chiron to help the Argonauts navigate the Mediterranean. But Newton and his contemporaries knew—as the early Greeks did not—that star maps suffer from built-in

---

[250] RS/MS/790/John Richardson the Younger to Folkes, 17 March 1736/7, Royal Society Library, London. The letter is in a folder alphabetized with the surname of correspondence but is not numbered.

[251] G. C. B. Robert, 'Historical Argument in the Writings of the English Deists', DPhil Dissertation (Worcester College, Oxford, 2014), p. 6. See also J. P. Ferguson, *An Eighteenth Century Heretic: Dr Samuel Clarke* (Kineton: Roundwood Press, 1976), pp. 120–36, pp. 210–26, and James E. Force, *William Whiston: Honest Newtonian*, pp. 15–16.

[252] Force, *William Whiston: Honest Newtonian*, p. 136.

obsolescence, because the entire sphere of the heavens, as it appears to an earth-bound observer, moves one degree westward every 72 years. Newton could therefore calculate the original date of the Greek map as 936 BC, which gave him a date for the Argonauts setting out to sea.[253]

Newton's initial decision to publish his *Chronology* was in response to the illegally published *Abregé de la Chronologie* (1726), an abstract of his studies that Newton had initially written for Princess Caroline in 1717, 'in that shape the properest for Her Perusal'.[254] After Newton sent the abstract to her, the Princess lent it to natural philosopher and writer Abbé Antonio Schinella Conti, who had come to London to observe the total solar eclipse and to visit Newton. As Newton's friend Zachary Pearce related, 'the Abbé, without the Princess's consent (as he believed) took a Copy of it; and…some time after, when he was in France, to which he went from England, a Translation of it in French was published at Paris, without Sir Isaac's Approbation or Knowledge'.[255] By May 1725, the Italian philosopher Antonio Cocchi intimated to Conti that Newton had lost interest in publishing his *Chronology*, only interested at that point in creating another edition of the *Principia*. Pearce affirmed that Newton was determined 'not to enter into Controversy with any man about any of the particulars of…[the Chronology], at his time of life, when he was so far advanced in years'.[256]

However, by 6 March 1726 Newton was finally persuaded by his friends to publish his *Chronology*, although he did not live long enough to see the publication come to pass.[257] On Newton's death, his heir John Conduitt commissioned the physician Thomas Pellett to go through his papers and decide what was worthy of publication. Conduitt ultimately sold the manuscript of the *Chronology* to printer Jacob Tonson, and gave the task of overseeing its publication to Pellett and Folkes who had to make a coherent work out of a manuscript that, as Pearce stated, left out 'some of the authorities and references, upon which [Newton] had grounded his opinion'.[258]

Folkes's close professional relationship to the ageing Newton as Vice President of the Royal Society made him an obvious choice to edit his posthumous works. A letter Folkes wrote to Sloane on 27 November 1726, a few months before

[253] Jonathan Rée, 'I tooke a bodkine', *London Review of Books* 35, 19 (10 October 2013). https://www.lrb.co.uk/the-paper/v35/n19/jonathan-ree/i-tooke-a-bodkine [Accessed 22 December 2019].

[254] Zachary Pearce, *A Commentary with Notes on the Four Evangelists and the Acts of the Apostles* (London: E. Cox, 1778), vol. 1, p. xli.

[255] Pearce, *A Commentary*, vol. 1, p. xli.

[256] Antonio Conti, *Lettere scelte di celebri autori all'Ab. Antonio Conti* (Venice: Domenico Fracasso, 1812), p. 23. [non credo ch'ei pensi all edizione della sua Cronologia, ma piuttosto ad una nuova de suoi Principii]; Jed Z. Buchwald and Mordechai Feingold, *Newton and the Origins of Civilization* (Princeton: Princeton University Press, 2012), p. 329.

[257] Buchwald and Feingold, *Newton and the Origins of Civilization*, p. 329.

[258] Pearce, *A Commentary*, vol. 1, p. xli.

Newton's decease, shows he was often a valued go-between for the Society's administrative business. Folkes wrote:

> I waited upon Sir Isaac Newton our President to show him the extract of the Societys accounts, and at the same time to know if he had thought of any persons to recommend for the Council of the ensuing year upon which he lookt over the list and pitch'd upon the enclosed names, to be offer'd to the approbation of your self, and the council who meet before the Election.[259]

When Newton was alive, Folkes even presented the third edition of his *Principia*, 'richly bound in morocco Leather', to the Royal Society in Newton's stead.[260] The frontispiece of the third edition of the *Principia* was John Vanderbank's 1725 engraving of Newton; Folkes would later model his own portrait on the same image (see chapter five). Newton also used data about the precession of the equinoxes to formulate his new chronology, and Folkes's astronomical expertise and access to Newton's manuscripts was useful in this regard. As we will see in chapter five, Folkes himself dated the Farnese Globe, the only large celestial globe that survives from antiquity, by analysing the stellar positions in the constellations carved on his surface.

In the auction catalogue of Folkes's books after his death, item 5114 was: 'An Abstract of Chronology of Sir Isaac Newton, which Mr Folkes sent to Sir Isaac, he not being at that time able to find his own, and which he returned with corrections in his own hand, *bound in Russia*, 4to'.[261] This copy was obtained from France 'by one who had it from Monsr Conti's MS';[262] Folkes wrote in the introduction to the manuscript that in 1726 he had lent his copy to Newton himself,

[259] BL MS Sloane 4048, f. 223, Folkes to Sloane, 27 November 1726, British Library, London. The list of names is not extant.

[260] RS/JBC/12, 31 March 1726, Royal Society Library, London. Folkes also had a copy of the first edition of the *Principia* in his library, originally the property of William Sancroft, Archbishop of Canterbury. As Munby stated, 'This copy, having subsequently passed through the hands of Sir Robert Peel, was bought by the late Lord Keynes, and bequeathed with his great Newton collection to King's College, Cambridge, in 1946'. See A. N. L. Munby, 'The Distribution of the First Edition of Newton's *Principia*', *Notes and Records of the Royal Society of London* 10, 1 (1952), pp. 28–39, on p. 38.

[261] *A Catalogue of the Entire and Valuable Library of Martin Folkes, Esq.*, p. 155. It sold for 14 shillings.

[262] Folkes provided the copy owned by the Lincolnshire antiquary, Maurice Johnson of the Spalding Gentlemen's Society, to friends in England. This 'Short Chronicle from the first memory of things in Europe to the conquest of Persia by Alexander the Great' has recently been sold at Sotheby's on 17 December 2019, lot 1. As the catalogue notes: 'Newton and Johnson were correspondents; and given their mutual interest in history and antiquities—Newton was in fact an honorary member of the antiquarian society Johnson headed—it would have been entirely natural for Newton to gift Johnson a copy of this text. https://www.sothebys.com/en/auctions/ecatalogue/lot.1.html/2019/science-technology-online-n10172 [Accessed 26 December 2019]. See Scott Mandelbrote, 'Newton and the Evidences of the Christian Religion', in Scott Mandelbrote and Helmut Pulte, eds, *The Reception of Isaac Newton in Europe*, 3 vols (London: Bloomsbury Academic, 2019), vol. 2; Saint-Omer: Bibliothèque d'agglomération du Pays, MS 786, 'An Abstract of Chronology by Sir Isaac Newton'. https://ccfr.bnf.fr/portailccfr/ark:/06871/004D03011645 [Accessed 2 July 2019]. My thanks to Scott Mandelbrote for this reference.

who had mislaid it among his papers. Newton then returned the manuscript to Folkes 'with a few corrections under his own hand not a month before his decease'. As Folkes indicated,

> some days after he asked me for it again telling me he had up and down corrected such passages as occurrd to him, but that he would now collate it throughout with the original which he had found, and correct whatever errors were yet remaining in it. I accordingly carried it to him the next time I waited upon him, and the last I had the Honour of seeing him, but found him already taken ill, so that I brought it back again in the same condition: and have since caused it to be bound up to keep by me whilst I live as a valuable token of the ffriendship that Great man was pleased to honour me with.[263]

In a private paean to Newton, Folkes then recorded Newton's time of death on the manuscript, including two laudatory poems by Lucretius (*De Rerum Natura*), a poem loved by the freethinking Folkes, and a verse by Edmond Halley, *An Ode on This Splendid Ornament of Our Time and Our Nation*, about his mentor. From Lucretius:

> Vivida Vis Animi pervicit, et extra
> Processit longe flammantia Moenia Mundi
> Atque omne Immensum peragravit Mente Animoque
>
> The living force of his soul gained the day
> on he passed far beyond the flaming walls of the world

and traversed throughout in mind and spirit the immeasurable universe.[264] From Halley, the poem printed in the 1687 edition of the *Principia* as its opening ode:

> Talia monstrantem mecum celebrate Cameonis,
> Vos qui Caelesti gaudetis Nectare vesci,
> NEWTONUM sancti reserantem scrinia Veri,
> Newtonum Musis carum, cui pectore puro
> Phaebus adest, totoque incessit Numine Mentem:
> Nec fas est propius mortali attingere Divos
>
> Oh, you heavenly ones who make merry on nectar,
> Celebrate with me in song the revealer of these things,

---

[263] Saint-Omer: Bibliothèque d'agglomération du Pays, MS 786.

[264] Saint-Omer: Bibliothèque d'agglomération du Pays, MS 786. Titus Lucretius Cari, *De Rerum Natura Libri Sex, with a translation and notes*, ed. and trans. H. A. J. Munro (Cambridge: Deighton Bell and Co; London: Bell and Daldy, 1900), vol. 1, book 1, lines 72–4, p. 4.

Newton to the Muses dear,

Newton who unlocked the barred treasure-chest of Truth:

Phaebus is in his pure breast, and enters his mind with his own divinity.

Nearer the Gods no mortal may approach.[265]

Halley's *Ode* implies that Newton's discovery of God's truth in the form of eternal celestial mechanics meant that 'by implication, one can look back to the origins of time and see the same natural phenomena described by Newton at work in the past'.[266] Past astronomy and chronology could thus inform present investigations, whether in natural philosophy or theology.

Folkes passed along to Charles Morgan, the master of Clare Hall, a manuscript copy of portions of this chronology, including sections on the ancient year and a 'computation of the times of burning the first temple and building the second'; Folkes remarked to his former teacher,

> I am so glad you have been so well entertained by Sr Isaac's book, and at the same time to find my own opinion of it so entirely confirmed...but indeed I have had that satisfaction from several hands, and I even here your Neighbour of the great College [Richard Bentley, Master of Trinity] who spoke very slight-ingly of the performance before it appeared begins not to talk so magisterially as he did before.[267]

Folkes also gave Morgan part of a manuscript copy of the second half of Newton's *Treatise of the System of the World*, which was based upon a series of lectures Newton gave in 1687 entitled *De Motu Corporum Liber*.[268] In this work, he postu-lated that 'all bodies have a disposition to gravitate, but it is only activated in vir-tue of them having this common nature', which is a relational account of action at a distance mutually shared between bodies.[269] Newton intended this work to comprise the third book of Newton's *Principia*, but decided to suppress this material, explaining:

---

[265] The translation is from A. Rupert Hall, *The Scientific Revolution 1500–1800: The Formation of the Modern Scientific Attitude* (London: Longmans, Green and Co, 1954), p. 246, footnote 2.

[266] Alessio Mattana, 'Antiquitas Non Fingo: Newton, the Moderns and the Science of Ancient History', *Journal of Eighteenth-Century Studies*, forthcoming. My thanks to Dr Mattana for sharing his work with me ahead of publication. It has since been published in vol. 43, 4 (2020), pp. 447–61.

[267] Charles Morgan Papers, MS G.3.13, Clare College Archives, Cambridge; College Letterbook, Clare College Archives, Folkes to Morgan, 6 January 1727/8, Letter 89.

[268] I. B. Cohen, ed., *A Treatise of the System of the World by Isaac Newton* (Philadelphia: American Philosophical Society, 2004), p. vii.

[269] Eric Schliesser, 'Newton and Newtonianism in Eighteenth-Century British Thought', in James A. Harris, ed., *The Oxford Handbook of British Philosophy in the Eighteenth Century* (Oxford: Oxford University Press, 2013), DOI: 10.1093/oxfordhb/9780199549023.013.003.

Upon this subject, I had indeed compos'd the third book in a popular method that I might be read by many. But afterwards considering that such as not sufficiently enter'd into the principles, could not easily discern the strength of the consequences, nor lay aside the prejudices to which they had been many years accustomed; therefore to prevent the disputes which might be rais'd upon such account, I chose to reduce the substance of that book into the form of propositions (in the mathematical way) which should be read by those only, who had first made themselves masters of the principles establish'd in the preceding books.[270]

As I. B. Cohen has surmised, 'we do not know whether Newton had any particular person or persons in mind when he wrote this statement, or whether he was merely aware that objections to his new book [from Cartesians] might be based more on prejudice than on mathematical argument'.[271] We also do not know whether Folkes had Newton's own manuscript, or a transcript, but we do know that Newton's reluctance to popularize the *Principia* led to Folkes's later comment that 'After Sir Isaac printed his *Principia*, as he passed by the students at Cambridge said there goes the man who has writ a book that neither he, nor anyone else understand'.[272] However, Folkes understood more than he let on. In his *System of the World*, Newton developed his idea that the ancients possessed a cosmology which he thought contained the centre of the original true religion. It was in this work that he 'intended to make public his views on ancient science' that was of use to Folkes when he edited Newton's *Chronology*, for included in the *System of the World* was Newton's belief that the heliocentric system was known by the ancient Greeks and Egyptians.

In addition, Folkes sent to Morgan some other mathematical papers that in their methodology illuminated Newton's own techniques in creating his chronology out of a discordant set of astronomical observations. Buchwald and Feingold indicated that Newton, convinced that the senses were not reliable enough to generate knowledge,

developed a way to overcome this limitation by taking the extraordinary tack of increasing the number of measurements without discarding any. He alone forged a trustworthy resultant out of a discrepant set of numbers, each of which was inherently unreliable, by taking an average among them. Instead of discarding every measurement other than the one thought to be the very best, as his

---

[270] Isaac Newton, *The mathematical principles of natural philosophy. By Sir Isaac Newton. Translated into English by Andrew Motte…in two volumes* (London: Benjamin Motte, 1729), introduction to book III.

[271] Cohen, ed., *A Treatise of the System of the World*, p. ix and p. xi.

[272] Keynes MS 130, f. 5. Martin Folkes's recollections of Newton, Cambridge University Library, Cambridge.

contemporaries usually did, Newton kept them all, thinking that a good number could be produced by combining a multitude of bad ones.[273]

In the papers he provided to Morgan, Folkes had transcribed in full a manuscript from Roger Cotes's posthumous work, *Aestimatio Errorum*, which accomplished the same thing mathematically. Cotes was the first Plumian Professor of Astronomy, appointed in 1707, and assisted Newton with editing the second edition of the *Principia*, which addressed Leibnizian criticism to provide more of a philosophical frame for the work through the addition of his 'General Scholium'. Cotes's *Aestimatio Errorum* explored the calculus of difference equations, and the weighted 'least squares' method. Cotes considered more generally:

> four observations p, q, r and s which may not be equally reliable, of the position of a point. He proposes, as the most probable true position, a weighted average with weights P, Q, R and S which are inversely proportional to the spread of errors to which the respective observations are subject. This proposal represents one of the earliest attempts to determine an average which uses all the observations but does not assign equal weights to all of them.[274]

Folkes must have studied this work carefully, along with the other papers he sent to Morgan when editing the *Chronology* in order to understand Newton's method of reconciling astronomical data.

Folkes's and Thomas Pellett's edition of the *Chronology*, while pragmatic, was respectful and rapidly published, appearing less than ten months after Newton's death. That did not mean it was not problematic. As Newton's friend Zachary Pearce lamented, it was a pity that Newton

> took so much of the same method in his Chronology which he took in his Principia, etc. concealing his proofs, and leaving it to the sagacity of others to discover them. For want of these, in some instances, what he says on Chronology does not sufficiently appear at present to rest upon anything but his assertions, and the want of these was thought so great by the editors (Martin Folkes, Esq. and Dr Pellet) that they or one of them, as I have been informed, did in some places put References to Authors in the margin of the Work; which are printed now as Sir Isaac's References, though not his, and not perhaps always referring to the very same places, upon which he founded his assertions.[275]

---

[273] Buchwald and Feingold, *Newton and the Origins of Civilization*, p. 5.

[274] H. Leon Harter, 'The Method of Least Squares and Some Alternatives—Part I', *International Statistical Review* 42, 2 (August 1974), pp. 147–74, on p. 148.

[275] Pearce, *A Commentary*, vol. 1, p. xlii.

The Chronology was indeed only edited slightly. As Schilt has noted:

> its 92 folios contained a neat text, with only minor deletions and interlinear additions, and just a handful of more substantial additions on the versos. In the chapter on the Temple, Folkes made emendations where he thought the main text was unclear or, in his view, plainly wrong. In a description of the dwellings of the priests, Newton wrote that these had 'cloysters on the outside [a]nd the pavement & buildings upon it were encompassed on the outside with a marble rail'. Folkes underlined most of this passage and added his corrections in the margin, changing the line to 'cloysters under them: and the pavement was faced on the inside with a marble rail'.[276]

Folkes and Pellet would take their red pencil and apply the same pragmatism of the *Chronology* to Newton's *Observations on the Prophecies of Daniel and the Apocalypse of St John* (1733). It seems they corrected factual inaccuracies and ignored Newton's own proposed section and chapter structure for the sake of coherence.[277] Indeed, Chapter X 'Of the Prophecy of the seventy Weeks' was a fabrication by Folkes compiled from the 'first pages of a work titled by Newton "The History of the four monarchies compared with Sacred Prophecy"', and other paragraphs taken from other manuscripts.[278] Folkes 'even deleted a passage where Newton, having discussed the application of the period of seven weeks mentioned in Daniel 9:25 to the time of Christ's second coming'.[279] Folkes wrote that 'This clause…I had rather leave to be explained by time then venture upon a rash interpretation of what I do not yet understand'.[280] The second part of the *Observations*, where Newton's views of the Apocalypse were represented, was also severely curtailed. There were also just practical matters, such as Folkes's suggestion to place Newton's extensive marginal notes 'in a somewhat less character on the bottom of the page'.[281] When Newton attempted to reorder the very events of Christ's life, make such pronouncements as

> All those things therefore which Christ did in the 4th, 5th 6th 7th and 8th chapters of Matthew, between the imprisonment of John and this feast, must be referred to the Summer and spring before, beginning at the winter solstice, when Christ upon the news of John's imprisonments went from Judaea through Samana into Galilee, and there began to preach,

---

[276] Cornelis J. Schilt, 'Of Manuscripts and Men: The Editorial History of Isaac Newton's Chronology and Observations', *Notes and Records: The Royal Society Journal of the History of Science* 74, 3 (2020), pp. 387–403, on p. 389.

[277] Schilt, 'Of Manuscripts and Men', p. 390.

[278] Schilt, 'Of Manuscripts and Men', pp. 392–3.

[279] Schilt, 'Of Manuscripts and Men', p. 393.

[280] MS Add 3989, f. 48r., Observations upon the Prophecies of Daniel and the Apocalypse of St John, Cambridge University Library, as quoted in Schilt, 'Of Manuscripts and Men', p. 393.

[281] MS Add 3989, f. 47r., Cambridge University Library.

an editor wrote laconically on the bottom margins of the page: 'I imagine all this page is to be out'.[282]

Folkes's careful attention to Newton's texts about the Temple in the *Chronology*, however, may have been because it was one of the pieces of biblical wisdom which interested him from the viewpoint of Freemasonry and as a mathematician and astronomer. Namely, Folkes was a devotee of metrology, whether it was for clarifying the scripturally ambiguous dimensions of the sacred cubit of the Hebrews used to build Solomon's Temple or the dimensions of the Roman foot.[283] He even had a 'curious model of the Temple at Jerusalem, in a glass case' in his house in Queen Square.[284] Folkes took as his motto (as did Newton) for *alba amicorum* (friendship books): '*omnia mensura et numero et pondere diposuisti*' or 'thou has ordered all things in measure and number and weight' from the apocryphal Book of Wisdom, 11:21.[285] Although Wisdom 11:20d was 'quoted by almost every Christian author between Augustine and Pascal writing about the importance of mathematics', it is suggestive of Folkes's (and Newton's) heterodoxy that the Book of Wisdom was non-canonical for the Church of England.[286]

And Folkes's interests reflected the current intellectual milieu. Speculative masonry emphasized 'a mathematical deity' which 'accorded well with the Newtonian conception of the universe, with a grand mathematician or architect at the centre'.[287] Previously, Newton had written a dissertation upon the sacred cubit, which was also republished in 1737 by antiquary Thomas Birch along with the works of John Greaves, who wrote the *History of the Royal Society* (1757). And, as Haycock noted,

> in 1723 the English publication of the French priest Bernard Lamy's *Apparatus Biblicus*, which included detailed engravings of both the Tabernacle and Temple, was published. The following year a grand model of the Temple, 'lately brought over from Hambourg', was put on display for the public in London, whilst in

[282] MS Add 3989, f. 55v., Cambridge University Library.

[283] The specification of the cubit was ambiguous: 'in the man's hand a measuring reed six cubits long, of a cubit and a handbreadth each: so he measured the thickness of the building, one reed; and the height, one reed' (Ezekiel 40:5). The Hebrew text reads: 'a measuring reed, six cubits by the cubit and a handbreadth'. 'It was unclear whether the six cubits marked on the reed were particularly large, each outsizing the regular cubit by a handbreadth; or whether the reed as a whole measured six cubits and one extra handbreadth. This was to become one of the key problems of much Temple scholarship'. See Jetze Touber, 'Applying the Right Measure: Architecture and Philology in Biblical Scholarship in the Dutch Early Enlightenment', *The Historical Journal* 58, 4 (2015), pp. 959–85, on p. 964.

[284] *A Catalogue of the Genuine and Curious Collection*, p. 5.

[285] See, for instance, Folkes's signature in the Album for the Egyptian Society, BL Add MS 52362, f. 4r., British Library, London.

[286] Jens Høyrup, *In Measure, Number and Weight: Studies in Mathematics and Culture* (SUNY Press, Albany, 1994), p. xv; See also, George Gömöri and Stephen Snobelen, 'What he May Seem to the World: Isaac Newton's Autograph Book Epigrams', *Notes and Records: The Royal Society Journal of the History of Science* 74, 3 (2020), pp. 409–52, on p. 431.

[287] Elliott and Daniels, 'School of True, Useful and Universal Science', p. 220.

1726 William Whiston advertised 'a small, but curious Course of Lectures' on astronomical subjects which also included 'Sacred Architecture past'...as well as 'Sacred Architecture Future'...It was also in this very same period that Freemasonry was first taking off in London, with prominent Newtonians... playing an active role.[288]

Folkes's attention to these matters was simply reflective of a zenith of interests in metrology and Freemasonry, and his editions of Newton's works on chronology an act of devotion to his mentor, and his own means of promoting Newtonianism and himself. Folkes may not have been able to have been Royal Society President just yet, but he could say he had edited and published the posthumous works of the most famous former holder of the office. As we will see in chapter five, he continued to venerate his mentor in the 1730s during his Grand Tour, as an ambassador of Newtonianism abroad.

[288] Haycock, Stukeley: p. 155.

5

Taking Newton on Tour

5.1 The Grand Tour

Qui sera sera, 'Who or What will be, will be' was the opening phrase that Martin
Folkes (1690–1754) chose as his personal motto and inscribed in his travel diaries
of his Grand Tour to France, Germany, and Italy from 24 March 1732/3, returning to London on 3 September 1735[1] (see figure 5.1). War between France and
England had been nearly continuous from the late seventeenth century, with the
War of the League of Augsburg and the War of the Spanish Succession; the 1720s
and 1730s provided a brief interlude where freedom to travel increased dramatic-
ally. Folkes recorded the Italian leg of his peregrination from Padua to Rome in
his journal.[2] His travel diary, like Folkes himself, is little known in the historical
literature, and when it is assessed, it is to provide examples of historical topog-
raphy and a more general commentary of the import of the Grand Tour, or analysed with regard to Folkes's interests in the collection and iconography of coins
and medals (see chapter six).[3]

In some ways, the journey was typical in its route (see figure 5.2), if not in its
travel mishaps; Folkes's Italian tour, with Lucretia and their three children, a
parrot, a Dutch dog, and a monkey, took the family from Ferrara to Ravenna and

[1] Part of this chapter appeared as: Anna Marie Roos, 'Taking Newton on Tour: The Scientific
Travels of Martin Folkes, 1733–35', *The British Journal for the History of Science* 50, 4 (2017),
pp. 569–601. Reprinted with permission, Cambridge University Press. Bodl. MS Eng. misc. c. 444,
Martin Folkes, 'Journey from Venice to Rome', Bodleian Library, Oxford; Martin Folkes Esq. *Memoranda
in case he should go abroad again, respecting his Estates, foreighn money & Places*, f. 18r., NRS 20658,
Norfolk Record Office, Norwich. The motto also appears on the mezzotint copy of the 1718 portrait of
Folkes by Jonathan Richardson. See National Portrait Gallery, London, NPG D36990. The exact translation of this phrase is difficult, because the exact phrase would not be used with the pronoun 'qui'
alone in French, instead beginning with 'ce qui'. The only modern French phrase that gets as close as
possible to this fatalistic meaning would be 'Ce qui doit arriver arrivera', which translates 'what needs
to happen will happen'. Used with the verb 'être'/to be instead of 'arriver'/to happen, an older form of
this phrase would be 'Ce qui doit être sera' (= what must be, will be). This rendition is very close to
Folkes's phrase, where the 'ce' in 'ce qui' might have been forgotten by the author, in which case it
could become 'Ce qui sera sera' (= whatever will be, will be). Folkes, however, was a near native
speaker of French. See also Lee Hartman, '"Que sera sera": The English Roots of a Pseudo-Spanish
Proverb', *Proverbium* 30 (2013), pp. 51–104. My thanks to Charlotte Marique for this explanation.
[2] Martin Folkes Esq. *Memoranda in case he should go abroad again*, f. 5v., NRS 20658, Norfolk
Record Office, Norwich. His account book indicates each stage of his journey and how many weeks he
stayed in each place.
[3] John Ingamells, *A Dictionary of British and Irish Travellers in Italy, 1701–1800: Compiled from the
Brinsley Ford Archive* (New Haven: Yale University Press, 1997), p. 365; Eisler, William Eisler, 'The
Construction of the Image of Martin Folkes (1690–1745), Part I', *The Medal* 58 (2011), pp. 1–29.

Martin Folkes (1690–1754): Newtonian, Antiquary, Connoisseur. Anna Marie Roos, Oxford University Press (2021).
© Anna Marie Roos. DOI: 10.1093/oso/9780198830061.003.0005

Fig. 5.1 John Smith, *Martin Folkes* after Jonathan Richardson Senior, mezzotint, 1719, 339 mm × 247 mm, 1902,1011.4540AN121298001, ©Trustees of the British Museum, Folkes's motto is in the centre bottom.

then to Loretto where he lamented of the beggars: 'walking a hundred yards in the street it goes beyond all human imagination, and is irksome beyond expression'.[4] Folkes recorded buying large purple grapes for his children from an itinerant merchant when they were travelling from Ferrara, and he rose early 'before the rest of the family was up' to go with his son Martin to Ancona to show him the sights.[5] He also reported on their way to Narni,

> I had the bad luck to lose my Dutch Dog, of which I have been able to hear no news tho I sent two servants back and offered a very great reward, as it has really given me far more uneasiness than I think such a sort of thing should but I have that weakness and am really fretted and went about it beyond measure.[6]

[4] Folkes, 'Journey from Venice to Rome', f. 74r.
[5] Folkes, 'Journey from Venice to Rome', ff. 57 and 69.
[6] Folkes, 'Journey from Venice to Rome', f. 84r.

Fig. 5.2 Folkes's Grand Tour keyed to the list of places he indicated he visited in his Travel Account Book in the Norwich Archives, map by the author.

Happily, the dog was recovered, having wandered back to Venice in search of Folkes, and was 'almost starved', Folkes commenting that 'the Dog moans still and seems to attempt at telling his misfortune'.[7]

---

[7] Folkes, 'Journey from Venice to Rome', f. 88r.

According to William Stukeley, Folkes went on his journey because he was 'baffled' by his loss of the election for the Royal Society presidency to Sir Hans Sloane. However, Folkes's Grand Tour was five years after his loss of the Royal Society election. In the interim, Folkes had also half-heartedly reconciled with Sloane and was elected one of his Vice Presidents on 8 February 1733; a number of letters between them indicated at least a functional professional relationship on an official level, if not *bon amie*. It seems the two gentlemen had decided to 'bury the hatchet', perhaps closing ranks in response to the external criticism Sloane (and the Royal Society) continued to receive for his curiosity mongering.[8] Folkes may have also pragmatically realized that without Sloane's assistance, he would never attain his ultimate ambition of the Royal Society presidency. So, to curry his favour, Folkes requested Sloane's medical advice for Richard Laughton at Cambridge,[9] and on another occasion he wrote concerning bidding on Sloane's behalf at auctions to build his collection of paintings.[10] Folkes had also distributed a preliminary list of Fellows whom Sloane desired to be on the Council to his friends and colleagues, to preempt any opposition to his choices.[11] He later served as an executor for Sloane's will.

Folkes, however, did not attend the Council meeting to be sworn in as Vice President.[12] He wrote to Sloane immediately afterwards:

I am quite ashamed of my self for my neglect of coming to the Council this afternoon to be sworn, and hereby missing an occasion of acknowledging my obligation for the honour you have done me in nominating me for one of your Vice Presidents, and I can plead nothing in my excuse but so absolute a want of memory that I did not in the least recollect the council till I heard my clock strike six, tho I yesterday put off an appointment on purpose to attend this afternoon, this being Sir the truth of the case I rather chuse to Expose the infirmity of my forgetfulness, than labour under the suspicion of any want of respect or acknowledgement for your favour; and therefore could not avoid giving you this immediate trouble to desire you will pardon.[13]

---

[8] Barbara Benedict, 'Collecting Trouble: Sir Hans Sloane's Literary Reputation in Eighteenth-century Britain', *Eighteenth-Century Life* 36, 2 (Spring 2012), pp. 111–42, on pp. 125–32.

[9] BL MS Sloane 4058, f. 342, Folkes to Sloane, 'Monday: July 15', British Library, London.

[10] BL MS Sloane 4058, f. 344, Folkes to Sloane, 'thurs: night', British Library, London.

[11] BL MS Sloane 4058, f. 338, Folkes to Sloane, 'Saturday morning 5 o'clock', British Library, London. The dating is surmised to be the 1730s, as Folkes mentioned assuring Dr Mortimer that Sloane's wishes would be followed, and Mortimer began as secretary of the Royal Society in 1730, with Sloane resigning the presidency in 1741.

[12] George S. Rousseau and David Haycock, 'Voices Calling for Reform: The Royal Society in the Mid-Eighteenth Century: Martin Folkes, John Hill and William Stukeley', *History of Science* 37, 118 (1999), pp. 377–406, on p. 381.

[13] BL MS Sloane 4058, f. 340, Folkes to Sloane, 'Tuesday, half an hour after six', British Library, London.

This may have been a snub or a genuine excuse, but whatever Folkes's reason for not attending, there must have been another reason for his journey abroad than being 'baffled' from losing the Royal Society election for the presidency. Some of it may have been personal circumstance, as Folkes's wife Lucretia had for some time exhibited symptoms of mental illness. As Marland has commented, in the Georgian period,

> many medical treatises, as well as novels and plays of the eighteenth century, depicted how social expectations inflamed or frustrated the fairer sex, and led to mental breakdown—unrequited love, inappropriate romances, the difficulties of matrimony or failure to marry at all, inability to conceive, or the sadness of being left a widow.[14]

Playwright Henry Fielding remarked of a character suffering mental illness in his play *Amelia*:

> These Fatigues, added to the Uneasiness of her Mind, overpowered her weak Spirits, and threw her into one of the worst Disorders that can possibly attend a Woman: A Disorder very common among the Ladies...Some call it Fever on the Spirits, some a nervous Fever, some the Vapours, and some the hysterics.[15]

Although Lucretia would have been considered fortunate for having married into money, her relative lack of education, the loss of her professional identity as an actress, and no doubt some social isolation among Folkes's social circles due to her rapid social mobility would have placed her under strain. Her manifestation of mental illness also coincided with the time she would have been undergoing the peri-menopause or menopause. However, although women were often, at least in the nineteenth century, confined to asylums for gynaecological disorders, this was not the case in the early modern period as 'there was not a widespread view of menopause as pathological'.[16] Although women were 'as silent on the effects and implications of menopause as they were about menstruation', there could have been some attendant anxiety about having left it too late to have another child.[17]

---

[14] Hilary Marland, 'Women and Madness', Centre for the History of Medicine, University of Warwick. https://warwick.ac.uk/fac/arts/history/chm/outreach/trade_in_lunacy/research/womenandmadness/ [Accessed 28 December 2019].

[15] Henry Fielding, *Amelia* (London: A. Millar, 1752), book 3, chapter seven.

[16] Sara Read, *Menstruation and the Female Body in Early Modern England* (London: Palgrave Macmillan, 2013), p. 178. See also Sara Mendelson and Patricia Crawford, *Women in Early Modern England, 1550-1720* (Oxford: Clarendon Press, 1998); Susannah R. Ottaway, *The Decline of Life: Old Age in Eighteenth-Century England* (Cambridge: Cambridge University Press, 2004), esp. pp. 38–40.

[17] Read, *Menstruation*, p. 179.

Whatever the reason, during their tour of Italy in the 1730s (see chapter five), Lucretia was manifesting symptoms of erratic behaviour. Tom Hill, Lennox's steward, wrote to his lord:

> With much ado I obtain'd leave to transcribe the following account relating to Mrs. Folkes out of a letter that came from abroad, having first sworn not to tel the person that sent it.

> There is come hither a Lady with her husband, three children, and a monkey, who are no more exempt from obedience to her, one than another, and all seemingly fellow-sufferers alike. I happen'd to be at a visit when she came in. In all my life did I never hear such an insupportable creature, nor so much nonsense in so small a space of time. You will be surpris'd when I tell you the husband is reckon'd as clever a man as any in England. His name is Folkes (Martin Folkes as she cals him) who used to be very much with the Duke of Richmond. The lady he married is very wel known in England. He designs making the tour of Italy and France, by which time I don't doubt but she wil turn out the most accomplisht of fine Ladys. She did think indeed of bringing a little dog and a cat to keep poor pug company, but that they could not possibly find more room in the coach. Such characters are no where to be met abroad, whatever they may be in England, and even there I never saw one come up to this.

> This is al that was read to me out of the letter. I could not help saying, what I fancy you'l join with me in, Poor Martin! In an evil hour didst thou take to thy bosom this Lady Mar-all.[18]

One could attribute this report to jealousy or nasty gossip, but Stukeley noted that, while in Rome, Lucretia Folkes 'grew religiously mad'.[19] Stukeley's assessment was confirmed by sobering letters written between Martin Folkes and his friend, the Italian natural philosopher and physician Antonio Cocchi, which indicated the severity of Lucretia's disorder. When Folkes left Florence to visit Pisa and Lucca in July 1735, Cocchi's family took care of his two daughters, who were left behind due to Lucretia's state of mind.[20] In November 1735, at which point Folkes had returned to London with his family, he sent Cocchi an abridged *Philosophical Transactions of the Royal Society* as a thank-you gift, and wrote,

[18] Charles Henry Gordon-Lennox, 8th Duke of Richmond and Earl of March, *A Duke and His Friends: The Life and Letters of the Second Duke of Richmond* (London: Hutchinson and Company, 1911), p. 253.
[19] William Stukeley, *The Family Memoirs of the Rev. William Stukeley* (London: Surtees Society, 1882), vol. 1, pp. 99–100.
[20] Leghorn MS Facs 589, 584B, Martin Folkes to Antonio Cocchi, 15 July 1735, British Library, London. These are a series of photographs from manuscripts in the family archive of Conte Enrico Baldasseroni in Florence, a descendant of Antonio Cocchi.

I must freely tell you the situation of my own affairs is on many account so dis-agreeable to me I know not what in world to do; my wife is nothing so outra-geous as when abroad, which lays me under really an incapacity to make my self easie by confining her, and on the other hand it is not possible for me to live with her or easie if she has it in her power to make me and all my ffriends uneasie.[21]

Perhaps Folkes had thought the travel would do her some good or halt gossip in his social circles in London.

From the perspective of intellectual history, Charles Weld in his 1848 *History of the Royal Society* claimed the journey was so Folkes could 'improve himself in classical antiquity'.[22] And that was also true. Lennox provided Folkes with a number of rather blustery letters of introduction, one to Alessandro Albani, the leading collector of antiquities in Rome, whom Lennox described as 'a very odd Curr, Ignorant enough, & proud as Hell; butt has the finest library, one of them, in Europe, & without exception ~~one of~~ the very best collection of bustos, in the World. You must flatter him upon his learning'.[23] Lennox also wrote a letter of introduction for Folkes to Teresa Grillo Pamphili (d. 1763), the Genoese poet who led an important literary salon (the Arcadia) in Rome, referring to her rather uncharitably as the 'Ugliest Bitch, in the World, Damn'd proud also, & stark staring mad, butt a Develish deal of Witt, some knowledge, & altogether *Une Maitresse ferme, sans un brin de religion* [a resolute woman without a touch of religion]'.[24] Lastly, Folkes thanked Lennox for his introductions to the notables of Venice: 'I have also here had the great-est civility on your Graces account both from the Resident and Mr Smith with whom he being known to you has set me on a very different foot from the generality of travelers that come here'.[25] The Resident was the English ambas-sador to Venice, Colonel Elizeus Burges, and 'Mr Smith', Thomas Smith, the art collector and connoisseur, banker to the British community at Venice, and noted host to those on the Grand Tour.

In a more serious vein, Folkes's faithful friend James Jurin wrote a letter for Folkes to Giovanni Poleni, the Italian natural philosopher and expert in

---

[21] Leghorn MS Facs 589, 584B, Martin Folkes to Antonio Cocchi, 6 November 1735, British Library, London.
[22] Weld, Charles Weld, *A History of the Royal Society*, 2 vols (London: J. W. Parker, 1848) vol. 1, p. 480.
[23] RS/MS/865/12, Charles Lennox to Martin Folkes, 31 July 1733, Royal Society Library, London.
[24] RS/MS/865/12, Charles Lennox to Martin Folkes, 31 July 1733, Royal Society Library, London. The letter itself to Pamphili is RS/MS/865/13. For Pamphili's role as host of the Arcadia, see Elisabetta Graziosi, 'Women and Academies in Eighteenth-Century Italy', in Paula Findlen, Wendy Wassyng Roworth, Catherine M. Siena, eds, *Italy's Eighteenth Century: Gender and Culture in the Age of the Grand Tour* (Stanford: Stanford University Press, 2009), pp. 103–19.
[25] Goodwood MS 110, Martin Folkes to Charles Lennox, 8 July 1733, West Sussex Record Office.

architecture and classical antiquity (who had also been engaged in the *vis viva* dispute with Samuel Clarke and other Newtonians).[26] Jurin stressed that Folkes was an

> excellent gentleman of high achievement in every kind of literature, especially
> physics and mathematics; whom Newton valued so highly that, not content to
> consider him worthy of his own close friendship, he voluntarily chose him to
> carry out the duties of Vice-President at the formal meetings of the Society in
> his own absence.[27]

Folkes subsequently met Poleni in Venice on 18 July 1733, with whom he discussed meteorology and the state of learning in England, and subsequently recommended him to Sloane for election to the Royal Society.[28]

Intellectual and aesthetic development was thus part of the reason for Folkes's tour, just as it was for natural philosopher George Berkeley before him, who, in his case, cultivated a taste for Greek Doric Architecture. In 1713, Berkeley wrote to Sir John Percival, stating that he was interested in the 'agreeable effects' of such buildings on his eyes, an interesting remark 'given the subjectivist theory of perception he recently articulated in his *New Theory of Vision* (1709)'.[29]

Like Berkeley, Folkes melded aesthetic and natural philosophical concerns, his tour one of scientific peregrination, or 'science on the move'—a concept that was prevalent from 1650–1750.[30] The origins of the scientific peregrination were in the '*peregrinatio academica* of early modern aspiring scholars' and/or the *peregrinatio medica*, as medicine was the subject for which foreign travel was most valuable; medical students brought back new techniques, knowledge, and *materia medica* to their homeland.[31] Indeed, in the 1670s Thomas Bartholin, when recalling his own *peregrinatio medica*, noted, 'Today there are many travellers; indeed, it seems as if the whole of Europe is on the move'. Young medical students travelled abroad to earn professional credentials and

---

[26] The *vis viva* dispute lasted from 1686 until the 1720s and concerned whether the *vis viva*, or momentum, was conserved in the universe. The dispute started between Descartes and Leibniz and resulted over confusion between momentum (mv) and kinetic energy ($1/2mv^2$). See Carolyn Iltis, 'Leibniz and the vis viva controversy', *Isis* 62, 1 (Spring 1971), pp. 21–35.

[27] For the letter of introduction, see Andrea Rusnock, ed., *The Correspondence of James Jurin (1684–1750): Physician and Secretary of the Royal Society* (Amsterdam and Atlanta, GA: Rodopi, 1996), p. 391.

[28] Folkes, 'Journey from Venice to Rome', f. 25r.; BL MS Sloane 4058, f. 60, Folkes to Sloane, 1 October 1733, British Library, London.

[29] Edward Chaney, *The Evolution of the Grand Tour: Anglo-Italian Cultural Relations Since the Renaissance* (London and Portland: Frank Cass, 1998), p. 328.

[30] Mark Greengrass, Daisy Hildyard, Christopher D. Preston, and Paul J. Smith, 'Science on the Move: Francis Willughby's Expeditions', in Tim Birkhead, ed., *Virtuoso by Nature: The Scientific World of Francis Willughby FRS (1635–72)* (Leiden: Brill, 2016), pp. 142–226, on p. 145.

[31] Zur Shalev, 'The Travel Notebooks of John Greaves', in Alistair Hamilton, Maurits H. Van Den Boogert, and Bart Westerweel, eds, *The Republic of Letters and the Levant* (Leiden: Brill, 2005), pp. 77–103, on p. 85. See also Ole Peter Grell and Andrew Cunningham, eds, *Centres of Medical Excellence? Medical Travel and Education in Europe, 1500–1789* (Aldershot: Ashgate, 2010).

attain accomplishments to become gentlemen of quality, with a cosmopolitanism to add polish to their English education. But by the time of Folkes's journey, scientific tours apart from the reasons of *politesse* and medical pedagogy had long developed as a separate enterprise. Previous works such as botanist John Ray's later published travelogue of his journey on the Continent (editions in 1693 and 1738) reinforced the image of 'diligent natural philosophers, engaged in the pursuit of activities conducive to the public [and scientific] good'.[32] In 1698, physician and naturalist Martin Lister (1639–1712) wrote a guidebook of his *Journey to Paris*, intending it specifically to appeal to fellow natural philosophers.[33] Lister directed the reader to his interests in natural history using his mature judgment and own eyes, offering 'clean Matter of Fact, and some short notes of an unprejudiced Observer'.[34]

Although not a doctor or a naturalist like Lister or Ray, Folkes had similar sentiments reflecting his specialism in mathematics and interest in architectural measurement, his journey designed to answer particular questions of natural philosophy. His diary reflected 'antiquarian science', a form of peregrination in which he used metrology to understand not only the aesthetics but the engineering principles of antique buildings and artefacts, as well as their context and place in the Italian landscape. Although Folkes followed in the tradition of past natural philosophers such as John Greaves, the Savilian Professor of Astronomy who desired to standardize and synchronize the 'weights and measures of all ancient and modern nations', we will see that Folkes also wished to verify Greaves's measurements of the Roman foot as an expression of his status as a Freemason and his current interests promoted in the Society of Antiquaries.

Using Folkes's diary, his account book of his journey, now held in the Norwich archives, and accompanying correspondence with other natural philosophers such as Francesco Algarotti (1712–64), Anders Celsius (1701–44), and Abbé Antonio Schinella Conti (1667–1749), we will determine to what extent this journey established Folkes's reputation as an international broker of Newtonianism, particularly in the field of optics and the subsequent application of his mentor Newton's natural philosophy to geodesy, the ultimate form of metrology. We will see that Folkes was successful in demonstrating the overall primacy of English scientific instrumentation to Italian virtuosi. His journey was an expression of the motto scrawled on his diary, symbolic of who and what he would become as President of the Royal Society and the Society of Antiquaries.

---

[32] Greengrass, Hildyard, Preston, and Smith, 'Science on the Move', p. 152.
[33] Raymond Phineas Stearns, ed., *Martin Lister, A Journey to Paris in the Year 1698* (Urbana, Chicago, and London: University of Illinois Press, 1967), p. 2.
[34] Lister, *A Journey to Paris*, p. 3.

5.2 Metrology

As Shalev has claimed, the study of early modern metrology or ancient weights and measures was:

> a central preoccupation of antiquarians, theologians, and natural scientists in the early modern period…The list of central figures who devoted full tracts to the subject—from Budé through Mariana and Scaliger to Newton, to name but a few—could amount to an introduction to early modern scholarship. With chorography, genealogy, and etymology, metrology provided local antiquaries and érudits a physical link to the past.[35]

Pastorino also noted that, 'Johann Caspar Eisenschmid, a French mathematician and cartographer from Strasbourg of the second half of the seventeenth-century, summarized well the goals of metrologists'. Eisenschmid stressed the importance of the precise knowledge of ancient measures for 'theologians, lawyers, physicians, and those who study the ancient texts of philosophers, historians, poets, and especially geographers, architects and writers of agriculture'.[36]

We have seen that Folkes's work on Newton's *Chronology* led to his interest in using biblical evidence to clarify the scripturally ambiguous dimensions of the sacred cubit of the Hebrews used to build Solomon's Temple, and to assess the dimensions of the Roman foot.[37] These interests, in turn, gave impetus to his interest in determining ancient measures to assess architecture in a critical fashion. It was a means of empirically assessing the material culture and aesthetics of past civilizations, part of Folkes's (and other antiquaries') methodology.

Therefore it is not surprising that one of Folkes's first tasks when he arrived in Venice was to measure the height of the Campanile by its shadow, concluding it was 140 feet high and then measuring again to confirm his conclusions; to do this, Folkes figured out the conversion factor between Venetian palms and English feet. Persevering despite suffering from a heat rash called *scotture*, Folkes proceeded to measure the Rialto Bridge in Venice to confirm his determination

---

[35] Zur Shalev, 'Measurer of All Things: John Greaves (1602–52), the Great Pyramid, and Early Modern Metrology', *Journal of the History of Ideas* 63, 4 (2002), pp. 555–75, on pp. 567–8.

[36] Johann Caspar Eisenschmid, 'Praefatio ad Lectorem', in *De ponderibus et mensuris…De valore pecuniae veteris* (Strasbourg: Henr. Leo Stein, 1708), quoted in Cesare Pastorino, 'Measuring the Past: Quantification and the Study of Antiquity in the Early Modern Period', unpublished paper.

[37] The specification of the cubit was ambiguous: 'in the man's hand a measuring reed six cubits long, of a cubit and a handbreadth each: so he measured the thickness of the building, one reed; and the height, one reed' (Ezekiel 40:5). The Hebrew text reads: 'a measuring reed, six cubits by the cubit and a handbreadth'. 'It was unclear whether the six cubits marked on the reed were particularly large, each outsizing the regular cubit by a handbreadth; or whether the reed as a whole measured six cubits and one extra handbreadth. This was to become one of the key problems of much Temple scholarship'. See Jetze Touber, 'Applying the Right Measure: Architecture and Philology in Biblical Scholarship in the Dutch Early Enlightenment', *The Historical Journal* 58, 4 (2015), pp. 959–85, on p. 964.

and comparison of past standards of measurement, recording on 8 July 1733 that 'writers not agreeing perfectly with one another about its dimensions, I went this morning to take its dimensions my self, and in order to be sure that I did I made the scetches on the other side to show the lines I really took the measures of'.[38] Folkes used packthread and a plumb bob, and as he was in Lennox's lodgings nearby, he was able to take several measurements.[39] Folkes carefully compared his measures of the Antonio da Ponte's Bridge to those taken by Francesco Sansovino, the son of Jacopo Sansovino (1486–1570), the *proto* or chief architect of Venice who was responsible for the rebuilding of the *Fabbriche Nuove di Rialto* complex (seat of the magistrates in charge of customs and duty).[40] Disturbed that his measurements of the bridge were different from Sansovino's, he wrote, 'I would readily think may be partly owing to defect in my own measure, partly perhaps an inaccuracy in the building the bridg itself to the intended model, and perhaps again in the not having to perfect these proportions between the English and Venetian foot'.[41] After making some enquiries after the original standard of the Venetian measures, which he found to be fruitless, he 'took some measures of St Geminianos Church in order to have made something out by them'. However, upon comparing the measures he had taken, Folkes found the church 'was not built by the Venetian foot but by the Roman Architectonick palm whose length comes out very exactly by it'.[42]

It was a perceptive comment. St Geminiano was in the Piazza San Marco, part of the complete remodelling of the piazza, the piazzetta, and surrounding buildings by Sansovino in Renaissance Venice in the classical style; as historian Fiona Kisby has noted, the 'effect was to superimpose an evocation of ancient Rome on the existing Byzantine elements'.[43] St Geminiano, directly across the piazza from San Marco, was a focal point of the architectural complex designed by Jacopo Sansovino in Vitruvian proportion to evoke the ancient Roman forum. Hence, Folkes suspected and proved that Sansovino used Roman measurements to create his building, particularly as Vitruvius in his third book of *De Architectura* gave a thorough description of the Roman foot or *pes* and palms (one-quarter of the foot)[44] (see figure 5.3). In this manner, principles of measure and geometry would

[38] Folkes, 'Journey from Venice to Rome', f. 20r.
[39] Earl of March, *The Duke and his Friends*, vol. 1, p. 251.
[40] Folkes referred to Francesco Sansovino, *Venetia, città nobilissima, et singolare: descritta in XIIII. Libri* (Venice: Steffano Curti, 1663), p. 365.
[41] Folkes, 'Journey from Venice to Rome', f. 20r.
[42] Folkes, 'Journey from Venice to Rome', f. 26r. Folkes also routinely complained in his journal that finding exact calculations for traditional Venetian measures was difficult.
[43] Fiona Kisby, *Music and Musicians in Renaissance Cities and Towns* (Cambridge: Cambridge University Press, 2005), p. 31; See also Deborah Howard, *Jacopo Sansovino: Architecture and Patronage in Renaissance Venice* (New Haven and London: Yale University Press, 1975).
[44] 'For when two palms are taken from the cubit, there is left a foot of four palms, and the palm has four fingers. So it comes that the foot has sixteen fingers, and the bronze denarius as many asses.' [E cubito cum dempti sunt palmi duo, relinquitur pes quattuor palmorum, palmus autem habet

Fig. 5.3 Canaletto, etching of Piazza San Marco, *c.* 1707–58, 144 mm × 211 mm, object number RP-P-OB-35.577, Rijksmuseum, Amsterdam, public domain.

ensure harmony and aesthetic appeal in the building's design. In other words, Folkes was investigating architectural metrology to provide what we would term archaeological information about the architects and builders of Renaissance Venice and, as we will see, ancient Rome.

Queries about ancient metrology began in 1573 with the publication of *De Mensuris et ponderibus Romanis et Graecis* by Lucas Paetus. That publication spawned an active debate about the exact length of the *pes*. Dissatisfied with inconsistencies in textual evidence, John Greaves, Professor of Astronomy at Oxford, visited Rome in 1639 to 'examine as many ancient measures, and monuments, in Italy, and other parts, as it was possible', and then compare 'these with as many Standards, and Originals, as I procure the sight of'.[45] He thus measured brass measuring rods in Roman ruins and the foot measures on the tomb of Titus Statilius Aper and on the statue of Cossutius. Greaves then noted with some pleasure that these foot measures exactly corresponded with the distance between the milestones on the Appian Way.[46] He then made comparisons between the

---

quatuor digitos. Ita efficitur uti pes habeat XVI digitos, et totidem asses aereos denarius]. Vitruvius, *On Architecture*, trans. Frank Granger, 2 vols (London: William Heinemann, Ltd, 1934), vol. 1, Book III, c. I, pp. 164–5.

[45] John Greaves, *A Discourse of the Romane Foot, and Denarius: from when, as from two principles, the Measures, and Weights, used by the Ancients, may be deduced* (London: M.F, 1647), p. 20.

[46] Greaves, *A Discourse of the Romane Foot*, pp. 20, 23.

Roman foot and the iron standard of the English foot in the London Guildhall and published his findings in his *Discourse on the Romane foot and Denarius* in 1647, concluding that the Cossutian foot was the 'true' Roman *pes*. Greaves also claimed that he thought he could find a relationship between the *pes* and the sacred cubit, something which Newton also noted in his own treatise on the cubit.

Folkes followed the same methodology as Greaves in his own travels. The diary of Sir Andrew Mitchell (1708–71), politician and fellow traveller in Rome, recorded that on 1 June 1733, he, Folkes, and Anders Celsius (who he had met in Florence, see section 5.4) went to see the Marquess Caponi whose palace court-yard had a similar standard stone 'engraved with the tools of a [masonic] work-man' with a bottom rule that was '969 parts of the English foot' (see figure 5.4). Celsius brought a measure of the Swedish foot for their comparison, which by tradition was said to adhere to Roman measures.[47] In 1736, Folkes identified the Root Canonical Greek foot as 1.0057142 times the English measure, and the Roman *pes* of .966 times the English foot, noting them as being engraved on a Standards stone tablet at the Roman Capitol that might have been the actual record of the *pes monetalis*. His paper about the stone was subsequently published in the *Philosophical Transactions of the Royal Society*; Folkes recorded 'setting the point of my Compasses' in the lines in the stone that represented the measures and noted that 'my chief Attention was given to the Roman foot, as of greater Consequence than the other measures'.[48]

In making his conclusion, Folkes considered previous measurements made by Fabretti, but recorded in his diary that he also measured the height of the Trajan Pillar 'using an exact two-foot rule brought from London'. Folkes was particularly interested in the column because the height of its shaft from the plinth of the base to the abacus of the column was said to measure exactly one hundred Roman feet, a *columna centenaria*.[49] The column was thus ideal to determine the dimensions of the Roman foot.[50] Now, of course, studies of metrology and measurement sur-veys of ancient buildings constitute some of the most important work in archae-ology.[51] He found the column, 'from the Ground to the Top of the Cimatium of the Capitol, to be "115 feet 10 inches 5/8"'; this height, divided by 120, gives nearly 966 for the quotient, thus confirming that the Roman foot was 0.966 the English foot and used frequently in Roman architecture. Ironically, Folkes's exac-titude of measurement has not been borne out in modern studies, as there appear

[47] BL Add MS 58318, f. 109r., Diary of Andrew Mitchell, Travels in Italy, Rome, 1733, British Library, London.

[48] Martin Folkes, 'An Account of the Standard Measures Preserved in the Capitol at Rome', *Philosophical Transactions* 39, 442 (1735–6) pp. 262–6, on pp. 262–3.

[49] G. Boni, 'Trajan's Column', *Proceedings of the British Academy* 3, 1–6, (1907), pp. 93–8, on p. 96.

[50] Folkes, 'An Account of the Standard Measures', p. 266; Raphaelis Fabretti, *De Columna Traiana Syntagma* (Rome: Nicolai Angeli Tinassij, 1693).

[51] David Soren and Noelle Soren, *A Roman Villa and a Late Roman Infant Cemetery: Excavation at Poggio Gramignano Lugnano in Teverina* (Rome: L' Erma di Bretschneider, 1999), p. 182.

in measuring the figures on the column to assess and compare the Roman artist's use of visual perspective as opposed to the use of perspective in modern artistic practice.[53] Their conversation comparing ancient and modern techniques was an extension of the literary quarrel of the 'ancients and moderns' that raged among French and English intellectuals in the seventeenth and early eighteenth centuries. Intellectuals who sided with the 'ancients' argued that antique literature offered the sole models for literary excellence; the 'moderns', on the other hand, challenged the supposed supremacy of the classical writers. This scholarly debate was extended to assessing technological and aesthetic achievements.[54] Folkes was particularly interested in discussing with Algarotti to what extent measurement could answer questions about the relative merits of ancient technology and artistic production.

Algarotti had associated himself with 'radical conversazioni and early masonic lodges' at some point during his tours of Padua and the Veneto, then Florence and Rome in the early 1730s, where he met Folkes.[55] Algarotti was best known for his popularization of Newton's precepts in his *Newtonianism for Ladies*, but he also wrote a number of letters and essays on aesthetics, one in which he analysed the helical frieze on Trajan's column depicting the two Dacian Wars. In a letter to Jean Paolucci, Algarotti recalled that he speculated with Folkes whether the bas relief on the Trajan column gave them any reason to 'presume that the ancients had not the slightest knowledge of the proportions of perspective'.[56] Folkes noted in an article he later published about the column that:

> it has been said by some, that the bas-reliefs on the shaft of this pillar increase in size upwards, in order to appear of the same size below; but this is not true; and I had an opportunity of satisfying myself from the plaister-cast of the whole pillar, kept at the French academy of painting and sculpture in Rome, where…
> I measured several of the fairest figures.[57]

It appeared from Folkes's observations of the column that the Roman artists who carved the frieze on the column did not use traditional perspectival techniques in their portrayal of the Dacian Wars.

Algarotti also recalled discussing with Folkes the arguments of Charles Perrault, who rose up through the French Academy to champion modern authors in the

[53] Francesco Algarotti to Jean Paolucci, 20 May 1763, in *Oeuvres de comte Algarotti traduit de l'italien*, 7 vols (Berlin: G.J. Decker, 1772), vol. 6, pp. 208–14.

[54] Excellent analyses of the ancients-versus-moderns debate may be found in Joan DeJean, *Ancients Against Moderns: Culture Wars and the Making of a Fin de Siècle* (Chicago: The University of Chicago Press, 1997); Joseph M. Levine, *The Battle of the Books: History and Literature in the Augustan Age* (Ithaca, NY: Cornell University Press, 1991).

[55] Massimo Mazzotti, 'Newton for Ladies: Gentility, Gender and Radical Culture', *British Journal of the History of Science* 37, 2 (June 2004), pp. 119–46, on p. 143.

[56] Algarotti to Paolucci, 20 May 1763, in *Oeuvres de comte Algarotti*, vol. 6, p. 210.

[57] Martin Folkes, 'On the Trajan and Antonine Pillars at Rome: Read 5 February 1735–6', *Archaeologia* 1, 2 (January 1779), pp. 117–21, on p. 120. Contrary to Folkes, the bas relief figures actually do slightly increase from 60 to 80 cm in height.

quarrel between the ancients and moderns. Folkes and Algarotti noted that Perrault was of the opinion that while the ancients excelled in sculpture (a genre that in Perrault's eyes demanded little 'reflection' and abstract thought), bas reliefs and painting required a deeper understanding of perspective and spatial relations that the ancients did not have. Perrault thought his opinion was supported by the 'casts of the Trajan Column' he saw in Paris, which to him were naive in their understanding of perspective space.[58] However, Folkes and Algarotti were skeptical about Perrault's claim, Algarotti writing 'the defects which are believed to [have] been seen in the ancient bas-reliefs, and especially in the Trajan column, prove nothing against the knowledge of the ancients in regard to perspective'. Rather, Algarotti and Folkes realized the artists who carved the column used a small number of emblematic forms to represent the events of the war rather than hundreds of individual participants; perspectival tricks could not assist the artist in emphasizing 'certain figures, groups, or parts of the composition'. Indeed, the composition is characterized by diagonals, so the 'figure of the emperor is emphasised along with the direction of gazes and actions of the figures surrounding him'.[59] To make the figures and iconography intelligible to a spectator who was on the move and at a great distance, the composition required a multi-perspectival approach, and simple converging and diverging orthogonals were not appropriate. Trajan's Column was in fact designed so that viewers could comprehend the story of the Dacian Wars from bottom to top standing in one place rather than circling the column twenty-three times as the carved spiral frieze does. A viewer could see key events of the war from two main vantage points. Far from being naive, the designer of the column's frieze was more like a great captain employing a stratagem of sophistication.

5.3 Venice and Newtonianism

Folkes's conversation with Algarotti reflected his pervasive interest in optics, optical illusion, and perception, topics which came to the fore when he was in Venice. Although the Venetians were aware of the theories of Newtonian optics, in the 1730s there was a degree of scepticism among natural philosophers there about the validity of the refrangibility of light in the visible spectrum. As a result,

---

[58] Charles Perrault, *Parallèle Des Anciens et Les Modernes En Ce Qui Regarde L'Éloquence*, 4 vols (Paris: J. B. Coignard, 1688–97), vol. 1, p. 190 and pp. 193–4: 'Trajane où il n'y a aucune perspective n'y aucune dégradation. Dans cette colonne les figures sont presque toutes sur la même ligne; s'il y en a quelques-unes sur le derriere, elles sont aussi grandes et aussi marquées que celles qui sont sur le devant; ensorte qu'elles semblent estre montées sur des gradins, pour se faire voir les unes au-dessus des autres'. See also Larry F. Norman, *The Shock of the Ancient: Literature and History in Early Modern France* (Chicago: University of Chicago Press, 2011), p. 235, note 10 and Anne Betty Weinshenker, *Falconet, His Writings and His Friend Diderot* (Geneva: Librairie Droz, 1966), p. 89.

[59] Tina Bawden, et al., 'Early Visual Cultures and Panofsky's *Perspektive als 'symbolische Form'*, *eTopoi: Journal for Ancient Studies* 6 (2016), pp. 525–70, on p. 551.

when he was in the Veneto, Folkes cultivated fellow mathematicians and astronomers to gain adherents of the Newtonian research programme in the Royal Society.

Before he left for his journey, and during his time on the Council of the Royal Society, he was exposed to a constant stream of accounts 'of the present state of Learning and Experimental Philosophy in Italy' from the English Jacobite Thomas Dereham, their contact in Florence. Dereham was the translator into Italian of *Astro-Theology*, a five-volume set of the *Philosophical Transactions of the Royal Society* (published by Felice Mosca between 1729 and 1734), and a work by Newtonian physician George Cheyne.[60] Dereham relied on luminaries such as Eustachio Manfredi, the Chair of Mathematics in Bologna, discoverer of a comet and asteroid, and observer of the transit of Mercury, for natural philosophical news to pass on to the Royal Society.[61]

Venice, in particular, was also a favourite haunt of Newtonian mathematicians and their intellectual circles, and Folkes went there to convince their virtuosi of the truth of Newton's work in optics. Twenty years previously, Newtonians had fought a campaign against Italian supporters of Leibniz in the calculus dispute. In Padua and Venice, after 1710, Jakob Hermann and Nicolas Bernoulli were 'promoters of a fierce campaign against Newton's methods of fluxions in the calculus'.[62] Hermann had taught differential and integral calculus using Leibniz's method at Padua from 1707 to 1713. However, Newton had friends as well, and hoped to diminish the prestige of Leibniz through the offices of Abbé Antonio Schinella Conti (1667–1749), who was born and educated in the Veneto but spent several years abroad before returning to Venice in 1726; Conti was in London between 1715 and 1718 to become part of Newton's intellectual circle, where he met Folkes. In his travel diary of 1733 Folkes noted, 'I had known the Abbé very well in England above 16 years ago, I had a very agreeable conversation with him as he is a Gentleman of great knowledg and politeness, and indeed the most knowing man of all the Italians of his side the country'.[63] Conti would also serve as an intermediary between Leibniz and Newton in the calculus affair, and through Conti's offices Newton thought he could lessen Leibniz's prestige in Italy, which had grown during the calculus dispute.[64] As we have seen, Folkes had also been tutored by Abraham De Moivre, who would subsequently be an intermediary in another mathematical dispute between Newton and Bernoulli, so Folkes was intimately au fait with the history of the Newton Wars.

---

[60] RS/JBO/14, 19 January 1726, p. 37, Royal Society Library, London; Harold Samuel Stone, *Vico's Cultural History: The Production and Transmission of Ideas in Naples* (Leiden: Brill, 1997), pp. 278–80.

[61] RS/JBO/14, 21 March 1727, p. 199, Royal Society Library, London.

[62] Paolo Casini, 'The Reception of Newton's Opticks in Italy', in J. V. Field and Frank A. J. L. James, eds, *Renaissance and Revolution: Humanists, Scholars, Craftsmen and Natural Philosophers in Early Modern Europe* (Cambridge: Cambridge University Press, 1997), pp. 215–29, on p. 217.

[63] Folkes, 'Journey from Venice to Rome', f. 37r.   [64] Casini, 'Reception', p. 217.

Before his journey, Folkes had attended a series of Royal Society meetings in which the Society responded to the refutation of Newton's optical work by Giovanni Rizzetti (1675–1751), a Paduan nobleman who belonged to the circle of the mathematician Jacopo Riccati (1676–1754) of Castelfranco Veneto. The Royal Society Journal Book for 27 June 1728 reported that the society received Rizzetti's *De Luminis Affectionibus Specimen Physico Mathematicum*; the tract was referred 'to Dr Desaguliers which he promised to make an extract of'.[65] In *De Luminis*, Rizzetti discussed refrangibility, the degree to which light refracts when passing from one medium into another, or, as Newton wrote, a 'predisposition, which every particular Ray hath to suffer a particular degree of Refraction'.[66] In particular, Rizzetti claimed he could not replicate the experiments in Book 1 of the *Opticks*, in which Newton used a candle to illuminate a two-coloured card with black silk threads wrapped around it, demonstrating the principle of chromatic aberration.[67]

Newton had placed a glass lens at a distance of six feet from the card, and used it to project the light coming from the illuminated card onto a piece of white paper which was at the same distance from the lens on the other side. He moved the piece of white paper back and forth, taking note where and when the red and blue parts of the image were most distinct. The black threads indicated distinctness of the image (when the lines created by the thread were sharpest). Newton found it 'impossible to focus the image of black silk-thread lines upon a blue background at the same distance as when the same lines are placed against a red background; the distance between the two positions of sharp focus was as much as one and half inches'.[68] In other words, to get a distinct red image, the paper had to be held 1.5 inches further away than to obtain a distinct blue image. Newton thus concluded that the blue light was refracted more by the lens than the red, and was more refrangible.[69]

Red and blue light were not strictly homogeneous, and so not all the blue light was more refrangible than all the red light, something Newton admitted. However, these experiments demonstrate a general effect: 'But these Rays, in proportion to the whole Light, are but few, and serve to diminish the Event of the

[65] RS/JBO/14, 21 March 1727, 10 April 1728, 27 June 1728, and 31 October 1728, p. 191, p. 235, pp. 249–50, Royal Society Library, London.
[66] Rupert Hall et al., *The Correspondence of Isaac Newton*, 7 vols (Cambridge: Cambridge University Press, 1959–81), vol. 1, p. 96.
[67] Aberration occurs as lenses have a different refractive index for each different wavelength of light.
[68] A. Rupert Hall, *All Was Light: An Introduction to Newton's Opticks* (Oxford: Clarendon Press, 1993), p. 96.
[69] Hall, *All Was Light*, pp. 96–7. See also Kirsten Walsh, 'Newton's Epistemic triad', PhD Dissertation (University of Otago, 2014), p. 104.

Experiment, but are not able to destroy it'.[70] Newton thus described an ideal experiment.

The inability to create these ideal conditions was the point upon which Rizzetti had seized. The popular lecturer John Theophilus Desaguliers (1683–1744) refuted Rizzetti's claims in a series of optical experiments performed at the Royal Society in 1728 in response to Rizzetti's *De Luminis Affectionibus*. In his article published in *Philosophical Transactions*, Desaguliers noted 'we hear indeed in a letter from Sir Thomas Dereham to Sir Hans Sloane, President of the Royal Society, that now Signior Rizzetti alledges, that he was deceived in his experiments by reason of the badness of his prisms which he had from Venice'.[71] In other words, Rizzetti claimed he did not have ideal experimental conditions due to the quality of his prisms.

As Schaffer has shown in his essay 'Glass Works', the quality of the prisms was important; Francesco Algarotti's attempts in Bologna in 1726 to replicate the *experimentum crucis* were described by Rizzetti as failures, as he could not isolate a single red ray of light; Algarotti eventually realized that his prisms were defective, and he had success at replicating the *experimentum crucis* when he tried again with some top-quality English prisms.[72] Schaffer shows us how and why acceptance or denial of the specific experimental apparatus became a flashpoint in disputes over knowledge and method; we see the beginnings of the establishment of standards for professional instrumentation, or what was termed the 'sociology of calibration'.

Schaffer did not analyse in detail the debate over Newton's *Opticks* in Italy, which continued long after Desaguliers's 1728 demonstrations in the Royal Society.[73] Long after Isaac Newton's fame was recognized, Venetian natural philosophers had an attitude of 'critical acceptance, if not open aversion' towards Newtonianism.[74] Part of the reason for Folkes's travels in Venice was so he could serve in Desaguliers's stead as the Newtonian demonstrator, and the quality of Folkes's prisms were still the focus of inquiry for Venetian intellectuals. It seemed particularly important that Folkes first prove the truth of Newtonian optics in the

---

[70] Isaac Newton, *Opticks* (New York: Dover, 1952), p. 26; Hall, *All Was Light*, p. 97.

[71] J. T. Desaguliers, 'Optical Experiments made in the beginning of August 1728, before the President and Several Members of the Royal Society, and Other Gentlemen of Several Nations, upon Occasion of Signior Rizzetti's Opticks…', *Philosophical Transactions* 35, 406 (1727–8), pp. 596–629, on pp. 598–9.

[72] Simon Schaffer, 'Glass Works: Newton's Prisms and the Uses of Experiment', in David Gooding, Trevor Pinch, and Simon Schaffer, eds, *The Uses of Experiment: Studies in the Natural Sciences* (Cambridge: Cambridge University Press, 1989), pp. 67–104, *passim*; Casini, 'Reception', p. 222.

[73] To be fair, Schaffer does acknowledge the controversy over the nature of Newton's genius and his posthumous legacy in his 'Fontanelle's Newton and the Uses of Genius', *L'Esprit Createur* 55, 2 (Summer 2015), pp. 48–61.

[74] Vincenzo Ferrone, *Intellectual Roots of the Italian Enlightenment* (Atlantic Highlands, New Jersey: Humanities Press, 1995), p. 95. See also Massimo Mazzotti, 'Newton in Italy', in H. Pulte and S. Mandlebrote, eds., *The Reception of Newton in Europe*, 3 vols (London: Bloomsbury, 2019), vol. 1, pp. 159–78, esp. pp. 164–5.

city whose glass had been the 'standard against which other glass was to be compared', and second, demonstrate the inherent virtue of the English prisms and instrumentation.[75] Folkes's status as a protégé of Newton (his vice-presidential office conferred by Sloane six weeks before his travels) and his fluency in French gained during his tutelage with De Moivre and cognate knowledge of Italian were also helpful assets.

On 9 July 1733, Folkes received a letter from Abbè Conti which began, 'I beg you to bring these Italian manners to an end and to treat me as a friend, that is the first favour I am asking you'.[76] Conti related that his colleague, Giovan Bernardo Pisenti (d. 1742), having heard of the success of past demonstrations of Newtonian optics, requested 'to see demonstrated geometrically the order of the colours of the image'.[77] Pisenti was a Somasco cleric educated by the Jesuits who also studied with Eustachio Manfredi in Bologna and served as a personal tutor to Girolamo Ascanio Giustiniani; Folkes's diary shows that he spent much of his time in Venice at the Giustiniani Palace with Girolamo's family, so it is likely he met Pisenti there.[78] Pisenti was also the translator of George Berkeley's *An Essay towards a New Theory of Vision* (1709) into Italian, and dedicated his translation to Berkeley's friend Sir John Percival, with whom he was also acquainted.[79] The initiative to publish Berkeley's work was probably born in the wake of the Newtonian optics controversy aroused in the Venetian world from the attacks of Rizzetti.[80]

Although many scholars have seen *The New Vision* as a precursor to Berkeley's immaterialist principles formulated in his *Principles of Human Knowledge* (1710), this would be an incomplete interpretation. As Atherton has shown, Berkeley in *The New Vision* was primarily querying the validity of the geometrically based account of vision that he saw as being held by writers on optics such as Malebranche and Descartes.[81] In Berkeley's examination of visual distance, position, magnitude, and associated problems of touch and sight, he refuted an account of distance vision which requires tacit geometrical calculations, positing

[75] Schaffer, 'Glass Works', pp. 72, 99.

[76] RS/MS/790/28, Letter from Antonio Conti to Martin Folkes, 9 July 1733, Correspondence of Martin Folkes, Royal Society Library, London.

[77] T. E. Jessop, *A Bibliography of George Berkeley*, 2nd ed., International Archives of the History of Ideas, vol. 66 (New York: Springer, 1973), p. 9. Pisenti's book was the first translation of Berkeley's work on vision into any language: George Berkeley, *Saggio d'una nuova teoria sopra la visione…ed un discorso preliminare al Trattato della cognizione*, trans. Giovanni Pisenti (Venice: Francesco Storti, 1732). Pisenti also translated Berkeley's work on cognition.

[78] Don Gasparo Leonarducci, *La Provvidenza. Cantica seconda. I primi quattro canti inedita del P. don Gasparo Leonarducci della Congregazione di Somasca* (Venice: dalla Tipografia di Alvisopoli, 1827), p. 5. This is a history of the Somasco congregation.

[79] Giovanni Antonio Moschini, *Della letteratura veneziana del secolo XVIII fino a' nostri giorni*, 4 vols (Venice: Dalla Stamperia Palese, 1806), vol. 1, p. 169.

[80] Casini, 'Reception', p. 217; Antonella Barzazi, *Gli Affanni dell'erudizione: studi e organizzazione culturale degli ordini religiosi a Venezia tra Sei e Settecento* (Venice: Istituto Veneto di Scienze, Lettere ed Arti, 2004), p. 181.

[81] Margaret Atherton, *Berkeley's Revolution in Vision* (Ithaca: Cornell University Press, 1990), pp. 4–16.

that we perceive space by lines and angles. Instead, Berkeley offered a new theory that saw perception as the empirical integration of two senses: sight and touch.

Malebranche and Descartes argued that distance was judged by the geometry of angles between the eyes and the perceived object, or via the angles of light rays that fell upon the eye.[82] One thus judges distances by the optic axes of the eyes which form an angle at the object; we judge whether the object is far or near based on the size of this angle. Berkeley rejected those accounts and was quite negative about using Euclidean geometry of the visual world as a basis for visual perception. Berkeley argued against the classical scholars of optics by claiming that space is perceived by experience. As Berkeley later stated in his *Philosophical Commentaries*, 'The common Errour of the Opticians, that we judge of Distance by Angles strengthens men in their prejudice that they see things without and distant from their mind'.[83] In Berkeley's view, the visual perception of distance is explained by the correlation of ideas of sight and touch, and this associative approach challenged appeals to geometric calculation that profess to explain completely monocular vision and the Moon illusion: 'anomalies that had plagued the geometric account'.[84]

Visual perception of distance was also of great interest to Folkes. We recall his conversations with Algarotti about the friezes on Trajan's Column. Folkes also gave advice to Robert Smith FRS, Plumian Professor of Astronomy at the University of Cambridge, in his *A Compleat System of Opticks in Four Books* (1738). Cantor has judged that the work was the most 'famous book dealing with the subject' after Newton's own *Opticks* (1704).[85] Smith acknowledged Folkes for his 'curious remarks' on fallacies in vision, applying optics to explain phenomena such as the Moon illusion, 'the Sun's apparent distance, on the apparent figure of the sky, on the apparent curvity of the sides of long walks and ploughed lands, and the changes of curvity by the observer's motion'.[86]

[82]   Daniel E. Flage, 'George Berkeley (1685–1753)', *Internet Encyclopedia of Philosophy*. http://www.iep.utm.edu/berkeley [Accessed 3 December 2016]. This is a peer-reviewed and scholarly encyclopedia.

[83]   Casini, 'Reception', p. 217; Antonella Barzazi, *Gli Affanni dell'erudizione: studi e organizzazione culturale degli ordini religiosi a Venezia tra Sei e Settecento* (Venice: Istituto Veneto di Scienze, Lettere ed Arti, 2004), p. 181. Margaret Atherton, *Berkeley's Revolution*, p. 3, and chapters two and three, *passim*. Atherton argues that Berkeley primarily had in mind the work of Descartes and Malebranche in his refutation of the geometric theory of vision; George Berkeley, 'Philosophical commentaries', in *The Works of George Berkeley, Bishop of Cloyne*, ed. T. E. Jessop and A. A. Luce, 9 vols (London: Nelson, 1948–64), vol. 1, section 603.

[84]   Flage, 'George Berkeley (1685–1753)'. The Moon illusion is that the Moon appears much larger when just above the horizon than when it is overhead.

[85]   Geoffrey Cantor, *Optics after Newton: Theories of Light in Britain and Ireland, 1704–1840* (Manchester: Manchester University Press, 1983), p. 19.

[86]   Robert Smith, 'Preface' in *A Compleat System of Optics in Four Books* . . . (Cambridge: Printed For the Author, 1738), pp. i–vi, on p. vi.

Folkes had also spoken to Algarotti about Molyneux's problem in some depth.[87] The Irish natural philosopher William Molyneux (1656–98) had asked John Locke whether, if a blind person could suddenly see, they would be able to recognize by sight an object's shape previously only known by touch. Presented with a globe and a cube, could such a person determine which was which just by looking? An affirmative answer meant that one believed in a rationalist and innate conception of space that is common to touch and sight. A negative answer indicated an empiricist view, that this is a relationship that we learn only through experience. Algarotti noted that Locke responded in his *Essay Concerning Human Understanding* that the formerly blind person would not be able to say with certainty 'which was the globe, which the cube...though he could unerringly name them by his touch'; the connection between the senses was learned.[88] Berkeley also offered a negative response: that a perception of distance was an 'act of judgment grounded on experience'.[89]

However, well before Diderot's famous essay *Letter on the Blind* (1749), Folkes spoke to Algarotti about Molyneux's problem in reference to the case of Nicholas Saunderson (1682–1739), the blind Lucasian Professor at Cambridge and fervent Newtonian who lost his sight at twelve months from smallpox and who could purportedly judge the size of a room and his distance from the wall by sound (see figure 5.5). Saunderson made Newton's *Opticks* the basis of several lectures that he gave at Cambridge, including some on the theory of vision.[90] Folkes was a close colleague of Saunderson and commissioned a painting of him. Saunderson affirmed he could have distinguished the objects with his sight restored, namely because he innately understood the mathematical definition of a sphere and cube and could identify them from the number of vertices they would present to him, as well as the shape of their cast shadows. This was an argument drawn out of Saunderson's *Treatise of Algebra*, in which he proved that the cube could be divided into six equal pyramids having their vertices at the centre of the cube, each with a volume of one-third of a prism having the same base and height.[91] Folkes, as a Newtonian mathematician, believed there was an innate geometrical

---

[87] Mazzotti, 'Newton for Ladies', p. 143. Francesco Algarotti, 'Al Signor Abate Ortes a Venezia, 18 Ottobre 1750', in *Opere Varie Del Conte Francesco Algarotti*, (Venice: Giambiatista Pasquali, 1757), vol. 1, p. 316.

[88] 'Al Signor Abate Ortes a Venezia', 18 Ottobre 1750', in *Opere Varie Del Conte Francesco Algarotti*, vol. 1, p. 316.

[89] George Berkeley, 'Essay Towards a New Theory of Vision', in the *Works of George Berkeley*, section 41, p. 186; John W. Davis, 'The Molyneux Problem', *Journal of the History of Ideas* 21, 3 (July–September 1960), pp. 392–408, on p. 396.

[90] 'Dialoghi Sopra l'Ottica Neutoniana', in *Opere Varie Del Conte Francesco Algarotti*, vol. 1, p. 316. Nicholas Saunderson, *The Elements of Algebra, in Ten Books* (Cambridge: Cambridge University Press, 1741), p. vi.

[91] Denis Diderot, 'Letter on the Blind', in M. J. Morgan, *Molyneux's Question: vision, touch, and the philosophy of perception* (Cambridge: Cambridge University Press, 1977), p. 42.

Fig. 5.5 Gerard Van der Bucht, *Nicholas Saunderson*, engraving after John Vanderbanks's 1719 portrait painted for Martin Folkes, 1740, 221 mm × 144 mm, Wellcome Collection, Attribution 4.0 International (CC BY 4.0).

quality to vision and was a very supportive subscriber to Saunderson's *Treatise of Algebra*, putting money in for eight books, '4 Royal and 4 common paper'.[92]

In order to create a demonstration to directly counter Rizzetti's disavowal of Newtonian optics and tacitly counter Berkeley's rejection of optics as a means of explaining vision, Conti indicated to Folkes that he made a figure 'to show visibly the different relative refrangibility', showing 'the different perpendiculars with which are made the angles of the refraction of the exiting rays; these angles are as small as the others that are made with the internal perpendicular are large'.[93] He then indicated to Folkes: 'You would do me a favour if you told me whether you imagine that in the same way. The obscurity of Mr Newton's book consists in failing to give the Theory of the refractions of the rays exiting a prism, that is what he had to start with in order to remove any doubt'.[94]

Conti was right. Newton had conjectured only sparsely in the *Opticks* about these issues; for instance, while analogizing about the behaviour of light and matter, he really did not even assert the materiality of light. As Cantor has shown, later popularizers of Newton such as Desaguliers, Smith, and later Willem 's Gravesande (1688–1742) interpreted Newton's *Opticks* as if it were 'an appendix to the *Principia*', in which forces were employed to explain all those phenomena in which light deviated from its natural rectilinear path'.[95] In this form of projectile optics, light was postulated to travel in a straight line with finite velocity, and

[92] Saunderson, 'Index of Subscribers', *The Elements of Algebra*.
[93] RS/MS/790/28, Royal Society Library, London.
[94] RS/MS/790/28, Royal Society Library, London.   [95] Cantor, *Optics after Newton*, p. 26.

refraction was explained by the influence of attractive forces. One could say, for instance, that refractive dispersion was due to a particulate model of light, the rays of bodies being of different sizes, the violet-making rays the smallest and the red-making ones the largest and heaviest.[96]

Conti then went on to mention the work of Paris lawyer and Anglophile Nicolas Gauger, who in 1727 produced a long account defending Newtonian optics. In July 1727, Gauger had written Conti a twenty-page letter which was publicly printed, addressing the danger of Rizzetti's attacks on the Newtonian system, with particular mention of the two-colour card experiment.[97] But five years later, Conti lamented to Folkes 'although Mr Gauger has given it quite amply…I don't know if his book is still being printed'. He went on, 'It is certain that if the ray was not divided into several threads, so they made different angles, you would see the image differently from how we see it. I am waiting for your decisions on this'.[98] But such clear, if sometimes superficial, explanations by Newtonian pedagogues had not reached or convinced all Venetian ears; hence Conti's plea to Folkes for a proper diagram and clarification of refrangibility.

Shortly after this exchange, on 5 August 1733, Folkes met Rizzetti, recording that Father Giovanni Crivelli 'calld on me with Sigr. Rizzetti author of a book which he gave me of the nature of light and colours against Sir Isaac Newton'.[99] Crivelli had written *Elementi di fisica* in 1731, a textbook which was a reinterpretation of recent discoveries in natural philosophy within an Aristotelian framework. Historian Vincenzo Ferrone considered Crivelli's work to be a 'meaningful record of the attitude of a large segment of society in the Veneto with regard to Newton's works'.[100] The first volume offered the most complete, if dispassionate, explanation of Newtonian gravity that appeared in Italy (for example, he rejects Newton's concept of a vacuum in space). Crivelli presented Cartesian and Newtonian optics alongside the objections of Rizzetti that subtly neutralized them, and did the same thing in his coverage of astronomy, describing Copernican, Tychonic, and Ptolemaic hypotheses.[101] Although one receives the impression from reading his manual that there was wide-ranging knowledge of Newtonian theory in the Veneto, one also 'clearly gets an image of local scholars rejecting what we might define any sort of Newtonian orthodoxy'.[102] In this light, Crivelli's visit with the antagonistic Rizzetti in tow may be interpreted as a challenge to the Newtonian Folkes.

---

[96] Cantor, *Optics after Newton*, p. 31.
[97] Nicolas Gauger, 'Lettre à M. l'abbé Conti, juillet 1727', *Continuation des Mémoires de Littérature et d'Histoire* 5, 1 (March 1728), pp. 10–51.
[98] RS/MS/90/28, Royal Society Library, London.
[99] Folkes, 'Journey from Venice to Rome', f. 30r.   [100] Ferrone, *Intellectual Roots*, p. 98.
[101] Ferrone, *Intellectual Roots*, p. 99.   [102] Ferrone, *Intellectual Roots*, p. 99.

After Rizzetti's visit and a short period of illness (Folkes and his wife contracted a kind of heat rash termed *scotture*), Folkes met Conti on 23 July. The latter had just come to town and, upon hearing about Rizzetti's visit to Folkes, proposed he and Folkes act and publicly 'make some of Sir I[saac] Newton's experiments on fryday next'. Folkes subsequently indicated in his travel diary on Friday, 27 August 1733: 'I was of the Palace Justinian in the morning, to have made some of Sir Isaac Newtons optical experiments but the Sun not stirring I could no more than get things in order for another day. The apparatus belonged to Abbe Conti who was home and he had brought some very good prisms from England'.[103] Clearly on the hunt for apparatus, Folkes had three days earlier indicated 'I was also at another person who is a practical Mathematician to see some instruments but nothing was particular what he had best were some English things'.[104]

On 28 August 1733, Folkes recorded triumphantly that 'I was again at the Giustiniani Palace and made most of the experiments of the first book of Opticks with very good success and to the satisfaction of all that were here which were…men of the first people of Venice'.[105] Folkes was rewarded by Giustiniani with a present of 'spanish Chocolate'.[106] On 2 and 7 September, Folkes then indicated:

> I was again to prepare some of the Experiments of the Opticks at Sr. Giustinianis, our Experiments there have made a good deal of noise and many of the chief Nobility are very desirous to see some if them which I have promised to repeat on ~~Saturd~~ tuesday next though it is but in a gross manner I can make them want of a convenient apparatus. They are very satisfactory and carry demonstration to all that are capable of comprehending them…I made the Experiment I had intended at the Justinian Palace. I have now shewn all that are considerable in the Book of Opticks in the first book, and though my apparatus was none of the best I was sufficiently satisfyd with them and am told it is what was never done with any success here before, there were several of the Chief nobility and some others here present. and every one that has any sense seems thoroughly satisfyd as far as he understands. tho I was a little surprised at some mistakes Crivelli fell into and was pretty obstinate…a great while which showed me he is not so perfect master of the philosophy as I thought before.[107]

Considering Crivelli's neutral stance in presenting Newton's optics in his *Elementi*, his obstinate behaviour during the demonstration was not surprising. Rizzetti also continued his anti-Newtonian campaign until the 1740s, with diminishing success.

---

[103] Folkes, 'Journey from Venice to Rome', f. 39r.
[104] Folkes, 'Journey from Venice to Rome', f. 39r.
[105] Folkes, 'Journey from Venice to Rome', f. 39r.
[106] Folkes, 'Journey from Venice to Rome', f. 39r.
[107] Folkes, 'Journey from Venice to Rome', ff. 39r.–40r.

Nonetheless, Folkes's performance had the required effect. When he left Venice, Folkes noted that his experimental success showed that he had 'conversd with many of [the Venetians] in a manner no English man almost has done before me'.[108] In a letter from the physician Pietro Michelotti to John Machin, secretary of the Royal Society, Michelotti indeed indicated that Folkes had stayed several days with him in Venice, indicating close familiarity.[109] Several of the Newtonian adherents that Folkes indicated in his diary that he had met in Venice, including Ludovico Riva (1698–1746), Professor of Astronomy at Padua, were subsequently put forward for election in the Royal Society and became Fellows.[110] Although Folkes worried about his apparatus, his use of Conti's 'very good' English prisms and his expertise in his demonstrations vindicated, or so he was told, Newtonian optics in the Veneto.

Folkes continued his Newtonian campaign when he travelled to Florence, but he still worried about the quality of his equipment. In a letter of c. 1734 to Sir Philip Stanhope (who was on his own Grand Tour at the time) Folkes wrote:

> The morning promising very finely Dr Cocchi has just sent to me to desire if I could to try to make some to Sr Isaacs Prismatic Experiments at his house; at which both he and my self should esteem your Lordships company a great honour...I can make them but in a bungling sort of a manner by reason the tackle is very indifferent.[111]

Folkes's son, also named Martin, was sent as a companion to Earl Stanhope to attend an experimental performance at the house of Antonio Cocchi. Cocchi was a well-known physician and defender of Newton whom Folkes met in 1723.[112] It

---

[108]   Folkes, 'Journey from Venice to Rome', f. 44r.

[109]   RS/EL/M3/32, Pietro Antonio Michelotti to John Machin, 6 October 1733, Royal Society Library, London. Michelotti noted, 'I have often understood from that most noble and intelligent gentleman Mr. Martin Folkes, a man furnished with every kind of virtue, who is still staying with me, that you have the most clear-sighted opinions in respect of the disciplines of mathematics and physics (in which I too take enormous delight)'. Folkes appended a letter of introduction for Michelotti to Machin to this piece of correspondence.

[110]   Folkes nominated Riva by letter on 1 October 1733 to be a Fellow of the Royal Society, as a 'Person of great Modesty and knowledge in his way and Author of Several Works, which he hath given me to be presented to the Society, one of which is a dissertation on certain fiery Meteors, that have lately appeared and done a pretty deal of Mischief. He is likewise strongly recommended by the Marquis Poleni and Dr Michaelotty'. Riva was elected on 24 January 1733/4. See EC/1733/07, The Royal Society, London. The dissertation that was mentioned was 'an account of some surprizing Meteors appearing from time to time in the Province or Trevigiana (in the dominions of Venice) described and explained by Signor Ludovico Riva in his Miscellanies in Latin'. Read to the Royal Society on 5 December 1734; see Register Book Original, RBO/19/5, The Royal Society, London.

[111]   U1590/C21/6a, Stanhope Papers, Kent Record Office, Canterbury.

[112]   For a comprehensive biography of Cocchi, see Luigi Guerrini, *Antonio Cocchi naturalista e filosofo* (Florence: Polistampa, 2002); Cocchi and Folkes were also both Freemasons, and Cocchi later became Master of the Florentine Lodge. See Nicholas Hans, 'The Masonic Lodge in Florence in the Eighteenth Century', *Ars Quatuor Coronatorum* 61 (1958), pp. 109–12. Folkes's son also attended several meetings at the Royal Society when he returned from England, as a means of furthering his education.

does not seem that Folkes's comments about his prisms were the result of a false modesty, and he was clearly worried because the presentation was important; Folkes had invited 'Mr Rhodes', secretary to Charles Fane, the British minister in Tuscany and 'Mr Mann' or Sir Horace Mann, 1st Baronet (1706–86), acting British Minister in Florence from 1738 and Minister from 1740–86.[113] (Folkes and Mann had initially met at Clare Hall, Cambridge as students).[114]

This presentation of Newtonian optics was much less public than the first, probably because outside the safety of the Venetian Republic the Inquisition was active and dangerous; it placed Locke's *Essay on Human Understanding* on its Index of Forbidden Books in 1734. Algarotti's *Il Newtonianismo per le Dame* had also been condemned as Masonic, and the Inquisition denounced Newtonian Antonio Conti whom Folkes had met in Venice. Open acceptance of Newtonian gravitation was impeded because the Catholic Church still banned belief in the Earth's motion, and refutation of Newtonian colour theory in itself was an implicit opposition to a moving Earth.[115] Folkes's demonstration probably impressed his audience, evidenced by his election to the Florentine Academy (Accademia del Disegno) on 9 January 1735.[116] Upon Folkes's return to England, when he informed Cocchi of his intention of visiting Italy again, Cocchi noted he was 'overjoy'd' and reminded him to bring a telescope and microscope from England. Conti added, 'I should very proud if you would habilitate my Terrace with Some of your Astronomical Observations…and other physical experiments which would be very much admired here'.[117] In Venice and Florence, Folkes thus successfully demonstrated Newtonian optics, cultivated fellow natural philosophers and architectural critics, and fashioned himself as an ambassador of Newtonianism.

5.4 Venice and Celsius

During his journey, Folkes established the primacy of Newtonian optics as well as awareness of standards of professional equipment; he also had another encounter that would further contribute to the primacy of Newtonianism and English instruments. On 18 September 1734, when he was in Venice, Folkes recorded in

---

[113] Frank Salmon, 'British Architects and the Florentine Academy, 1753–94', *Mitteilungen des Kunsthistorischen Instituts in Florenz* 34, 1/2 (1990), pp. 199–214, on p. 200.

[114] In a letter to Walpole, Mann recalled speaking with Folkes, and the editor speculated that Mann and Folkes may have met in Italy. Indeed they did. Horace Mann to Horace Walpole, 19 April 1777, in *Horace Walpole's Correspondence* (New Haven: Yale University Press, online edition), vol. 24, pp. 289–90, note 3. http://images.library.yale.edu/hwcorrespondence/default.asp [Accessed 2 November 2016].

[115] Casini, 'Reception', p. 220. See also Paolo Casini, *Hypotheses Non Fingo: Tra Newton E Kant* (Rome: Edizioni di storia e letteratura, 2006), p. 108.

[116] Salmon, 'Florentine Academy', p. 200. See Accademia del Disegno, Florence Archivio di Stato, MS 18, Vacchetta…1723–37, *ad diem*.

[117] RS/MS/790/26, Cocchi to Folkes, 3 March 1736, Royal Society Library, London.

his diary a fortuitous meeting with '2 Swedish Gentlemen. Mr Schelsius Prof. of Astronomy at Upsal, and a young Gentleman who travels at the Kings expense to study Architecture. they are very well bred understanding men and make a handsome figure. Mr Schelsius presented me with a book he has published on the Aurora borealis'.[118]

'Schlesius' was of course Anders Celsius; his work on the *aurora borealis*, the *Nova Methodus distantiam solis a terra determinandi* (New method for determining the distance from the Earth to the Sun), was first read to the Royal Society on 8 April 1736 and printed in *Philosophical Transactions*. Celsius was also one of the first to suggest a connection between aurorae and changes in the Earth's magnetic fields, and he would also make observations of sightings in England.[119] As we recall (see chapter two), Folkes would have been very interested to talk to Celsius, as he made some of the earliest sightings of aurorae from King's Lynn, not too far from his country seat at Hillington Hall.[120]

Folkes then noted the next day,

> I had invited the 2 Swedish Gentlemen, and I passd the day with them, they gave me several curious accounts both of their own country and their observations in Germany, they also gave me the measure and weights of their own country in which they have been both been very curious, and I am so well satisfied of their acquaintance that tis a satisfaction they are going the same way and will be of service to me observing the same things as my self.

Not only did Celsius address Folkes's metrological interests but, more importantly, Folkes became involved in the scientific work of Celsius, assisting him with his geodesic expedition with Maupertuis to measure the shape of the Earth. The argument over the Earth's shape involved predictive calculations by Newton, and thus its outcome was crucial to the primacy of Newtonianism. When Celsius came to London to procure scientific instrumentation for the expedition, he was introduced by Folkes to key members of the Royal Society and to Fellows of the Society of Antiquaries, as Folkes promoted close ties between the two organizations which had common interests in metrology.

When he met Folkes, Celsius had just been to Paris on a grand tour of European observatories and had met Algarotti in Italy.[121] And when they arrived in Paris

---

[118] Folkes, 'Journey from Venice to Rome', f. 42r.

[119] The record of the reading of Celsius's book is in the Register Book Original, RBO/19/57, Royal Society Library, London; see also Anders Celsius, 'Observations of the Aurora Borealis Made in England by Andr. Celsius, F. R. S. and Secr. R. S. of Upsal in Sweden', *Philosophical Transactions* 39, 441 (1735), pp. 241–4.

[120] Mike Lockwood and Luke Barnard, 'An Arch in the UK: Aurora Catalogue', *A & G: News and Reviews in Astronomy and Geophysics* 56 (August 2015), pp. 4.25–4.30, on p. 4.26.

[121] See also N. V. E. Nordenmark, *Anders Celsius: Professor iUppsala 1701–1744* (Luleå: Almqvist & Wiksell, 1936).

together at the end of the summer of 1734, Algarotti introduced Celsius to Maupertuis and his colleagues, all Newtonians interested in geodesy. Here the idea for the Lapland expedition to measure the true shape of the Earth was born, the expedition given a royal warrant five months later to find and provision a ship to take the travellers to Stockholm.[122] The expedition would help settle a dispute that Maupertuis had with the French cartographer Jacques Cassini over whether the Earth was prolate (longer along its axis from north to south), or oblate (broader along its diameter at the equator). Cassini believed that the Earth was prolate. Maupertuis, from his expedition, and Newton, through predictive calculation, proved it was oblate. The expedition to Lapland, lasting from July 1736 to March 1737, measured an arc of the meridian from Torneå to Kittis. The scientific party consisted of Maupertuis, Clairaut, Camus, Le Monnier, the Abbé Outhier, and of course Celsius, who would measure the meridian arc near the North Pole. Another expedition organized by the French to South America (1735–43) near the equator to measure the meridian arc would eventually demonstrate that Newton was correct about the Earth's shape.

The travellers would use a fixed-length pendulum clock, a theodolite, and a zenith sector that could measure the shape of the Earth in tropical, temperate, and Arctic latitudes. The first method depended on the comparison of the periods of pendulum clocks at different latitudes, eliminating a temperature-dependent variable (clocks went a second a day slower for every two divisions of temperature-rise on the thermometer).[123] Newton noted in his *Principia* that astronomers found pendulum clocks move more slowly near the equator, thus supporting his prediction that their 'increase in weight from the equator to the poles is…as the square of the…sine of the latitude'.[124] Other than comparing the period of pendulum clocks, one could also use a theodolite and surveyors' rods to measure out a large distance (sixty to seventy miles) and, using a zenith sector, measure the degree of arc that distance covered: a more direct means of measuring the Earth's surface.

As Celsius was on his way to London, Maupertuis asked him to obtain several instruments by George Graham.[125] Maupertuis desired a model of Graham's portable zenith sector for observing the transits of fixed stars that allowed for the calculations of latitude; it was by using this sector to measure Gamma Draconis

---

[122] Mary Terrall, *The Man who Flattened the Earth: Maupertuis and the Sciences in the Enlightenment* (Chicago: University of Chicago Press, 2002), p. 102. See also Rob Iliffe, 'Aplatisseur du Monde et de Cassini'; Maupertuis, Precision Measurement, and the Shape of the Earth in the 1730s', *History of Science* xxxi (1993), pp. 335–75.

[123] This explanation is adapted from Sorrenson, 'George Graham: Visible Technician', pp. 210–11.

[124] Sorrenson, 'George Graham: Visible Technician', p. 211.

[125] Terrall, *Flattened Earth*, p. 102. In his letter to Celsius, Maupertuis indicated he wished for a model of James Bradley's instrument for observing the transits of fixed stars, as well as Graham's astronomical pendulum clock. See RS/EL/M3/24, Extract of a letter from Pierre Maupertuis, dated at Paris, to Andreas Celsius, 22 November 1735, Royal Society Library, London.

that James Bradley had discovered the aberration of starlight in 1728. Gamma Draconis has a high northerly position of 51.5 degrees north of the celestial equator, which takes it through the zenith as seen from London. The atmosphere of the Earth causes refraction of light which makes stars appear higher than they really are, but using a star that passes directly overhead like Gamma Draconis solves such problems. It was thus used to measure stellar parallax. However, in his work with Samuel Molyneux, Bradley detected a small apparent motion of Draconis not due to parallax (the annual timetable for Draconis's parallax was different), but instead resulting from the finite speed of light and the orbital motion of the Earth.[126]

Maupertuis and Folkes had become friends in 1728 during the Frenchman's own Grand Tour. Folkes was also a chief patron of Graham, having nominated him to the Fellowship of the Royal Society on 8 December 1720.[127] Folkes also appended to his travel journal a list of prices for Graham's watches, often securing them as gifts for his scientific correspondents.[128] As we recall (see chapter four), Folkes and Graham were close friends, recorded in John Byrom's diary as often in various taverns such as the Sun after Royal Society meetings.[129] When Celsius was in London in 1735, he observed a lunar eclipse from Graham's house using some of Bradley's new telescopes.

Celsius was also invited to attend meetings at both the Royal Society, where he discussed the *aurora borealis* he had seen in Sweden, and the Society of Antiquaries, showing his deep connections between the two organizations. For instance, on 13 November 1735, Celsius presented at Antiquaries 'a draught of two kinds of Runic Characters, namely the Vulgar and Helsingic together with some remarks on them', runes on the Malsta stone in Rogsta, Hå/lsingland, Sweden[130] (see figure 5.6). This paper on runes was published by the *Philosophical Transactions of the Royal Society* in 1737 with an attempt to date the stone by examining the genealogy of the family it described and applying the rationale used to date events in Newton's Biblical *Chronology of Ancient Kingdoms*

---

[126] James Bradley, 'A Letter from the Reverend Mr. James Bradley Savilian Professor of Astronomy at Oxford, and F.R.S. to Dr Edmond Halley Astonom. Reg. &c. giving an Account of a new discovered Motion of the Fix'd Stars', *Philosophical Transactions* 35, 406 (1727–8), pp. 637–61.

[127] RS/JBO/13/18, 8 December 1720, Royal Society Library, London.

[128] Folkes also gave Graham's watches to his aristocratic patrons. On 16 October 1642, Montagu wrote to Folkes to thank him 'for the watch which I have receiv'd safe and lyke it mytely'. RS/MS/790, Royal Society Library, London.

[129] Byrom, *Diary*, vol. 1, p. 109: 1725. Tuesday, 6th April…'to Paul's Church Yard, where Mr. Leycester and I went, Mr. Graham, Foulkes, Sloan, Glover, Montagu…There was a Lodge of Freemasons in the room over us, where Mr. Foulkes, who is Deputy Grand Master, was till he came to us'.

[130] SAL/MS/264 B, Anders Celsius, '"Lapis Maltadensis"'. Transcription and translation of a runic inscription on the Malsta stone in Rogsta, Hå/lsingland, Sweden by Prof. Anders Celsius, Hon. FSA'. Register transcribed by Joseph Ames of papers read or submitted to the Society, vol. 2, p. 309, Society of Antiquaries Library, London. See also Sven B. F. Jansson, *Runes in Sweden* (London: Phoenix House, 1962), pp. 79–80.

Fig. 5.6 Watercolour sketch of Anders Celsius's discovery of runes on the Malsta stone in Rogsta, Hälsingland, Sweden, which was engraved and published in the *Philosophical Transactions*, CLP/16/51, © The Royal Society, London.

Amended, which Folkes of course had edited.[131] In Folkes's presence, Celsius attended the Society of Antiquaries again on 9 March 1735/6 and on 18 March 1735/6, where Celsius 'delivered an account of an ancient Alabaster vase found in Sweden…at the Antiquarian College of Stockholm…which was read, likeways presented the Society with a wooden print of the said vase'.[132] As Celsius received his instruments, Folkes's friends received an antiquarian present, part of a time-honoured exchange of gifts characteristic of the Republic of Letters. The ties between the Royal Society, Antiquaries, and Newtonianism strengthened.

Folkes also gave advice to Celsius when he was on expedition. In December 1736 Celsius wrote to Folkes, stating he was 'one thousand times grateful for the courtesies you did for me during my stay in London'. He continued, 'We have observed the difference of latitude between Torne and Pello with Mr Graham's instrument. The observations were made in September in Pello, having not found

---

[131] Anders Celsius, 'An Explanation of the Runic Characters of Helsingland [sic]', *Philosophical Transactions* 40, 445 (1737/8), pp. 7–13, on p. 13. Celsius wrote, 'if we suppose Frumunt (the creator of the monument) to have been thirty years of Age when he erected this Monument for his Father, and, with Sir Isaac Newton, allow thirty Years for each Generation, we shall find three hundred and thirty Years from the Death of Fifiulsi to the Birth of Fidrasiv, who is the Stock of these Generations'.

[132] SAL/MS/264 B, vol. 2, pp. 164–5, Society of Antiquaries Library, London. As Celsius 'presented' the Society with the print, rather than 'showing' it or exhibiting it, it is presumed it was a gift.

any suitable star other than the Dragon. We then observed it in October here in Torne'.[133]

In other words, Graham's sector measured the angle at which a particular star, in this case Draco, or the Dragon, near the zenith passed the meridian (the star's altitude). The expedition then moved to another mountaintop at Pello and repeated the same measurement. The difference between the star's altitudes was due to their having travelled south (changing latitude). As Sorrenson stated, they then had to 'establish the distance between the two mountains by surveying a series of triangles and measuring several of the baselines with wooden poles'.[134] Once they had the distance between the two points and the difference in angles, they 'calculated that the length of one degree of arc near the polar circle was longer than at Paris', and concluded the Earth was oblate, vindicating Newton. There was one problem. As Celsius continued,

> and since this star [Draco] is close to the ecliptic pole, the declination changed in the meantime.[135] That is why I beg you, Sir, to send me the rule according to which one calculates the change of declination from one month to the next. We don't have here in memory what Mr Bradley gave about this in the Philosophical Transactions. And I would like to know as well if Mr Bradley has established anything yet about the irregularities that he found in the stars located around…the equinoxes, since our star of the Dragon is very close to that location'.[136]

As mentioned, Bradley's observations of the motion of Draco led to his discovery of stellar aberration. As Celsius was using Draco as a reference point for his triangulation measurement, he was concerned that this aberration would affect his measurements. Folkes's reply to Celsius's query is not extant, but we know that Bradley's calculations regarding stellar aberration were communicated. Maupertuis, in his publications about the exhibition, incorporated Bradley's data about declination, and used Bradley's precision in measuring stellar positions as an example of new standards of astronomical practice only possible with the best instruments. Maupertuis stated, 'Mr Bradley very kindly shared with me his latest discoveries on the motions of the stars, and communicated to me the necessary correction to the arcs we observed'. As Terrall has shown, this precision of English instruments was used as a weapon against Cassini's claims for a prolate Earth. When Maupertuis's book about the shape of the Earth (*Le figure de terre*) finally did appear in 1738, he wrote to Folkes:

---

[133]   RS/MS/790/21, Letter from Celsius to Folkes, 3 December 1736, Royal Society Library, London.
[134]   Sorrenson, 'George Graham: Visible Technician', p. 214.
[135]   A star's declination changes gradually due to precession of the equinoxes and annual parallax.
[136]   RS/MS/790/21, Letter from Celsius to Folkes, 3 December 1736, Royal Society Library, London.

It is my turn now to thank you after the obliging letter that you did me the honour of writing me. The most flattering reward I could get from a work that has faced many difficulties and risks is to see it has the approval of an excellent judge such as you. Such approval buttresses me against the injustice of those who were not pleased by our measures, and their conspiracy. It is true that it is something singular and that it will be difficult to understand that the Astronomer [Cassini][137] who has the highest reputation in France was mistaken 5 different times in the measures he undertook and that he has always found that the Earth is elongated in 1701, 1713, 1714, 1733, and 1734....But I am nevertheless convinced that the Earth is flattened.[138]

He then continued, 'It has been very sweet to me, Sir, to find myself still in the honour of your memory after such a long time...I [send you my book] to you as one of the first scholars in Europe'.[139]

Folkes's interest in stellar parallax, aberration, and the modification of the constellations over time was demonstrated by a model he had made of the globe which he demonstrated to the Society of Antiquaries on 7 July 1736 when he returned to England (see figure 5.7). It was a plaster of Paris copy of the Farnese Globe, the oldest surviving depiction of this set of Western constellations; Folkes recorded in his travel diary 'Diam of the Farnesian Globe. 2f. 5 inch. 1/10 English Measure', likely indicating he saw and measured in it Rome.[140] The Globe puts the celestial figures against a grid of circles comprising the celestial equator, the colures, the ecliptic, the tropics, and the Arctic and Antarctic Circles. This allows for the constellations to be positioned accurately. It was noted at the Antiquaries' meeting that 'the colure of this globe passes by those parts of the asterism by which it is said to have past in the days of Hipparchus but the intersection of the Equator and Eccliptic is not at the Colure'. This was a reference to how the 26,000-year precession, or axial wobble, on the Earth that Newton discovered shifts the positions of the celestial circles over time, so the observed positions on the Farnese Globe can be traced to a particular date in celestial time. 'The declination of the Arctic and Antarctic Circles will correspond to a particular latitude for the observer whose observations were adopted by the sculptor...a detailed analysis of the globe will reveal the latitude and epoch for the observations incorporated in the globe'.[141] The Antiquaries entry noted that its age could be determined to be during the time of the Antonine emperors.

---

[137] 'Astronomer' is written with a capital A in French in the original letter.
[138] For English instrumental precision, see Terrall, *Flattened Earth*, p. 137.RS/MS 790/42, Letter from Pierre Maupertuis to Martin Folkes, 26 July 1738, Royal Society Library, London.
[139] RS/MS/790/42, Royal Society Library, London.
[140] Folkes, 'Journey from Venice to Rome', f. 98v. This was noted by Kristen Lippincott, 'A Chapter in the Nachleben of the Farnese Atlas', p. 287; SAL/MS/264 B, vol. 2, pp. 201–2.
[141] B. E. Schaefer, 'The Epoch of the Constellations on the Farnese Atlas and their Origin in Hipparchus' Lost Catalogue', *Journal for the History of Astronomy* xxxvi (2005), pp. 167–96, on p. 196.

Fig. 5.7 Stereographic projection by Martin Folkes of the Farnese Globe in Richard Bentley, ed., *M. Manilii Astronomicon* (London: Henrici Woodfall; Pauli et Isaaci Vaillant, 1739), p. xvii, 25.9 × 52.1 cm. The Library, All Souls College, Oxford.

Geodesy also continued to be a preoccupation of Folkes when he eventually realized his goal of being appointed President of the Royal Society and then the Society of Antiquaries, nearly a decade after his initial Italian visit. Folkes became one of the strongest supporters in the Royal Society of John Harrison and his longitude clock, convinced of Harrison's 'strong impulses of a natural and uncommon genius'.[142] And, in 1745, Folkes intervened when Antonio de Ulloa, who went on the Peruvian expedition to measure the meridian arc near the equator, was captured by the English as he sailed home when the expedition was completed, as at the time Spain and Britain were at war. Ulloa's calculations were confiscated and he languished in prison in Portsmouth, until he was granted leave to travel to London at the request of the Royal Society. Folkes mediated, and on 8 May 1746 used his knowledge of Spanish to summarize the details of Ulloa's expedition in thirty-three pages of manuscript which he presented to the Society later that month;[143] Folkes reproduced Ulloa's triangulation and its results, which confirmed Newton's predictions of the shape of the Earth (see figure 5.8). Folkes

[142] RS/JBC/20/184, Royal Society Library, London. This was from the address Folkes gave when presenting the Copley Medal to Harrison. See Jim Bennett, 'James Short and John Harrison: Personal Genius and Public Knowledge', *Science Museum Journal* 2 (Autumn 2014), http://dx.doi.org/10.15180/140209.

[143] RS/L&P/1/479, Paper, 'Journal of Observations Made in Peru; and Account by Martin Folkes' by Antonio de Ulloa, 1746, Royal Society Library, London. For an approachable treatment of Ulloa's voyage, see Larrie D. Ferreiro, *Measure of the Earth: The Enlightenment Expedition that Reshaped Our World* (New York: Basic Books, 2013); See also Francisco de Solano, *Don Antonio de Ulloa, Paradigma del Marino Científico de la Ilustración Española* (Coimbra: Universidade de Coimbra, 1990) which has a nice description of Ulloa's mineralogical work and discovery of platinum.

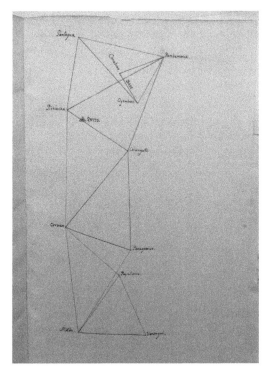

Fig. 5.8 Martin Folkes's diagram of Antonio de Ulloa's Triangulation for the Peruvian Expedition, Letters and Papers, Decade 1, Volume 10A, 5 December 1745 to 1 May 1746, MS no. 479, © The Royal Society, London.

noted if the 'proportion resulting of the Axis of the Earth' from the Maupertuis expedition 'to its Equatorial Diameter' of Ulloa's measurements 'will be nearly 223 to 222 surprizingly near to that given from calculation only, by the great Sir Isaac Newton, of 229 to 230, in the last edition of his Principia.'[144] Rather ironically, Folkes noted that 'Dr Bradley's aberration of the fixed stars from the successive propagation of light is not brought into the account'.

Folkes also recorded their barometric observations made in the Andes, as well as the heights of the mountains, and included Ulloa's ethnographic information about the Peruvians and descriptions of the flora and fauna, including the deadly coral snake. Folkes noted that he did not want the material to be made public, as Ulloa intended to publish a book about his exploits when he returned to Spain (accomplished in 1748, with twelve subsequent editions in French, English, Dutch, and German); Folkes noted that he was unwilling to publicise Ulloa's

[144] RS/L&P/1/479, pp. 30–1.

findings because of the pleasure the public would 'without doubt receive' from Ulloa's book.[145] Apart from the evident esteem Ulloa and Folkes developed for each other, vindication of Newtonianism was all the more powerful from an outside entity. Ulloa was made an FRS, most of his papers were returned to him by order of the Lords Commissioners of the Admiralty, and it was arranged that he would get travel to Portugal, which was a neutral in the War of the Austrian Succession, so Ulloa could get back to Spain. However, Folkes kept one set of manuscripts. In the posthumous auction catalogue of Folkes's collection, it was recorded that he had held back 'an abstract of Don Antonio de Ulloa's Journal, made by the French Astronomers and himself at Peru' in memory of his friend.[146]

5.5 Folkes as Newton?

Whether in defence of Newtonian optics, or in support of Maupertuis's flattened Earth, Folkes's travel diary reveals that he made a case for the primacy and precision of English instruments and measurement in natural philosophy. This precision of measurement would also extend to antiquarian questions in which conversion factors were sought between the English and Roman foot to comprehend Roman architectural engineering. His diary also represented, as Feingold reminds us, the 'confabulatory life' of the scholar, the diffusion of scientific knowledge through informal discussion with colleagues.[147] Henry Guerlac has commented that 'as historians of ideas we are happiest when we can navigate from the firm ground of one document to the next, and we are prone to forget how great a part travel, gossip and word-of-mouth have played in the diffusion of scientific knowledge, indeed of knowledge of all sorts'.[148] In this manner, the journal of Folkes's scientific peregrination demonstrates to us the parameters of Georgian 'antiquarian science' and Newtonianism.

The diary also represented the manifestation of Folkes as an international broker of Newtonianism, something also reflected in his role as a connoisseur and collector in the Royal Society after his tour. On 13 April 1738, Sloane informed the Society that William Freman FRS 'had purchased a fine Marble Bust of the late Sir Isaac Newton, with an intention of making a Present of it to the Society: and therefore, as it would be proper to consider beforehand of a suitable place in

---

[145] RS/L&P/1/479, p. 1. Ulloa published his account in he and Don Jorge Juan's *Relación histórica del viaje a la América Meridional*, 4 vols (Madrid: Antonio Marin, 1748). See de Solano, *Don Antonio de Ulloa*, pp. 335–6 for a discussion of Ulloa's editions.

[146] *A Catalogue of the Entire and Valuable Library of Martin Folkes, Esq.* (London: Samuel Baker, 1756), p. 155.

[147] Mordechai Feingold, 'Confabulatory life', in P. D. Omodeo and K. Freidrich, eds, *Duncan Liddel (1561–1613): Networks of polymathy and the Northern European Renaissance* (Leiden: Brill, 2016), pp. 22–34.

[148] Henry Guerlac, *Newton on the Continent* (Ithaca: Cornell University Press, 1981), p. 46.

the Meeting Room, to set it up in, he proposed that Mr Folkes with any other Gentlemen he be pleased to join with him, might be desired to consider of it, and report their opinion to the Society'.[149] On Folkes's advice, the Society subsequently bought a pedestal for the bust for £2.7s, approving payment on 19 June[150] (see figure 5.9). Folkes also had in his personal collection a large plaster bust of Newton as well as one of the Earl of Pembroke; Roubiliac, as he did for Newton, sculpted a bust of Pembroke, and he was known to make plaster copies as well as terracotta studies for his work.[151]

The previous year Folkes had a portrait of himself painted by Vanderbank, which John Faber did in mezzotint. Vanderbank painted Folkes seated next to a portrait bust of Newton that looks similar to that which Freman donated to the Society, including the pedestal which was subsequently purchased; if so, Vanderbank/Faber's portrait of Folkes may be the earliest portrayal of a Roubiliac bust, as well as this particular sculpture of Newton, and Folkes had access to it when it was in Freman's possession[152] (see figure 5.10). We also see that Folkes

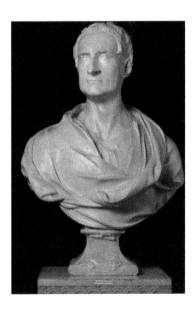

Fig. 5.9 Louis François Roubiliac, Sculpture Bust of Sir Isaac Newton, Marble, 1737–8, S/0018, © The Royal Society, London.

[149] RS/JBO/17/231–2, Meeting of 13 April 1738, Royal Society Library, London.

[150] RS/CMO/3/79, p. 194, Meeting of 19 June 1738, Royal Society Library, London.

[151] David Bridgewater, '18th-Century Portrait Sculpture'. http://english18thcenturyportraitsculpture.blogspot.com/2016/05/bust-of-isaac-newton-by-roubiliac-at.html [Accessed 12 January 2020]; Milo Keynes, *The Iconography of Sir Isaac Newton to 1800* (Woodbridge: The Boydell Press, 2005), pp. 36–7. The auction catalogue of Folkes's collection lists under 'plaster figures': 7th May lot 4, a large bust of the earl of Pembroke, on a painted deal term, lot 5 ditto of Sir Isaac Newton on a ditto.

[152] David Bridgewater, 'Portraits of Martin Folkes'. http://bathartandarchitecture.blogspot.com/2015_05_21_archive.html [Accessed 12 January 2020]. The Vanderbank portrait was in the possession of the O'Briens on the Monare Estate on Foynes Island in County Limerick, Ireland until 1992. See ICON Notes for Martin Folkes, National Portrait Gallery, London.

Fig. 5.10   John Faber Jr, *Martin Folkes,* mezzotint after John Vanderbank, 1737 (1736), 353 mm × 253 mm, Yale Center for British Art, Paul Mellon Collection.

had himself depicted with drapery around his soldiers, his normally rotund face leaner and more resolute, just like the bust of Newton behind him.

This attempt by Folkes to resemble Newton physiognomically demonstrates the contrast in the portraiture he commissioned of himself before and after his Grand Tour. His 'Kit-Cat' portrait by Jonathan Richardson painted in 1718 (see figure 4.6) shows no indication of Folkes's association with Newton, or even of his interests in astronomy, lacking the typical props of books or scientific instruments. Richardson instead painted this portrait as a character study to commemorate his and Folkes's friendship and Folkes as an epicure, not as an official image.

However, the 1736 portrait of Folkes, as well as another 1739 portrait commissioned from John Vanderbank, very much shows him in the image of Newton (see figures 5.11 and 5.12). Here we see Folkes's self-fashioning as the heir apparent to the Royal Society presidency, something he achieved in 1741 after Sloane's death. Folkes's journey to Italy was an expression of the motto scrawled on his travel diary, symbolic of who and what he would become: a natural philosopher who would manifest several elements of the Newtonian programme during his tenure as President of the Royal Society.

All of these artworks also slyly demonstrate one of Folkes's interests developed when he was on tour—what Malcolm Baker has termed investigative viewing and the process of perception.[153] Baker has pointed out that in 1731 John Conduitt commissioned a terracotta bust of Newton from Roubiliac that served as a study for the marble sculpture donated to the Royal Society.[154] The terracotta was 'made under the eyes of Mr Conduitt and several of Sir Issac Newton's particular friends…from many pictures and other busts, and esteemed more like than anything extant of Sir Issac.'[155] On Newton's death, Conduitt allowed John Michael Rysbrack to take casts of Newton's face, and Roubiliac used these to create the terracotta bust. The bust was later owned by surgeon John Belchier FRS, who in 1785 bequeathed it to the Royal Society with instructions that it should be placed

Fig. 5.11 John Vanderbank, *Martin Folkes*, oil on canvas, 1739, private collection, courtesy of Christopher Foley FSA, Lane Fine Art.

---

[153] Malcolm Baker, 'Making the Portrait Bust Modern: Tradition and Innovation in Eighteenth-Century Sculptural Portraiture', in Jeanette Kohl and Rebecca Müller, eds, *Kopf/Bild Die Büste in Mittelalter und Früher Neuzeit* (Munich and Berlin: Deutscher Kunstverlag, 2007), pp. 347–66, on p. 358. For the bust itself, see 'Isaac Newton (1642–1727), Object ID ZBA1640', Royal Museums Greenwich. https://collections.rmg.co.uk/collections/objects/220530.html [Accessed 7 January 2020].

[154] Milo Keynes, *The Iconography of Sir Isaac Newton*, p. 75.

[155] 'Minute Book, 18 August 1785', Royal Society London, as quoted in Keynes, *The Iconography of Sir Isaac Newton*, p. 8.

Fig. 5.12   John Vanderbank, *Three-quarter-length portrait of Sir Isaac Newton, aged eighty-three years,* oil on canvas, 1725–1726, 1270 mm × 1016 mm, © The Royal Society, London.

in the Royal Observatory at Greenwich. Belchier also stated in his will that, as a portrait, it was 'esteemed more like than anything extant of Sir Isaac'.[156] Conduitt wished the terracotta and marble portrait busts of Newton to display his powers of concentration, 'intentness of thought', perseverance, and 'patience equal to his sagacity and invention', along with the highly accurate empirical detail of his visage.[157] 'One way in which this is achieved, is…through the active engagement of the viewer in the discernment of the subtle difference in surface…the play of the eye between more broadly worked areas and those passages that are particular and specific in their detail…a Lockean mode of empiricist viewing, in effect'.[158] After all, 'the image was created under the eyes of a Fellow of the Royal Society who was right at the heart of that circle concerned to disseminate and promote the new empiricism…it is as if the viewer is invited by these sculptural prompts

---

[156]   'Isaac Newton (1642–1727)', Object ID ZBA1640, Royal Museums Greenwich [Accessed 7 January 2020].

[157]   Baker, 'Making the Portrait Bust Modern', p. 362.

[158]   Baker, 'Making the Portrait Bust Modern', p. 362.

to emulate that "intentness of thought"—or at least intentness of observation that Newton possessed".[159]

Such a sculpture that promoted such Lockean empiricist viewing would have appealed to Folkes, who on his Grand Tour demonstrated his own highly developed powers of observation in optics and in the details of architecture. After all, one of Folkes's prize possessions was a 'seal with Locke's head, on a cornelian'.[160] Folkes did not just emulate Newton's 'intentness of thought' by looking at the busts, but also had his own painted portraits visually echo that 'intentness of thought' that the bust of Newton displayed. At the same time, his portraits served as sly optical illusions. Folkes was not Newton, but he could look like him. He could appropriate his iconography and physiognomy two-dimensionally, the closest he could get to appropriating Newton's three-dimensional identity and legacy. For Folkes, the Vanderbank portraits were evidently a source of aesthetic pleasure and iconographic self-promotion as an international broker of Newtonianism, a campaign that ultimately resulted in him securing the Royal Society presidency in 1741.

---

[159] Baker, 'Making the Portrait Bust Modern', p. 362.
[160] *A Catalogue of the Genuine and Curious Collection of Mathematical Instruments...of Martin Folkes, Esq* (London: Samuel Langford, 1755), p. 6.

6

Martin Folkes, Antiquary

In the study of Antiquities, as in all others, judgment, genius is necessary

—William Stukeley[1]

6.1 Numismatics

In the 1730s, in addition to his Newtonianism Folkes continued to pursue his metrological and antiquarian work, or 'scientific antiquarianism', developing a specific expertise in numismatics, an interest which preoccupied him for the rest of his life. When Folkes embarked on numismatics, Maurice Johnson, President of the Spalding Gentlemen's Society, predicted that Folkes, whom he described as 'that judicious Gent', would create 'the most accurate account extant' of English coinage.[2] Johnson was correct in his assessment, as Folkes's *Table of Silver Coins* (1745), and his posthumous *Table of Silver and Gold Coins* (1763) became the standard numismatic texts for a century. Part of the reason was Folkes's diligence in searching Robert Cotton's library (which formed the core of the British Library) resulting in his rediscovery in 1747 of its extensive coin collection; 'in the report of the Committee of the House of Commons' the collection was thought to have been lost in 1731 in the fire where it was housed at Ashburnham.[3] Part of the reason was that Folkes could consult, over time, his own vast numismatic library of over 300 works, containing most of the major works devoted to English and

---

[1] Part of section 6.2 concerning The Egyptian Society will appear as Anna Marie Roos, 'The First Egyptian Society', in Anna Marie Roos and Vera Keller, eds, *Collective Wisdom. Collecting in the Early Modern Academy*, Techne Series (Turnhout: Brepols, 2020), forthcoming. William Stukeley, 'Preface', to *The Medallic History of Marcus Aurelius Valerius Carausius*, 2 vols in 1 (London: Charles Corbet, 1757), vol. 1, p. 1.

[2] Bodl. MS Eng. misc. *c.* 113, f. 344v; Maurice Johnson to William Stukeley, 17 March 1744, Bodleian Library, Oxford, in Diana and Michael Honeybone, *The Correspondence of William Stukeley and Maurice Johnson, 1714–54* (Woodbridge: Boydell Press, 2014), p. 208.

[3] Bodl. MS Eng. misc. *c.* 113, ff. 343r–344v, Bodleian Library, Oxford, printed as 'Maurice Johnson, Jr to William Stukeley, 20 June 1747', in William Stukeley, *The Family Memoirs of the Rev. William Stukeley* (London: Surtees Society, 1882), vol. 3, p. 354; See also Hugh Pagan, 'Martin Folkes and the Study of the English Coinage in the Eighteenth Century', in R. G. W. Anderson, M. L. Caygill, A. G. MacGregor, and L. Syson, eds, *Enlightening the British: Knowledge, Discovery and the Museum in the Eighteenth Century* (London: The British Museum Press, 2003) pp. 158–63, on p. 158.

Martin Folkes (1690–1754): Newtonian, Antiquary, Connoisseur. Anna Marie Roos, Oxford University Press (2021).
© Anna Marie Roos. DOI: 10.1093/oso/9780198830061.003.0006

European coinage.[4] He was also a collector of coins and medals, the twenty-six-page auction catalogue of his estate indicating that it took nearly a week to disperse his numismatic specimens, as well as his English, Swiss, Roman, papal, and Danish medals.[5] Part of the reason was his keen visual apprehension, Folkes even having his own portrait medal struck by Ottone Hamerani (1695–1761) when he travelled to Rome (see figure 6.1). Its iconography integrated his interests in numismatics, Masonic, and, as we will see in this chapter, Egyptian iconography and culture. As William Eisler noted and as we have seen, 'Folkes...employed various forms of portraiture to promote his endeavours internationally'.[6]

The publication of John Evelyn's *Numismata* (1697) as well as Joseph Addison's *Dialogues upon the Usefulness of Ancient Medals* (1726) led to 'a growth in enthusiasm for "medallic history"', as well as an interest in physiognomy, the portrait considered reflective of moral character and identity.[7] The 'ideal of the purity of the ancestral face or head came to preoccupy Early Georgians.... Anything which disguised the shape of the head was, by the standards of the "medallic" historians, considered tantamount to an attempt to deceive history'.[8] Folkes's medallic portrait was thus unadorned and inspired by Roman Republican verism and coinage, connecting him to his civic role as an officer in the Royal Society, as well as to the virtues of Roman stoicism. Folkes also had a portrait gem of himself done by Lorenz Natter, who had begun copying ancient gems for Grand Tourists under the influence of Baron Philipp von Stosch in Florence, and who moved to England in 1739; Folkes's portrait was done shortly afterwards in 1741, modelled after Jacques Antoine Dassier's medal of him (see below). Although previous gems cut for Roman tourists were done in a periwigged 'contemporary' style, Natter introduced a classical aesthetic for his portrait of Folkes.[9]

[4] George Kolbe, 'Godfather to all Monkeys: Martin Folkes and his 1756 Library Sale', *Asylum* (April–June 2014), pp. 38–92, on pp. 45–6.

[5] *A Catalogue of the Genuine, Entire and Choice Collection of Coins, Medals, and Medallions in Gold, Silver and Bronze of the Learned and Ingenious Martin Folkes, Esq* (London: Samuel Langford, 1756).

[6] William Eisler, 'Paul Mellon Centre Rome Fellowship: The Medals of Martin Folkes: Art, Newtonian Science and Masonic Sociability in the Age of the Grand Tour', *Papers of the British School at Rome* 78 (November 2010), pp. 301–2.

[7] Matthew Craske, *The Silent Rhetoric of the Body: A History of Monumental Sculpture and Commemorative Art in England, 1720–1770* (New Haven: Yale University Press, 2007), p. 79; Malcolm Baker, 'A Genre of Copies and Copying? The Eighteenth-Century Portrait Bust and Eighteenth-Century Responses to Antique Sculpture', in Tatjana Bartsch, Marcus Becker, Horst Bredekamp, and Charlotte Schreiter, eds, *Das Originale der Kopie: Kopien als Produkte und Medien der Transformation von Antike* (Berlin and New York: Walter de Gruyter, 2010), pp. 289–312, on p. 302.

[8] Craske, *The Silent Rhetoric of the Body*, p. 79.

[9] J. Boardman, J. Kagan, C. Wagner, and C. Phillips, *Natter's Museum Britannicum. British Gem Collections and Collectors in the Mid-eighteenth Century* (Oxford, Archaeopress in Association with the Hermitage Foundation UK and the Classical Art Research Centre, 2017), p. 10; Elizabeth Nau, *Lorenz Natter (1705–1763): Gemmenschneider und Medailleur* (Biberach an der Riß: Biberacher Verlagsdruckerei, 1966), Medal 83, p. 98. An example is in the Hermitage Museum.

Fig. 6.1 Ottone Hamerani, Martin Folkes, 1742, bronze, 37 mm, M.8467,
© The Trustees of the British Museum, Creative Commons BY-SA 4.0.

There was also a spate of works on coinage produced in the early eighteenth century. While there had been several more general numismatic works published in the previous century, such as Andreas Helvetius (Morellius's) *Specimen Universae Rei Nummariae Antiquae* [An Essay Towards an Universal History of Ancient Coins and Medals] (1683) attempts in the early part of the eighteenth century to create a history of British coinage were dominated by churchmen. It was also a process characterized by delays in publication. John Sharp, Archbishop of York (1645–1712), was the first to attempt a work of British numismatics, although his book 'was not published in its entirety until 1785'.[10] In turn, Bishop Nicholson utilized Sharp's work in his *English Historical Library* (1696–9), and Bishop Fleetwood 'gave an account of English money in his *Chronicon Preciosum* (1707)', reprinted with plates in 1745.[11]

In 1707, when the Society of Antiquaries of London was reconstituted, Humphrey Wanley proposed one of its tasks would be a 'Historical account of the Coin', particularly as there was no national numismatic collection extant, the Royal Collection not containing 'more than a handful of English coins predating the reign of James I'.[12] There was no effort made until 3 January 1721, when the Society of Antiquaries again proposed that it 'would be much for the honour of the Kingdom, and particularly of the Society, to attempt a complete description and history of all the Coins relating to Great Britain'.[13] From 1721–4, different antiquaries collaborated on the project, including Maurice Johnson, the founder of the Spalding Gentlemen's Society, who divided 'all the Legends and accounts of Coins that relate to Britain into five areas' for an overarching *Metallographia Britannica*.[14] Stukeley, for instance, examined all the British coins in Sloane's collection, Wanley analysed Saxon numismatics, and Roger Gale[15] concentrated on Roman coinage that related to Britain. Peter de Neve, the Society of Antiquaries President, was to analyse 'the whole of the medieval and later English coinage'.[16] In May 1722, Stukeley showed the Society, 'What he had done towards collecting the coyns of Great Britain from the earliest times according to the respective tasks assumed to themselves some time ago. He showed a great number of British coyns drawn out in some order, when he proposes to demonstrate their great Conformity with the old Greek and Punic'.[17] Unfortunately, the task proved overwhelming and was not achieved; only Stukeley produced anything substantive, engraving twenty-three plates which were published by his executor Richard

---

[10] Robin J. Eaglen, 'The Illustration of Coins: An Historical Survey, Part I', Presidential Address 2009', *British Numismatic Journal* 80 (2010), pp. 140–50, on p. 143. https://www.britnumsoc.org/publications/Digital%20BNJ/pdfs/2010_BNJ_80_7.pdf [Accessed 7 September 2019].

[11] Eaglen, 'The Illustration of Coins', pp. 143–4.   [12] Pagan, 'Martin Folkes', p. 158.

[13] Roger Ruding, *Annals of the Coinage of Great Britain and its dependencies: from the earliest period of authentic history to the reign of Victoria*, 3 vols (London: the author, 1812), vol. 1, p. xi.

[14] John Nichols, *Literary Anecdotes of the Eighteenth Century*, 6 vols (London: Nichols, Son, and Bentley, 1812), vol. 6, p. 157. The Society of Antiquaries Minute Books has several versions of this passage under the dates 3 January 1721/23, 30 May 1722, and 1 April 1723/4.

[15] Roger Gale (1672–1744), antiquary and numismatist, was FRS (1717), serving as its treasurer, and elected FSA the same year, eventually becoming Vice President. An MP for Northallerton, Yorkshire, he was also commissioner for excise from 1715 to 1735.

[16] Pagan, 'Martin Folkes', p. 158.

[17] Society of Antiquaries Minute Books, 30 May 1722, Society of Antiquaries Library, London. See also D. F. Allen, 'William Stukeley as a Numismatist', *The Numismatic Chronicle* 10 (1970), pp. 117–32, on pp. 119–20. Stukeley's original numismatic notebook is in the archives of Corpus Christi, Oxford: 'Brittish coins drawn by William Stukeley 1720: Britannia Metallica'.

Fleming in quarto.[18] There were also isolated plates that the Society of Antiquaries had engraved that documented scarce medals and coins such as the silver crown of Henry VIII (*c.*1545), which featured later in Folkes's *Table of English Silver Coins*.[19]

Stephen Martin Leake, who had been an esquire with Folkes at the installation of the Order of the Bath, was the first to write a book on the English coinage: *Nummi Britannici Historia, or An Historical Account of English Money from the Conquest* (1726). However, Leake was not a specialist, his selection of coins was limited, and his work was 'founded chiefly upon the authority of printed books, instead of original Records'.[20] Leake's work also contained only eight plates, the other six added to the enlarged 1745 and 1793 editions.[21]

As Leake's work was unsatisfactory, in March 1732 Folkes succeeded in taking over the project with the Society of Antiquaries of London to produce a definitive account of English coinage. On 10 February 1731, Folkes brought to the Society 'an account of the English Gold and Silver Coins examined by the Balance and compared with the Standard weight', no doubt using the several sets of scales and standard weights he kept at Queen Square, along with a 'hydrostatical balance' that determined the 'density and controlled purity of different substances'.[22] Folkes hired George Vertue (1684–1756), the first official engraver to the Society of Antiquaries since its revival in 1717, to engrave the numismatic plates; the only illustrations of coinage that the Society produced were for its *Vetusta Monumenta* [Ancient Monuments] which was only sporadically published and in a large format inconvenient for study.[23] By July 1732, Vertue was touring England examining Folkes's coins and medals collection in Norfolk,[24] and the collections of other antiquaries like Browne Willis ('very perfect' and in 'large number'), as well as those belonging to Lord Pomfret, who had bought the Arundel collection.[25] According to Crystal B. Lake, Willis published *A Table of The Gold Coins of the*

---

[18] Ruding, *Annals of the Coinage*, vol. 1, p. xi.
[19] Crystal B. Lake 'Plate 1.20: Medals of Henry VIII, Edward VI, Elizabeth I, and James I', *Vetusta Monumenta: Ancient Monuments.* https://scalar.missouri.edu/vm/vol1plate20-sixteenth-century-english-medals [Accessed 2 January 2019].
[20] Ruding, *Annals of the Coinage*, vol. 1, p. ix.
[21] Eaglen, 'The Illustration of Coins', p. 144.
[22] Minutes of the Society of Antiquaries, vol. I, p. 282, Society of Antiquaries, London; *A Catalogue of the Genuine and Curious Collection*, p. 3, p. 7; Anders Lungren, 'The Changing Role of Numbers in 18th-Century Chemistry', in Tore Frängsmyr, J. L. Heilbron, and Robin E. Rider, eds, *The Quantifying Spirit in the 18th Century* (Berkeley: University of California Press, 1990), pp. 246–67, on p. 247.
[23] Pagan, 'Martin Folkes', p. 158, p. 161, note 5. The plates are 37 and 38 in vol. 1 of the *Vetusta*, published in 1747.
[24] BL Add MS 5833, Copy of letter from Browne Willis to Folkes of 28 July 1732, where Willis mentions Vertue 'design'd himself the pleasure of waiting on you', f. 152r., British Library, London. The letters between Folkes and Willis were copied by antiquary William Cole, who bitterly complained about Folkes's illegible handwriting and idiosyncratic spelling.
[25] Crystal B. Lake, 'Plate 1.20: Medals of Henry VIII, Edward VI, Elizabeth I, and James I', *Vetusta Monumenta: Ancient Monuments*, trans. Raymond Marks. A Digital Edition. https://scalar.missouri.edu/vm/vol1plate20-sixteenth-century-english-medals [Accessed 21 November 2020].

Kings of England (1733) in one-sheet folio 'based on his private collection, and, according to Nichols, he was responsible for the "making" of some of the plates' of tables of coins that the Society of Antiquaries intended for their *Metallographia Britannica*, which eventually appeared in their *Vetusta Monumenta*.[24] Willis was described as 'extremely well versed in coins, he knows hardly anything of mankind', and in 1741, he donated his entire cabinet of English coins, 'at that time looked upon as the most complete collection in England', to the University of Oxford, 'with the stipulation that it be displayed and visited annually on the 19th of October' on St Frideswide's Day. The university, due to the lavishness of the bequest, ended up buying the collection for 150 guineas, and the collection is still extant.[26]

During his tour, Vertue borrowed coins from Willis for Folkes's use, including a 'small gold' piece from the reign of James I, as they had the whole and half piece, 'but want[ed] the quarter'.[27] It was painstaking work. Vertue was initially reluctant to take the commission, being unacquainted with numismatic Latin abbreviations and the ensuing potential for error, but 'a greater price was proposed to him [by Folkes] than any other engraver would demand'.[28] Folkes clearly wanted the best illustrations for his work so it could be readily used as a reference guide, and he had worked extensively with Vertue before.

Vertue had painted watercolours and engraved two plates for the Society of Antiquaries' *Vetusta Monumenta* featuring fragments of the Cotton Genesis, a sixth-century manuscript of the Book of Genesis in Greek housed in the Cottonian Library, which was badly damaged in the Ashburnham House fire in 1731.[29] One of the most important Early Christian manuscripts, no other exerting 'a wider influence during the Middle Ages', the work featured approximately '330 miniatures interspersed through the Genesis text on some 215 folios'.[30] After the fire, only eighteen badly charred scraps of vellum were left.

In the explanatory material for the plates in the *Vetusta*, it was noted:

> Since, therefore, there was little doubt but that fragments of this precious codex, which had escaped the destructive force of flames in some measure, gratified with [its] most agreeable appearance not only the eyes but also the minds of all lovers of antiquity, a decision was made to publish, arranged on a pair of plates,

[26] Crystal B. Lake, 'Plate 1.20: Medals of Henry VIII, Edward VI, Elizabeth I, and James I'; Nichols, *Literary Anecdotes*, vol. 6, p. 191, p. 205. Willis's grandfather was the physician Thomas Willis, and Browne Willis donated several of his grandfather's manuscripts to the Bodleian Library.

[27] Letter from George Vertue to Maurice Johnson, 29 July 1732, in Nichols, *Literary Anecdotes*, vol. 2, pp. 248–9.

[28] Letter from George Vertue to Maurice Johnson, 29 July 1732, in Nichols, *Literary Anecdotes*, vol. 2, p. 249.

[29] Now BL Cotton MS Otho B VI.

[30] Kurt Weitzmann and Herbert Kessler, *The Cotton Genesis: The Illustrations in the Manuscript of the Septuagint, Volume 1. British Library, Codex Cotton Otho B. Vi.* (Princeton: Princeton University Press, 1987), p. ix; Kurt Weitzmann, ed., *Age of Spirituality: Late Antique and Early Christian Art, Third to Seventh Century; Catalogue of the Exhibition at the Metropolitan Museum of Art, November 19, 1977, Through February 12, 1978* (New York: Metropolitan Museum of Art, 1979), p. 457.

the best parts of those fragments, which were able to be recovered from the shipwreck, as it were, and to be reproduced with illustrations. Martin Folkes, esquire, acting as President of the Society of Antiquaries, agreed to perform the arduous task of collecting and arranging them and of deciphering the fading contours of their letters.[31]

Folkes rearranged the fragments to try to put them in chronological order and filled in the gaps caused by fire damage with 'excerpts from Lambert Bos's edition of the Codex Vaticanus published in 1709. Members of the Antiquaries consulted Grabe's 1703 collations as well as Peter Lambeck's published commentary on the Vienna Genesis (1665–1671)'.[32]

As an editor of Newton's biblical chronologies and student of James Cappel, Folkes at least had the expertise in classical Greek (if not Byzantine Greek) to accomplish such a task, and the Society approved of his and Vertue's efforts—but William Stukeley was dismissive. He wrote that Folkes conducted 'the work, at the engravers' so as the fragments 'appear crouded in the plates, and much of their true beauty is lost'.[33] Folkes had provided Stukeley with a set of the plates out of friendship,[34] which Stukeley proceeded to cut and paste to create his own manuscript booklet, giving each fragment 'a distinct page ... and short account of each picture, for we cannot show too much respect, for so invaluable a curiosity'.[35] Stukeley was particularly interested in the codex fragment of the ark, convinced it was key to showing Noah's original design, and then how it was nailed, caulked, and pitched when constructed, another drawing of which he provided with its measure in cubits, another example of the pervasive interest in metrology in this era.[36] As for Vertue, he must have enjoyed the process of working for Folkes and engraving fragments and coins, or at least found it lucrative. In 1753, he subsequently engraved the coin collection of the Earl of Pembroke as well as the *Medals, coins, great seals, and other works* (1753) of antiquary Thomas Simon, which must have been one of the last numismatic works Folkes collected for his library.[37]

---

[31] *De Codice Genesos Cottoniano Dissertatio Historica*, Explanatory Account to Plates 1.66–1.61.68, Portrait of Robert Cotton with Genesis Fragments, *Vetusta Monumenta: Ancient Monuments*, trans. Raymond Marks. A Digital Edition. https://scalar.missouri.edu/vm/vol1plates66-68-account [Accessed 16 January 2020].

[32] Crystal B. Lake and Benjamin Weichman, Plates 1.66–1.68, Portrait of Robert Cotton with Genesis MS Fragments, *Vetusta Monumenta: Ancient Monuments*, trans. Raymond Marks. A Digital Edition. https://scalar.missouri.edu/vm/vol1plates66-68-robert-cotton-portrait-with-genesis-ms [Accessed 28 August 2020].

[33] Wellcome MS 4726, f. 2r., Wellcome Library, London.

[34] Wellcome MS 4726, Wellcome Library, London. There is a note by Stukeley on the first (unnumbered) leaf: '13. Aug. 1749. Martin Folkes Esq. LL.D. gave me the set of prints done by the Antiquarian Society'.

[35] Wellcome MS 4726, f. 2r, Wellcome Library, London.

[36] Wellcome MS 4726, ff. 7v and 8r, f. 26r, Wellcome Library, London.

[37] George Vertue, *Medals, coins, great seals, and other works of Thomas Simon, engraved and described by George Vertue* (London: J. Nichols, 1753, 2nd ed., 1780).

Later in 1732, several letters about English coinage then ensued between Browne Willis and Folkes. Folkes reported to Willis his analysis of the numismatic collection at Trinity College, Cambridge with regard to the gold coins produced during the reigns of Henry V and VI, thinking the presence of absence of annulets on their surface a distinguishing characteristic. Folkes stated, 'on my return to London...I hope to get some insights from some old verdicts of the Pyx which I was promised access to, which I could not get access before'.[38] Folkes also informed Willis that he had Vertue engrave a noble of Henry IV 'from the Earl of Pembroke's cabinet' that 'Mr [Hugh] Howard furnished me with the quarter of the same thing', and he gave Willis prints after the proofs were completed.[39] Willis in return gave Folkes a series of drawings of coins from his own collection. Folkes enthused that Howard's collection had a 'variety of Irish money the greatest I have met with', and that the collections of John Sharp, Archbishop of York, were also available for viewing. He then lamented that the gold coinage of Philip of Spain and Mary I 'is yet a desideratum', and the collections of his friend James Granger had been sold, which 'will be an irreparable loss to me if it is gone where I cannot come at it'.[40] Willis and Folkes continued to correspond for years, Willis writing to Folkes in 1743 that he was in a 'quest' in searching collections in Bedfordshire and Huntingdonshire for the half noble of Richard II, the individual in possession 'tempted to part with it by an exchange of an Elizabeths Angel'.[41] Just as natural historians like Hans Sloane traded specimens and locality records of birds and butterflies by post before the establishment of the British Museum in 1753, antiquaries like Folkes and Willis exchanged coins, drawings, and locations of numismatics, 'as there was no public or private collection that, by itself, contained a good enough selection of such coins'.[42]

What is less explicable is why Folkes took on his numismatic project a mere eight months before he left for his Grand Tour. It may have been simply that his Tour was something he embarked upon rather suddenly to help his wife Lucretia; after all, Folkes admitted to Antonio Cocchi, 'my wife is nothing so outrageous as when abroad'.[43] Perhaps it was because Folkes could gather comparative

---

[38] BL Add MS 5833, f. 152v, British Library, London. The word 'pyx' comes from 'pyxis', or a small box used to transport coins. The Assay Office has an annual trial that randomly tests a selection of coins for metallic content, weight, and measure to see if the Royal Mint has met requirements.

[39] BL Add MS 5833, f. 152v, British Library, London.

[40] BL Add MS 5833, Folkes to Willis, 17 September 1732, 19 October 1732, f. 153r., f. 155r., British Library, London.

[41] RS/MS/250/4/24, Willis to Folkes, 6 October 1743. Royal Society Library, London. Another letter much in the same vein from Willis to Folkes is in RS/MS/250/4/19, Royal Society Library, London.

[42] Pagan, 'Martin Folkes', p. 159. See James Delbourgo, *Collecting the World: The Life and Curiosity of Hans Sloane* (Cambridge: Harvard University Press, 2018).

[43] BL MS Facs 589, 584B Martin Folkes to Antonio Cocchi, 6 November 1735, British Library, London.

numismatic information while travelling. He recorded in his diary: 'I have employd my self some days in understanding the Venetian money and the changes that have been made in it as also the affairs of their Bank', noting 'they keep all their accounts in the Roman figures which seems an argument of its antiquity'.[44] He went to the Palazzo della Zecca, a building designed by Sansovino that housed the Venetian mint, recording the weights and measures of the bars used to make ducats.[45] He returned on another occasion to witness the coin stamping, observing they usually gave 'about 12 strokes to a half Ducat and 14 to a ducat', but deciding in comparison to English and particularly Newtonian coinage that 'the money is very ill stampt'.[46] Folkes did calculations for fellow virtuosi on the value of Florentine specie, spoke with the Procurator of Venice about the art of engraving coins, and borrowed a gold ducat from the Venetian consul to study.[47]

When he was in Nuremberg, Folkes met with a local salon headed by mathematician, astronomer, and translator Johann Gabriel Doppelmayr (1677–1750) and gathered information about the local coinage; his account book for his journey had a printed list of gold monies and their relative weights[48] (see figures 6.2 and 6.3). Doppelmayr had written the *Historische Nachricht* (1730), a history of mathematicians and artists of Nuremberg, which included illustrations of medallion portraits of his countrymen, a topic of conversation he may have had with Folkes. Doppelmayr would later correspond with Folkes concerning his atlas of cosmological maps known as the *Atlas coelestis* (1742), and his writing of the first German work on electricity.[49] While on tour, Folkes wrote to Sloane to recommend him for election to the Royal Society.[50]

On Folkes's return from his Tour, he resumed his correspondence with Browne Willis, writing in September 1736 and apologizing for 'being so rude for not having answered sooner'. Folkes read parts of his *Table of English Gold Coins* to the Society of Antiquaries on 15 and 22 January 1736, following with another paper on the weights and value of ancient monies such as the denarius and 'Greek Didrachmas', including those he 'had seen in cabinets in Italy', as well as in the Earl of Pembroke's collection.[51] Folkes 'expressed his thanks to Pembroke 'both to the Late Noble Collector and the present Possessor for the great freedom and Goodness with which theyr have always admitted his access to it'.[52] On 29 April

[44] Folkes, 'Journey from Venice to Rome', f. 36r.
[45] Folkes, 'Journey from Venice to Rome', f. 32r–33r.
[46] Folkes, 'Journey from Venice to Rome', f. 38r.
[47] Folkes, 'Journey from Venice to Rome', f. 36, f. 41r–42r.
[48] NRS 20658, inside back cover, Norfolk Record Office, Norwich.
[49] RS/MS/250/26, Martin Folkes; RS/MS/790, Letters 36–9, Royal Society Library, London.
[50] BL MS Sloane 4058, f. 60, Folkes to Sloane, 1 October 1733, British Library, London.
[51] Minutes of the Society of Antiquaries, vol. II, pp. 139–40.
[52] Minutes of the Society of Antiquaries, vol. II, pp. 139–40.

Fig. 6.2 Christoph Jacob Trew (1695–1769), friendship album, MS 1471, fol. 83r., University Library Erlangen-Nürnberg.

1736, the Society subsequently 'order'd that Mr Folkes Dissertation on the Gold Coins of England be printed at the Expense of the Society and that 500 Copies thereof be printed'.[53] Folkes claimed of his subsequent *Table of Gold Coins* (1736) that it was 'published for no other view' than 'in Hopes the observations of my Friends may correct my mistakes'.[54] Indeed, Browne Willis had observed that Folkes had omitted a large sovereign of James I in his work, as well as failing to indicate the different mints in which the coinage of the same value and design was produced; Folkes also admitted he had not seen the noble of Richard III. Folkes concluded that he hoped to 'go on with the gravings; which I hope to do soon... But I am ashamed to own my being abroad so many years from my

[53] Minutes of the Society of Antiquaries, vol. II, p. 174.
[54] Martin Folkes, *A Table of English Gold Coins: from the eighteenth year of King Edward III, when gold was first coined in England: with their several weights and present intrinsic values* (London: Society of Antiquaries, 1736); BL Add MS 5833, Folkes to Willis, 8 September 1736, f. 157r., British Library, London; Pagan, *Martin Folkes*, p. 159.

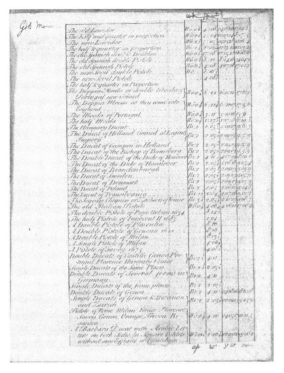

Fig. 6.3 Martin Folkes' memorandum book, coinage weights pasted inside back cover, (NRO, NRS 20658), Norfolk Record Office, Norwich.

papers, have so addled my hand from them, that I am frequently at a loss to make out things I had formerly done'.[55]

Despite Folkes's reservations, as Crystal B. Lake and David Shields have stated, in 1744 the Society of Antiquaries 'agreed to help Folkes produce a project not unlike the *Metallographia Britannica* by promising to pay for the engravings; Stukeley reportedly spoke out against this plan, and Folkes revised his tables again in 1745 and showed the Society of Antiquaries some drafts of material he was preparing for his omnibus publication.[56] On 13 February 1745 it was recorded in the Antiquaries Minute Books that Folkes:

pursuant to a proposition agreed to in December last, brought a Scheem of the English Gold & Silver Coins disposed in the Order & Size of the Book of Coins he has Published, to shew the manner he intended they might be done by the

---

[55] BL MS 5833, Folkes to Willis, 8 September 1736, f. 157v., British Library, London; Pagan, *Martin Folkes*, p. 159.

[56] Crystal B. Lake and David Shields, 'Plates 1.37–1.38: Tables of English Coins', *Vetusta Monumenta: Ancient Monuments*, https://scalar.missouri.edu/vm/vol1plates37-38-table-of-english-coins [Accessed 2 January 2020].

Society, which he computed in that method, to take up near fifty Plates. He also offer'd his care and study to add to each Plate, one printed Leafe, as an explanation of the coins therein represented: which altogether with some later discoveries of coins and remarks on them, would make another such Vol. as is already done by him. Added to this he Generously offer'd considering the great expense of such a Work, over and above, his pains in that affair, that he would at his own Expence make up those he had already done ten Plates, and Present them to the Society in part of the Work. For which kind offer he had the Unanimous Thanks of the Members of the Society present.[57]

In 1745, Folkes's first *Table of English Silver Coins* was also published. Although the work had no illustrations, it was a reasonably thorough history of the coinage from the Norman Conquest to George II, using archival sources, documents from the mint, coins borrowed from other antiquaries such as James West, and Folkes's own empirical observations of weights and measures. Folkes also consulted his extensive numismatic library, such as James Anderson's *Selectus Diplomatum et Numismatum Scotiae Thesaurus* (1739), for an analysis of the Scottish recoinage in 1707, as well as 'groupings of monographs related to the controversies surrounding the re-coinage of English silver coins under William III' in his own library.[58] Folkes gave a bound copy of his book to each of the Fellows of the Society of Antiquaries 'who took it as an Instance of his sincere Affection to them; and returned Him their hearty thanks'.[59]

Why was Folkes so intent on his project? Although historian Hugh Pagan noted that there was no evidence of why 'Folkes suddenly took an interest in coins in the early 1730s', after he lost the Royal Society presidency in 1727, in this period Folkes more generally moved from his pursuit of astronomy and mathematics to antiquarianism.[60] As Rebekah Higgitt has shown, Folkes was intimately involved in the creation and design of the Royal Society's Copley Medal, first issued in 1731 to Stephen Gray.[61] It also seems Folkes's close working relationship with Newton when Newton was Warden of the Mint may have been the impetus behind his numismatic interests. Just as Folkes promoted Newtonianism in the Royal Society and then abroad, writing an extensive history of numismatics with the Mint's archival material would honour Newton. Conduitt, the husband of Newton's step-niece and his executor, succeeded Newton as Master and Warden of the Mint on his decease, and did the rough

[57] Minutes of the Society of Antiquaries, vol. V, p. 54, pp. 55r and v.
[58] Martin Folkes, *Tables of English Silver and Gold Coins: First published by Martin Folkes, Esq; and now Re-printed, With Plates and Explanations* (London: Society of Antiquaries, 1763), p. 154; Kolbe, 'Godfather to all Monkeys', p. 46.
[59] Minutes of the Society of Antiquaries, vol. V, p. 15.   [60] Pagan, 'Martin Folkes', p. 159.
[61] Rebekah Higgitt, 'In the Society's Strong Box': A Visual and Material History of the Royal Society's Copley Medal, c.1736–1760, *Nuncius: Journal of the Material and Visual History of Science* 34, 3 (2019), pp. 284–316.

classification of Newton's mint papers before the archives descended through the Earls of Portsmouth.[62] Folkes, in his numismatic works, used evidence gathered from Conduitt to identify previously unknown Masters of the Mint, as well as the Mint's administrative structure, to supplement evidence from the coins themselves. We see, for instance, among the manuscripts in Folkes's own collection: 'An Abstract of the Accounts of Thomas Neale'. Neale was Master of the Mint from 1686 to 1699 preceding Newton.[63] Folkes wrote to Browne Willis, 'what I covet from every-body, is as many names of Mint-Masters, from the Reverses of each sort [of coin], as I am able'.[64]

As many of the mint records were destroyed during the English Civil War, Conduitt's summaries of past comptrollers' accounts also allowed Folkes to calculate total coinage produced in the reign of James I and during the interregnum.[65] Folkes's careful attention to these amounts in his *Table of English Silver Coins* (1747) was possible because the past fluctuation of specie produced and its value had been of primary interest to Newton. During his tenure in the Mint, Newton had to ensure that coinage reflected the current value of precious metal, and to make his determination he employed historical evidence. Newton paid the Exchequer and Peter de Neve, the President of the Society of Antiquaries, £27 for copies of records from 1603 to 1701 about the Trial of the Pyx, past weights of coin denominations, and totals of coinage minted.[66] These records, in turn, were excerpted by antiquaries, such as Folkes, later in the eighteenth century in an attempt to reconstruct Britain's numismatic history.[67]

So Folkes was simply carrying out Newton's past investigations for less pragmatic and more antiquarian reasons. For instance, in his analysis of the total amount of coinage produced year by year, Folkes procured a memorandum written by Newton used to support his 1717 devaluation of the guinea to 21s from 21s 6d to bring it nearer to its value in silver bullion. As Folkes related in his *Table of English Silver Coins*, Newton realized the coin, the Louis d'or, which was 'worth but 17s. and three farthings a piece, passed in England at 17s.6d. I gave notice thereof to the lords commissioners of the treasury, and his late majesty put out a

[62] MINT 19/6, 'Newton manuscripts in the library of the Royal Mint', The National Archives, Kew, p. 5.

[63] *A Catalogue of the Entire and Valuable Library of Martin Folkes, Esq.* (London: Samuel Baker, 1756), p. 156.

[64] BL Add MS 5833, Folkes to Willis, 19 October 1732, f. 154v., British Library, London.

[65] Folkes, *A Table of English Silver Coins from the Norman Conquest to the Present Time: with their weights, intrinsic values, and some remarks upon the several pieces* (London: Society of Antiquaries, 1745), p. 71, p. 90.

[66] MINT 19/6, vol. 4, 127a, 'Newton manuscripts in the library of the Royal Mint', The National Archives, Kew.

[67] See Ga 12885, 'Notebook containing Extracts taken from amongst Sir Isaac Newton's papers relating to directions about the trail of the monies of gold and silver in the pix and other technical notes relating to the Mint', 18th century, Department of Manuscripts and Special Collections, Nottingham University Library.

proclamation that they should go but at 17s and thereupon they came to the mint, and 1,400,000 were coined out of them'.[68] Folkes then noted Newton's decision accounted for the 'large coinages of gold about that time'. Folkes also featured especially the silver crown of Henry VIII (c.1545) in his *Table of English Silver Coins*, 'an illustration of Henry VIII's fraught attempts to manage the nation's finances by minting new coins and determining their valuation', a reflection of both Folkes's and Newton's interest in numismatic debasement.[69] Just as Folkes had edited Newton's *Chronology* as Conduitt wished for posthumous publication, he also felt it necessary to document Newton's technological achievements at the Mint. In this manner, Folkes was one of the first to place Newton's life and letters in historical context.

Similarly, Folkes also composed a manuscript: 'An Account of the Royal Society from it's [sic] First Institution'.[70] It was a detailed history of the Society's early years until 1742, recounting its incorporation, noting the differences in the list of Fellows between the first and second Charters and the presentation of the mace to the Royal Society by Charles II, and the circumstances of the donation of the Arundel Library; for instance, if the society failed, Arundel stipulated the books be reverted to him. He made note of William Dugdale's separation in 1678 of the part of the library concerning the antiquities of Norfolk, of personal interest to Folkes, as his estates were in that county. Folkes also transcribed each of the orations on the election of the monarchs to the society, a list of Fellows' elections and decease until 1749, rules of election, and by-laws. It was an antiquarian history of the Royal Society for reference, but also to preserve its heritage.

Lastly, the metrological aspects of numismatics—indeed Folkes's first *Table of Gold Coins* (1736) was purely concerned with their weights and values instead of their physical appearance—would also have appealed to his interests in weights and measures more generally. Folkes weighed English coins himself with surprisingly varied results, weighing 'no fewer than eight examples of the rare surviving groats of Edward I'.[71] 'The majority were from 80 to 85 troy grains, but three of them weighed 92, 116, and 138 grains respectively. From this diversity of weight, it should seem that they were trial-pieces'.[72] In 1732, Folkes repeated his experiments at the Society of Antiquaries, bringing 'an account of the English Gold & Silver coins examined by the balance & compared with the standard weight'.[73] Folkes also had his own 'hydrostatical ballance' at Queen Square, as well

[68] Folkes, *Table of English Silver Coins*, p. 130.
[69] Crystal B. Lake, 'Plate 1.20: Medals of Henry VIII, Edward VI, Elizabeth I, and James I', *Vetusta Monumenta: Ancient Monuments*, https://scalar.missouri.edu/vm/vol1plate20-sixteenth-century-english-medals [Accessed 2 January 2019].
[70] RS/MS/702, Martin Folkes, "An Account of the Royal Society from it's [sic] First Institution', Royal Society Library, London.
[71] Pagan, 'Martin Folkes', p. 159.   [72] Ruding, *Annals of the Coinage*, vol. 1, p. 206.
[73] Society of Antiquaries Minute Books, vol. 1, p. 282, as quoted in Crystal B. Lake and David Shields, Plate 1.69, Standard of Weights and Measures, *Vetusta Monumenta*. A Digital Edition. https://scalar.missouri.edu/vm/vol1plate69-weights-and-measures-1497 [Accessed 28 October 2019].

as two sets 'of scales and weights by Read', and another pair of assaying scales and weights by the same instrument maker.[74]

As Pagan pointed out, even Folkes's *Table of English Silver Coins* (1745), also published by the Society of Antiquaries, 'makes much use of data of a metrological character'.[75] Just as Folkes attempted to get the measurements of the Roman foot when in Italy, he analysed in his book the evolution of the use of the Troy Weight from the 'Pound of the Tower, or the Moneyers pound' after the Norman Conquest, comparing weights of the Troyes pound, the English pound and the *colonia* ounce., concluding the English Troy Weight was not used at the Mint until the 18th regnal year of Henry VIII.[76] As David Shields has indicated:

Following Folkes, who noted the official conversion to troy weight in the reign of Henry VIII, metrologists have since argued that Henry VII's Standard of Weights and Measures represents a working standard of the 'ancient British pound' rather than the Troy pound. However, by formalizing an equivalence between ancient practices, the crown's 'Tower Standard', and troy weight, Henry VII anticipated both the replacement by his son in 1527 of the Tower Pound with the troy, and the administrative actions and enforcement mechanisms that changing the King's standard would necessitate.[77]

Folkes's exactitude made his work a reference point for popularized manuals and plates of weights and measures.[78] On 11 November 1742, he communicated to the Royal Society 'an account of the Proportions of the English and French Measures and Weights, from the Standards of the same kept at that Society'.[79] This was followed by a set of weighing and measuring experiments by members of the Royal Society such as George Graham and Joseph Harris, one of the Masters of the Mint, of the standard of yard and the several standard weights of the Troy and Avoirdupois Pounds housed in several institutions such as the Mint, Guildhall, Tower of London, and the Founders Company, as well as a series of French institutions.[80] This set of trials even occasioned a group visit to Folkes's house in

---

[74] *A Catalogue of the Genuine and Curious Collection*, p. 3 and p. 7.
[75] Minutes of the Society of Antiquaries, vol. IV, p. 189. 'On 2 February 1744, the Vice President [Folkes] was pleased to read the remaining part of his Dissertation on our Silver Coins, whereon a motion was made and secured that he will suffer it to be printed at the expence of this Society now, which he was pleased to consent too, and Mr Bowyer a Member of this Society is desired to attend'.
[76] Folkes, *Table of English Silver Coins*, p. 3.
[77] Lake and Shields, 'Plate 1.69, Standard of Weights and Measures'.
[78] See, for instance, J. Millan, *Coins, Weights and Measures. Ancient and Modern. Of all Nations. Reduced into English on above 100 Tables. Collected and Methodiz'd from Newton, Folkes, Arbuthnot, Fleetwood...* (London: J. Millan, 1749).
[79] John Nichols, *Literary Anecdotes*, vol. 2, pp. 583–4.
[80] 'An Account of a comparison lately made by some gentleman of the Royal Society, of the standard of a yard, and the several weights lately made for their use; with the original standards of measure and weights in the exchequer, and some others kept for public use, at Guild-hall, Founders hall, the tower, &c', *Philosophical Transactions* 42, 470, (1743), pp. 541–56. George Graham's brass standard

Queen Square, in which the Standard Yard was delivered from the Exchequer by George Graham, where all the measuring trials were collated and found to agree with each other.[81]

In 1747, based upon this research, Folkes would go on to write a manuscript treatise about the true weight of the unit of the carat employed in assessing the renowned Pitt or Regent diamond, brought back from Madras by its governor Thomas Pitt and sold in 1717 to the French Regent, Philippe II, Duke of Orléans for £135,000. The stone was set in Philippe II's coronation crown, and described in an engraving of it to weigh 547 grains, which most readers would understand to imply the stone was 136 carats.[82] Folkes, however, noted these sensationalistic assumptions were based on a misunderstanding. The grains described were of 'the French or Paris ounce; for Carat or Diamond grains are no aliquot parts of the gros or ounce'[83] and that jewellers' carats were something different entirely, based traditionally upon 'the Venetian ounce, divided into 144 Carats, each Carat equal to 3 Troy Grains and two tenths'.[84] To complicate the matter further, 'the Carat weight now in use among Jewellers' Folkes found was a 'small matter lighter' than the original Venetian weight. From his calculations, Folkes demonstrated the weight of the diamond was substantially less than advertised, 'in Troy grains 433 and a half, or one grain and a half more than 18 penny weight', very close (28.061) to the modern weight of the diamond assessed at 28.128 grams.[85]

In 1746, the 'Standard of Weights and Measures produced on order of Henry VII in 1497' was engraved by George Vertue as Plate 1.69 for the Society's *Vetusta Monumenta* (see figure 6.4). Vertue's engraving was based upon an 'original parchment that was pasted on an oak table formerly in the Treasury of the King's Exchequer at Westminster and in the collection of Edward Harley, 2nd Earl of Oxford and Mortimer, at the time the print was engraved'.[86] As David Shields has indicated, the plate presented 'a dated index of data relevant to the Society… [of Antiquaries']…investigation and analysis of English coinage and medals', that dovetailed with its programme of support for Folkes's numismatic work. In 1750, Folkes's metrology became subject to further research by Samuel Reynardson, who presented a comprehensive historical study of weights and measures in

---

yard and diagonal scales from the Royal Society collection are still extant. See Science Museum Group. Graham's brass standard yard. 1900–159, Pt2 Science Museum Group Collection Online. https://collection.sciencemuseumgroup.org.uk/objects/co8028873/grahams-brass-standard-yard-measuring-yard [Accessed 4 August 2020]; Science Museum Group. Diagonal scale, screwed to mahogany base. 1900–161, Science Museum Group Collection Online. https://collection.sciencemuseumgroup.org.uk/objects/co60115/diagonal-scale-screwed-to-mahogany-base-drawing-instruments-diagonal-scales [Accessed 4 August 2020].

My thanks to Alison Boyle and Rebekah Higgitt for this information.

[81] 'An Account of a comparison', p. 550.

[82] MS 2391/4, Martin Folkes, 'Account of the Pitt diamond, 1747', Wellcome Library, London, p. 1.

[83] MS 2391/4, pp. 1–2.   [84] MS 2391/4, pp. 3–4.   [85] MS 2391/4, p. 4.

[86] Shields, 'Plate 1.69, Standard of Weights and Measures'.

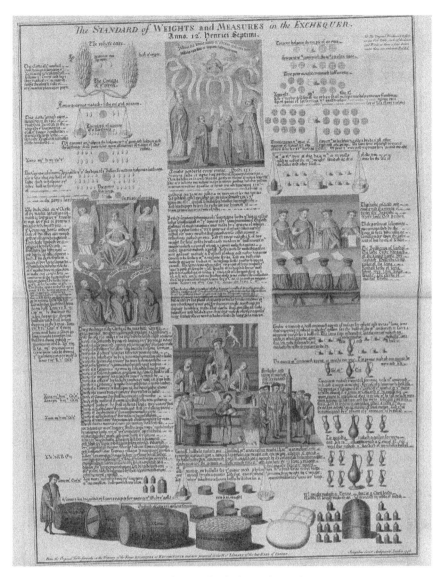

Fig. 6.4   Plate 1.69: Engraving of a Standard of Weights and Measures, 1497, *Vetusta Monumenta* (1746), vol. 1, Getty Research Institute, archive.org.

English history to prove that the 'present Avoirdupois Weight is the legal and ancient Standard for the Weights and Measures of this Kingdom.'[87] Characteristically, he then calculated the exchange rate between the Indian

[87] Samuel Reynardson, 'A State of the English Weights and Measures of Capacity', *Philosophical Transactions* 16, 491 (1749), pp. 54–71. Samuel Reynardson (1704–1797) was of Holywell Hall, Lincolnshire. In the 1760s he commissioned astronomer and designer Thomas Wright of Durham to lay out the landscape park and ornamental gardens.

pagoda and the English pound when Pitt bought the gem, noting 48,000 pagodas at 8 shillings and 6 pence each was equivalent to £20,400.[88]

In this period, Royal Society Fellows were also preoccupied with creating standards for weights and measures, including standardized tables of specific gravities of metals, so all of these concerns dovetailed in Folkes's numismatic work. There was in fact so much work in compiling specific gravity tables that critic John Hill lampooned the Royal Society, stating of one article in the *Philosophical Transactions*:

> Dr Davies, before he delivers his tables, gives some account of the authors to whom he has been obliged for the experiments of which they consist…he mentions particularly among the members of the royal society who have given tables of this kind, Mr Ellico, Mr. Graham and others…It is observable, Dr Davies had candour enough to acknowledge, that in many of the particular relating to these, there is some fallacy, when he repeated the experiments with the utmost caution and exactness, was not able to make them answer in the same manner.…[These observations] were improperly introduced among the transactions of a society whose name should give a sanction to the truth of what is ushered into the world under it, and whose motto, Nullius in verba, expresly declares against taking things on trust.[89]

If Hill understood the nature of statistical work in the Royal Society, such as Newton's and Cotes's methodology of weighted averages which Folkes employed when reconciling Newton's astronomical data in the *Chronology*, he may have realized that his critique was without merit (see chapter four). The new statistics forged a trustworthy result out of a discrepant set of numbers, precisely for such data as specific gravity measurements.

And Hill was hardly one to criticize Folkes or the societies of which he was President for lack of accurate empirical observation. A spirit of precise empiricism was even borne out in Folkes's final numismatic publication, the Posthumous *Tables of Silver and Gold Coins* (1763), which employed engraver Francis Perry to alter and supplement those of George Vertue; the work was edited by Andrew Gifford and John Ward, and overseen by Lord Willoughby. The engravings portrayed over 450 coins from Folkes's own collection and those of his fellow antiquaries. Joseph Ames, the secretary of the Antiquaries, sent each member of the Society's Council 'a Set of the Plates already engraven' to compare with Folkes's earlier edition.[90] In an act of numismatic crowd-sourcing, a committee of Fellows

---

[88] MS 2391/4, p. 5.

[89] [John Hill], 'Art. CXXXIX. Philosophical Transactions, Number 488. For the Month of June, 1748', *Monthly Review* 2 (1749–50), pp. 466–75, on pp. 470–1.

[90] The Society purchased his forty-four plates and copy for £120 after his death, which had been the cost of engraving, and William Folkes gave them some of his brother's manuscripts. See John Nichols,

of the Society were told that 'whatever observations or Additions' they would 'judge proper to be made' should be communicated in writing to Ames, to 'compleat and Perfect the Work.'[91] George North (1707–72), one of the Society's best numismatists, kept several of the replies in his own copy of Folkes's earlier edition, collating coins and their typologies between collections.[92] The final edition was quite successful, subsequently used by Ruding in his *Annals* in 1817 as well as in Seaby's *Standard Catalogue of Coins of Great Britain and Ireland* (1952); forty-four of the engraved plates of the English silver coins with specimens through the protectorate of Oliver Cromwell were even plagiarized in a numismatic publication by Thomas Snelling.[93]

Although Vertue's and Perry's engravings tended to 'humanise facial features' not present in the coins themselves—not entirely surprising, as Vertue's primary occupation was as a portrait engraver—Folkes, as a connoisseur of portraits (see chapter five), had also had some good insights noted by others about the visual portrayals of monarchs on the coins[94] (see figure 6.5). Maurice Johnson wrote to Stukeley 'Last Thursday I continued my historical Notice under King H[enry] VII and took notice of a Remark made by Mr Folkes his tables of this being the first of our Kings whose Coin bears a true Similitude to his Pictures, and that done en Profile.'[95] As Lake and Shields noted, 'Coins' depictions of monuments, historical dress, and symbolic objects as well as their inscriptions meant that they could be consulted in the service of a variety of historical inquiries. Antiquaries throughout the long eighteenth century, therefore, consulted coins in the course of developing histories of architecture, fashion, and literature, among other subjects.'[96]

Because of Folkes's expertise and sensitivity to visual evidence, he was asked to correct the plates for the *Numismata Antiqua* (1747), a set of engravings from the extraordinary collection of Thomas, 8th Earl of Pembroke, which included North America's first coins; this was a frequent reference source for Folkes's own

Biographical and Literary Anecdotes of William Bowyer, Printer, FSA (London: John Nichols, 1782), p. 179.

[91] The letter is pasted in the inside cover of George North's copy of the plates. See Arch.Num. X. 28,28*, [44 plates illustrating English gold and silver coins, circulated for observations to the committee of the Society of Antiquaries, in preparation for the edition of 1763, Tables of English silver and gold coins], Bodleian Library, Oxford. The Fellows included 'Lord Willoughby, Robert Bootle, Esq., Dr Chauncy, Mr Serjeant Ayre, Dr. Lyttleton, James West, Esq., Mr North, Dr Gifford, Dr Ducarel'.

[92] See Arch.Num. X. 28,28*, Bodleian Library, Oxford.

[93] Pagan, 'Martin Folkes', p. 160; Eaglen, 'The Illustration of Coins', p. 145; Nichols, *Literary Anecdotes*, p. 179.

[94] Eaglen, 'The Illustration of Coins', p. 148.

[95] Bodl. MS Eng. misc. c. 113, 24 October 1747, Johnson to Stukeley, ff. 345r–346v, Bodleian Library, Oxford; Martin Folkes, *A table of English silver coins from the Norman conquest to the present time: with their weights, intrinsic values, and some remarks upon the several pieces* (London: Printed for the Society of Antiquaries, 1745), p. 16.

[96] Crystal B. Lake and David Shields, 'Plates 1.37–1.38: Tables of English Coins'.

Fig. 6.5 George Vertue, engraved coins featuring Charles I, in Martin Folkes, *Tables of Silver and Gold Coins*, (London: Society of Antiquaries 1763–6), plate XIII, New York Public Library, Hathi Trust, Public Domain.

numismatic work[97] (see figure 6.6). The Earl of Pembroke died in 1733, and his son Henry memorialized his collection with the *Numismata*'s publication. In his own *Tables of English Silver and Gold Coins*, Folkes recorded and illustrated the very rare silver money coined in 1652 in New England with its motif of the American pine, as well as coinage minted by Lord Baltimore, the governor of Maryland, from his own collection, including 'Lord Baltimore's shilling, sixpence and groat'.[98]

There were other means of memorializing the friendship of Folkes and Pembroke. Folkes was a frequent guest at the 8th Earl of Pembroke's home at Wilton, and he and Sir Andrew Fountaine, numismatist and warden of the Royal Mint, 'were closely involved in the arrangement and cataloguing' not only of

---

[97] *Numismata Antiqua in tres partes divisa. Collegit olim et aeri incidi vivens curavit Thomas Pembrochiae et Montis Gomerici Comes* (London, n.p., 1747). There is reference to the Pembroke collection in the reprint of Folkes's *Tables of English Silver and Gold Coins* (1763) on pp. 142, 185, 203, 205, 206.

[98] Folkes, *Table of English Silver and Gold Coins Reprinted*, p. 98; *A Catalogue of the Genuine, Entire and Choice Collection of Coins*, p. 14.

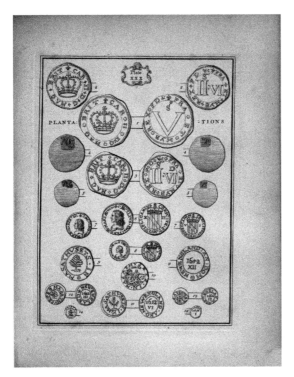

Fig. 6.6   Plate XXX, New England Shillings, figures 4, 5, 9, 11–14 in Martin Folkes, *Tables of English Silver and Gold Coins* (London: Society of Antiquaries 1763–6), New York Public Library, Hathi Trust, Public Domain.

Pembroke's coin collection, but also his collection of antique sculptures.[99] In commemoration of their efforts, Wilton's Great Room or Double Cube Room, the most important of the public rooms, displayed busts by Roubiliac of Folkes as well as Fountaine, part of a large assortment made up of marbles from the Arundel, Guistiniani, and Mazarin collections.[100] The busts sat on tables of Egyptian granite and lapis lazuli. One of the several eighteenth-century guidebooks about Wilton said of the 8th Earl, 'Bustoes he was particularly fond of, as they expressed with more strength and exactness, the lineaments of the face'.[101] Henry, the 9th Earl of Pembroke, commissioned the busts in 1751 'for placing in the new library', where they were ultimately moved. Both the 8th and 9th Earls relied on Fountaine and Folkes for advice about their collection, 'but at

[99]   Malcolm Baker, 'Attending to the Veristic Sculptural Portrait in the Eighteenth Century', in Malcolm Baker and Andrew Hemingway, eds, *Art as Worldmaking: Critical Essays on Realism and Naturalism* (Manchester: Manchester University Press, 2018), pp. 53–69, on p. 57.
[100]   Malcolm Baker, 'A Genre of Copies and Copying', p. 294.
[101]   James Kennedy, *A Description of the Antiquities and Curiosities in Wilton House, Salisbury* (London: E. Easton, 1769), pp. iv–v, as quoted in Baker, 'A Genre of Copies and Copying', p. 295.

the heart of the friendship between the three men lay their antiquarian interests and their continuing dialogue with antiquity'.[102] The portrait busts themselves, as Baker has indicated, were also modelled on Roman images which were intently naturalistic; Fountaine was in classical dress, and although Folkes was represented in a fashionable soft cap and fur collar that was strikingly contemporary, the detailed (and unflattering) portrayal of his face resembled veristic portraiture of the Roman Republic.[103] George Vertue remarked upon seeing it: 'having lately seen a Marble bust of Martin Folkes esq. cut in Marble by Mr. Rubillac—Sculptor, which I think is a most exact likeness of him—his feature strong & musculous, with a Natural & Just air of likeness—as much as any work of that kind ever seen—equal to any present or formal ages'.[104]

Folkes's and Fountaine's numismatic interests were commemorated on their portrait busts themselves; 'the cartouche or name plate was a direct adaptation of the reverses of Jacques Antoine Dassier's medals of Fountaine and Folkes, imitating Roman coinage'[105] (see image 6.7). Dassier struck Folkes's medals in February 1740/1, George Vertue remarking in his Notebooks that:

Dassier has published proposals for cutting several medals or Dies—the portraitures of famous men Living in England. Martin Folkes Esq. is done very like him...The subscription is—four Guineas for thirteen medals...[They are] done from life and are free and boldly cut but not so elaborately nor so high finish. As others, there appears a little of the fa-presto [sic] (see figure 6.8).[106]

Other individuals featured included De Moivre and John Montagu, 2nd Duke of Montagu, his star from the Order of the Bath on his shoulder. Folkes was very familiar with the renowned Dassier, as the medallist's series of twenty-four small medals, the *Hommes illustres* of Reformation figures such as John Calvin and Philip Melanchthon, was accompanied by a dedicatory medal honouring Folkes's uncle William Wake, Archbishop of Canterbury.[107] 'In a letter to the medallist, William Wake commends him for applying his great artistry to the immortalization of heroes of the mind and spirit, rather than the cruel and violent men customarily depicted in such objects'.[108] Folkes, not surprisingly, had a complete set of Dassier's seventy-four medals of 'illustrious persons...in four boards', and in

[102] Malcolm Baker, "'For Pembroke Statues, Dirty Gods and Coins'": The Collecting, Display, and Uses of Sculpture at Wilton House', *Studies in the History of Art* 70 (2008), pp. 378–95, on pp. 387–8.

[103] Baker, 'Attending to the Veristic Sculptural Portrait', p. 57.

[104] '[George] Vertue's Note Book B.4 [British Museum. Add. MS. 23, 704]', *The Volume of the Walpole Society* 22 (1933–4), pp. 143–62, on p. 152.

[105] Baker, 'For Pembroke Statues', p. 388.

[106] '[George] Vertue's Note Book B.4', p. 101 and p. 104.

[107] William Eisler, *Lustrous Images from the Enlightenment: The Medals of the Dassiers of Geneva*, ed. Matteo Campagnolo (Geneva: Skira, 2010), p. 36.

[108] Eisler, *Lustrous Images*, p. 38.

Fig. 6.7 Louis François Roubiliac, Bust of Martin Folkes, 1747, Marble. Earl of Pembroke Collection, Wilton House, Image: Conway Library, The Courtauld Institute of Art.

his house in Queen Square displayed a bust of the Earl of Pembroke 'on a painted deal term'.[109] Folkes's coin collection, his portrait busts, and the busts of his friends all had intertwined iconographic references to show his status as an antiquary and natural philosopher, an 'intellectually relevant means...to show their mutual esteem and broad affinities'.[110]

As for Henry Pembroke, in a dedicatory copy to Folkes of the *Numismata Antiqua* Henry wrote, 'My much honoured Friend Martin Folkes Esquire is desired to accept of this Copy of my father's medals and coins for poor

---

[109] *A Catalogue of the Genuine, Entire and Choice Collection of Coins*, p. 3; *A Catalogue of the Genuine and Curious Collection*, p. 3.

[110] Douglas James, 'Portraits in Medical Biography: Alexander Pope (1688–1744), Poet, Patient, Celebrity', *Journal of Medical Biography* 21, 4 (2013), pp. 200–8, on p. 207.

Fig. 6.8a, b Jacques-Antoine Dassier, Medal of Martin Folkes, Bronze, 1740, 54 mm, obverse and reverse. Accession Number 2003.406.17, Gift of Assunta Sommella Peluso, Ada Peluso, and Romano I. Peluso, in memory of Ignazio Peluso, 2003, Metropolitan Museum of Art, New York, Public Domain.

Acknowledgement for the great Trouble he has had in correcting the plates for the press'.[111] As it was a memorialization in the form of a visual statement, it was not an entirely practical work; antiquary Joseph Ames subsequently created and printed an index of four leaves for private circulation 'to the gentlemen who have Lord Pembroke's Book' which transcribed the names of the coins portrayed on the plates. Ames advised to 'number your increasing Leaves with a Pencil or Ink, from 1 to 308 inclusive; then you will readily find the printed part, and Tables answer to it'. The emphasis on plates was reminiscent of publications of natural history collections of the previous century such as Martin Lister's *Historiae Conchyliorum* (1685–92), in which rare or wondrous items from the natural world, like shells, were engraved, with very little explanatory text.[112] In eighteenth-century numismatics, the visual taxonomy as interpreted by the artist or engraver still played a paramount role.

There were also attempts by antiquaries and natural philosophers to provide more exact representations of ancient coins, combining empirical observation and technology with the growing interest in the medallic image in the early eighteenth century. In an article appended to his *Employment for the Microscope* (1753), Henry Baker included a set of 'directions for obtaining an exact Representation or Picture of any Coin or Medal'. It involved making a wax impression of the coin, inking the wax seal with copperplate printer's ink, and

---

[111] 'Pembroke Book 1747'. Holabird-Kagin Americana, 22 February 2014 Auction, Lot 1362. https://www.liveauctioneers.com/item/24254201_1747-pembroke-book [Accessed 4 September 2019].
[112] Anna Marie Roos, *Martin Lister and his Remarkable Daughters: The Art of Science in the Seventeenth Century* (Oxford: Bodleian Library Publishing, 2018), pp. 123–4.

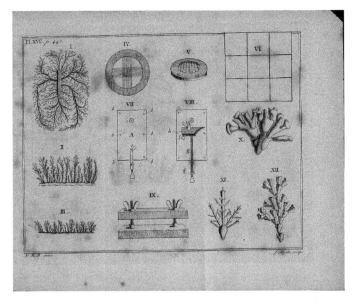

Fig. 6.9   Henry Baker, *Employment for the Microscope* (London: R. Dodsley, 1753), facing p. 441, plate IX. Columbia University Libraries, Archive.org, Public Domain.

then inking a paper by using a screw press with iron planks[113] (see figure 6.9). Antiquaries were interested in the processes of the creation of coin and seal impressions; Richard Rawlinson amassed a huge collection of medieval seal matrices and in 1738 Hans Sloane devoted one of the registers in his collection to *Impressions of Seals*.[114]

Baker's article was an adaptation of one published in the *Philosophical Transactions* in 1744.[115] Using Baker's method,, one could 'procure a noble Collection of genuine Prints or Medals, which may be placed in Books, in orderly Series': a collection-by-proxy. When Baker read his article on 19 April 1744, he demonstrated an impression made of an ancient silver medal, a half crown, and, in a bit of self-promotion, the 'picture and Relievo' of the Copley Medal recently awarded to him, an act that encompassed the intertwined relationship between natural philosophy and antiquarianism. Technology, metrology, and aesthetics were inherent to numismatics, a discipline that encompassed the 'scientific antiquarianism' that characterized Folkes's work and that of the Royal Society.

---

[113]   Henry Baker, *Employment for the Microscope* (London: R. Dodsley, 1753), pp. 436–42.
[114]   John Cherry, *Richard Rawlinson and his Seal Matrices* (Oxford: Ashmolean, 2006), p. 19.
[115]   Henry Baker, 'An Easy Method of Procuring the True Impression or Figure of Medals, Coins, &c. Humbly Addressed to the Royal Society', *Philosophical Transactions* 43, 172 (31 December 1744), pp. 135–7.

6.2 The Egyptian Society, 1741–3

Folkes's interest in coins and medals also dovetailed with his interests in Egyptian antiquities, and he was a very early member of the Egyptian Society (1741–3). The following analysis of the Society helps our understanding of Folkes's *social* milieu, *apart* from his involvement in the Royal Society and Antiquaries. However, it was a club that evinced an empirical and object-oriented approach to artefacts very similar to that employed in the Royal Society and the Society of Antiquaries, and thus assists in comprehending Folkes's *intellectual* world. As Young has commented,

> Fellows of the early Royal Society attended experimental demonstrations less to learn how to produce effects for themselves, than to access empirical material from which causal and axiomatic principles could be derived through discursive practice. At society meetings, members could access this material in a variety of ways; experimental demonstrations existed alongside the reading of experimental reports, accounts of lay empirical practices and travellers' reports.[116]

The Egyptian Society Fellows followed the same methodology to discover principles of sociocultural and religious practices in Egypt, as well as to provide 'object biographies' of artefacts that belonged to members that they examined with speculations on manufacture and use. The Society was another example of 'scientific antiquarianism' that was typical to Folkes's interests, and, as we will see, Folkes was also involved in numismatic research for the Society during its tenure. When Folkes signed the *album amicorum* of the Society, his mot was the same as Newton's: *Numero Pondere et Mensura, Deus Omnia fecit* (God creates all things through number, weight, and measure), reflecting his research in mathematics and his interest in the metrology of ancient architecture and artefacts.[117] This marriage of natural philosophy and antiquarianism was reflected in the methodological practices as recorded in the minutes of the Egyptian Society, which also had several FRS and FSA among its membership.

The Society, the first of its kind, was created for the purpose of 'promoting and preserving Egyptian and other antient learning'.[118] Renaissance humanists such as Athanasius Kircher in his *Oedipus Aegyptiacus* (1652–4) proclaimed that Egypt

---

[116] Mark Thomas Young, 'Nature as Spectacle: Experience and Empiricism in Early Modern Experimental Practice', *Centaurus* 59 (2017), pp. 72–96, on p. 91.

[117] BL Add MS/52362, f. 4r, Minute Book of the Egyptian Society, kept until 28 May 1742 by the Rev. Jeremiah Milles, as 'Reis Effendi', and from 5 Nov. by the Rev. Richard Pococke; 11 Dec. 1741–16 Apr. 1743, British Library, London.

[118] SGS No 80, MB5, Letter from William Stukeley to Maurice Johnson, 16 June 1750, f. 59A, Spalding Gentlemen's Society Library, Spalding, reprinted in Honeybone and Honeybone, *Correspondence*, pp. 155–7.

was the cradle of civilization. 'Another humanist scholar, Giordano Bruno, pushed such claims further, arguing that Christianity had perverted the true, original teachings of the Egyptians, who had worshipped God in the sun'.[119] Egyptian tablets with hieroglyphics were donated to the earliest Ashmolean Museum in the seventeenth century, and there was a view of 'some Egyptian buildings' in the Tradescant Collection at Lambeth.[120] However, it was not until *The Travels, or Observations relating to Several Parts of Barbary and the Levant* (1738) by Thomas Shaw (1694–1751), Chaplain to the English Factory in Algiers, that interest in Egyptology in the eighteenth century was revived. By the late 1730s, British travellers such as Richard Pococke began to go further south into Egypt than Saqqara.[121] Jean Terrasson's bestselling *Séthos, histoire ou vie tirée des monuments, anecdotes de l'ancienne Égypte* (1731), described the 'life of Sethos, an Egyptian prince who undergoes various trials to prepare him for initiation into the mysteries of Isis'.[122]

Terrason's work was translated into English and German in 1732, and was also well known in European Masonic circles, for whom 'Egypt remained the crucible of knowledge and wisdom'.[123] The oldest Masonic documents, the 'Manuscript Old Charges' claimed that when Abraham went into Egypt he taught the Egyptians the liberal arts and sciences, including the 'science of Geometry in practice, for to work in stones all manner of worthy words that belonged to...all manner of buildings'. Euclid himself was said to have given Egyptian masons a series of rules of conduct.[124] James Anderson's Masonic *Constitutions* (1723) claimed that a Jewish exile in Egypt 'provided the chosen people with the opportunity to become "a whole Kingdom of Masons", with Moses himself as their "Grand Master"'.[125] Egyptology, as interpreted in the eighteenth century, reflected Folkes's Masonic beliefs and interest in metrology, and as he was a dedicated attendee; it is worth delineating its activities to understand his intellectual milieu.

Extant for two years (1741–3), the Egyptian Society's antiquarian members either travelled to the Land of the Pharaohs as part of their Grand Tour, or were simply interested in Egypt's material, linguistic, or hermetic connections to current problems in intellectual history and theology. Like other similar

---

[119] Dan Edelstein, 'The Egyptian French Revolution: Antiquarianism, Freemasonry and the Mythology of Nature', in Dan Edelstein, ed., *The Super Enlightenment: Daring to Know Too Much* (Oxford: Voltaire Foundation, 2010), pp. 215–41, on p. 217.

[120] John Nichols, ed., *Bibliotheca topographica Britannica*, 8 vols (London: J. Nichols, 1780–90), vol. 2, p. 97, p. 125.

[121] James Stevens Curl, *The Egyptian Revival: Ancient Egypt as the Inspiration for Design Motifs in the West* (Abingdon and New York: Routledge, 2005), p. 144.

[122] John Hamill and Pierre Mollier, 'Rebuilding the Sanctuaries of Memphis: Egypt in Masonic Iconography and Architecture', in Jean-Marcel Humbert and Clifford Price, eds, *Imhotep Today: Egyptianizing Architecture* (London: UCL Press, 2016), pp. 207–20, on p. 209.

[123] Edelstein, 'The Egyptian French Revolution', p. 219.

[124] Hamill and Mollier, 'Rebuilding the Sanctuaries of Memphis', pp. 209–10.

[125] Edelstein, 'The Egyptian French Revolution', p. 219.

Fig. 6.10   Andrew Miller, *Lebeck*, mezzotint after Sir Godfrey Kneller, mezzotint, 1739, 353 mm × 250 mm. Medical Historical Library, Harvey Cushing/John Hay Whitney Medical Library, Yale University.

organizations such as the Royal Society Dining Club, it had its origins in a dinner at the Lebeck's Head Tavern, Chandos Street, Charing Cross on 11 December 1741.[126] Lebeck was one of the most 'noted tavern keepers of his day', his rather beefy countenance painted by Sir Godfrey Kneller, wearing a linen cap and holding a glass of beer[127] (see figure 6.10). He 'distinguished himself by providing the best food and most delicate wines for his clientele at extravagant prices', his tavern a virtual temple of Epicureanism fit for the first meeting of the Society for its 'inaugural feast of Isis'.[128] Although the Society usually met in London, it is possible that it also met at Boughton House, belonging to John, 2nd Duke of Montagu who was a member. There is an Egyptian Hall at Boughton, and as members were accustomed to 'dining in each other's houses', this 'presumably

[126] A point noted by Warren R. Dawson, 'The First Egyptian Society', *The Journal of Egyptian Archaeology* 23, 2 (Dec. 1937), pp. 259–60, on p. 259.

[127] Jaap Harskamp, *Streetwise: Art at Heart of the City, Streetscapes from Lorenzetti to Mondrian* (Armorica Editions, 2015), p. 280. http://issuu.com/bookhistory/docs/streetwise_f53b04df393fff [Accessed 28 August 2018].

[128] Dawson, 'The First Egyptian Society', p. 259.

gave rise to the hall's name. No doubt evocative of the spirited gatherings in the room, Chéron's ceiling portrays a rotund Bacchus riding on an ass with accompanying putti and bacchanalian accoutrements'.[129] Just as in Masonic Lodges or the Thursday Club for the Royal Society, the Egyptian Society practised ritual feasting.

The first four members of the Society were Captain Frederick Lewis Norden, Reverend Richard Pococke, John Montagu, 4th Earl of Sandwich, and physician Charles Perry, all of whom had travelled to Egypt.

Norden was the King's official representative on the first Danish expedition to Egypt under King Christian VI in 1737–8, his interest in Egypt stimulated by the German diplomat Baron Philipp von Stosch, whom he had met on his Grand Tour.[130] Norden was a talented draftsman and cartographer, and in 1739 he came to London with other naval officers such as Ulrich, Count of Danneskiold-Samsøe, who were asked to 'serve under the British flag'. During this time, Norden worked on his Egyptian material, wrote a critique of John Greaves's *Pyramidographia* (1646), a study of the metrology of the Egyptian pyramids, and sent his remarks from Portsmouth to Folkes on 11 October 1740, 'whom he had met already', and in whose care he had left part of his archives from Egypt'.[131] Folkes spoke of Norden's work with 'praise to several connoisseurs', and the latter was elected FRS on 8 January 1740/1, the certificate of election written and signed by Folkes.[132] In London, Norden met Richard Pococke, 'who had passed the Danish expedition on the Nile', and the idea for the Egyptian Society was born.[133] In 1741, to acknowledge his election to the Royal Society, Norden published an illustrated essay, 'Drawings of some Ruines and Colossal Statues at Thebes' with 'four of his hundreds of drawings'.[134] Norden's introduction to his essay was addressed to Folkes, emphasizing that he followed the Society's maxim of *nullius en verba*, his 'fixed resolution to pursue no other rule than that of delivering with truth and simplicity such accounts as I am able to give of those places'.[135]

---

[129] Richard, Duke of Buccleuch and Queensberry, *Boughton: The House, Its People and Its Collections* (Hawick: Caique Publishing, 2006), p. 110. The Egyptian Hall also features a model bridge in the gothic style made by Stukeley for John, 2nd Duke of Montagu.

[130] Marie-Louise Buhl, Erik Dal, and Torben Holck Colding, *The Danish Naval Officer: Frederik Ludvig Norden* (Copenhagen: The Royal Danish Academy of Sciences and Letters, 1986), pp. 11–12; See also Paul John Frandsen, *'Let Greece and Rome Be Silent': Frederik Ludvig Norden's Travels in Egypt and Nubia, 1737-1738* (Denmark: Museum Tusculanum Press, 2019). My thanks to Richard Parkinson for alerting me to this source.

[131] Buhl, Dal, and Colding, *Norden*, p. 36. John Greaves, *Pyramidographia; Or, a Description of the Pyramids of Egypt* (London: George Badger, 1646). The remarks are published in the first volume of Norden's posthumously published *Voyage d'Egypte et de Nubie* (Copenhagen: Maison Royale des Orphelins, 1755), pp. 89–101.

[132] RS/EC/1740/21, Royal Society Library, London; 'Preface', *Voyage d'Egypte et de Nubie*, p. xxiv.

[133] Buhl, Dal, and Colding, *Norden*, p. 36.

[134] Paul Frandsen, *Let Greece and Rome Be Silent: Frederik Ludvig Norden's Travels in Egypt and Nubia, 1737-1738* (Copenhagen: Museum Tusculanum Press, 2019), p. 25.

[135] Frederik Ludvig Norden, *Drawings of some Ruines and Colossal Statues at Thebes in Egypt, with an account of the same in a letter to the Royal Society* (London: n.p., 1741), p. 4; Frandsen, *Let Greece and Rome Be Silent*, p. 166.

Sadly, the publication of Norden's more complete travel accounts was posthumous, as he died on 22 September 1742 of tuberculosis in Paris where he had travelled to recover. Norden was 34 years old, and Montesquieu noted to Folkes, 'I am very angry; he was a man of merit whom we relied on for his spirit and knowledge. The poor man had a very happy ending, considering himself lost only an hour before being lost'.[136] In October 1748, when Folkes was President of the Royal Society, he endorsed the publication by subscription of the 'valuable' work: 'The Travels of our late Worthy Member Captain Norden'; this work eventually became the *Voyage d'Egypte et de Nubie* published by the Royal Danish Academy in 1755, engraved by Carl Marcus Tuscher.[137]

In 1738–9, Lord Sandwich had visited Greece, Turkey, and Egypt in his Grand Tour and met Perry in the Levant.[138] Several other leading antiquaries,, nobles, and explorers who had not visited the Nile joined later including Folkes and Folkes's fellow Mason, John, 2nd Duke of Montagu. Folkes and Pococke had probably met in Italy, as in a letter to his mother (Mrs Elizabeth Pococke) from Rome, dated 30 April/11 May 1734, Pococke stated: 'Went to see Mr Fowkes [sic] a very ingenious fellow of the Royal Society, he has his family here'.[139] Pococke's cousin, Jeremiah Milles, who became Egyptian Society secretary, was in Italy with him. Another member of the Society (*inter alia*) was Stukeley, who, influenced by the work of Athanasius Kircher, was interested in finding common origins between early Hebrew and Egyptian religious practices.[140]

The Egyptian Society and its remaining Minute Book (BL Add MS 52362) has previously been characterized in the context of the Grand Tour and popularity of travel writing.[141] Certainly there was a fashion in this period for Egyptian travel

---

[136] Montesquieu to Folkes, 27 September 1742, *Correspondance de Montesquieu*, vol. 1, p. 375. [Le pauvre capitaine Norden' est mort, Monsieur,j'en suis très fâché; c'était un homme de mérite, et nous comptions beaucoup son esprit et son savoir. Le pauvre homme a eu une fin très heureuse; il ne se jugeait perdu qu'une [heure] avant que de l'être].

[137] RS/CMO/4/5, Royal Society Library, London; Frandsen, *Let Greece and Rome Be Silent*, p. 167. The English translation was by Peter Templeman, and published in 1757.

[138] David Boyd Haycock, 'Ancient Egypt in 17th and 18th Century England', in Peter Ucko and Timothy Champion, eds, *The Wisdom of Egypt: Changing Visions Through the Ages* (London: UCL Press, 2003), pp. 133–61, on p. 148.

[139] BL Add MS 19939, Original letters of Dr. Richard Pococke [Bishop successively of Ossory and Meath] to his mother, written during his travels in Italy, Germany, etc.; 30 Nov/10 Dec. 1733–Aug. 1737; and containing minute accounts of the places visited by him, f. 126; BL Add MS 22978 (copy letter made by Mrs Pococke), Tour through France and North Italy, including Rome, in the form of letters to his mother; 28 Aug./8 Sept. 1733–1730 June/11 July, 1734; with corrections in Pococke's hand, Italy: Travels in, by Dr. Pococke: 1733–1737, f. 1, British Library, London. My thanks to Rachel Finnegan for this information.

[140] The Society was eventually limited to thirty members, 'those that have been in Egypt excepted'. BL Add MS 52362, f. 33r., British Library, London.

[141] There is one other brief account of the Society: M. Anis, 'The First Egyptian Society in London (1741-3)', *Bulletin de l'Institut français d'archéologie Orientale* 50 (1950), pp. 99–105. For themes of travel and Georgian sociability, see the short discussion of the Society in Rachel Finnegan's excellent study of the Divan Club: 'The Divan Club, 1744–46', *EJOS* IX (2006), pp. 1–86, esp. pp. 15–16; Rachel Finnegan, *English Explorers in the East (1738–1745): The Travels of Thomas Shaw, Charles Perry and Richard Pococke* (Leiden: Brill, 2019), pp. 105–10, 284–5.

and antiquities, a journey there seen as going beyond the usual vapid Grand Tour venues.[142] As Charles Perry noted in his preface to his *A View of the Levant* (1743):

> Indeed, the Turkish Empire...at least the more central Parts of it (such as Constantinople, the Archipelagean Isles, the Sea-coasts of Asia Minor, and of Syria) are now become pretty trite Subjects...But none of our own Countrymen...have yet wrote fully, and circumstantially, upon the Government, Politics, Maxims, Manners, and Customs of the present People of Egypt.[143]

Perry dedicated the book to Lord Sandwich, and used his anecdotes of travel in his publication.

The Society has also been studied as an exemplar of Georgian sociability, a drinking club akin to the Dilettanti or the Divan Club, and certainly tomfoolery abounded within its meetings and associated social visits. Stukeley reported when he first met Lord Sandwich, 'then lately come home from his Egyptian travels', that the nobleman (in an act of casual racism), put on the habit of the Arabs inhabiting those:

> oriental countrys...'.Tis called camissa, a black short gown, with open sleeves, loose; a slit on the breast, for convenient putting on; reaching down only to the knees, the body and legs otherwise naked.[144]

Not a habit to repeat (see figure 6.11).

Lord Sandwich also comically (at least to him and his fellow members) referred to himself as the 'Sheik' when chairing meetings, while the secretary was given the name 'Reis Essendi', the treasurer (Andrew Mitchell), the 'Hasneda'r', the collector of the fees (W. F. Fauquier), the 'Mouhasil',[145] and the inspector of Egyptian medals (Smart Lethieullier), the 'Gumrocjee'.[146] The Arabic, Ottoman Turkish, or Persian names given to the offices may have been due to Folkes: as mentioned in chapter two, Folkes was a correspondent of the Moroccan Ambassador Mohammed Ben Ali Abgali and hosted Abgali as a member of the Royal Society Council when he spent eighteen months in England in 1725–7.

---

[142] Major works on early modern Egyptology include: Brian Curran, *The Egyptian Renaissance: The Afterlife of Ancient Egypt in Early Modern Italy* (Chicago: University of Chicago Press, 2007); Daniel Stolzenberg, *Egyptian Oedipus: Athanasius Kircher and the Secrets of Antiquity* (Chicago: University of Chicago Press, 2013); Stephanie Moser, *Wondrous Curiosities: Ancient Egypt at the British Museum* (Chicago: University of Chicago Press, 2006).

[143] Charles Perry, Preface to *A View of the Levant: Particularly of Constantinople, Syria, Egypt and Greece* (London: T. Woodward, 1743), p. 1.

[144] SGS No 80, MB5, Letter from William Stukeley to Maurice Johnson, 16 June 1750, f. 59A, Spalding Gentlemen's Society Library, Spalding, reprinted in Honeybone and Honeybone, *Correspondence*, p. 156.

[145] An Ottoman Tax Collector. See Finnegan, 'The Divan Club', p. 16.

[146] Anis, 'The First Egyptian Society', p. 102.

Fig. 6.11 Sketch of a camisa, in a letter from William Stukeley to Maurice Johnson, 16 June 1750, SGS No 80, MB5 fol. 59A, by kind permission of the Spalding Gentlemen's Society, Spalding, Lincolnshire.

Abgali was made FRS in 1726, and was also a friend of John, 2nd Duke of Montagu, and Lennox who also belonged to the Society.[147]

The Egyptian Society has also been dismissed by historians as only promoting traditional humanist philology, linguistics, and textual criticism in the context of its Egyptian studies. Levitin stated, for example, that it 'never seemed to go beyond discussion of the objects that were brought in to broader claims about ancient history, contenting themselves with the standard explanations from Herodotus, etc (at least as far as the minutes allow us to see)'.[148] However, a detailed

---

[147] RS/MS/790, Letter from Abgali to Folkes (In Arabic with accompanying translation in French): 'You are aware that said Mr Russel had come with me for a job that was to earn him 250 pounds sterling a year; but since he had to go back to England because of a setback, I felt obliged to write you to ask you to recommend him to my honoured friends the Duke of Richmond, and the Duke of Montaigue, so that through them he may have some advantage that could make him go through this life more comfortably. It is your praiseworthy habit, as it is theirs, to please all those who bond with you. I beg you to forgive me for the trouble I am giving you, being convinced that my friendship alone for you has made me do this'.

[148] Dmitri Levitin, 'Egyptology, the Limits of Antiquarianism, and the Origins of Conjectural History, c. 1680–1740: New Sources and Perspectives', *History of European Ideas* 41, 6 (2015), pp. 699–727, on p. 726.

analysis of the Minute Book, as well as its associated contextual sources, shows to the contrary, that the Egyptian Society was more than a travellers' association, a social club, or one dedicated to tried and true historical analysis.[149] As James indicated, 'the enthusiasm which had led to the formation of the Egyptian Society represented a growing interest in ancient Egypt and its physical remains of a quality quite different from that demonstrated by the random collection of Egyptian antiquities by Sir Hans Sloane and other general collectors'.[150]

Stukeley even dismissed the scholars' penchant for applying 'immediately to Herodotus, as the oracle, the spring when all antient learning must needs flow'.[151] Rather, Folkes and the other members of the Society were engaged in comparative analysis of language, customs, and cultural institutions that certainly included, but went beyond, 'standard explanations from Herodotus' using their own experiences of travel to provide the cultural and historical context of material culture. These sources evince an approach in the Egyptian Society to textual sources and artefacts that characterized early modern antiquarianism and the development of early archaeology, one that also had a close relationship with the methodology of natural philosophers.[152] As Stukeley stated, 'In the study of Antiquities, as in all others, judgment, genius is necessary'.[153]

6.3 The Egyptian Society Minute Books

The first page of the Society's Minute Book displays a finished drawing of a bronze Egyptian Sistrum, a symbol of Isis, and musical rattle used in processions and religious ceremonies such as the festival of Pasht at Bubastic[154] (see figure 6.12). 'Plutarch and others have mentioned the sistrum as having been intended to frighten away Typhon, or the evil spirit; and in describing the instrument, mentions that a cat with a human visage is depicted on it'; the cat is Pasht's representative. The sistrum served as the icon of the society, appearing on its summons to membership and meeting invitations.[155] Its supposed properties to drive off evil spirits made the sistrum a popular item to collect, the virtuoso Sir Hans Sloane also having one in his possession.[156] The sistrum which the Egyptian Society

---

[149] William Stukeley, MS: On Egyptian Antiquitys. 1742. Freemasons' Hall Library, London.
[150] T. G. H. James, *The British Museum and Ancient Egypt* (London: British Museum Press, 1981), p. 5.
[151] William Stukeley, MS: On Egyptian Antiquitys. 1742. Freemasons' Hall Library, London, f. 2.
[152] For more on innovations in antiquarianism, see Lydia Janssen, 'Antiquarianism and National History: The Emergence of a New Scholarly Paradigm in Early Modern Historical Studies', *History of European Ideas* 43, 8 (2017), pp. 843–56.
[153] Stukeley, *Medallic History*, vol. 1, p. 1.
[154] BL Add MS 52362, f. 1, British Library, London.
[155] Stukeley pasted an invitation with the seal of Lord Sandwich into the first page of his notebook dedicated to Ashtaroth, ABRAHAM. Freemasons' Hall Library, London.
[156] Stukeley, *Medallic History*, vol. 1, p. ix.

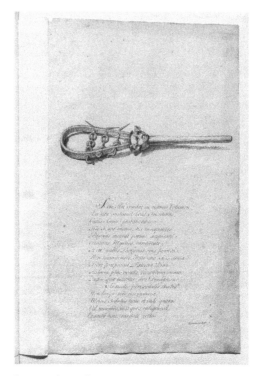

Fig. 6.12 Sistrum drawing from the Egyptian Society Minute books with opening poem, BL MS Add. 52362, f. 1r., © The British Library Board.

owned also served as Sandwich's staff of office, a token of its purpose and its jollity, and its cultural appropriation.

This sociability was also indicated by the next two pages, featuring an admonitory neo-Latin poem (one could even say neo-humanist poem) composed by Thomas Dampier, serving as a warning against interlopers into the Society's business.[157] Dampier was Lower Master at Eton from 1745–67, and a beneficiary (*inter alia*) of the Advowson of the Rectory and Parish Church of Eynesbury under Lord Sandwich's gift.[158] He also was described as a *bon vivant* and highly popular in society, if somewhat indulgent to his pupils. 'A boy under him, fonder of cricket than of Latin and Greek, boasted he was sure of getting his remove— "Dampier loves a good glass of wine, I'll write to my father to send him a hamper of claret, and mark, if I do not soon swim in the Upper School".'[159]

---

[157]   BL Add MS 52362, ff. 1–2, British Library, London.

[158]   MS HINCH 2/69, Deed of Indemnity. 1. Rt. Hon. John, 5th Earl of Sandwich. 2. William Palmer of Brampton, co. Huntingdon, Esq. Huntingdonshire Library and Archives, Huntingdon, Cambridgeshire.

[159]   Sir H. C. Maxwell Lyte, *A History of Eton College* (London: Macmillan, 1911), p. 336.

Nonetheless, Dampier was a skilled classicist, as the fifty-two verses in the Egyptian Society Minute Book were written in hendecasyllables, a rigid metrical scheme of eleven syllables per line, which is fiendishly difficult to compose.[160] He was in good company, as several other members of the society were also skilled composers and recipients of joking neo-Latin verses.[161] Antiquary Daniel Wray, for instance, was so known for his 'vivacity and laughing air', and for his comic poetry, that Richard Roderick, Fellow of Magdalene College, Cambridge at one point dedicated to him the verse: 'Capricious Wray A Sonnet needs must have'.[162] The poem for the Egyptian Society was well received, and it was ordered to be appended to the Minute Book in a fine calligraphic round hand, probably executed by the writing master and engraver George Bickham (1683/4–1758) or his son of the same name; one of the numismatic drawings in the Minute Book was signed *Bickham, delin* (*delineat* or drawn).

Dampier's verse was read at the meeting of 22 January 1742 by Reverend Jeremiah Milles (1714–84), the secretary who would go on to become President of the Society of Antiquaries of London (1768). Dampier's poem began:

> If this volume should fall into the hands of anyone who takes it up amiss without being properly instructed in the initiation of the followers of Isis, then I urge him to halt here, and not turn over the following pages, for he will be making trial of mysteries with an ill-omened curiosity, which no amount of tears, alas, will expiate. For Isis did not place the 'sistrum' at the head of this book for no reason; it is her weapon, a thing to be feared for its penetrating rattle more than that which the altar of Dindymene[163] shakes.

The sistrum played a key role in Society meetings that went beyond the ceremonial or convivial. Its reputed power to drive off spirits was a topic of discussion on 11 December 1742, the winter solstice, when the Society met to elect John, 2nd Duke of Montagu as a member.[164] William Stukeley delivered a paper on the origins of the sistrum's magical qualities, a paper he repeated when he gave it on 16 January 1751/2 at the Spalding Gentlemen's Society in Lincolnshire.[165]

---

[160] My thanks to Tom Holland for this point. The verses were presented to the Egyptian Society on 22 January 1741/2 and ordered to be put into the Minute Book. BL Add MS 52362, f. 17r., British Library, London.

[161] BL Add MS 4456, Verses, etc., almost entirely by Birch or addressed to him, ff. 42, 53, 70, 176–7, British Library, London.

[162] BL MS 4456, f. 42, British Library, London. Justice Hardinge, 'Biographical Anecdotes of Daniel Wray, Esq. FRS and FSA', in John Nichols, ed., *Illustrations of the Literary History of the Eighteenth Century*, 8 vols (London: for the Author, 1817–58) vol. 1, pp. 5–118, on p. 117.

[163] 'Dindymene' is another name for Cybele, the great mother goddess, who was worshipped with drums and pipes.

[164] Montagu was proposed for membership by Martin Folkes. BL Add MS 52362, f. 15v.

[165] Minute for 16 January 1751/2, SGS Minute Books, Spalding Gentlemen's Society, Lincolnshire.

Stukeley believed that the animal sacrifices performed by the prophet Abraham in the Old Testament involved a form of sistrum, so the prophet could, as the scripture noted: 'drive away the birds of prey coming down upon the sacrificed animal, in order to feed on them'.[166] Here [says Stukeley] 'then is the true foundation of the famous Egyptian Sistrum, a religious utensil properly speaking, pertaining to the act of sacrificing...the memorial of it, preserv'd by the Egyptians, tho' they knew not the true original of it; but in their way attributed to it, the power of driving off every evil, spiritual or temporal'.[167] He continued, 'their notion of its prophylactic virtue remained notorious, whence in Statues of Isis, she holds a Sistrum in one hand, a pitcher or water vase in another, which intimated an invocation to the divine Being, to protect, and preserve the regular flux of the Nile'.[168] Lastly, the sistrum could be seen in the constellation Engonasis, showing Adam 'kneeling before an altar, in the act of sacrificing two pigeons', and scaring off 'two ravenous birds attempting to steal doves from the altar' with sistrum.

Such ties between the Old Testament prophets, ancient astronomy, and Egyptian gods and their symbols were not unique to Stukeley, but were a penchant among natural philosophers, as we have seen in others, including his fellow Lincolnshire man and friend Sir Isaac Newton, and of course, Folkes. Newton claimed in his *Chronology of Antient Kingdoms Amended* based upon some fairly specious evidence, that 'Pharaonic Egypt was not a kingdom of major significance until after the time of King Solomon', and that, contrary to the Bible, all civilizations, including Egyptian, stemmed from the Israelites.[169] He also argued in his work, (heavily indebted to Vossius' *De Theologia Gentili*), that Egyptian religion was post-diluvian idolatry, a corrupted form of the true religion begun by Noah: Newton said 'it's certain that ye old religion of the Egyptians was ye true [Noachian] religion tho corrupted before the age of Moses by the mixture of fals[e] Gods with that of ye true one'.[170] According to Newton, 'each tribe or people had come to worship its own images, turning their ancestors into deities and naming stars and constellations after them. They had worshipped what Newton tellingly referred to as "hieroglyphical figures" as well as stones, statues and sculptures'.[171] Clearly, the Egyptian Society was not just beholden to Herodotus in its interpretations of material culture and artefacts, but set these interpretations into a much richer intellectual context involving chronology and biblical exegesis.

[166] Stukeley, 'Dedication', *Medallic History*, vol. 1, p. xii.

[167] Stukeley, 'Dedication', *Medallic History*, vol. 1, p. xii.

[168] Stukeley, 'Dedication', *Medallic History*, vol. 1, p. xii.

[169] Ernest Davis, 'Review of Newton and the Origin of Civilization, by Jed Z. Buchwald and Mordechai Feingold', https://cs.nyu.edu/davise/papers/Newton.pdf [Accessed 4 September 2018].

[170] Newton Yahuda MS 41, f. 5, as quoted in David Haycock, *William Stukeley: Science, Religion and Archaeology in Eighteenth-Century England* (Woodbridge: The Boydell Press, 2002), p. 151.

[171] Haycock, *William Stukeley*, p. 151.

After his first verse about the power of the sistrum, Dampier continued his hendecasyllabic poem, advising:

If you are eager to become our fellow-devotee, I do not demand that you make a long voyage, constantly seeking out from all places merchandise, or coins, save for those which venerable antiquity commends, and which are readily identified because of their rusty and well-known rarity. You may well search for these everywhere with admirable acquisitiveness and with unusually keen enthusiasm, and should not shrink from stirring the hidden gods of the underworld, nor from cheating Charon of his coin [i.e. engaging in archaeological research and excavations].

Although the Grand Tour was alluring, 'collections of classical antiquities made by wealthy Grand Tourists such as the Second Earl of Arundel also inspired the collection and appreciation of those remains at home'.[172] Sir William Fermor, Baron Lempter (d. 1711), who bought the Arundel marbles, had his county seat Easton-Neston, Northamptonshire radically altered to house this collection of ancient Greek, Roman, and Egyptian statues. As Haycock has shown, Stukeley himself believed that the pursuit of such collecting and learning in England had become 'so universal' that it was 'a necessary qualification for a Scholar & a Gentleman. it seems to me to carry with it a great force in enlarging the understanding of cultivating that Manly greatness of Spirit peculiar to the English'.[173]

However, as Dampier advised, if it was possible, it was advisable to go to Greece and then to Egypt, considered by many early modern scholars as the oldest civilization in the world and at the basis of philosophy, as well as the mathematical disciplines of the *quadrivium*. His poem continued:

But while you rival the fame of Pythagoras, take care that, with boldness begotten of ignorance, you do not ignore the health-giving soil of Attica, and Aeolus, god of the winds. When the god so orders, his temple *[in Athens]* is not buffeted by either north or south wind, but when it looms on the horizon it should be treated with reverence by passing sailors. If you do this, you will secure the god's favour and safely arrive on the shores of Egypt, and the kindly nursery of gentle Pallas. O land of Egypt, holy mother and nurse-maid of the arts, who in rivalry with her own river Nile pours forth far and wide the wealth of knowledge. No age, and no place remains silent about you, and none ever will. After so many centuries, the Dead One *[Memnon]* arises once more, with limbs unscathed, and

[172] Haycock, *William Stukeley*, p. 151.
[173] Bodl. MS Eng. misc. c. 323, f. 81r., Bodleian Library, Oxford, as quoted by Haycock, *William Stukeley*, p. 115.

gives you praises which increase day by day, and his marble memorial recalls with fine words the honours due to you. If you have any intelligence, my friend, you will range over this land with observant steps, and gather abundant harvests of Learning. Thus you will win for yourself a glory to be recorded on a bright page.

But we implore you, Isis, and your Osiris, and whichever gods and goddesses dwell in Egypt, to protect this book with a more watchful divine power, lest wicked insects and little worms destroy it with eager teeth, perchance preventing later ages from enjoying your teachings, and preventing any from reading that my name is linked to such great names?[174]

And who were these 'great names' with whom Dampier would have wished to have been associated? The next few pages of the Minute Book provided a clue, serving as a relatively infrequent example of an English *album amicorum*, a humanist genre much more prevalent in German-speaking Protestant lands in the early modern period than in England, although it is true English *alba* became more prevalent through the eighteenth century[175] (see figure 6.13). Students on their academic peregrinations would collect signatures of famous scholars as souvenirs, proof of contacts and/or credentials, illustrating their participation in scholarly networks. To some degree, these books were also part of self-fashioning one's scholarly or gentlemanly image, as signatures, though they could stand alone, were often accompanied by *sententiae* or a mot.[176]

Not surprisingly, Lord Sandwich was first to sign as the 'Sheik', proclaiming in humorous Latin that he was lucky to have penetrated the recesses of the Egyptian pyramids with his body (a slender physiognomy borne out by his portrait made

---

[174] BL Add MS 52362, ff. 1–2, British Library, London.

[175] *Alba* were, however, popular in Scotland. See James Fowler Kellas Johnstone, *The Alba amicorum of George Strachan, George Craig, Thomas Cumming*, Aberdeen University Studies 95 (Aberdeen: University of Aberdeen, 1924); Jan Papy, 'The Scottish Doctor William Barclay, His Album Amicorum, and His Correspondence with Justus Lipsius', in Dirk Sacré and Gilbert Tournoy, eds, *Myricae: Essays on neo-Latin Literature in Memory of Jozef Ijsewijn*. Supplementa humanistica Lovaniensia 16 (Leuven: Leuven University Press, 2000), pp. 333–96; J. K. Cameron, 'Leaves from the Lost Album amicorum of Sir John Scot of Scotstarvit', *Scottish Studies* 28 (1987), pp. 35–48. My thanks to Vera Keller for these points and references.

[176] English scholars often did not keep *alba*, but they did sign them, even the relatively unsocial Sir Isaac Newton. See George Gömöri and Stephen D. Snobelen, 'What He May Seem to the World: Newton's Autograph Book Epigrams', *Notes and Records: The Royal Society Journal of the History of Science* 74, 3 (2020), pp. 409–52. See also Max Rosenheim, 'The album amicorum', *Archaeologia* 62 (1910), pp. 251–308; M. A. E. Nickson, *Early Autograph Albums in the British Museum* (London: Trustees of the British Museum, 1970); Franz Mauelshagen, 'Networks of Trust: Scholarly Correspondence and Scientific Exchange in Early Modern Europe', *The Medieval History Journal* 6 (2003), pp. 1–32; and Bronwen Wilson, 'Social Networking: The "album amicorum" and Early Modern Public Making', in Massimo Rospocher, ed., *Beyond the Public Sphere: Opinions, Publics, Spaces in Early Modern Europe* (Bologna: Società editrice il Mulino, 2012), pp. 205–23.

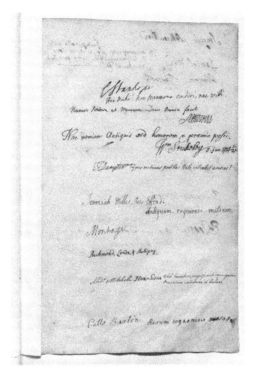

Fig. 6.13 Friendship album page from the Egyptian Society meeting minutes, with Folkes's motto taken after Newton, BL MS Add. 52362, f. 4r, © The British Library Board.

in 1740 to celebrate his voyage)[177] (see figure 6.14). During his travels in the Levant, Sandwich took lessons in local languages, including Turkish lessons in Istanbul to understand local customs, and visited several iconic Egyptian sites such as Alexandria, Giza, Saqqara, and Heliopolis. His joking comment was based on personal experience, as he remarked in his memoirs when exploring the interior of Giza:

At the bottom of the partition, which stops the passage, is a small hole, just big enough for a middle-size man to creep through upon his belly, not without some

[177] *Sandwich Sheik pyramidum recessus tenui corpore felix penetravi.* BL Add MS/52362, f. 3r. Joseph Highmore, 'John Montagu, 4th Earl of Sandwich, 1740', 48 in x 36 in, National Portrait Gallery, NPG 1977. For the portrait, which celebrates his voyage to the Middle East and shows his drawings of the pyramids, see https://www.npg.org.uk/collections/search/portraitExtended/mw05613/John-Montagu-4th-Earl-of-Sandwich [Accessed 27 July 2019].

Fig. 6.14 Joseph Highmore, *John Montagu, 4th Earl of Sandwich*, oil on canvas, 1740, 48 in. × 36 in. (1219 mm × 914 mm), © National Portrait Gallery, London.

pain and difficulty: in this manner you writher yourself along like a serpent, for the space of eight feet, after which you come into a breathing-place, where you have the satisfaction of standing upright, and in that manner repose yourself after the fatigue of the entrance.[178]

In his memoirs, Sandwich made several observations about Egyptian religion. On the one hand, they were not based upon Stukeley's premise that it was post-diluvian idolatry, but on the other, they were not completely reliant upon Herodotus. Sandwich remarked:

Nothing certainly would afford a more copious subject to an author than the religion of the Aegyptians…in the explanations of the many mysteries and

enigmas…But such an undertaking…is liable to the very obvious objection of such an explanation's being the pure invention of the author; who, being sensible that it would not be easy to contradict him, might be concluded to

---

[178] John Montagu, 4th Earl of Sandwich, *A Voyage Performed by the Late Earl of Sandwich Round the Mediterranean in the Years 1738 and 1739 Written by Himself* (London: T. Cadell, 1799), p. 454.

have given an entire scope to his imagination…And such I make no sort of doubt is the foundation, on which all modern writers have built.[179]

Although Sandwich uses Herodotus at points in his discussion of Egyptian culture, as well as Lucian and Pliny, most of his work is characterized by clear, matter-of-fact observation, scepticism, and the weighing of evidence based upon what he has seen. It was a practical empirical sense sharpened by his tenure as a military commander and understanding of the local landscape not dissimilar to that of a natural historian. He dismissed the argument that the pyramids were temples as well as being sepulchres, simply because of the difficulty of passage for corpulent Egyptian priests! He also asked with reference to Egyptian religious rituals and hieroglyphs, 'how is it possible to explain mysteries, which were never made public even in the nation in which they were in use?'

His cautious scepticism, however, did not forestall his curiosity or his ruthless acquisitiveness and cultural appropriation of archaeological finds. On his return to England in 1739, Sandwich brought with him 'two mummies and eight embalmed ibises from the catacombs of Memphis; a large quantity of the famous Egyptian papyrus, 50 intaglios, 500 medals…and a very long inscription, as yet undecyphered, on both sides of a piece of marble of about two feet in height…[and] plans and drafts of the pyramids'.[180] William Stukeley recorded that on 23 February 1741/2, 'Lord Sandwich dissected an Egyptian Mummy at Mr Ffolkes. The linnen foldings were unwrapt and the flesh was not discernable; being quite chang'd into dust and gum, together with that part of the Linnen which was next the bones'.[181]

Folkes's house served as a laboratory, not only for the Royal Society, but also for the Egyptian Society. Likewise, personal acquisitions of artefacts often shaped the Egyptian Society meetings, which took an approach very similar to that of the Royal Society sessions. As we have seen (see chapter four), Royal Society fellows carried out experimental tests at home, borrowing appropriate instruments, and repeating the experiments in the demonstration room of the Repository Museum.[182] As Marples has noted, 'the minutes of the ordinary meetings of the Society reveal that, more often than not in this period, objects were "shewn" and then not deposited into the Repository but taken home again—still, the showing of them by the collector or finder meant that Fellows knew where to find them, if

---

[179] Sandwich, *A Voyage Performed*, pp. 418–19, as quoted in Haycock, 'Ancient Egypt', p. 149.
[180] John Cooke, 'Memoirs of the Noble Author's Life', in *A Voyage Performed by the Late Earl of Sandwich Round the Mediterranean in the Years 1738 and 1739 Written by Himself* (London: T. Cadell, 1799), p. iii.
[181] William Stukeley, *Memoirs of the Royal Society*, vol. 11, f. 23, Spalding Gentlemen's Society, Spalding, Lincolnshire. Stukeley gave Maurice Johnson, President of the Spalding Gentlemen's Society, a copy of his memoirs so he knew more about London intellectual life.
[182] Anita McConnell, 'L. F. Marsigli's Visit to London in 1721, and His Report on the Royal Society', *Notes and Records of the Royal Society of London* 47, 2 (1993), pp. 179–204, on p. 191.

required'.[183] The Egyptian Society Fellows followed the same methodology, to dis-cover principles of sociocultural and religious practices in Egypt, as well as to provide 'object biographies' of artefacts that belonged to members that they examined with speculations on manufacture and use.

On 22 January 1741/2, Sandwich presented an ibis to the Society for inspection; Sandwich gave another specimen to Sir Hans Sloane for his collection.[184] In the last millennium BC, 'the mummification of animals...assumed enormous proportions', the ibis cult of the god Thoth (god of writing, scribes, and wisdom) having been firmly established during the Ptolemaic and Roman periods.[185] Most of the birds were votive animals given by pilgrims to temples and buried subse-quently in the temple catacombs, with 1.75 million ibis remains at Saqqara and four million at the Tuna-el-Gebel catacombs.[186] Sandwich indeed reported both in the meeting minutes, which later shaped his published writings. He witnessed at Saqqara several rooms of '20, or 30 feet long' with 'very capacious passages, on each side of which are several large square niches, filled with earthen pots of a conic figure, in each of which is contained an embalmed Ibis...[they] are fre-quently found so entire, that not only the bones, but very often even the plumage remains in its original perfection, but at the touch moulders away into powder'.[187] Like he did about the pyramids, Sandwich cautiously equivocated considering the opinions of Herodotus and Strabo about ibises and Egyptian worship, but also went beyond them, stating 'their superstition was not so gross as is commonly imagined, for it was not the animals themselves to which they paid their homage, but the particular deity, which they imagined symbolized in some peculiar

---

[183] Alice Marples, 'Scientific Administration in the Early Eighteenth Century: Reinterpreting the Royal Society's Repository', *Historical Research* 92, 255 (February 2019), pp. 183–204, on p. 199.

[184] BL Add MS 52362, ff. 16–17. This presentation of an ibis 'in a long earthen vessel...as long as one's arm' was also recorded by Stukeley, *Memoirs of the Royal Society*, vol. II, f. 8, Spalding Gentlemen's Society. On Sandwich's gift to Sloane, see John and Andrew van Rymsdyk, *Museum Britannicum, Being an Exhibition of a Great Variety of Antiquities* (London: I. Moore, 1778), p. 78. It is also possible to see the ibis in Sloane's inventories in his Catalogue of Miscellanea, f. 174v, item 1119: 'An ibis preserved by the Ægyptians in an earthen red cylindricall pott sealed up wt. a white cement. In those are contained sometimes hawks given me by my Lord Sandwich who brought it from the pyramids of Ægypt'. *Sir Hans Sloane's Miscellanea which comprises his catalogues of Miscellanies, Antiquities, Seals, Pictures, Mathematical Instruments, Agate Handles and Agate Cups, Bottles, Spoons.* Digital Edition, ed. Kim Sloan, Alexandra Ortolja-Baird, Julianne Nyhan, Victoria Pickering, and Martha Fleming. 2019. https://enlightenmentarchitectures.reconstructingsloane.org/cataloguemiscellanies/index.html [Accessed 28 July 2020]. It is unclear which species of ibis it was, whether the white-bodied Sacred Ibis, Ibis religiosa (*Threskiornis aethiopicus*) with its black head, neck, and black wing feather tips, or the Glossy Ibis (*Plegadis falcinellus*), which has more iridescent plumage.

[185] H. te Velde, 'A Few Remarks Upon the Religious Significance of Animals in Ancient Egypt', *Numen* 27 (1980) pp. 76–82, on p. 81; Ahmed Tarek, Mohamed Abdel-Rahman, Nesma Mohamed, and Ahmed Abdellatif, 'Study of Some Types of Different Wrappings on Ibis Mummies from Catacombs of Tuna-el-Gebel, Hermopolis', 2016, Poster session, https://isaae2016.sciencesconf.org/89426 [Accessed 29 July 2019].

[186] Tarek et al., 'Study of Some Types of Different Wrappings'. Thomas Greenhill also mentions mummified birds in his *[Nekrokedia]: Or, the Art of Embalming* (London: Printed for the author, 1705), p. 328.

[187] BL Add MS 52362, f. 17 r. and v. Sandwich, *A Voyage Performed*, p. 469.

quality of the animal, under whose form they worshipped'.[188] We now know because the difference 'between man and animals was not so absolute but regarded as relative, since men and animals both together are living beings', Egyptian 'gods could be represented as humans, as animals and in composite human-animal form'.[189]

In the subsequent meeting of 5 February 1741/2, Richard Pococke removed one of Sandwich's mummified ibises from its earthenware red cylindrical pot and dissected it, drawing it during the process, measuring the pots, counting the folds of linen, and examining the skeleton; for example, 'no. 7 represents the Ibis after so much of the crust is taken off as to discover the wing and leg bones, whose position is represented…The wing bones marked A and the tibia mark'd B are 5 inches long, the other bones in proportion. I could discover no traces of the head or Bill tho it is most probable that the whole bird was embalmed entire'.[190] Attendees of the meeting discussed which particular species of ibis served as the votive animals, making comparisons in the appearances between several candidates. Pococke's detailed empirical observation was reminiscent of matter-of-fact descriptions in the Royal Society's *Philosophical Transactions* by Edward Tyson, William Cowper, and Nehemiah Grew of the dissection and comparative anatomy of exotic animals that had been recently living, such as the opossum or rattlesnake.[191] Like Sandwich's observations later appeared in printed memoirs, Pococke's detailed drawings later appeared in Pococke's publication of his travels, the *Description of the East*, with his description in the Minute Book providing the text to accompany the image[192] (see figure 6.15).

Pococke, who would become Bishop of Meath, brought the same sort of detailed empiricism to Society meetings. In 1737, he went on a three-year tour of the East and started from Egypt, subsequently visiting the Holy Land, Cyprus, Crete, the Aegean islands, Asia Minor, and modern-day Continental and Northern Greece.[193] Like Sandwich, he accumulated, according to Stukeley, 'innumerable Egyptian antiquities' such as 'two stone Carvings, Vases in a human body, male and female. Osiris and Isis'.[194] Pococke signed the Minute Book indicating that during his travels he saw the colossal statues of Memnon in the ruins

---

[188] BL Add MS 52362, f. 17v.; Sandwich, *A Voyage Performed*, p. 416.

[189] H. te Velde, 'A few remarks', p. 79.   [190] BL Add MS 52362, f. 19r.

[191] Edward Tyson, 'Carigueya seu Marsupiale Americanum Masculum, or the Anatomy of a Male Opossum: in a letter to Dr. Edward Tyson, from Mr William Cowper', *Philosophical Transactions* 24, 290 (1704), pp. 1565–90.

[192] Rachel Finnegan, 'The Travels and Curious Collections of Richard Pococke, Bishop of Meath', *Journal of the History of Collections* 27, 1 (2015), pp. 33–48, on p. 40. Richard Pococke, *A Description of the East and Some Other Countries*, 2 vols in 3 (London: W. Bowyer, 1743), vol. 1, p. 233.

[193] Ioli Vingopoulou, 'Pococke, Richard. A Description of the East', Travelogues, Aikaterini Laskaridis Foundation. http://eng.travelogues.gr/collection.php?view=219 [Accessed 26 August 2018].

[194] Stukeley, *Memoirs of the Royal Society*, vol. 2, ff. 37–8, Spalding Gentlemen's Society, Spalding, Lincolnshire.

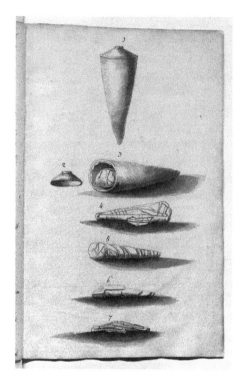

Fig. 6.15 Mummified Ibis from Egyptian Society Minute Book, BL MS Add. 52362, f. 18r, © The British Library Board.

of Thebes, spending 'above half a day', but did not hear the legendary voice sound emanating from the northern colossus, a sign of favour from the gods.

> The statue on the north side lost much of its torso from the waist up during a powerful earthquake in the region in 27 or 26 BCE and soon became famous for the eerie high-pitched sound it produced early in the morning, a phenomenon now understood...to have been caused by the heat of the rising sun warming the stone base...In antiquity, it gradually came to be understood as the voice of Memnon, who as the son of Eos, was closely connected with the dawn.[195]

Memnon was a running theme, and even joke, in the *album amicorum*. William Lethieullier noted '*Dimidio magicae resonant ubi Memnone chordae*', i.e. 'Where the magic notes reverberate from the broken Memnon' from Juvenal. Charles Stanhope FRS (1673–1760), secretary of the Treasury and a politician best known

---

[195] Patricia A. Rosenmeyer, *The Language of Ruins: Greek and Latin Inscriptions on the Memnon Colossus* (Oxford: Oxford University Press, 2018), p. 4.

for his transactions in the South Sea Bubble, indicated 'Woe is me, I did not hear or see Memnon'.

Like he did with the mummified ibis, Pococke documented the 'measurement and angles of the huge seated statues [of Memnon], focusing on the one to the north … With scientific accuracy, he recorded statistics for each part of its body: the distance from foot to knee (19 feet), foot to ankle (2 feet, 4 inches), bottom of foot to top of instep (4 feet), the foot itself (5 feet wide), and the depth of the leg (4 feet)', as well as the hieroglyphs on the statues.[196] When Pococke brought a mummy back from his travels for the delectation and delight of the Egyptian Society, it would receive a similar detailed description.

He unwrapped the mummy during a Society meeting held at Richmond House, the Whitehall residence of Folkes's dear friend Lennox, which 'commanded the relatively public area of the royal garden and the Banqueting House'.[197] Richmond House, a wonder of Palladian architecture, was most famous for providing a backdrop to a magnificent firework display. In April 1749, the *Music for the Royal Fireworks* by Handel was commissioned to celebrate the Peace of Aix-la-Chapelle negotiated to end the War of the Austrian Succession, accompanied by a fireworks display in Green Park. Stukeley, who received tickets from John, 2nd Duke of Montagu, described the 'variety of beautiful forms of wheels, stars, globes, fountains, as is inconceivable … joined with thundrings & a great variety of rockets' & thought it 'the most amazing sight I ever saw'.[198] Despite Stukeley's approbation, while the music was successful, the fireworks were not: it rained, most of the fireworks refused to light, a pavilion burnt down and members of the audience were injured by a stray rocket. Horace Walpole complained, 'The fireworks by no means answered the expense, the length of preparation, and the expectation that had been raised'.[199]

On 15 May 1749, the opportunistic Lennox bought the remaining fireworks left over and staged his own display for his friends (including Folkes) in the gardens at Richmond House and on the adjoining Thames. Lennox's display was not only done under the pretence of the Duke of Modena being in London, but also to assuage the Royal Society's embarrassment at the abortive Green Park event; the burning down of the Temple Pavilions was supposed to have been

---

[196] Rosenmeyer, *The Language of Ruins*, pp. 1–2.

[197] See Rosemary Baird, 'Richmond House in London: Its History, Part I', *British Art Journal* 8, 2 (Autumn 2007), pp. 3–15, on p. 5. 'Charles Lennox, 1st Duke of Richmond and Lennox (1672–1723), was the natural son of Charles II and his aristocratic French mistress, Louise de Keroualle (1649–1734). As befitted their close relationship to the Crown, the first three Dukes of Richmond and Lennox (of this fourth creation of the Richmond title) lived near to what had been the old Whitehall Palace', p. 3. Stukeley remarked the mummy was 'open'd at the late Duke of Richmonds house in Whitehall'. Stukeley, Ashtaroth, ABRAHAM, f. 19. Freemasons' Hall Library, London.

[198] Bodl. MS Eng Misc. e. 128, f. 52r., Stukeley diary, Bodleian Library, Oxford. My thanks to Liam Sims for this quotation.

[199] 'Fireworks', in Christopher Hibbert, Ben Weinreb, John Keay, and Julia Keay, eds, *The London Encyclopaedia*, 3rd ed. (New York: Pan-Macmillan, 2011), p. 294.

Fig. 6.16 *A View of the Fire-Workes and Illuminations at his Grace the Duke of Richmond's at White-Hall and on the River Thames*, on Monday 15 May, 1749. © Victoria and Albert Museum, London.

prevented by Stephen Hales FRS's 'experiments to secure buildings from fire by layering earth on floors and roofs to slow the progress of the flames', an approach vouched for by Royal Society secretary Cromwell Mortimer.[200] Lennox's display, by contrast, was a very different affair. 'A contemporary engraving shows sixteen different types of fireworks, some of which were discharged from barges on the river, with the caveat that such prints were often made in the anticipation of these ephemeral events. But the event was certainly a success, Horace Walpole enthusing: 'Whatever you hear of the Richmond fireworks, that is short of the prettiest entertainment in the world, don't believe it'[201] (see figure 6.16).

Lennox became a member of the Egyptian Society because of his noble status, and he was proposed by his good friend Folkes; in September 1740 Lennox hosted Norden at his home and begged Folkes for a copy of Greaves's work on the pyramids as 'wee are all Egiptian mad'.[202] Lennox had a 'musaeum' or 'Curiosity Room' of natural history and antiquarian objects at Richmond House; in 1741, John Cheale, the Norroy King at Arms,[203] complemented the Duke on its contents: 'Your curiositys are far before those in ye Royall Society'.[204] His museum

[200] Simon Werrett, *Fireworks: Pyrotechnic Arts and Sciences in European History* (Chicago: University of Chicago Press, 2010), pp. 154–5.

[201] 'The Renaissance Duke', catalogue to 2018 exhibition, Goodwood House, Sussex, https://www.goodwood.com/globalassets/venues/goodwood-house/summer-exhibition-2017.pdf [Accessed 9 October 2019].

[202] BL Add MS 52362, f. 16r.; RS/MS/865/20, From Lennox to Folkes, 19 September 1740, Royal Society Library, London.

[203] The office of the Norroy King at Arms, dating from the thirteenth century, is a Provincial King at Arms at the College of Heralds whose domain is England north of the Trent. Others who held this honorary office were Sir William Dugdale from 1660 to 1677, and Peter Le Neve from 1704 to 1729.

[204] John Cheale to Charles Lennox, 17 November 1741, in Charles Henry Gordon-Lennox, 8th Duke of Richmond and Earl of March, *A Duke and His Friends: The Life and Letters of the Second Duke of Richmond*, 2 vols (London: Hutchinson and Company, 1911), vol. 2, p. 383.

must have been quite a sight for the Egyptian Society members to examine; from its collections, he gave the Society a 'fine balloting box made after a Canopus brought from Egypt'.[205] It does not seem surprising that Lennox would host a mummy dissection in his premises.

As with the ibis, the description of the mummy was published in Pococke's *Description of the East*, the elaborate measurement of the skeleton characteristic of 'virtual witnessing' in a *Philosophical Transactions* paper, so as to produce in a reader's mind such a complete image of an experimental scene that the need for either direct witnessing or replication was eliminated (in this case not possible, as each mummy is unique).[206] Pococke also commented on the type of wooden coffin his four foot, ten-inch (about 1.5 metres) mummy was in, comparing it to simpler presentations where 'the bodys rapped up in linnen were cover'd over with a think plaister; and painted according to the pattern they chose'.[207] He speculated that these mummies may have been put into reed coffins, or even more humble remains put into palm boughs, remnants of which he saw in the Egyptian catacombs.[208] Indeed, more recent archaeological surveys have confirmed that papyri and coffins ranged from 'custom-made to mass-produced products with blanks for the patron's name'.[209]

He then made notes on the types of pigments being utilized, thinking the blue was akin to 'strow[n] smalt', or powdered blue potassium glass cast or strewn over a mordant like gold size or white lead base, an alternative to precious lapis lazuli.[210] Pococke was not far removed from the truth, as it is thought the origins of smalt were in the 'blue pigment used by the ancient Egyptians, known to us as 'Egyptian blue' or frit; Egyptian blue was the first artificial pigment, and while 'both pigments are made from glass that has been coloured blue and both are also used as glazes on ceramics…Egyptian blue contains copper, whereas smalt derives its colour from cobalt'.[211] Pococke also surmised that patterns on the mummies were created using stencils, and it is thought that as the 'training of scribes and painters in ancient Egypt was based largely on repetition', copybooks of motifs were certainly in use.[212]

---

[205] BL Add MS 52362, f. 29v., British Library, London. He later also gave the balloting balls.

[206] Simon Schaffer and Stephen Shapin, *Leviathan and Air Pump: Hobbes, Boyle, and the Experimental Life* (Princeton: Princeton University Press, 1985).

[207] BL Add MS 52362, f. 22v.   [208] BL Add MS 52362, f. 23r.

[209] Melinda Hartwig, 'Method in Ancient Egyptian Painting', in Valérie Angenot and Francesco Tiradritti, eds, *Artists and Colour in Ancient Egypt: Proceedings of the Colloquium Held in Montepulciano, August 22nd–24th, 2008* (Montepulciano: Missione Archaeologica Italiana a Luxor, 2016), pp. 28–56, on p. 31.

[210] BL Add MS 52362, f. 22v.

[211] 'The Changing Properties of Smalt Over Time', Tate Britain, January 2007, https://www.tate.org.uk/about-us/projects/changing-properties-smalt-over-time [Accessed 26 August 2018].

[212] Melinda Hartwig, *Tomb Painting and Identity in Ancient Thebes, 1419–1372 BCE* (Turnhout: Brepols, 2004), p. 19.

Far from adhering to Herodotus with his speculations on pigments and techniques, Pococke went further than the Greek historian. In fact, throughout his work he uses not only Strabo's *Geography*, Pliny's *Historia naturalis*, and Ptolemy's *Geographia*, but his personal library also included a raft of more contemporary published travel journals including Reland's *Palaestina* (1714) and Shaw's *Travels* (1738).[213] Pococke was likely using a mixture of ancient and modern sources, with a critical eye to supplement his own observations.[214]

Pococke also criticized Herodotus in the Minute Book for his thoughts concerning embalming: 'There are two other ways of embalming spoken of by Herodotus. He does not mention that the bodies were wrapped in lin[n]en, or the heads filled with Bitumen, and in the Catacombs I saw several heads that had been filled with Bitumen. Tis probably these bodies furnish'd the Mummy of dried flesh used in Medicine…for which the Catacombs were searched at the time the Venetians traded to Egypt when that medicine was much in vogue'.[215] This is of course a reference to the medicine *mumia*, an ancient Egyptian and then early modern panacea thought to prevent infection and stop bleeding; common cemeteries were regularly raided for this lucrative trade.

Pococke was not alone in his scepticism about Herodotus, or in his trying to understand techniques of mummification. In 1705, English surgeon Thomas Greenhill made a survey of all literature on the practice, advocating that the English should 'adopt the practice for hygienic reasons, and that surgeons, rather than undertakers, were best qualified to perform it for aristocratic clients'.[216] Like Pococke, he read the ancients like Herodotus or Diodorus Siculus, and after a process of comparison he aimed to 'suggest some new Thoughts, as plausible'.[217] He questioned Herodotus' account of extracting the brains from corpses through the 'Nostrils with a crooked Iron…as being it self unpractible and ridiculous, which any one skill'd in Anatomy will ready agree to. But grant it could be done, the afore-said extraction of the Brain thro' the Nostrils, must nevertheless so dilacerate the cartilaginous parts of the Nose, that the carnous and cutaneous parts would sink, and thereby render the Face deform'd'.[218]

For his time and trouble, two weeks after the Egyptian Society was founded, Pococke was 'introduced by Martin Folkes to the Society of Antiquaries', where he

---

[213] Adrian Reland, *Palaestina ex monumentis veteribus illustrata* (Utrecht: Willem Broedelet, 1714); Thomas Shaw, *Travels or Observations Relating to Several Parts of Barbary and the Levant* (Oxford: Printed at the Theatre, 1738).

[214] John R. Bartlett, 'Richard Pococke in Lebanon, 1738', *Archaeology and History in Lebanon* 16 (Autumn 2002), pp. 17–33, on p. 29.

[215] BL Add MS/52362, f. 22v.

[216] Thomas Greenhill, *[Nekrokedeia]: Or, the Art of Embalming* (London: Printed for the author, 1705). Christina Riggs, *Egypt: Lost Civilisations* (London: Reaktion Books, 2017), p. 126. Hans Sloane was one of the subscribers to Greenhill's work.

[217] Greenhill, *[Nekrokedeia]*, p. 240.   [218] Greenhill, *[Nekrokedeia]*, p. 249.

showed the Fellows Egyptian coins. Five weeks later, he was then elected to the Royal Society on 12 November 1741, with Folkes and Newton's physician and collector Richard Mead (who also had a mummy in his own museum[219]) supporting his nomination, showing the clear links and overlapping interests between the societies.[220]

Egyptian Society member Charles Perry (1698–1780), physician and medical writer, who signed the Minute Book with a mot mentioning his tour of Thebes,[221] in the next meeting displayed a cross-shaped 'ornament of glass beads, that he found lying upon the breast of the Mummy... of three colours viz Green, red, & white'.[222] The artefact would have been part of a bead net, which enveloped contemporary mummies, sometimes as a girdle, apron, or skirt of glass or faience beads. Perry would later go on to write *A View of the Levant* (1743), describing his tour of France, Italy, Egypt, Palestine, Constantinople, and Greece; he dedicated the work to Lord Sandwich.[223] In his Egyptian travels, Perry went up the Nile to Aswan, providing 'the earliest description of the Temple of Isis at Behbeit El-Hagar, and the frescos in the tombs of the Beni Hasan necropolis'.[224] The mummy with its beaded network was late Dynastic or early Ptolemaic, its case hewn out of a solid piece of sycamore. It was found in the Sakkarah Catacombs by Arab traders, purchased by Perry, and sent to London in 1741.[225]

The beaded net on the mummy inspired William Stukeley to make further connections between the Old Testament prophets and ancient Egypt, as described in his discourse in a manuscript dedicated to the biblical patriarchs. In Genesis, we are told that although Jacob died in Egypt, he wished to be buried in Canaan with his wife and ancestors. Jacob's son Joseph, an Egyptian official at the time of Jacob's decease, had his father's body embalmed to keep it from decay during the long journey. Stukeley speculated that the crossed arms of the mummies were a

---

[219] Alexander Gordon, *An Essay Towards Explaining the Antient Hieroglyphical Figures on the Egyptian Mummy, in the Museum of Doctor Mead, Physician in Ordinary to his Majesty* (London: For the author, 1737).

[220] Finnegan, 'Travels and Curious Collections', pp. 40–1.

[221] *Lector, Thebam vidi centum obrutam portis* [Reader, I have seen Thebes buried with her one hundred gates]. The line is reminiscent of Juvenal Satire 15:6: '*vetus Thebe centum iacet obruta portis*' [ancient Thebes lies buried with her one hundred gates], but Perry's modification spoils the hexameter versification.

[222] BL Add MS 52362, f. 23r.

[223] Charles Perry, *A View of the Levant: Particularly of Constantinople, Syria, Egypt and Greece...* (London: T. Woodward and C. Davis, 1743).

[224] Sotheby's, *The Library of a Greek Bibliophile, Aldines and an important Qur'an*, 28 July 2020, Lot 85, Perry: *A View of the Levant*. https://www.sothebys.com/en/buy/auction/2020/the-library-of-a-greek-bibliophile-travel-books-aldines-and-an-important-quran/perry-a-view-of-the-levant-particularly-of [Accessed 28 July 2020].

[225] Charles Perry, *A View of the Levant*, p. 470. When Perry died, the mummy passed into the hands of Richard Cosway (1740–1821), after which it was purchased by Thomas Pettigrew and dissected again, though at that point it proved to be in poor condition. See Warren R. Dawson, 'Pettigrew's Demonstrations upon Mummies: A Chapter in the History of Egyptology', *The Journal of Egyptian Archaeology* 20/3–4 (November 1934), pp. 170–82, on p. 170.

superstition, the adulteration of a biblical event in which Joseph crossed his hands blessing Jacob on his deathbed (Genesis XLVIII), from which the Egyptians:

> took a high notion of the form of the cross; as we see it frequently in their monuments; and placed Jacobs arms across in preparing his body; supposing a very great sanctity inherent therein; that is was prophylactic; to keep off all evil powers, with corruption, too much heat or moisture. So that the body should be preserv'd through all ages, to meet the soul in its future state. and we find accordingly, they may be preserv'd for ever. hence the origins of Mummys, and the constant custom of crossing their arms.[226]

Stukeley thought that the resurrection of the soul in Christianity was corrupted into Egyptian funerary practice to preserve the physical body. Lastly, Stukeley thought that the beaded ornament on the mummy with its 'venerable cross' was really a shepherd's pouch, again transmuted from vague memories of the Old Testament prophets who kept sheep.[227] Stukeley stated, 'the shepherds crook, the whip, the pouch were taken by the Egyptian priests into the number of their sacred instruments; held highly prophylactic, and able to drive off evil powers'.[228]

Certainly, not everyone accepted Stukeley's views, particularly Alexander Gordon (c.1692–1755), who had published two treatises to decipher hieroglyphics and to illustrate and interpret the iconography of all Egyptian mummies extant in England.[229] 'In his essays, Gordon speculated on the meaning of some of these symbols, drawing on the writings of earlier scholars who recognized that they offered clues to a greater understanding of Egyptian religion, art and society'; one of his plates of hieroglyphics (Plate XXV) was dedicated to Folkes, who was a patron of the volume.[230] In his published work and in his manuscripts, Gordon 'assiduously defended the precedence of Egyptian antiquity over Jewish, and held the deist view that Egyptians had worshipped one god before the coming of idolatry'.[231] Gordon's precedence of ancient Egypt over the Judaic Old Testament

---

[226] Stukeley, Ashtaroth, ABRAHAM, f. 8. Freemasons' Hall Library, London.

[227] Stukeley, Ashtaroth, ABRAHAM, f. 24. Freemasons' Hall Library, London.

[228] Stukeley, Ashtaroth, ABRAHAM, f. 18. Freemasons' Hall Library, London. Some of these ideas were derived from Athanasius Kircher's *Egyptian Oedipus* (1655).

[229] Alexander Gordon, *An Essay Towards Explaining the Hieroglyphical Figures on the Coffin of the Ancient Mummy Belonging to Captain William Lethieullier* (London: Printed for the Author, 1737); *Twenty-Five Plates of All the Egyptian Mummies and Other Egyptian Antiquities in England* (c. 1739)—both in folio. William Lethieullier (1701–56) was the brother of antiquary Smart Lethieullier.

[230] Nicola J. Shilliam, 'Mummies and Museums: Egyptology in Britain in the Early Eighteenth Century', Princeton University Library. https://library.princeton.edu/news/marquand/2019-03-18/mummies-and-museums-egyptology-britain-early-eighteenth-century [Accessed 14 January 2020].

[231] Stephen W. Brown and Warren McDougall, eds, *Edinburgh History of the Book in Scotland: Volume II, Enlightenment and Expansion: 1707–1800* (Edinburgh: Edinburgh University Press, 2011), pp. 264–5. BL Add MS, Alexander Gordon, 'An Essay towards Illustrating the History, Chronology, and Mythology of the Ancient Egyptians (1741)', 8834, British Library, London.

prophets was precisely what Stukeley was against; Stukeley's mot in the *album amicorum*, taken from Horace's *Epistulae*, stated 'It annoys me that something is faulted not because it is obtusely or inelegantly wrought but because it is recent, while not indulgence but praise and prizes are sought for old things [*indignor quidquam reprehendi, non quia crasse/compositum illepideve putetur, sed quia nuper/nec veniam antiquis, sed honorem e praemia posci*].[232] Despite Stukeley's pronouncements, it is clear that Gordon, as did many natural philosophers, con- sidered Egypt to surpass India or China in 'discourses on the origin and diffusion of arts and science and of idolatry. It was also central to arguments about the Judaic tradition', within and without the bounds of the Egyptian society.[233] Again, considerations of material culture entered intellectual debates that were well beyond simple reliance upon Herodotus.

Some of the last sets of artefacts to be examined by the Society were coins and medals. On 2 August 1741, Lord Sandwich wrote to Folkes from his estate at Huntingdon,

> I have just now received an account of a great acquisition I have made as a vir- tuoso a box full of Egyptian medals by the Ruxley Galley from Alexandria. The box is now unopened at the custom house and as I know not whether they may not \meet/ with some ill usage should they fall into any profane hand. I should take it as a particular favour if you would trouble to withdraw them from thence and keep them in your hand till I have the pleasure of seeing you again.[234]

At Sandwich's request Folkes made his appraisal of the medals, of their 'species and value', which Lord Sandwich was 'not much disappointed at', reporting he had 'great hopes from another parcel which are to be brought to me by Pocock from Athens of 236 brass ones 45 silver ones'.[235]

On 19 March 1741/2, it was recorded that 'a motion was made that Mr Folkes be desired to receive examine & keep a Register of such Egyptian medals as shall be brought to him by members of this Society'.[236] A committee was duly appointed to examine Egyptian medals, and later, in April 1742, Pococke produced a design of a copperplate for a 'series of Egyptian medals proposed to be engraved by the Society', reminiscent of Folkes preparing engravings for his own numismatic works.[237] Pococke also had an interest in medals, having Folkes's *Table of English Silver Coins* in his library, as well as a 'copper medal struck in his honour'.[238] In the summer of 1742, Pococke, Charles Perry, and Smart Lethieullier were then

---

[232] BL Add MS 52362, f. 4r.   [233] Brown, *Edinburgh History of the Book*, p. 264.
[234] RS/MS/790/72, Royal Society Library, London.
[235] RS/MS/790/73 and 74, Royal Society Library, London. Letter from Sandwich to Folkes, 22 September 1741 and 11 October 1741.
[236] BL Add MS 52362, f. 25v.   [237] BL Add MS 52362, f. 27r.
[238] Finnegan, 'Travels and Curious Collections', p. 35.

ordered to 'enquire after proper persons to draw and engrave the said meals and to make a report of the prices demanded for the said work', and Lethieullier was appointed 'gumroue Gi' or 'Inspector, Controller, and Examiner of the Egyptian medals'.[239] On 31 January 1742/3, Pocock wrote to Folkes:

> I have had the Egyptian medals of five collections drawn, and I have a person not employed in drawing them, so that if you would be pleased to look out what Egyptian medals you have of the Imperial series, I should be obliged to you if you would be pleased to trust me with them for a short time…and will send my servant to them tomorrow morning.[240]

Ultimately, one Mr Tuscher was ordered to be paid £13.14s to engrave the Society's collection. Several pages of drawings for these engravings (ninety-six medals) were in the last pages of the Minute Book, where the medals were identified by their owner's initials on the drawings in the Minute Book: Mr. F (Folkes); S.L. (Smart Lethieullier); C.P. (Charles Perry) (see figure 6.17).

At the same time the Egyptian Society was examining its medals, Folkes was having a medal of his own struck in Rome with Egyptian motifs (see figure 4.4). Although detailed iconographic analysis of this piece has been done by William Eisler in the context of what he claims were Folkes's Jacobite sympathies, the production of the medal could be explained as simply being connected to Folkes's activities in the Egyptian Society and his interest in masonry.[241] On the reverse, the medallion had symbols 'taken directly from the seal of the Neapolitan "Perfetta Unione" [masonic] lodge: the pyramid (but here transformed into the pyramid of Cestius at Rome), pillars of the porch of the Temple of Solomon, sphinx with crescent moon, acacia branch, and sunburst'.[242]

This iconography indicates that Folkes had established ties with the lodges there during his previous journey there, as well as with the lodge in Florence when he visited his friend Antonio Cocchi (see chapter four). Folkes's medallion is dated on orthodox principles as Anno Lucis 5742, the year of the world or of Masonry 5742, as the Masons thought the creation was 4000, or 4004 years before the Christian era; the date of the medal is generally agreed to be 1742.[243] The phrase on the medal, *sua sidera norunt*—His own constellations have acknowledged him—is from the sixth book of the *Aeneid*, where Virgil described

---

[239] Probably derived from the Ottoman Turkish for 'Customs Officer'. See Finnegan, 'The Divan Club', p. 17.

[240] RS/MS/250, Letter 17, Royal Society. See also Finnegan, *English Explorers in the East*, p. 107.

[241] Eisler, 'The Construction of the Image of Martin Folkes (1690–1754) Part II', pp. 4–16.

[242] Alastair Small and Carola Small, 'South Italy, England and Elysium in the Eighteenth Century', *Antiquaries Journal* 79 (1999), pp. 333–42, on p. 335.

[243] Martin Folkes Medal, object M.8466, British Museum, London. https://www.britishmuseum.org/collection/object/C_M-8466 [Accessed 1 November 2019].

Fig. 6.17 Page of Coins and Medals from Egyptian Society Minute Book with identifying initials for various numismatic collections, BL MS Add. 52362, f. 35r, © The British Library Board.

an encounter in Elysium between Aeneas and his father Anchises. The inscription also indicates that the medal may have been struck at Rome to commemorate Folkes's successful journey there to spread Newtonian and Masonic principles, as well as 'to show the high esteem in which Folkes was held in the city of antiquities...about the time that he was elected a member of the French Academy'.[244]

Examining Egyptian coins and medals was the last major venture of the Society, its final meeting being held on 16 April 1743. Its demise may have been due to the unpopular and self-serving insistence (repeated in several of the meeting minutes) by Charles Perry that every member had to purchase a copy of another member's book, as the publication of his *View of the Levant* was imminent.[245] It also may have been that the Society simply ran out of Egyptian objects to exhibit and discuss, as in their last few sessions they also examined Greek antiquities, such as a 'very curious foot of bronze of Greek workmanship...found [by Lord Sandwich] in the bottom of a well in the Island of Mycone', of interest because it 'admirably well represents the ligature of the

---

[244] Martin Folkes Medal, object M.8466, British Museum, London. https://www.britishmuseum.org/collection/object/C_M-8466 [Accessed 1 November 2019].

[245] Finnegan, 'The Divan Club', p. 17.

ancient sandal'. But whatever the real reason, the Egyptian Society represented Georgian scientific antiquarianism very well, with its interdisciplinary blend of classicism, humanism, empirical observation, metrology, early archaeology, and biblical exegesis, all areas reflective of Folkes's intellectual interests. The Egyptian Society was representative of the complexity of the intellectual currents of the early Enlightenment that characterized early modern antiquarianism and natural philosophy (*pace* Herodotus).

7

Martin Folkes and the Royal Society Presidency

Patronage, Biological Sciences, and Vitalism

I assure you Sir and I think I already told you, but I will repeat it to you many times again, that I care more, if possible, about your kindness and your sensitivity than I care about the beauty of your mind and the extent of your enlightenment and of your knowledge
—Madame Geoffrin to Folkes, 15 March 1743

7.1 Folkes as Administrator

The 1740s was the decade in which Folkes came into his own as a statesman of antiquarianism and natural philosophy, achieving his goal of being Royal Society President on the Society's Anniversary Day elections on 30 November 1741. In their study of scientific networks and innovation, Bruno Latour, Hal Cook, and David Lux have demonstrated that 'new information and ideas tend to come from people with many weak social bonds'.[1] This chapter analyses how Folkes's large number of far-flung social contacts during his presidency engendered scientific creativity in the Royal Society, challenging the long-standing mistaken impression among scholars that the organization was in decline in the eighteenth century. Evidence also shows that before he was incapacitated by a paralytic stroke in 1752, Folkes was an excellent administrator.

At the beginning of Folkes's presidency, James West continued as Treasurer, and John Machin and Cromwell Mortimer as secretaries. Charles Lennox wrote to Folkes a week before the election, 'If Sr. Hans [Sloane] abdicates, I promise you my vote for president, preferable to any man in the World, because I am sure there is not a man in the world so fitt for it in every respect as yourself, & I will

---

[1] David S. Lux and Harold J. Cook, 'Closed Circles or Open Networks: Communicating at a Distance During the Scientific Revolution', *History of Science* 36, 2 (1998), pp. 179–211, on p. 182; Bruno Latour, *Science in Action: How to Follow Scientists and Engineers Through Society* (Cambridge, MA: Harvard University Press, 1987).

Martin Folkes (1690–1754): Newtonian, Antiquary, Connoisseur. Anna Marie Roos, Oxford University Press (2021).
© Anna Marie Roos. DOI: 10.1093/oso/9780198830061.003.0007

certainly be in town & attend you for that purpose'.[2] He hoped 'for the sake of the learned world' that Folkes would preside in that chair as long as you live', a prediction that very nearly came true.[3]

On 16 November 1741, the week before the elections, it was recorded in the minutes that 'the vice President [Folkes] said he was charged with a Message from the President this morning, who desired him to bear his Respects...and to acquaint them that the weakness in his Limbs...and the precarious State of his Health...will render it impractible for him to give that attention to the Society which his Office requires and therefore he desires them to think of some other proper person for that Office'.[4] Folkes, James Lowther, James West, and John Hadley then went to Sloane's house in Bloomsbury to 'consult with him some measure to reconcile, if possible, his holding that Office in some manner as not be injurious to his health'. They reported on the day of the Anniversary Day elections that Sloane ordered them to proceed and elect a new President, and Folkes was duly elected, with his first meeting as President on 16 December 1741.

Before Folkes's election to the presidency, Sloane and Folkes continued their rapprochement. Sloane appointed Folkes on 14 January 1740/1 to chair a crucial finance committee. In 1728, Sloane as President had instituted a series of proposals passed by the Council for the admission of Fellows and to rectify financial arrears. The first required that each candidate had to be nominated by three or more Fellows, the second stipulated that Fellows in arrears with their 'subscriptions and had signed the obligation to pay them should be sued in legal form', and the third exempted foreign members from payment.[5] Through these methods, £1,600 was raised, and in 1732 the money was used to buy an estate of forty-eight acres at Acton in Middlesex for the Society.[6] Despite Sloane's efforts, by 1741 the Society was again in arrears, as 'expenditure had regularly begun to exceed income in the late 1730s'.[7] In 1739, James West, the treasurer, reported that 'his receipts were £95 short of his expenses; in the following year the deficit increased to £240, and stringent measures had to be taken. The cause was that many Fellows were remiss in paying their subscriptions, which should have been remitted quarterly'.[8] The Society Minutes recorded:

    [2] RS/MS/865/21, Lennox to Folkes, 18 November 1741, Royal Society Library, London.
    [3] RS/MS/865/25, Lennox to Folkes, 5 December 1742, Royal Society Library, London. Folkes had a stroke in 1752 and had to resign the presidency.
    [4] RS/MS/645, Minutes of Meetings of the Council of the Royal Society, f. 34, Royal Society Library, London.
    [5] Henry G. Lyons, 'The Officers of the Society (1662–1860)', *Notes and Records: The Royal Society Journal of the History of Science* 3 (1940–1), pp. 116–40, on pp. 127–8.
    [6] Lyons, 'The Officers of the Society', p. 128.
    [7] Aileen Fyfe, Julie McDougall-Waters, and Noah Moxham, '350 Years of Scientific Periodicals', *Notes and Records: The Royal Society Journal of the History of Science* 69 (2015), pp. 227–39, on p. 230.
    [8] Henry G. Lyons, 'Two Hundred Years Ago: 1739', *Notes and Records: The Royal Society Journal of the History of Science* 2, 1 (1939), pp. 34–42, on p. 38.

Mr Treasurer West spoke concerning the present State of the Society's affairs with regard to their Revenues; and represented that the Charges and Expenses had of late exceeded the Income, and thereby incumbered the Society with several Debts, so that it is now become necessary to consider of some speedy Remedy to recover the Society out of its present state…whereas many Debts are owing to the Society as well from living Members as from the Executors of those that are deceased…we did therefore take occasion to lay before them the list of the arrears that are due for Contribution.[9]

As Sloane discovered, Folkes's mathematical gifts lent themselves to a talent for accountancy, and his interests in antiquarianism and detail made him an excellent compiler and organizer of financial records. Folkes subsequently compiled a list of those members who had not paid, those who had been abroad, as well as those 'Gentlemen, who through misfortune or otherwise may seem intitled some Tenderness from the Society', or those who had done service and might expect abatement of dues.[10] Among these was Folkes's old friend John Byrom, who had not attended regularly since Folkes's defeat for the presidency in 1727 and was £34.9s in debt to the Society: 'Mr Byrom have liberty to take up his Bond upon making up his past Payments of twenty five points against Christmas next'.[11] Richard Pococke, Folkes's fellow member of the Egyptian Society was also given another year to pay his debts.[12] Folkes also was frank in his committee report to Sloane, 'the whole Revenue of the Royal Society amounts only to £232 per Annum…after the Deduction of Taxes, necessary Repairs, Charges of getting in, and other Contingencies…it appears your annual Expences can hardly be computed at less than £380 yearly…the getting in your Contributions is at this time a matter of the Greatest Consequence'.[13] He then suggested a quarterly payment of dues, and, if necessary, a visit by a 'proper Officer' to the residence of the Fellows to collect the monies; some Fellows, such as Richard Bentley, nearing the end of his life and suffering a loss of £4,000 in the South Sea Bubble, were as much as £75 in arrears.[14] Due to Bentley's distress, on 23 February 1741/2 Folkes wrote to him personally, suggesting:

[9] RS/MS/645, Minutes of Meetings of the Council of the Royal Society, f. 1., Royal Society Library, London.
[10] RS/MS/645, Minutes of Meetings of the Council of the Royal Society, f. 5., Royal Society Library, London.
[11] RS/MS/645, Minutes of Meetings of the Council of the Royal Society, f. 31 at a meeting of 16 November 1741, Royal Society Library, London.
[12] RS/MS/645, Minutes of Meetings of the Council of the Royal Society, f. 31 at a meeting of 16 November 1741, Royal Society Library, London.
[13] RS/MS/645, Minutes of Meetings of the Council of the Royal Society, f. 7., Royal Society Library, London.
[14] Hugh de Quehen, 'Bentley, Richard (1662–1742), Philologist and Classical Scholar', Oxford Dictionary of National Biography. https://www.oxforddnb.com/view/10.1093/ref:odnb/9780198614128.001.0001/odnb-9780198614128-e-2169 [Accessed 27 Nov. 2019].

that upon your paying to the Treasurer of the Society the Summ of five pounds before the 25th day of the next month of March, you will be entitled to be discharg'd from all arrears due to the Society, to have your bond delivered up and be excused from all future payments to the same. This herefore of it is their request that you will comply with as the Summ tho a trifle is of consequence to the Society whose well being in great measure depends on their steeling the affairs of their little finances on a regular foot without distinction of persons.[15]

Folkes also proposed a yearly audit, with a report of arrears and contributions to the auditor. Francis Hauksbee, serving as clerk and housekeeper of the Society, subsequently sent a letter to 'every member in arrear, signifying the Society's demands upon him', as well as to 'Executors of such Members who died in Arrear, and who Bonds have not been yet discharged'; twenty-one days were given until proceedings were started. These Orders and Resolutions were also hung in the Meeting Room of the Society, with a copy enclosed with every letter.[16]

These actions began to bring the Society back to financial health; Hauksbee's report on the day of Folkes's presidential election on 30 November 1741 revealed that the Society now had a positive balance of £297.9s.3d, and this report likely secured the election for him.[17] On 16 December 1741, his first meeting as President, Folkes also gave the Society £100 to 'assist them in the present low State of their Revenue'.[18] Overall, Folkes 'left the Society in a much more flourishing condition than when he was elected President; for, at the time of his resignation [in 1752], their funded capital amounted to £3,000'.[19]

On 29 October 1741, immediately before he was elected as President, Folkes also began keeping careful registers of Society business. This was done for three purposes. Firstly, Folkes produced a table of contents of the Journal or Minute Books that cross-referenced Guard and Register Books, as he wished to keep a record of which papers were produced at meetings and subsequently published, creating detailed tables to show where papers were in the publication process. Secondly, Folkes wanted a current list of all gentlemen proposed for Fellowship to assess their solvency, in order to prevent the financial problems that the Society had faced in the last years of Sloane's presidency. Thirdly, as part of the continual reform of the Repository, he also recorded all donations of books for the Library

---

[15] RS/MS/790/50, Folkes to Bentley, Royal Society Library, London.

[16] RS/MS/645/22, Minutes of Meetings of the Council of the Royal Society, Royal Society Library, London. This was Francis Hauksbee the Younger (1687–1763), the nephew of the elder natural philosopher of the same name. He was an experimental demonstrator and instrument seller.

[17] RS/MS/645, Minutes of the Royal Society, f. 35, Royal Society Library, London.

[18] RS/MS/645, Minutes of the Royal Society, f. 39, Royal Society Library, London.

[19] Charles Weld, *A History of the Royal Society: with memoirs of the Presidents*, 2 vols (Cambridge: Cambridge University Press, 2011; reprint 1848 ed.), vol. 1, p. 525.

or objects for the Repository, with the names of 'all those to whom the Society is obliged for each present'.[20]

Throughout the 1740s, he kept careful track of the *Philosophical Transactions*, endorsing by hand each paper that was read through and recording any responses he had to authors; for instance, he indicated when the author wished the paper to be returned for editing, if the author printed his work on his own accord, or, in the case of Desaguliers presenting a paper on 'the rise of vapours' on 21 April 1743, 'returned by desire, and the Dr dying no further account was given to the Society'.[21]

It was not until 1751 that the *Philosophical Transactions* fell into crisis. Folkes by that point was declining in health, having a stroke in 1752, and the journal fell two years behind in publication. Furthermore, as Fyfe, McDougall-Waters, and Moxham have rightly indicated, 'The journal itself had come under biting satirical attack by John Hill, an actor, apothecary, and naturalist who had been bitterly disappointed in his hopes of being elected to the Fellowship, and who took his revenge by publishing three works in two years ridiculing the Society and *Transactions*'.[22] In response, 'the post-1752 *Transactions* was to be edited by a standing Committee of Papers (in practice the Society's governing Council), who would use a secret ballot to generate a collective decision, thus avoiding any imputation that Committee members might have been bribed or coerced'; eventually the peer review practice would develop.[23] Hill's satire portraying Folkes as a 'drooling epicure' and the establishment of the Committee of Papers have given historians the impression that Folkes was a poor administrator but, that said, during most of his presidency the archival evidence indicates that Folkes had tended to journal business very well.

In 1742, Folkes also organized the striking of the Copley Medal with funds derived from a bequest made by Sir Geoffrey Copley in 1709. There were initial expenses and issues with casting, as well as the cost of the precious metal. The gold medal had the same value as the Copley fund's annual disbursement of £5,

---

[20] RS/MS/213; RS/DM/5/8/2, Chronological list dating from 1633–1722, of donations to the Royal Society, 'extracted out of Journal Books of the weekly meetings', Royal Society Library, London.

[21] RS/MS/213, Table of Contents of Journal Books, 1741–8, compiled by Martin Folkes, Royal Society Library, London. See also MS 5408, Wellcome Collection, London, representing sixty-five letters and papers sent to the Royal Society of London, 1745–8, and mostly published in the *Philosophical Transactions*. Most are initialled by Martin Folkes and endorsed with the dates when they were read to the Society. Another set from 1746–8, MS 2391, Wellcome Collection, consists of eleven short memoirs and papers sent to the Royal Society of London, including four holograph items by Folkes; all are initialled, signed, or endorsed by him.

[22] A. Fyfe, J. McDougall-Waters, and N. Moxham, 'Guest Editorial', *Notes and Records: The Royal Society Journal of the History of Science* 69 (2015), pp. 227–39, on pp. 231–2. This special edition of the journal examined the development and history of peer review as part of a 2015 commemoration of the foundation of the *Philosophical Transactions*. The satires that Hill wrote, the first two published anonymously, were: *Lucina Sine Concubitu* (1750), *A Dissertation on Royal Societies* (1750), and *Review of the Works of the Royal Society* (1751).

[23] Fyfe, McDougall-Waters, and Moxham, 'Guest Editorial', p. 232.

and there were fourteen medallions struck, as well as twelve silver and twenty-four copper medals, which were given as gifts or purchased by Fellows.[24] On 10 November 1736, when serving as Vice President under Sloane, Folkes had in fact been the first to suggest to the Society that the Copley donation, initially made to support an annual experimental demonstration, would be better employed by converting

> the value of it into a medal or other honorary prize, to be bestowed on the person whose experiment should be best approved; by which means he apprehended a laudable emulation might be excited among men of genius to try their invention, who in all probability may never be moved for the sake of lucre.[25]

Folkes's idea, no doubt partially spurred on by his interest in numismatics, was approved at the next month's Council meeting, with he, Samuel Gale, James West, and James Theobald ordered to 'consult together about a proper Medal'. Although 'three months later, in February 1737, the committee submitted to the Society a design for the medal by the engraver, George Vertue, and it was approved by the President and the Council', it would not be until 1742, during Folkes's presidency, that the medals were actually made and awarded to John Belchier, James Valoue, Stephen Hales, Alexander Stuart, and J. T. Desaguliers.[26]

Folkes's actions to regulate finance and procedural matters were also evidenced by his commission of a formal survey by the Library Committee, a committee which he had previously headed for the Society in the 1720s. On 12 July 1742, the committee recommended 'the future ordering and preservation of the Society's papers'.[27] On 14 June 1743, Folkes reported that 'the Library below stairs' needed furnishing 'for the better preservation and security of the Books'.[28] He not only provided chairs, tables, and other furniture the following year, but made a complete catalogue of the Society's library collection up to 1744 (3,250 books) to keep track of duplicates and identify which particular books were in need of purchase as well as donations, actions for which he received fulsome thanks from the Council.[29]

---

[24] M. Yakup Bektas and Maurice Crosland, 'The Copley Medal: The Establishment of a Reward System in the Royal Society, 1731–1830', *Notes and Records: The Royal Society Journal of the History of Science* 46, 1 (1992), pp. 43–76, on p. 47; See also Rebekah Higgitt, '"In the Society's Strong Box": A Visual and Material History of the Royal Society's Copley Medal, c.1736–1760', *Nuncius: Journal of the Material and Visual History of Science* 34, 3 (2019), pp. 284–316.

[25] RS/CMO/3/69, Royal Society Library, London.

[26] Bektas and Crosland, 'The Copley Medal', p. 47.

[27] Marie Boas Hall, *The Library and Archives of the Royal Society, 1660–1990* (London: The Royal Society, 1992), p. 9.

[28] Boas Hall, *The Library and Archives of the Royal Society*, p. 9.

[29] 'But more Especially desired to testify their gratitude to the President for all the great care he has been at in the new fitting up of the lower Library; and for the Singular mark of his favour in being at the pains of making out that Exact and regular Catalogue of all the books in it, which he lately made them a present of' RS/CMO/3/111, Minutes of the 30 November 1744 Council Meeting.

So many books were donated in the 1740s that by 19 July 1753 'the Secretaries were ordered to prepare a special list of those presented since the catalogue had been made by Folkes in 1744'.[30]

Folkes was a devoted bibliophile: after he died, his own enormous library at Queen Square took forty days to sell at auction, realizing £3,091 in 1756, and he lived proximally to other book collectors and collections.[31] Anthony Askew, bibliophile and protégé of Richard Mead, physician to Newton, and noted collector, was also an inhabitant of Queen Square.[32] Askew imitated his mentor's bibliophilia, noting, 'Our house in Queen Square was crammed full of books. We could dispense with no more. Our passages were full; even our garrets overflowed'.[33] Mead also lived near Folkes, his residence on Great Ormond Street in Bloomsbury featuring a library of 10,000 books. During the 1740s and 1750s, Folkes would dine with Mead at his home at an informal club, often meeting other bibliophiles such as Askew, Nicholas Mann (d. 1753), Master of Charterhouse, and Thomas Birch. Other guests included Richard Pococke from the Egyptian Society, and numismatist Browne Willis who assisted Folkes with his works on English coinage (see chapter six).[34]

Folkes's own library was not only large, but choice, featuring, for example, a sixteenth-century edition of the travel guide to Greece by the ancient geographer Pausanias (*fl.* mid-second century A D), bound in gold-tooled 'Turkish style' calf for Jean Grolier by the royal binder Gommar Estienne. This book was the first edition of the Latin translation by Romulus Amasaeus of Udine (1489–1552), the teacher of Cardinal Alessandro Farnese, papal legate, to whom the edition is dedicated; the binding was the 'only one executed for Grollier in Estienne's oriental manner', and in 2002 the book was auctioned in the Longleat House estate sale, selling for £41,825.[35] Folkes collected the best, his connoisseurship informed by a keen empiricism honed as a natural philosopher and virtuoso.

---

[30]   Boas Hall, *The Library and Archives of the Royal Society*, p. 9.

[31]   *A catalogue of the… library of Martin Folkes, esq., president of the Royal society… lately deceased; which will be sold by auction by Samuel Baker… To begin… February 2, 1756, and to continue for forty days successively (Sundays excepted)* (London: Samuel Baker, 1756). Folkes's art and instrument and coin collections were sold separately, taking fifty-six days for the entire collection to disburse. As a point of comparison, physician Richard Mead's collection of 10,000 books and his collection of sculpture, paintings, and applied arts took fifty-six days to auction at his death.

[32]   For correspondence between Askew and Mead, see, for instance, Askew's letter to Mead of 1745 concerning emendations to Hippocrates, MS 8068/1, Wellcome Library, London.

[33]   Simon Shorvon and Alastair Compston, *Queen Square: A History of the National Hospital and its Institute of Neurology* (Cambridge: Cambridge University Press, 2018), p. 52. See Anthony Askew, *Bibliotheca Askeviana sive catalogus librorum rarissimorum* (London: Baker and Leigh, 1775).

[34]   For a list of those who attended Mead's dining club, see A. E. Gunther, *An Introduction to the Life of Thomas Birch D.D., F.R.S. 1705–1766* (Suffolk: The Halesworth Press, 1984), p. 33.

[35]   Pausanias (*fl.* mid-2nd century A D) *Veteris Graeciae descriptio*. Translated from the Greek by Romulus Amasaeus of Udine (1489–1552). Florence: Lorenzo Torrentino, 1551, Christie's, Printed Books and Manuscripts from Longleat, 12 June 2002. https://www.christies.com/LotFinder/lot_details.aspx?intObjectID=3934596&cid [Accessed 3 August 2020].

During his presidency, Folkes's bibliophilia may have explained why he was more concerned in improving the Library than he was with restoring the Royal Society's Repository Museum; his past rivalry with Hans Sloane may have been another reason. In the 1730s, 'Sloane's collections featured increasingly strongly in the Society's meetings, when specimens were required to augment discussion'; in 1735, a mantis was shown from his collection along with two creatures brought by Peter Collinson, as well as a polypus in 1738.[36] Donors would also often give objects to the Society and to Sloane's personal collection, as both were held in high esteem by natural philosophers. In some cases, the 'written description [of the object] would be given to the Royal Society', but the actual samples of 'the natural phenomenon were forwarded to Sloane's collection'; the Royal Society was the scientific arbiter of the object's value, but the object itself went to Sloane's private repository in recognition of his status as master collector.[37] This division persisted under Folkes's presidency; Folkes's personal collection of objects could not rival that of Sloane's, and neither could that of the Repository. Folkes may have simply recognized that, through Sloane's collections and correspondence, the 'perceived public role of the Royal Society in the early eighteenth century' shifted 'to the facilitation of diverse forms of individual inquiry through the increased circulation and dissemination of knowledge materials, and the connection of private resources'.[38] This would be a process of knowledge creation that Folkes would continue.

It is thus clear that the Royal Society, during the majority of Folkes's presidency, was not moribund or administratively ill-managed, as some historians such as George Rousseau have claimed. Folkes was deeply concerned about the Society's finances, library, journal, and historical record, reflecting his past interests in collecting, connoisseurship, and antiquarianism. His establishment of the Copley Medal, furthermore, 'institutionalized public reward of excellence' in natural philosophy.[39]

Even William Stukeley, who had criticized Folkes for his 'infidel' beliefs, was well satisfied with the quality of Royal Society research throughout the 1740s; in July 1750, as Haycock has pointed out, Stukeley reflected on his good fortune at having escaped 'from the country indolence, to enjoy the pleasure of these meetings', and he described the Society as 'a most elegant and agreeable entertainment for a contemplative person; here we meet, either personally or in their works, all the genius's of England, or rather of the whole world, whatever the globe

---

[36] Jennifer M. Thomas, 'A "Philosophical Storehouse": The Life and Afterlife of the Royal Society's repository', PhD Dissertation (Queen Mary, University of London, 2009), p. 98; RS/JBO/16, 13 March 1735, p. 108 and RS/JBO/17, 1 June 1738, p. 271, Royal Society Library, London.

[37] Thomas, 'Philosophical Storehouse', p. 72.

[38] Alice Marples, 'Scientific Administration in the Early Eighteenth Century: Reinterpreting the Royal Society's Repository', *Historical Research* 92, 255 (February 2019), pp. 183–204.

[39] Bektas and Crosland, 'The Copley Medal', p. 67.

produces that is curious, or whatever the heavens present'.[40] Stukeley thought so much of the Royal Society meetings that he kept intermittent but detailed accounts of his attendance from 13 November 1740 to 22 February 1749/50, and gave four volumes of recollections to Maurice Johnson of the Spalding Gentlemen's Society.[41] Stukeley noted to Johnson that the meetings were 'one of the inducements, that drew me to Town again', eventually settling permanently in 1747 as rector of St George the Martyr in Queen Square.[42] Publicly, Folkes and Stukeley also remained cordial, Stukeley recording that the two went together on 4 June 1750 to hear the annual Fairchild Lecture by Archbishop Denne 'On the Beautys of the Vegetable World', and Folkes attended a Geological Soirée hosted by Stukeley to examine specimens that he had gathered and exhibited in his personal library.[43]

However, by 1751, Stukeley's private remarks about Folkes and the Royal Society became scathing. In the 1720s, before Stukeley moved back to Lincolnshire, he had been a regular attender of meetings. Although he had never been a member of Newton and Halley's predominant 'mathematical party' in the Royal Society, he did have a close relationship with Newton as a fellow Lincolnshire native. However, 'under Folkes's presidency...he found himself increasingly sidelined'.[44] Stukeley associated more and more with John Hill, the son of his old mentor Theophilus Hill, a Lincolnshire clergyman and doctor, and it seems came around to his views of the Society and of Folkes.[45] Stukeley's diary entries became more and more critical, particularly when Folkes would not countenance papers that contradicted the physics in Newton's *Principia* or those he considered nonsensical. As Haycock noted, on 31 October 1751, Stukeley recorded in his diary that a 'Dr Hickman sent a paper being a calculation of the quantity of matter in the planet Jupiter, of the celerity of his progressive motion'.[46] Hickman suggested that, contrary to Newtonian principles, 'the motion of planets is not owing to the principle of gravitation & progression: but rather only to the action of the Sun'.[47] Stukeley then noted that he 'was sorry to see in a numerous meeting, not one member able or willing to take this paper, & give an account of

---

[40] Bodl. MS Eng. misc. e. 129 f. 56; Weld, *History of the Royal Society*, vol. 1, p. 526, as quoted in Haycock, *William Stukeley*, p. 225.

[41] William Stukeley, 'Memoirs of the Royal Society', 4 vols, 1740–1750, Spalding Gentlemen's Society, Spalding, Lincolnshire.

[42] Stukeley, 'Memoirs of the Royal Society', vol. 1, pp. 1–2.

[43] Lyons, *The Royal Society*, p. 176.

[44] David Haycock, *William Stukeley: Science, Religion and Archaeology in Eighteenth-Century England* (Woodbridge: The Boydell Press, 2002), p. 226.

[45] Haycock, *William Stukeley*, p. 226.

[46] Stukeley, diary, 31 October 1751, Bodl. MS Eng. misc. e. 130, f. 39, as quoted in Haycock, *William Stukeley*, pp. 226–7.

[47] Stukeley, diary, 31 October 1751, Bodl. MS Eng. misc. e. 130, f. 39, as quoted in Haycock, *William Stukeley*, pp. 226–7.

it'.[48] Stukeley attributed this state of affairs to no one having scientific expertise, but more likely it was because Folkes considered Hickman's hypothesis unlikely.

Stukeley was also dismayed by Folkes's seeming censorship of particular topics proposed to the Society. On 14 March 1751, Stukeley recorded disapprovingly that 'a long l[ett]er with many drawings [was] deliver'd to the president. he thought, it proposd a squaring of the circle, a p[er]petual motion, or the longitude; & put it into his pocket for waste'.[49] There is little doubt there was some favouritism in Folkes's presidency—as a member of the Board of Longitude and supporter of horologist John Harrison in the 1740s, Folkes would not, for instance, countenance papers about longitude schemes that used the lunar method rather than timekeeping. On January 1741/2, Folkes with James Bradley examined the second version of Harrison's clock and both were 'unanimously of Opinion that so useful an Invention should be carried on to as great a degree of Perfection as the Nature of it will admit', giving Harrison another £500 to bring 'his said machine to perfection'.[50] No other approach would get past Folkes, and 'the automatic rejection of such material exasperated Stukeley, probably contributing to their growing rift'.[51]

Despite Folkes's partisanship, his term of office from 1741 until 1752 was characterized by some targeted and astute patronage of native English 'engineering'. Due to his interests in metrology, Folkes recognized the importance of the interaction between precision instrumentation, and applied and basic 'science'. In addition, while Folkes promoted natural philosophical work among aristocrats, as Royal Society President he also was a fervent and consistent sponsor of talent from humbler social backgrounds, such as Gowin Knight (1713–72), creator of artificial steel magnets, and Benjamin Robins (1707-51) who developed principles of ballistics and fundamental aerodynamic principles. These were the 'men of genius' whom he had in mind when proposing the scheme for the Copley Medal.

Furthermore, Folkes was European-facing in his research programme in the Royal Society, promoting research in biological vitalism, and served as a patron for Swiss naturalist Abraham Trembley (1710–84) who, like his English counterparts, was also of modest origins. Trembley, who was serving as a tutor for the children of Count Willem Bentinck van Rhoon at Sorghvliet in the Netherlands, performed experiments on the regenerative qualities of freshwater hydra or polyps, with Folkes himself participating in experimental trials with the creatures that Trembley sent. (The hydra were *Hydra viridissima* or *Chlorohydra viridissima*, a freshwater species of cnidarian in the north temperate zone.) In a related fashion, Folkes was also a benefactor of microscope maker John Cuff, whose

---

[48] Stukeley, diary, 31 October 1751, Bodl. MS Eng. misc. e. 130, f. 39, as quoted in Haycock, *William Stukeley*, pp. 226–7.

[49] Stukeley, diary, 14 March 1751, Bodl. MS Eng. misc. e. 130, f. 81, as quoted in Haycock, *William Stukeley*, pp. 226–7.

[50] Papers of the Board of Longitude, Royal Greenwich Observatory Archives, RGO 14/5, f. 14.

[51] Haycock, *William Stukeley*, p. 227.

instruments aided Trembley's research, as well as the popular writer Henry Baker, who also supported Cuff, but who also abused Folkes's trust by plagiarizing and publishing Trembley's experimental reports for financial gain.

Folkes's support of Knight, Robins, Trembley, Baker, and Cuff may have been due to his close and lasting friendships with his tutor Abraham De Moivre and the watchmaker George Graham, who both rose from humble beginnings. Having married an actress, Folkes was also not entirely one for the niceties of social position, and once he became Royal Society President he could use his power and his amiability to spot and support talent, resulting in the Society evincing considerable strengths in innovations in 'materials science', ballistics, microscopy, and investigations in the 'life sciences'.

7.2 Patronage of Gowin Knight

In January 1742, shortly after his assumption of the Royal Society presidency, Folkes was elected as one of the eight foreign members of the *Académie Royale des Sciences* in Paris to replace the deceased Edmond Halley. Montesquieu wrote congratulations from himself and Maupertuis, while wheedling for some publications of English natural philosophy, joking that such a present would be at the discretion of the President of the Royal Society.[52] After his election, Folkes sent in a paper about magnetism; the record indicated Folkes's acquaintance Maupertuis was in the chair reading, 'une Lettre de Mr Folkes, contenant des particularités intéressantes sur l'Aiman, et sur d'autres Sujets de Physique et de Litterature' (a Letter from Mr Folkes, containing interesting particulars on the Magnet, and on other Subjects of Physics and Literature).[53] Folkes also wrote Halley's memoir, which was adapted by J. J. d'Ortous de Mairan for his own 'Éloge de M. Halley' for the *Mémoires de l' Académie Royale des Science* (1742).[54] On 30 June 1746, Folkes was also elected 'Auswärtiges Mitglied' of the 'Academia Scientiarum Germanica Berolinensis', again contributing to one of the meetings by sending a letter about 'l'aiman' or the magnet which was read to the members on Thursday, 2 February 1747.[55]

---

[52] Montesquieu to Folkes, 27 September 1742, *Correspondance II, Oeuvres Complètes de Montesquieu*, ed. Philip Stewart and Catherine Volpilhac-Auger, Sociéte Montesquieu (Paris: Ens Éditions—Classique Garnier, 2014–), vol. 19, letter 529.

[53] My thanks to Christian Dekesel for this information.

[54] Eugene Fairfield MacPike, ed., *Correspondence and Papers of Edmond Halley* (London: Taylor and Francis, Ltd, 1937), pp. v. and vi.

[55] K. R. Biermann and G. Dunken, eds, *Deutsche Akademie der Wissenschaften zu Berlin. Biographischer Index der Mitglieder* (Berlin: Akademie Verlag, 1960), p. 35; E. Winter, *Die Registres der Berliner Akademie der Wissenschaften 1746–1766* (Berlin: Akademie Verlag, 1757), p. 108. See also *Registres de l'Academie, 1746–1786, Berlin-Brandenburgische Akademie der Wissenschaften.* https:// akademieregistres.bbaw.de/data/protokolle/0029-1747_02_02.xml [Accessed 30 November 2019]. My thanks to Christian Dekesel and Torsten Roeder for this information. Henry Pemberton (editor of the third edition of the *Principia*) and mathematicians James Bradley and James Stirling were also nominated to the Berlin Academy at the same time.

Some of the content of the letter was related to the discoveries of Gowin Knight, a son of a provincial clergyman who greatly impressed Folkes. Knight invented artificial steel magnets, as well as those made with linseed oil and powdered iron scales and then moulded into shape, a technique still in use. His magnets were more powerful than natural loadstones, which were literally worth their weight in silver due to their commercial usage to magnetize compass needles. Powdered loadstone was considered by Paracelsus to be of medical efficacy.[56]

Loadstones were also subjects of sensationalistic demonstrations. On 13 March 1728, the giant loadstone owned by the 2nd Duke of Devonshire was exhibited at the Royal Society in Folkes's presence. The magnet 'sustained a full 177lb', and a later demonstration on 9 April 1730 by Desaguliers saw it lift 175 pounds.[57] The Devonshire loadstone subsequently passed to James, Lord Cavendish, and on 25 April 1745 Folkes read a paper about it to the Royal Society, comparing it to another Chinese magnet owned by the King of Portugal. Both loadstones were set in mahogany cases by jeweller William Dugood, another Fellow of the Royal Society. Folkes noted that the Portuguese stone 'according to the account of it given under the print weighted naked 512 ounces which being as I suppose Portuguese weights, answer pretty nearly to 39 pounds and 5 ounces Troy. And it lifts as now armed 2880 of the same Portugal ounces'.[58] In a manner that was characteristic of his interest in metrology, Folkes included a comparison of the Portuguese Troy Weight to the English Troy in his report, the 'Portuguese ounce equivalent to 443.6 Troy grains'.[59]

An artificial magnet that could approach the Devonshire loadstone's power was thus on the list of *desiderata*. As Fara has noted, 'one of Knight's customers offered him £50 if he could make an artificial magnet rivalling' the strong loadstone belonging to the Portuguese King.[60] Knight, who kept his methods secret, employed the technique of 'divided touch', where two bar magnets were placed end-to-end, their opposite poles together. He then placed an unmagnetized steel bar on top, its middle over where the two bar magnets met. The magnetized bars were separated by sliding them apart in opposite directions along the length of

---

[56] And of course, the development of carbon-steel permanent magnets led to more investigations of their healing powers. Maximilian Hell, a Jesuit Priest and Professor of Astronomy at the University of Vienna treated patients with steel magnets, which later influenced Franz Anton Mesmer, and the idea of 'animal magnetism'. See James D. Livingston, *Driving Force: The Natural Magic of Magnets* (Cambridge, MA: Harvard University Press, 1996), p. 203.

[57] RS/JBO/13/314 and 454, Royal Society Library, London. 'Dr Desaguliers produced the famous Great Loadstone of his Grace the Duke of Devonshire and shewed the Society several experiments upon it. The stone weighed 105lb and was curiously filled up and Capt being supported in a fine Mahogany frame, adapted with screws to raise it'. The loadstone was later donated to the Ashmolean and now resides in the Museum of the History of Science, Oxford.

[58] RS/L&P/9/385, 'An Account of a Print showed to the Royal Society, 25 April 1745'.

[59] RS/L&P/9/385, 'An Account of a Print showed to the Royal Society, 25 April 1745'.

[60] Patricia Fara, *Sympathetic Attractions: Magnetic practices, beliefs, and symbolism in eighteenth-century France* (Princeton: Princeton University Press, 1996), pp. 48–9.

the unmagnetized steel bar. The unmagnetized steel bar was turned over to magnetize the opposite side, and the entire process was repeated until the steel bar
was sufficiently magnetized.[61]

On 11 February 1746/7, Folkes, accompanied by William Jones, attended
experimental demonstrations of Knight's steel magnets at his lodgings in
London.[62] Folkes reported 'he had seen Knight's magnets lift heavy iron keys
and weights' up to thirty-five pounds, as well as the magnetization of steel needles that lifted 'easily with either of its ends, the two iron weights...weighing
together 5 oz: 10 drachmas: 5 grains'.[63] He subsequently asked him to Queen
Square for similar exhibitions, with 'five pieces of iron for loadstones' at the
ready.[64] In 1745, Knight made formal demonstrations to the Royal Society.[65]
Folkes noted that:

> it was further proposed, that the temperd Needle having its virtue again
> destroyd, should be touched on the fined armed Terella belonging to the
> Society...the present of the Earl of Abercorn...The needle being touched there-
> with, was found to have acquired a strong potency, and to lift about the same
> weight, as when touched on the Doctors large and artificial magnet, that is to say
> about fifteen penny weight.[66]

Folkes subsequently wrote to naturalist Abraham Trembley (see section 7.4),
'what appears even more extraordinary is...[Knight]...is capable of fortifying
the natural magnet in nearly no time and of changing their poles whenever he
wanted to'. Folkes sent Trembley a present of small artificial magnet made by
Knight that could draw weights of five pounds. Knight went on to develop 'the
means of strongly magnetising the needles and bars by successively combining
bar magnets to reinforce them. His compass design included a single steel bar,
straight needle, a replaceable pivot and an agate bearing'.[67]

Folkes also claimed, correctly, that Gowin's improved azimuth compass would
enable the British 'to increase and promote greatly our foreign trade and commerce, whereby we are provided at home with the fruits, the conveniences, the

[61] H. H. Ricker III, 'Magnetism in the Eighteenth Century', *The General Science Journal* (5
December 2011), pp. 1–32, on p. 24. https://www.gsjournal.net/Science-Journals/Research%20Papers/
View/3817 [Accessed 19 January 2020].

[62] MS 2391/3, Royal Society Papers, Wellcome Library, f. 1. The account in Folkes's hand.

[63] Fara, *Sympathetic Attractions*, p. 40. MS 2391/3, Royal Society Papers, Wellcome Library, f. 3v.

[64] RS/JBO/12/203–4, Royal Society Library; *A Catalogue of the Genuine and Curious
Collection*, p. 3.

[65] Gowin Knight, 'A Letter to the President, Concerning the Poles of magnets being Variously
Placed', *Philosophical Transactions* 43, 476 (1745) pp. 361–3; RS/L&P/12/381.

[66] MS 2391/3, Royal Society Papers, Wellcome Library, ff. 6–7.

[67] 'The Compass', Museum of the History of Science, Oxford. http://www.mhs.ox.ac.uk/collections/
imu-search-page/narratives/?irn=10901&index=0 [Accessed 19 January 2020].

curiosities and the riches of the most distant climates'.[68] In 1749, the Royal Society discussed a compass that had been made useless when struck by lightning, and Knight saw a commercial opportunity. He cleverly magnetized a common steel sewing needle and used it as a compass needle, realizing it retained magnetism longer than one made of soft iron. He wrote that the smallest sewing needles, were 'strong enough to bear the Weight of a Card; and are neither so hard and brittle as to break; and they are generally better pointed than any that a common Workman could pretend to make *extempore*'.[69] Sewing needles had been used in the past to pin specimens and illustrations of specimens to organize them taxonomically, but this was a novel repurposing.[70] Most compasses had brass caps and points to reduce vibrations of the card and needle, but Knight thought this would be too expensive for common usage, so he experimented with making them out of glass and agate, finally hitting upon ivory caps 'with a small bit of agate at the top' to provide the needed stability and economy for his instrument. Ivory's lower price relative to brass was due to the expansion of Portuguese, Dutch, and English merchants into Southeast and Far Eastern Asia to trade in tusks and slaves, and the ubiquity of turning lathes made the idea practical.[71] Knight's steering compass, demonstrated to the Royal Society in July 1750, was thus the first major improvement of the mariner's compass in the eighteenth century.[72]

Knight's activities 'exemplify the commercial initiatives of natural philosophers which were affecting English society as profoundly as the widely-cited example of industrial manufacturing processes'.[73] Magnets were also commercial and aesthetic commodities, their strength as well as their setting symbols of natural philosophical power and might.

Though from humble origins, via his invention, patronage, and 'commercial opportunism', as well as his own initiative, Knight became the first Principal Librarian of the British Museum, writing with the General Committee the very plan 'for the general Distribution of Sir Hans Sloane's Collection of natural and other productions', noting that 'the greatest and most valuable part of this Collection consists of things relating to natural history. Wherefore that part will first claim our attention, and will merit a particular regard in the general

---

[68] Martin Folkes, RS/JBC/19/366 (20 November 1747), as quoted by Patricia Fara, '"Master of Practical Magnetics": The Construction of an Eighteenth-Century Natural Philosopher', *Enlightenment and Dissent* 14 (1995), pp. 52–87, on p. 56; Gowin Knight, 'A Description of a Mariner's Compass contrived by Gowin Knight, M.B. F.R.S', *Philosophical Transactions* 46, 495 (1750), pp. 505–12.

[69] Knight, 'A Description of a Mariner's Compass', p. 512.

[70] Anna Marie Roos, *Martin Lister and his Remarkable Daughters: The Art of Science in the Seventeenth Century* (Oxford: Bodleian Library Publishing, 2018), p. 130.

[71] Martha Chaiklin, 'Ivory in World History: Early Modern Trade in Context', *History Compass* 8, 6 (June 2010), pp. 530–42, on p. 537 and p. 540.

[72] 'Gowin Knight: Steering Compass', Royal Museums Greenwich. https://collections.rmg.co.uk/collections/objects/42659.html [Accessed 4 August 2020].

[73] Fara, 'Master', p. 59.

distribution.⁷⁴ Although Knight was elected to the Royal Society, and even a member of the Royal Society Council in 1751, he was unsuccessful in his bid to become secretary in 1752, defeated by Thomas Birch who was supported by a Whig Circle led by Philip Yorke, 2nd Earl of Hardwicke.⁷⁵ The naturalist and mineralogist Emanuel Mendes da Costa (1717–91) noted in a letter to William Borlase about Knight's candidacy:

> Dr Gowin Knight who first invented the Artificial Magnets & has enriched the World with usefull Discoveries on the Sailors compasses and Needles and who is an Oxonian Physician declared himself a Candidate. This gentleman was supported by the Physical party & a Great Number of the Working Members. Mr. Birch who was chosen was the other Candidate & was strongly supported by the Nobility & other honorary Members.⁷⁶

Despite Folkes's efforts at inclusiveness for those from different social backgrounds, 'working' members outside of a charmed circle still faced obstacles in the Royal Society.

7.3 Folkes and Benjamin Robins

Folkes was also the patron of ballistics expert Benjamin Robins. Robins was a military engineer, and like Knight, of a humble background, born to a tailor and practising Quaker in Bath of modest means. Recognized for his abilities by physician Henry Pemberton, who edited the third edition of the *Principia*, Robins was moved to London as an adolescent to develop his talents for mathematics and literature and became a fervent Newtonian.⁷⁷ In 1727, when he was twenty, he published a mathematical proof of the eleventh and last proposition of Newton's *Treatise of Quadratures* in the *Philosophical Transactions*, interestingly, using fluxions,⁷⁸ and the following year, an assured confutation of a dissertation by Jean Bernoulli concerning the laws of motion in bodies impinging on one another.

---

⁷⁴ Fara, 'Master', pp. 52–3; British Museum MS CE4/1, 'A Plan for General Distribution of the Sir Hans Sloane's Collection', 14 January 1757 in *Original Letters and Papers*, vol. 1, p. 51.
⁷⁵ David Philip Miller, 'The "Hardwicke Circle": The Whig Supremacy and Its Demise in the 18th-Century Royal Society', *Notes and Records of the Royal Society of London* 52, 1 (1998), pp. 73–91, on p. 76.
⁷⁶ BL Add MS/28535, ff. 70–3, Da Costa to William Borlase, 27 February 1752, as quoted in Miller, 'The "Hardwicke Circle"', p. 77.
⁷⁷ Biographical material has been taken from James Wilson, *Mathematical Tracts of the Late Benjamin Robins esq.*, 2 vols (London: J. Nourse, 1761); Brett D. Steele, 'Benjamin Robins', *Oxford Dictionary of National Biography* (Oxford: Oxford University Press, 2012), https://doi.org/10.1093/ref:odnb/23823 [Accessed 21 July 2020].
⁷⁸ Benjamin Robins, 'Demonstration of the Eleventh Proposition of Sir I. Newton's Treatise of Quadratures', *Philosophical Transactions* 34, 397 (1727), pp. 230–6.

Robins was approved as a candidate for election as a Fellow of the Royal Society on 9 November 1727 in a Council meeting where Folkes was present; as Robins and Folkes were approximately the same age and had similar mathematical interests, they were presumably well acquainted at this point.[79] Robins's subsequent publications were devoted to attacks on those opposed to Newton, such as Leibniz and Berkeley, most to do with the *vis viva* controversy, and some treatises defending geometrical approaches to the calculus and proposing the elimination of infinitesimals.

Robins's interest in geometry spawned his study of ballistics, including experimental measurements of air resistance and drag caused by different shapes of ammunition. His invention of a whirling arm apparatus demonstrated that air resistances at subsonic and supersonic speeds follow 'different rules and are associated with different phenomenological behaviour'.[80] Best known for his book *New Principles of Gunnery* (1742), Robins also developed the ballistic pendulum for measuring the speed of musket balls; projectiles were aimed from various distances at a square plate which pivoted on an axis, and the arc length described by the pendulum allowed for calculation of the speed of the bullet.[81] Ballistics expert Charles Hutton (1737–1823) would later build his own larger model for experiment with cannons.[82] Robins also anticipated research on the thermodynamics of the internal combustion engine by estimating the pressures and temperatures of gases created by combusting gunpowder in a gun barrel through which a ball was being shot.[83] Robins's *New Principles of Gunnery* was subsequently translated into German and expanded by Leonhard Euler; this went on to be used for over a century on the continent, and translated into French by Jean-Louis Lombard. Euler's commentary on Robins's book was subsequently re-translated into English and 'published by the order of the Board of Ordnance with remarks by Hugh Brown of the Tower of London'.[84]

Folkes's interest in Robins's work was due to Folkes's appointment as Master of the Royal Military Academy, Woolwich. Founded in 1741, Woolwich was modelled after the equivalent body in Caen, and served as a training school of military engineers and artillery officers. In 1740, John, 2nd Duke of Montagu, was

[79] RS/CMO/2/340, Minutes of a meeting of the Council of the Royal Society, 9 November 1727, Royal Society Library, London.

[80] W. Johnson, 'Benjamin Robins, FRS (17-7-1751): New Details of his Life', *Notes and Records: the Royal Society Journal of the History of Science* 46, 2 (1992), pp. 235–52, on pp. 235–36. Johnson's publications on Robins are collected in a volume: *Collected works on Benjamin Robins and Charles Hutton* (New Delhi: Phoenix Publishing House, 2001).

[81] RS/MS/130/11/11, Drawing of a ballistic pendulum by Benjamin Robins, Royal Society Library, London.

[82] Benjamin Wardhaugh, *Gunpowder and Geometry: The Life of Charles Hutton: Pit Boy, Mathematician and Scientific Rebel* (London: William Collins, 2019), pp. 88–9.

[83] Johnson, 'Benjamin Robins, FRS (17-7-1751)', on pp. 235–6.

[84] RS/MS/39, Papers of Benjamin Robinson, Royal Society Library, London.

appointed Master-General of the Ordnance, and on 3 September of that year, he
wrote to Folkes:

> Every thing relating to our academy at Woolwich is settled except the appoint-
> ing the chief master...if it be not disagreeable to you that you would let me
> make you the nominal master...your name the head of the thing would not
> only a great credit to it, but might be of great service in your overlooking now
> and then the ways of proceeding in the nature of the director of the accademy in
> France. I have mentioned this to nobody, so thus I beg you will sincerely and as
> a friend do on it as you lyke best.[85]

Although Folkes did not teach at the academy himself, with mathematician John
Muller (1699–1784) appointed as deputy headmaster to carry out his duties,
Folkes was interested in overseeing the transformation of the institution from a
mere school into a proper academy with a cadet company, as well as in relevant
military technology. Convinced of the efficacy of a military education, Folkes
visited the Caen Military Academy and sent his own son Martin there to be educated (see chapter eight).

Robins's Newtonianism and Euler's subsequent commentary on his book further recommended him to Folkes, who, in 1746, asked Robins to be on a special
committee organized in the Royal Society to answer questions from the
Committee of Common Council of the City of London. Other members were
James Jurin, William Jones, George Lewis Scott, and John Ellicott. The topic was
proposed alterations to London Bridge, and the Society's advice mostly had to do
with surveying the water depths at high and low tide under the drawbridge, and
'ascertain[ing] in how many of the arches the dripshot piles are driven; how close
together; and how far the tops of them are below low still water mark', mainly to
understand the forces of the water's rise and fall on the piers and piles according
to their shapes and sizes.[86] Robins's work with his whirling arm, which tested the
air resistance given by differently shaped objects, may have been seen as useful in
this regard, as well as the research he did in hydraulics as he 'investigated the
resistance of pyramids and inclined planes to further demonstrate the limitations'
of Newton's theories of fluid mechanics.[87]

The next year, Robins was invited by Folkes to give a series of papers and
experimental demonstrations at the Royal Society. On 2 June 1747, Robins presented to the Royal Society 'Of the Nature and Advantages of Rifled Barrel Pieces',
as well as another paper on 4 June 1747 concerning experiments on air

---

[85] RS/MS/790, Montagu to Folkes, 3 September 1740, Royal Society Library, London.
[86] Charles Hutton, *Tracts on Mathematical and Philosophical Subjects*, pp. 120–1; W. Johnson,
'Early Bridge Consultants, Benjamin Robins, F.R.S. and Charles Hutton, F.R.S. and Mis-judged Bridge
Designer, Thomas Paine', *International Journal of Mechanical Sciences* 41 (1999), pp. 741–8, on p. 742.
[87] Steele, 'Benjamin Robins', *Oxford Dictionary of National Biography*.

resistance.[88] In previous works, Robins had considered 'the numerous irregularitys which take place in most of the operations of Gunnery', employing his understanding of what we now call the 'Magnus effect' to explain why rifles, particularly those he had seen from Germany and Switzerland, were of greater accuracy than muskets. The 'Magnus effect' is seen when a moving spherical body has a spin, moving away from the intended direction of travel, a technique often employed by baseball players throwing curve balls or spin bowlers in cricket. The spin changes the airflow around the moving body, generating the Magnus force via conservation of momentum. However, as Robins pointed out, the direction of forward motion of a rifled bullet as opposed to a musket 'coincides with its axis of rotation... the flow of air passing around the bullet is uniform', and no lateral or pressure forces would be created to deviate the bullet shot from a rifle from its intended direction.[89] Robins also proposed egg-shaped rather than spherical shapes for bullets. The problem with using spherical bullets, in Robins's view, was that the 'axis of the sphere's spin would not necessarily coincide with the direction of motion'.[90] Instead, Robins claimed that 'making use of bullets of an egg-like form... will acquire by the rifles a rotation round its larger axis; and is constantly forced by the resistance of the air into the line of its flight. As we see, that by the same means arrows constantly lie in the line of their direction, however that line be incurvated'.[91]

Impressed with his demonstrations, Folkes subsequently wrote Robins a formal letter relating to a query from the Cavaliere Giuseppe Ossorio, chief minister and Savoyard representative of King Charles Emmanuel III of Sardinia. In 1748, the British government was also courting the Savoyard State due to the so-called 'Diplomatic Revolution', the 'alliance between the old enemies the Courts of Versailles and Vienna', so a sharing of military technology involving the Royal Society was strategic.[92] Ossorio was thus formally received by the Royal Society and duly elected a Fellow, his certificate signed by Folkes and the appropriately aristocratic Duke of Richmond, Duke of Montagu, and Philip Stanhope, 4th Earl of Chesterfield.[93] The Cavaliere in particular sought a copy of Robins's presentation about cannon charges made to the Royal Society a year previously, to satisfy his Sardinian Majesty's curiosity deriving from 'experiments made by his order at Turin'.[94]

---

[88] RS/MS/130/1; RS/MS/130/2, Royal Society Library, London.

[89] Brett D. Steele, 'Muskets and Pendulums: Benjamin Robins, Leonhard Euler, and the Ballistics Revolution', *Technology and Culture* 35, 2 (April 1994), pp. 348–82, on p. 364.

[90] Steele, 'Muskets and Pendulums', p. 364.

[91] Wilson, *Mathematical Tracts of the late Benjamin Robins*, vol. 1, p. 338.

[92] Christopher Storrs, 'The British Diplomatic Presence in Turin: Diplomatic Culture and British Elite Identity 1688-1789/98', in Paola Bianchi and Karin Wolfe, eds, *Turin and the British in the Age of the Grand Tour* (Cambridge: Cambridge University Press, 2017), pp. 73–90, on p. 78.

[93] RS/EC/1748/02, Royal Society Library, London.

[94] W. Johnson, 'Aspects of the Life and Work of Martin Folkes (1690–1754)', *International Journal of Impact Engineering* 21, 8 (1998), pp. 695–705, on p. 704.

Robins's answer to Folkes and Ossorio, read to the Society on 7 January 1747/8, is still extant in fair copy in manuscript.[95] In a series of tests, Robins fired a cannon with different charges into a bank of earth at fifty yards, showing that 'a piece fired at a given elevation and with the common best charge … ranged further than one shooting with larger charges', again noting the deflecting 'Magnus effect' on range of fire.[96] He recommended using one-third the weight of the cannonball in gunpowder as an optimum, or 'best charge', finding that using more than this weight was a waste of resources, and noting that this was the standard used in the French military. Robins wrote, 'every effect to be produced by use of firearms should be executed with the least powder possible, not only as hereby much ammunition may be saved, but also, because with smaller charges the piece is less heated and strained, the service of it is more prompt and the flight of the shot is much more steady'.[97]

For his efforts and his demonstrations in the Royal Society, in 1746 Robins was awarded the Copley Medal, Folkes writing and delivering the oration which confirmed Robins's scientific prowess and status as a Newtonian acolyte. Folkes wrote,

The great Sir Isaac Newton … investigated the laws of the resistances made to bodies in motion, during their passage through the air and other fluids, and, those upon different theories, and upon different suppositions. He also made experiments upon the resistance given to funipendulous bodies in their oscillations, and to other in their fall, which he dropped for that purpose from the highest part of the cupola of St. Paul's church: but he never had the opportunity of making trials, upon those much greater resistances, that shells and bullets are impeded by, in those immense velocities with which they are thrown from military engines.[98]

And Robins clearly did have such an opportunity, of which he made great use.

Robins would continue to be active in the Society, doing a set of calculations concerning the height and range of fireworks after observing the famous triumphal celebration of the Peace of Aix-la-Chapelle in Green Park. He gave a fixed quadrant to a friend stationed on 'top of Dr Nisbett's house in King Street near Cheapside, where he had a fine view of the upper part of the [pyrotechnics] building erected in the Green Park' so the degree of launch and height of ascent of the fireworks could be observed.[99] Robins concluded that while most rockets had

[95] RS/MS/139, Royal Society Library, London. The work was also published in Wilson, *Mathematical Tracts of the late Benjamin Robins*, vol. 1, pp. 295–305.
[96] Johnson, 'Aspects of the Life and Work of Martin Folkes', p. 704.
[97] RS/MS/139, Royal Society Library, London, pp. 12–13.
[98] Wilson, *Mathematical Tracts of the late Benjamin Robins*, vol. 1, pp. xxx–xxxi. Funipendulous bodies are those that hang from a rope or string.
[99] Wilson, *Mathematical Tracts of the late Benjamin Robins*, vol. 1, p. 317.

their heights 'limited between 400 and 600 yards', he believed they could be made to go further with more careful composition and manufacture and could be of military use for signalling.

Working with the Woolwich military academy, Robins then had some rockets made and fired from fixed locations—Godmarsham, Kent; Beacon Hill in Essex; London-Field in Hackney; and Barkway on the border from Hertfordshire—to see if they could be seen by different stationed observers. (His rocket trials may have taken a cue from Edmond Halley's previous crowd-sourced appeal in 1715 for observations to provide data for the publication of his map of the total solar eclipse.) On 27 September 1748 at 8 p.m., the firing of the rockets for 'geometrical use' took place; Robins and John Canton FRS measured their height, with observers also stationed in Trinity College, Cambridge. Robins repeated the trials on 12 October, noting 'the largest of the rockets, which were about two inches and a half in diameter' rose the highest and were most readily observed.[100] At this point, Robins ceased his fireworks experiments, as he had accepted a post as Director General of Engineering for the East India Company.[101]

In November 1748, however, an advertisement in the *Gentleman's Magazine* appeared, asking if 'ingenious gentlemen, who are within 1, 2 or 3 miles of the fireworks, would observe, as nicely as they can, the angle which the generality of the rockets shall make with the horizon, at their greatest height, this will determine the perpendicular ascent of those rockets to sufficient exactness'.[102] The crowd-sourced trials were carried out by one Samuel Da Costa of Devonshire Square, possibly related to naturalist Emanuel Mendes da Costa, 'Mr Banks a gentleman, who had for many years practiced making rockets', and two unidentified technicians; Canton again measured the heights the projectiles achieved.[103] Although there was little response to the advertisement in the *Gentleman's Magazine*, the experimental results were published in the *Philosophical Transactions*, the trials made in the presence of 'several worthy members' of the Royal Society, probably including Folkes. The experiments showed, according to Robins's prediction, that fireworks from 2.5 to 3 inches were 'sufficient to answer all the purposes they are intended for'; larger diameters and more gunpowder were not necessary. Similar to the cannon experiments Robins did for Ossorio, the practitioners were trying to find 'the sweet spot' of using just enough gunpowder to accomplish a military aim as well as an entertaining spectacle.

---

[100] Wilson, *Mathematical Tracts of the late Benjamin Robins*, vol. 1, p. 325.

[101] Wilson, *Mathematical Tracts of the late Benjamin Robins*, vol. 1, p. 325.

[102] 'A Geometrical Use proposed from the Fire-Works', *Gentleman's Magazine* 18 (November 1748), p. 488. Simon Werrett, 'Watching the Fireworks: Early Modern Observation of Natural and Artificial Spectacles', *Science in Context* 24, 2 (2011), pp. 167–82, on p. 178. Werrett identifies the advert as probably coming from Robins, but it may have come from subsequent experimenters like Da Costa.

[103] Wilson, *Mathematical Tracts of the late Benjamin Robins*, vol. 1, p. 325.

As for Robins, his career did not go off with a bang, but with more of a whimper. After one year in India, he died of malarial fever on 29 July 1751, unmarried, his grave unmarked. It was a sad end for Robins, described by a British artillery officer in 1789, as one who 'was in gunnery what the immortal Newton was in philosophy, the founder of a new system deduced from experiment and nature'.[104] Folkes was subsequently named an executor of his will and guardian of his scientific papers; in another unfortunate turn of events, Folkes's own death in 1754 precluded the publication of Robins's scientific works until 1761.

7.4 Folkes, Madame Geoffrin, and Abraham Trembley

Folkes did not confine his patronage to English experimental philosophers, but during his Grand Tour made and continued contact with several Italian, German, and French intellectuals such as Maupertuis, afterwards having Voltaire elected to the Royal Society in 1743. During his travels, Folkes also became acquainted with Madame Marie Thérèse Rodet Geoffrin who convened a famous salon on the Rue St-Honoré in Paris, and on his return they maintained an active correspondence.[105] Via Geoffrin's information networks, Folkes learned of the discoveries of Swiss naturalist Trembley and started an extraordinary correspondence with him.

From 1740 to 1744, Trembley discovered the regenerative properties of the hydra or polyp.[106] Folkes's friend Charles Lennox subsequently had Trembley serve as tutor to his oldest son, taking him on his own Grand Tour. As Ratcliff has eruditely shown, work on zoophyte polyps challenged taxonomic systems and the Aristotelian Chain of Being, which had dominated understanding of the natural world, and raised questions of vitalism, materialism, and deism.[107] For instance, did the polyp have a soul, and could that soul be split into two? Was the ability of the polyp to regenerate support for preformation, the idea that all living things pre-existed as invisible germs? Folkes and microscopist and FRS Henry Baker would subsequently discuss similar questions about parasitic twins.

[104] Alessandro V. P. d'Antoni, *A Treatise on Gunpowder, a Treatise on Fire-Arms, and a Treatise on the Service of Artillery in the Time of War*, trans. Captain Thomson (London: 1789), p. xvii, quoted in Steele, 'Benjamin Robins', *Oxford Dictionary of National Biography*.

[105] Harcourt Brown, 'Madame Geoffrin and Martin Folkes: Six New Letters', *Modern Language Quarterly* 1, 2 (1940), pp. 215–21. The letters were not translated in this article, so I have provided English translations.

[106] See Marc J. Ratcliff, 'Abraham Trembley's Strategy of Generosity and the Scope of Celebrity in the Mid-Eighteenth Century', *Isis* 95, 4 (December 2004), pp. 555–75; Mark J. Ratcliff, *The Quest for the Invisible—Microscopy in the Enlightenment* (Aldershot: Ashgate, 2009), pp. 103–24. While Ratcliff has analysed hydra as a means of scientific exchange and networking and in the context of development of microscopy, for the purposes of this biography I am more interested in analysing the personal relationship between Folkes and Trembley, and Folkes as a mentor to younger natural philosophers.

[107] Ratcliff, *The Quest for the Invisible*, p. 134, p. 257.

Why did the study of polyps become so central to the Royal Society? In the previous century, the study of insects (which according to the pre-Linnaean classification system included hydra and molluscs) had often provoked derision. Martin Lister, the first conchologist and arachnologist, admitted in the preface to his *Exercitatio Anatomica in qua de Cochleis* (1696), a comprehensive anatomical guide to land shells and slugs, that he was aware that his biological work might 'provoke the laughter of spectators'.[108] He also wrote to his friend Edward Lhwyd, the second keeper of the Ashmolean Museum, stating that there were 'censorious mouthes who think and say a man that writes on Insects can be but a trifler in Phisic'; Lister hoped he would be left 'alone to pursue Philosophie amongst the inferiour sort of beings'.[109] By the eighteenth century, however, the discussion of the polyps had become a major focus of research because it struck a chord as it raised questions about argument from design underlying what Alexander Pope described as 'A mighty maze! But not without a plan'. The sometimes puzzling behaviour or anatomical quirks of insects, molluscs, and hydra 'presented special challenges for those hoping to discover teleology in every corner of creation…[presenting] no apparent value in proclaiming the greater glory of God'.[110] As Folkes himself wrote in a letter to Trembley upon receiving his book on the hydra, 'how wrong were we until now to confuse the laws which perhaps concern only a small part of the creation with more general laws of nature'. Microscopists and naturalists 'also introduced the…problem of scale by opening questions as to why the great Creator would enclose so much information in so inaccessible a vault as a mite's wing or a beetle's body'.[111]

Folkes and Lennox performed several of Trembley's experiments themselves to answer such questions. Lennox participated for the entertainment, but Folkes was interested in the advantages and limitations of the instruments used to study the hydra, particularly John Cuff's (1707?–c.1772) innovative microscopes. As Folkes was a patron of horologist George Graham, extolling the merits of his instruments to fellow natural philosophers, so was he a patron of Cuff and his instruments. Ever the metrologist, Folkes even wrote a practical guide to using Cuff's new micrometer, appended to Henry Baker's *The Employment of the Microscope* (1753), which was written to encourage gentlemen naturalists to make their own observations. As we have seen, Folkes was an enthusiast of the instruments, having written a commentary on the microscope that Leeuwenhoek bequeathed to

---

[108] Martin Lister, 'Preface', *Exercitatio anatomica In qua de Cochleis, Maximè Terrestribus & Limacibus, agitur* (London: Samuel Smith & Benjamin Walford, 1694), p. 2.

[109] Bodl. MS Ashmole 1816, f. 176r., Bodleian Library, Oxford. Lister also expresses much the same sentiment in MS Ashmole 1816, f. 116r.

[110] Kevin L. Cope, 'Notes from Many Hands: Pierre Lyonnet's Redesign of Friedrich Christian Lesser's Insecto-Theology', in Brett C. McInelly and Paul E. Kerry, eds, *New Approaches to Religion and Enlightenment* (Vancouver and Madison: Fairleigh Dickinson University Press, 2018), pp. 1–34 on p. 5.

[111] Cope, 'Notes from Many Hands', p. 6.

the society, and he was the first to notice their bi-convex optics (see chapter four). Though it was not surprising that Folkes brought insights into the practicalities of using a micrometer and calculating the minute dimensions of specimens, he also was a perceptive enough administrator and natural philosopher to promote and patronise discoveries in natural philosophy on a grander scale.[112]

Part of the means by which Folkes found out and was able to promote discoveries in natural philosophy was due to Madame Geoffrin. Geoffrin has been characterized as the inventor of the Enlightenment salon, a private and more intimate affair than London coffee-house or tavern meetings. How did she run her salon? As Dena Goodman remarked,

First, she made the one-o'clock dinner rather than the traditional late-night supper the sociable meal of the day, and thus she opened up the whole afternoon for talk. Second, she regulated these dinners, fixing a specific day of the week for them. After Geoffrin launched her weekly dinners, the Parisian salon took on the form that made it the social base of the Enlightenment Republic of Letters: a regular and regulated formal gathering hosted by a woman in her own home which served as a forum and locus of intellectual activity.[113]

Although there are not many sources for the early days of Geoffrin's salon in the 1730s and 1740s, 'it seems clear that her guests were at first recruited from among writers and scholars (Fontanelle, Marian and Marivaux) present in Madam Tencin's salon. In the 1740s, the sciences and savants were in fashion'.[114] This was precisely when Geoffrin began corresponding with Folkes after he returned from his Grand Tour. Their correspondence ranged over a variety of subjects, but it is notable that Geoffrin advised Folkes about his health in a frank manner, suggesting they had close familiarity. On 16 January 1743, she wrote to Folkes:

Since you assure me Sir that the concern I have about your health will make it valuable to you, I repeat again that I am very interested in it, to force you not only to keep it good but also to prevent all pain thanks to a few precautionary cures such as bleeding, purges, it is always needed for someone that is as fat as

[112] Martin Folkes, 'Some account of Mr Leeuwenhoek's curious microscopes lately presented to the Royal Society', *Philosophical Transactions* 32, 380 (1723), pp. 446–53, on p. 449; James Hyslop, 'John Mayall and reproductions of early microscopes', Explore Whipple Collections, Whipple Museum of the History of Science, University of Cambridge, 2008. https://www.whipplemuseum.cam.ac.uk/explore-whipple-collections/microscopes/dutch-pioneer-antoni-van-leeuwenhoek/mayall-reproductions [Accessed 30 September 2019].
[113] Dena Goodman, *The Republic of Letters: A Cultural History of the French Enlightenment* (New York, Cornell University Press, 1994), pp. 90–1.
[114] Antoine Lilti, *The World of the Salons: Sociability and Worldliness in Eighteenth-century Paris* (Oxford: Oxford University Press, 2015), p. 72.

you are, and this fat is the reason my fear for what I had heard [about your health] was greatly increased.[115]

Folkes's ritual feasting had its disadvantages, not helped by his assumption in 1747 of the presidency of the Society's weekly Thursday's Dining Club known as the 'Royal Philosophers' held at the Mitre Tavern at 4 p.m. A Menu of 24 March 1747 consisted of:

2 dishes fresh Salmon, Lobster sauce
Cod's Head
Pidgeon Pie
Calves Head
Bacon and Greens
Fillett of veal
Chine of Pork
Plumb Pudding
Apple Custard
Butter and Cheese[116]

Nobility were allowed to be members of the club if they would donate game. On 3 May 1750, with Folkes in the Chair, it was resolved 'That any Nobleman or Gentleman complimenting this company annually with venison, not less than a haunch, shall…be deemed an Honorary Member and admitted as often as he comes, without paying the fine which those members so who are elected by Ballott'.[117] There were also presents of watermelons from Portugal and Spain. Along with the usual Fellows attending, such as James Bradley or Charles Lennox, Hogarth was invited twice, and an examination of one year's attendance in 1748 demonstrated the international 'flavour' of discourse Folkes promoted. There were guests such as Jean Jallabert, Chair of Experimental Physics at Geneva; Don Jorge Juan, a Spanish mathematician who measured territorial degrees in equatorial America; Johannes Nicolaas Sebastiaan Allamand, Professor of Philosophy and Natural History at Leiden who gave the first explanation of the Leyden Jar; as well as Cavaliere Giuseppe Ossorio, who dined on 16 February 1748, having, as we have seen, been made a Fellow in the previous year.[118] A more unfortunate symbol of internationalism was the inclusion of Nicholas de Montaudouin, part

[115] RS/MS/250, f. III, Letter 13, Geoffrin to Folkes, 15 December 1742, Royal Society Library, London.

[116] Sir Archibald Geikie, Annals of the Royal Society Club; the record of a London dining-club in the eighteenth and nineteenth centuries (London: Macmillan and co., 1917), p. 26.

[117] RS Minute Book One, Royal Society Dining Club, Royal Society Library, London; Geikie, Annals of the Royal Society Club, pp. 38–9.

[118] Geikie, Annals of the Royal Society Club, pp. 31–2.

of a prominent family who made their fortune in Nantes through the transatlantic slave trade.[119] It was not unusual for the Royal Society to have such associations, as part of Sir Hans Sloane's fortunes and collections had their origins in his wife's family's involvement in the slave trade in Jamaica.[120] Sloane's collector, naturalist James Petiver, also used networks in the slave trade to obtain specimens; as Kathleen Murphy has noted:

> The importance of slavery and the slave trade to Atlantic economic and social structures meant that the naturalist relied on the institutions, infrastructures and individuals of the slave trade and plantation slavery. [Petiver] exploited the commercial routes created by the slave trade to build his natural historical collections.[121]

The production of natural knowledge was fundamentally bound up with colonialism.

It was also entangled, as we have seen, with visual iconography and self-fashioning, particularly important in continental salon culture. Geoffrin reported being given a copy of Folkes's portrait medal struck by Italian medallist Ottone Hamerani (see chapter six) which was circulating around the French salons as a material dissemination of his erudition and Masonic beliefs:

> some time ago I received a letter from you as this man was with me, I asked him permission to read it, he saw from the movements of my face that it pleased me, he trifled with it, and he asked me who it was from, I told him your name, and how much I loved you... a few days later I went to the countryside, and upon my return, I found on the mantelpiece a small package of some weight, I wondered who brought it, my people told me they didn't know, it increased my eagerness to know what it carried, and I was very satisfied when I found your medal, so true to life that I screamed saying ah this is Monsieur Folkes, after surprise I felt joy, I actually feel much joy from having your portrait.[122]

Lastly, Geoffrin also passed along useful gossip about French *philosophes*, describing, for example, the animosity between Folkes's friend Charles de Secondat, Baron de Montesquieu and his son Jean-Baptiste de Secondat, Baron de La Brède.

[119] James A. Rawley and Stephen D. Behrendt, *The Transatlantic Slave Trade: A History, revised edition* (Lincoln and London: University of Nebraska Press, 2005), pp. 120–2.
[120] See James Delbourgo, *Collecting the World: The Life and Curiosity of Hans Sloane* (Cambridge, MA: Harvard University Press, 2018), particularly chapter two.
[121] Kathleen Susan Murphy, 'James Petiver's "Kind Friend" and "Curious Persons" in the Atlantic World: commerce, colonialism and collecting', *Notes and Records: The Royal Society Journal of the History of Science* 74, 2 (2020), pp. 259–74, on p. 261.
[122] RS/MS/250, f. II, Letter 70, Geoffrin to Folkes, 17 July 1743, Royal Society Library, London.

Montesquieu arrived in England in November 1729 and left in April 1731; through Philip Stanhope, 4th Earl of Chesterfield, Montesquieu met Folkes, who 'explained to him the procedure of the law courts'.[123] Folkes introduced Montesquieu to the Royal Society, where he met the Duke of Richmond and Montagu, who enchanted him. Montesquieu told their mutual friend Gaspero Cerati 'Je fus reçu il y a trois jours member de la Société royale de Londres' when he was elected FRS on 26 February 1730.[124] The two natural philosophers corresponded often and had a close friendship. Folkes even provided him (as well as Maupertuis) not only with scientific books, but also 'capotes anglaises', or condoms, via the post, as condoms were available over the counter in England and banned in France.[125] Maupertuis replied that he and Montesquieu shared out the precious prophylactics, although 'in good conscience, I believe I could have taken three-quarters... it is much to the shame of our nation that while it applies with such success all the frivolity in gallantry... we have to resort to strangers to defend us against the dangers to which our beautiful women expose us'.[126] In 1739, when Folkes was choosing a school for his son in France, the philosopher offered his son hospitality and mentoring in Bordeaux for a year; Montesquieu promised that 'we will put him in good company and do whatever possible to that one day he will be like his father... only being as libertine as a gallant man should be'.[127] Several years after Montesquieu's visit to London, he wrote to Folkes, 'Of all

---

[123] Ira O. Wade, *The Structure and Form of the French Enlightenment, Volume 1: Esprit Philosophique* (Princeton: Princeton University Press, 2015), p. 151.

[124] John Churton Collins, *Voltaire, Montesquieu and Rousseau in England* (London: Eveleigh Nash, 1908), p. 164. Montesquieu indicated to Cerati 'I was received three days ago as a member of the Royal Society of London', and later wrote to Folkes, 'Je vous prie, parlez un peu de moi à MM. les ducs de Richmond et de Montagu: le temps que j'ai passé à leur faire ma cour a été le plus heureux de ma vie'. [Please tell MM. the Dukes of Richmond and Montagu: the time I have spent paying court to them has been the happiest of my life.] Montesquieu to Folkes, 10 November 1742, *Correspondance II*. [Souvenez-vous que vous nous avez promis M, votrefils pour un an à Bordeaux; nous le mettrons en bonne compagnie et nous ferons tout ce qui sera en nous pour qu'il ressemble un jour à son père. Peut-être qu'une des grandes villes de province qu'il y ait [en France], pour un jeune homme qui trouve bonne compagnie d'honnêtes gens et des amis, vaut mieux que Paris même. Je vous réponds que j'aurai les yeux sur lui et qu'il ne sera libertin que comme le doit être un galant homme, et que je serai son mentor].

[125] See Jacques Battin, 'Montesquieu, les sciences et la médecine en Europe', *Histoire des Sciences Médicales* xli, 3 (2007), pp. 243–54, on p. 251.

[126] Maupertuis to Folkes, 6 January 1743, as quoted by Battin, 'Montesquieu, les sciences et la médecine en Europe', p. 251.

[127] Montesquieu to Folkes, 14 February 1744, *Correspondance II*; Robert Shackleton, *Montesquieu: A Critical Biography* (Oxford: Oxford University Press, 1961), p. 175. 'Souvenez-vous que vous nous avez promis M, votrefils pour un an à Bordeaux; nous le mettrons en bonne compagnie et nous ferons tout ce qui sera en nous pour qu'il ressemble un jour à son père. Peut-être qu'une des grandes villes de province qu'il y ait [en France], pour un jeune homme qui trouve bonne compagnie d'honnêtes gens et des amis, vaut mieux que Paris même. Je vous réponds que j'aurai les yeux sur lui et qu'il ne sera libertin que comme le doit être un galant homme, et que je serai son mentor'. [Remember that you promised us M[onsieur], your son for a year in Bordeaux; we will put him in good company and we will do everything so one day he is just like his father. Perhaps one of the great provincial towns [in France], for a young man who finds good company with honest people and friends, is better than Paris itself. I reply to you that I will have my eyes on him and that he will be a libertine only as a gallant man should be, and that I will be his mentor]. Antoinette Ehrard, with a study by William Eisler, *Portraits de Montesquieu: Répertoire Analytique* (Clermont-Ferrand: Presses Universitaires Blaise Pascal, 2014), p. 17.

people in the world your memory is dearest to me; I would rather live with you than with anyone. To live with you is to love you'.[128]

In 1741, Montesquieu arranged a marriage for his son against his will to Marie Catherine de Mons, and placed him in a civil service position, the *président à mortier* at the Bordeaux parlement; he also discouraged Jean-Baptiste from a career in natural philosophy.[129] This was a blow to Jean-Baptiste; he and Folkes had enjoyed a lively correspondence about scientific matters, Folkes sending him parcels of books, such as 'the best' edition of the *Principia*, or a copy of Roger Cotes *Harmonia Mensurarum*.[130] In 1741, a very miserable Jean-Baptiste subsequently wrote to Folkes, saying he could not 'renounce the inclination that I have for physics and natural history'. He asked Folkes if he could procure a hydrostatic balance on his behalf for his research, and noted that if he 'could succeed in anything it would be flattering for me to be able to present my work to the illustrious presence of the Royal Society', perhaps to convince his father to change his mind. In 1743, Jean-Baptiste published one paper on 'Of Stones of a Regular Figure Found at Bagneres' in the *Philosophical Transactions*, and was duly nominated and elected by Folkes to the Royal Society a year later.[131] Indeed, 'Jean-Baptiste de Secondat had no desire to carry out what his father had been unable to pursue and almost never took his seat in the city parlement of Bordeaux. At the death of Albessard, from whom the office was leased, Montesquieu finally had to sell his office on 4 August 1748 and Jean-Baptiste abandoned his post on 10 June 1748 to devote himself to his personal research', becoming a geologist and botanist.[132]

Just as Folkes would serve as a scientific mentor for Jean-Baptiste, he would serve as a mentor for the arguably more talented Abraham Trembley, with Geoffrin initially acting as the intermediary. Geoffrin wrote to Folkes,

You ask me Sir for some clarification about worms that you cut into pieces and who grow back the part that was cut off, I'll repeat to you what I think I already told you...We started talking about that here about two years ago, from a man named Mr Bonet from Genève who claimed he had discovered it, but it is said that the discovery was made by one Mr Tremblet from Holland, it kept Paris busy for a long time...It was said that the worms were quite big, in a way you could distinguish very easily what a surprising operation nature was making on

---

[128] *Lettres Familiares* xxx, February 1742, *Oeuvres*, vol. vii, p. 253, as quoted in Collins, *Voltaire, Montesquieu and Rousseau in England*, p. 164.

[129] RS/MS/250, f. III, Letter 93, 15 December 1742, Geoffrin to Folkes, Royal Society Library, London.

[130] RS/MS/790, 17 August 1739, Jean-Baptiste Montesquieu to Folkes, Royal Society Library, London.

[131] RS/EC/1744/06, Royal Society Library, London.

[132] François Cailhon, 'Secondat, Jean-Baptiste de', trans. Philip Stewart, dans *Dictionnaire Montesquieu* [online], directed by Catherine Volpilhac-Auger, ENS Lyon, September 2013. http://dictionnaire-montesquieu.ens-lyon.fr/fr/article/1376477218/en [Accessed 23 September 2019].

them, that [if] you cut their head and another head grew back at once, that if you cut them in half, a tail grew back on one end, and a head on the other, that if you cut them in half along their body, the half of the body grew back. I talked about this to Mr de Maupertuis who seemed very enthusiastic about this discovery, He told me that he considered these worms to be a species between plants and animals.[133]

'Mr Tremblet' was Abraham Trembley, and 'Bonet' was Charles Bonnet, Trembley's relative who introduced Trembley to the eminent biologist René-Antoine Ferchault de Réaumur. Trembley's experiments on 'polyps', or the freshwater cnidarian, *Chlorohydra viridissima* and *Hydra vulgaris*, were of great excitement to the natural philosophical world.[134] Trembley discovered animal regeneration, grafting, and asexual reproduction, for which he won the Copley Medal. Geoffrin went to Réaumur's study, where she saw the creatures for herself—she confessed, 'I couldn't tell head from tail, both extremities seemed to be ending in a point and that this point was a bit greenish'. Folkes quickly wrote to Montesquieu to get more news, stating:

We learn from the news that Mr. de Réaumur read his memoir on this insect which multiplies by division and one of our friends [Geoffrin] had given me some news some time ago; here we are passionately seeking some clarification on such an extraordinary fact. Shall we see soon this thesis? Where can you get us some light on this? Will it be necessary to say that the plants are animals of an imperfect class or that animals are only the most sublime of plants?[135]

Réaumur published Trembley's findings in the preface to the sixth volume of his 1742 *Mémoires* of the *Histoire des insectes* which were sent to Folkes; Folkes also wrote to Bentinck and Réaumur to get a copy of Trembley's abstract of his work.[136] Réaumur himself responded on 14 December 1742, sending 'the first one that comes out of my house. I have not been able to make others appear in

---

[133] RS/MS/250, f. III, Letter 13, Royal Society Library, London.
[134] *Hydra vulgaris* was discovered by Antonie van Leeuwenhoek in 1702.
[135] Folkes to Montesquieu, 23 November 1742/4 December 1742, *Correspondance II*. [Nous apprenons par les nouvelles que M. de Réaumur a lu son mémoire sur cet insecte qui se multiplie par la division et dont un de nos amis m'avait, il y a quelque temps, donné quelques nouvelles; nous souhaitons ici avec passion quelque éclaircissement sur un fait si extraordinaire. Verrons-nous bientôt ce mémoire? ou pouvez-vous nous procurer quelque lumière là-dessus? faudra-t-il dire que les plantes sont des animaux d'une classe imparfaite ou que les animaux ne sont que les plus sublimes des plantes?]
[136] L&P/1/174, 6 February 1742, Letter, 'Note on Réaumur's history of insects, 6th volume, being an extract from a letter referring to the account of Trembley's freshwater polyp' from Pierre Louis Moreau de Maupertuis to unknown recipient [Folkes], Royal Society Library, London.

public right now because the book will only be presented to the King next Sunday'.[137]

On 14 January 1742/3, Folkes initiated a remarkable series of correspondence with Trembley about his discoveries and made him better known to the Royal Society.[138] As Ratcliff has pointed out, 'Folkes had been privately informed about the existence and curious characteristics of the polyp by Georges Louis Leclerc du Buffon in July 1741', and Gronovius had published an article about them in the *Philosophical Transactions*.[139] This correspondence again speaks not of a decline in the Society, but a 'burgeoning sophistication developing then in biological studies, those on the lower organisms in particular'.[140] Trembley's views challenged the 'prevailing preformationist view of generation', suggesting 'matter and active forces alone were sufficient to explain the generation of living organisms'— beliefs appealing to a freethinker like Folkes.[141]

Folkes first congratulated Trembley on his discovery 'beyond such wonder, which seems to offer us some new enlightenment in one of the greatest mysteries of nature'; he urged him to publish a book about his discoveries, as he thought 'the beauties of nature are much more interesting than all the history of the actions of the greatest men'.[142] Folkes then mentioned that he had communicated Trembley's work to the Royal Society, which occurred on 20 January 1742/3. Trembley sent along a sample in a 'powder jar…put in a basket, which is well stuffed with hay', along with two bottles of water, 'imagining water from the Thames will be appropriate for them too'. A carafe of worms to feed them was included, although Trembley also fed them aquatic aphids, beef, mutton, and veal. He recommended to Folkes a small brush to rid the polyps of parasites, and also advised adhering a 'small pin to the end of a feather' to retrieve the hydra for the dissections and replication experiments.[143]

By the following month, Folkes reported to Trembley he had met with twenty Fellows of the Royal Society at his house at Queen Square to see the hydra with a

---

[137] RS/MS/250, f. 3, Letter 96, Royal Society Library, London.
[138] This correspondence is described briefly in Howard M. Lenhoff and Sylvia G. Lenhoff, 'Abraham Trembley and His Polyps, 1744: The Unique Biology of Hydra and Trembley's Correspondence with Martin Folkes', *Eighteenth Century Thought* 1 (2003), pp. 255–79. Sylvia Lenhoff generously shared copies of the entire Folkes/Trembley corpus with the author from the Trembley Family Archive in Geneva, and this chapter will analyse the letters in greater detail.
[139] Marc J. Ratcliff, 'Abraham Trembley's Strategy of Generosity and the Scope of Celebrity in the Mid-Eighteenth Century', *Isis* 95, 4 (December 2004), pp. 555–575, on pp. 559–60.
[140] Lenhoff and Lenhoff, 'Trembley', p. 266.
[141] Shirley A. Roe, 'John Turberville Needham and the Generation of Living Organisms', *Isis* 74, 2 (1983), pp. 158–84, on p. 158.
[142] Letter of 14 January 1742/3 from Folkes to Trembley, Trembley family archives, Geneva [Je vous félicite aussi, Monsieur, d'une découverte qui, outre le merveilleux qu'elle contient, semble nous offrir quelques nouvelles lumières dans un des plus grands mystères de la Nature…Je suis de ceux à qui les merveilles de la Nature sont bien plus intéressantes que toutes les histoires des actions des plus grands hommes].
[143] RS/MS/250, Letter of 12 March 1743, Trembley to Folkes, Royal Society Library, London.

magnifying glass and microscope, observing how they fed, and reported 'all that I have done was to convince them that the polyps were really animals'. In his demonstrations, Folkes had the assistance of optician and microscope artisan John Cuff, whose instrument shop on Fleet Street was three doors away from the Society's premises at Crane Court. Folkes wrote, 'Mr. Cuff, who has at my Request brought hither an excellent Microscope, will be so kind as to endeavour at shewing the Insect itself to such GENTLEMEN, as not having yet seen it, may now be willing to take a View of it in that Manner'.[144]

Cuff had been an attendant at Royal Society meetings since 1738, where he met Johann Nathanael Lieberkühn, a German physician. Lieberkühn had invented a solar microscope which projected images from the specimen stage on the wall, as well as a simple microscope fitted with a 'concave mirror or reflector to study injected animal specimens with epi-illumination. The Lieberkuhn reflector, or reflecting speculum... [was] made of silver or another highly polished metal, and increase[d] the amount of light illuminating a specimen'.[145] Lieberkühn's instruments became famous among continental natural philosophers, Count Gasparo Cerati of Pisa (1690–1769) reporting to the Royal Society during his Grand Tour—'I saw with very great satisfaction the inventor of the solar microscope Mr Liberküm and he shewed me the kindling of a very high sublimed spirit of Wine by the Effluvia of Electricity'.[146] After hearing about the instruments' reputations and watching Lieberkühn's demonstration at the Royal Society, Cuff 'took great pains' to improve his own microscopes and, in Henry Baker's words, 'bring them to perfection'.[147]

Folkes also asked Trembley for a small sample or 'vaccine' of the animals to be sent to him for further experiments, and noted that Lennox and Montagu were eager to see the hydra: 'the Duke [of Richmond] was at my house yesterday, when the basket arrived, however, we could not perform any experiments because we had to be at the Society'.[148] Lennox subsequently wrote to Folkes, 'I am glad you recommended the care of these creatures to [Trembley's emissary] Wolters. I hope they will arrive safe for I long to see them as much as you can... I

---

[144] [Martin Folkes], 'Some Account of the Insect called the Fresh-water Polypus, before-mentioned in these Transactions, as the same was delivered at a Meeting of the Royal Society by the President, on Thursday, March 24. 1742-3', *Philosophical Transactions* 42, 469 (1742-3, pub. 1744), pp. 422–36, on p. 434.

[145] Johann Nathanael Lieberkühn (1711–1756), Molecular Expressions™. https://micro.magnet.fsu.edu/optics/timeline/people/lieberkuhn.html [Accessed 25 September 2019].

[146] RS/L&P/1/360, Copy of Letter of Gasparo Cerati to Cromwell Mortimer, 1 January 1745, Royal Society Library, London.

[147] Deborah Warner, 'The Oldest Microscope in the Museum', Smithsonian, 13 May 2015, Smithsonian, https://americanhistory.si.edu/blog/microscope [Accessed 1 October 2019].

[148] Letter of 11 March 1743 from Folkes to Trembley, Trembley Family Archives, Geneva [Mais j'ai déjà eu avec moi vingt Messieurs de la Société Royale pour les voir chez moi ce matin; tout ce que je n'ai encour fait n'e été que ce sont convaincre tous ceux qui y ont été que ce sont vraiment des Animaux; et ça été une vraie satisfaction pour moi qu'aucun d'eux m'en a douté... ou il aurait dîné avec moi chez le Duc de Richmond Lundi dernier; le Duc était chez moi hier, quand le panier arriva, mais nous ne pûmes alore fait aucune expérience parce qu'il fallait se rendre à la Société].

long to hear of the arrival of the Pollipus's'.[149] Montagu also wrote to Folkes, 'if you can spare Mr Tremblays account of the Polipus I should be obliged to you if you would lend it me and let me take a coppy of it to carry with me into the Country to shew the Dutches of Montagu, and if you can let me have it this morning you shall have it again in the evening but I want [to] coppy it if you durst think it proper'.[150]

In a letter of 25 March, Folkes reported that he had received his 'vaccine', and showed the hydra to the Royal Society, as well as 150 other people at his home in Queen Square. Folkes wrote 'I had seen them perform an infinite amount of animal functions, and that I have actually done the experiments on the sectioning and the reproduction of the parts'. These experiments included Folkes feeding hydra still attached to their mothers, and realizing the mother and offspring had a shared gut: a correct observation.[151] As Lenhoff and Lenhoff pointed out, Folkes also questioned Trembley's assertion that the gut of the polyp 'is a caecum that has no natural opening at all on the bottom'.[152] He then asked about Trembley's microscopic technique and urged him to publish.

Folkes continued to do a series of microscopic observations, noting that the rings on the hydra were like the earthworm, appearing and disappearing when contracted, as well as describing their incredible ability to extend their mouths to feed. He noted the different sorts of worms that they fed upon, and that a person had found them in England near a dock next to the Thames. Folkes subsequently exhibited the polyps at the Royal Society, sending a draft summary about the experiments to Trembley for his approval, and noted to him that Henry Baker and physician James Parson 'found themselves quite often at my house' to see the hydra.[153] There was also a reading of Réaumur's *Mémoires* about the hydra at the Royal Society. Folkes reported 'so impatient that we prayed Mr Desaguliers, who was present at that time, to vividly read it out loud in English pretty well and on the following Thursday, I made Mr Zollman translate it', with Folkes making corrections. (Philip Henry Zollman (d. 1748) was the Foreign Secretary to the Royal Society.) Folkes also noted to Trembley that he was pleased that Lieberkühn was doing observations of the hydra, as 'his microscopes are the most perfect ones I have ever seen, and sent a drawing of a polyp that budded five to six times'.[154]

Lennox was also enthusiastically involved in the experiments. He was a patron of several natural historians such as Taylor White and George Edwards, who published the *Natural History of Uncommon Birds* between 1743 and 1751,

[149] RS/MS/865/28, 15 February 1742/3, Royal Society Library, London; RS/MS/865/29, 22 February 1742/3, Royal Society Library, London.
[150] RS/MS/790, undated letter 'Monday' from Montagu to Folkes, Royal Society Library, London.
[151] Letter of 25 March 1743 from Folkes to Trembley, Trembley Family Archives, Geneva.
[152] Lenhoff and Lenhoff, 'Trembley', p. 270. [Est-il vrai que pendant qu'ils sont attachés, le bateau de petit communique avec celui de la mère…].
[153] Letter of 19 April 1743 from Folkes to Trembley, Trembley Family Archives, Geneva.
[154] Letter of 25 March 1743 from Folkes to Trembley, Trembley Family Archives, Geneva.

as well as a two-volume French edition, *Histoire Naturelle de Divers Oiseaux*. Edwards dedicated the first volume (1745) to the Duke and the second volume (1748) was dedicated to Lennox's wife Sarah.[155] In 1733 Edwards became bedell of the Royal College of Physicians, in charge of the college's administration, library, and collections. This role permitted Edwards to be introduced to the most significant collectors of the day, like Lennox, who in turn bought his illustrations and books. For his books, Edwards drew and engraved illustrations of many animals and animal art in the collections of fellow virtuosi, such as a painting of a dodo owned by Sir Hans Sloane, making of them in his words, 'a natural and accurate portrayal'.[156] As Arthur MacGregor has noted, his 'practice of carefully acknowledging the source of each of his subjects sheds considerable light on the extent to which exotic birds and animals were to be found in the possession of a range of owners from wealthy grandees to humble citizens, as well as specialist traders who emerged to supply this growing market'.[157] As well as birds, Edwards illustrated some of the animals from Lennox's menagerie, such as some of the first images of Chinese goldfish, which Edwards tells us were increasingly common in the London area due to the trade of sea captains and the fashion for chinoiserie.[158] The title page of Edwards' 1751 edition portrayed an image of his Copley Medal, awarded to him in 1750 with the support of Folkes and Lennox.

The hydra thus could be considered another more minute member of Lennox's menagerie and reflective of his interests in natural history. The Duke reported he 'cut it [the hydra] from the posterior part, besides the usual head, as the detached arm seemed to have grown towards the middle'. Lennox also wrote to Folkes,

> My two English & one Dutch Pollipus are in good health, & have eaten up all the worms, so pray let me have some worms if you can spare any, I mean a good many, for they have damn'd good stomachs; & now let me aske you if you can spare me these three pollipus's, for if you could I should make Lady Young very happy with them.[159]

---

[155] 'The Renaissance Duke', catalogue to 2018 exhibition, Goodwood House, Sussex. https://www.goodwood.com/globalassets/venues/goodwood-house/summer-exhibition-2017.pdf [Accessed 9 October 2019].

[156] George Edwards, 'Preface', *Natural History of Uncommon Birds and of Some Other Rare and Undescribed Animals* (London, the author, 1743–51).

[157] Arthur MacGregor, 'Patrons and Collectors: Contributors of Zoological Subjects to the Works of George Edwards (1694–1773)', *Journal of the History of Collections* 26, 1 (1 March 2014), pp. 35–44, quote from article abstract.

[158] 'The Renaissance Duke', catalogue to 2018 exhibition, Goodwood House, Sussex. https://www.goodwood.com/globalassets/venues/goodwood-house/summer-exhibition-2017.pdf [Accessed 9 October 2019]. See also Anna Marie Roos, *Goldfish* (London: Reaktion Books, 2019).

[159] RS/MS/865/31, Lennox to Folkes, 29 April 1743, Royal Society Library, London.

Unfortunately, Lennox's luck ended shortly after:

> I am inconsolable for I have not only kill'd my worms, butt I believe my four
> pollipus's, by putting them for two hours only in the sun to day. So I wish you
> would dine with me to morrow to comfort me, & in case...my pollipus's should
> really be dead, if you could afford me one more you would make me very happy.
> The Duchess of Richmond also desires Miss Lucrece to dine here to morrow;
> pray bring the microscope that we may examine the dead bodys.[160]

Folkes also experienced unexpected difficulties with keeping the creatures in his
home, as in a case of domestic science gone awry, several suddenly 'perished in
the cabinet of my daughter who was taking care of them'. His daughter Lucretia's
cabinet was painted with white lead, and Folkes speculated that this was harmful
to the hydra, something Trembley remarked should be subjected to further
experiment.[161] Hugh Hume-Campbell, 3rd Earl of Marchmont (1708–94), had a
similar incident with his polyps, writing to Henry Baker that 'I suppose Mr Folkes
heard of the accident which befell some of my Polypes from Mr MacLauren...
about 12 Polypes I kept in town in the parlour whilst the painters were painting the rooms up on pair of Stairs, the Polypes all turned of a dark Colour and
drew in their arms as the Smell of the paint was very strong'. Campbell changed
the water, and they recovered and budded.[162] In present research, hydra have
been used as bioassays to test for environmental toxins; a small amount of lead
may act as a mitotic stimulator of all dividing cell types in *Hydra*, promoting
budding.[163]

As with the proceedings in the Egyptian Society, there were humorous Latin
verses written by the literati about the hydra experiments and their mishaps.
Among Folkes's papers at the Royal Society was a series of verses by neo-Latin
poet and MP Nicholas Hardinge called 'Wray's polyps', referring to Daniel Wray,
antiquary, Royal Society Fellow, and Egyptian Society member.[164] Wray was an
intimate of William Hogarth, and his tutor, the Reverend Thomas Birch, was also
Hogarth's and Folkes's friend.[165] Wray subsequently served Philip Yorke, Lord
Hardwicke, as Deputy Teller of the Exchequer, and they, with others such as

---

[160]   RS/MS/865/32, Lennox to Folkes, 1743, Royal Society Library, London.
[161]   Letter of 28 September 1743 from Folkes to Trembley, Trembley Family Archives, Geneva.
[162]   A. Hume-Campbell to Henry Baker, 19 September 1743, John Rylands Library, Manchester,
Henry Baker Correspondence, vol. 1, f. 241r.
[163]   C. L. Browne and L. E. Davis, 'Cellular Mechanisms of Stimulation of Bud Production in *Hydra*
by Low Levels of Inorganic Lead Compounds', *Cell and Tissue Research* 177, 4 (February 1977),
pp. 555–70; Matthew J. Beach and David Pascoe, 'The Role of Hydra Vulgaris (Pallas) in Assessing the
Toxicity of Freshwater Pollutants', *Water Research* 32, 1 (1998), pp. 101–6.
[164]   John Nichols, ed., *Illustrations of the Literary History of the Eighteenth Century*, 8 vols (London:
for the Author, 1817–58), vol. 1, pp. 40–3.
[165]   Robin Simon, *Hogarth, France and British Art* (London: Hogarth Arts, 2007), p. 26.

Birch, co-wrote the *Athenian Letters* (1741–3), considered the best commentary on Thucydides at the time. Wray was also a frequent composer of comic neo-Latin verse, and poets such as Thomas Edwards also dedicated works to him.[166] *Wray's polyps* was written in the fiendishly difficult metre of hendecasyllables, and closely modelled on Roman poet Catullus' famous verse about the death of his mistress' sparrow written in the same metre. Catullus's verse opened,

> Weep, ye Cupids and Venuses
> and whatever there is of rather pleasing men:
> the sparrow of my girlfriend has died,
> the sparrow, delight of my girl,
> whom she loved more than her own eyes

The first verse of *Wray's polyps* read:

> Weep, ye Mercuries and Apollos,
> And all men of the more erudite kind,
> The Polyp has fallen on his last day,
> With whom playful Wray had learned to play,
> And offer a feast when he sought it
> Of soft and tender little worms,
> Wray, that rival of all Folkes' witticisms.[167]

Although songbirds and sparrows were love-gifts in classical antiquity, and Catullus could have given his lover Lesbia a bird, the sparrow in particular had erotic connotations. 'Passer', Latin for sparrow, was a slang word for 'penis'. The allusion to 'Mercuries' in the first line of Wray's poem is likely an allusion to the use of mercury in treating venereal disease, the reference to the soft polyp a double entendre. The poem continued, indicating Wray had also lost his experimental specimens as well as his virility.

> The Polyp has fallen on his last day,
> Who Wray loved more than his companions.
> For he endured the knife, and drew strength
> From the very wound, and was wont
> To be reborn stronger, and from crumbs and little left-overs
> To draw mental strength and spirit,

[166] Edwards dedicated a Spenserian sonnet to Wray, published in the appendix to his *Canons of Criticism*. See David Hill Radcliffe, 'Spencer and the Tradition: English Poetry 1579–1830'. http://spenserians.cath.vt.edu/TextRecord.php?action=GET&textsid=34514 [Accessed 4 October 2019].
[167] RS/MS 250/2/6, Royal Society Library, London.

And not to die, after the fashion of us haughtier bipeds
Now, thanks to your efforts, my close friend,
The hapless Wray, complains of the absence
Of his guest and sweet little friend,
And requests some consolation for his grief,
And is now eager to pass the lingering night
In dining, jests and songs.[168]

Despite satires such as these, Folkes took seriously his role as patron to Trembley. As Lennox was preparing to serve as the British government's envoy to The Hague during the War of the Austrian Succession, Folkes gave him a small 'newly invented [Cuff] microscope, produced here, that has the merit of being portable' to take to Trembley.[169] Folkes also told Trembley that he would have news shortly of his election to the Royal Society, which happened on 20 May 1743 'to a very full meeting'. Lennox subsequently visited Trembley in Sorghvliet in late May 1743 to present the microscope, writing to Folkes that he saw in Trembley's room:

a Dozen large glasses of about a foot high, each of them holding about a gallon or six quarts of Water, & well stock'd with pollipus's. I believe he has several hundreds of them...I had almost forgott 'tho to acquaint you with one more very extraordinary observation of Mr. Trembley's, which is, that in the double headed pollipus...there was butt one comon gutt between them at first, so that the feeding of one head, had the same effect as feeding of both'.[170]

The latter observation confirmed Folkes's beliefs about the continuity of the gut between organisms that reproduce by budding.

Folkes subsequently reported to Trembley on 8 June 1743 that he observed that the polyps moved towards candlelight, extending their arms; hydra have extraocular sensitivity in their tentacles, despite their lack of any structure recognizable as a traditional photoreceptor.[171] Trembley confirmed these observations, putting a cardboard case over the glass where the hydra were kept, cutting a chevron shape in one side to let in light; the hydra assembled in that shape in response.[172] Lennox went again to see the polyps with his usual enthusiasm, Trembley

---

[168] RS/MS 250/2/6, Royal Society Library, London.
[169] Letter of 2 May 1743 from Folkes to Trembley, Trembley Family Archives, Geneva. See also Lenhoff and Lenhoff, 'Trembley', p. 271.
[170] RS/MS/865/34, Lennox to Folkes from Utrecht, 24 May/4 June 1743, Royal Society Library, London.
[171] Letter of 16 June 1743 from Folkes to Trembley, Trembley Family Archives, Geneva; S. Guertin and G. Kass-Simon, 'Extraocular spectral photosensitivity in the tentacles of *Hydra vulgaris*', *Comparative Biochemistry and Physiology Part A: Molecular & Integrative Physiology* 185 (June 2015), pp. 163–70.
[172] RS/MS/250, 7 June 1743, Trembley to Folkes, Royal Society Library, London.

reporting, 'I have only seen a few people observe them with as much taste and pleasure as he did. He has enjoyed seeing my long-armed polyps seize their prey, take it to their mouths and swallow it'.[173]

Trembley was also overwhelmed with Folkes's gift of the microscope, writing 'it only had one defect: it is infinitely too beautiful for me. The kindness, Sir, that you are showing me overwhelms me'.[174] Despite his flattery, Trembley was essentially pragmatic, finding Cuff's instrument very useful. As Warner has indicated,

> When Trembley visited London in 1745, he asked Cuff to make a microscope that would facilitate observations of aquatic creatures as they moved about. By 1747, Cuff was boasting of 'The AQUATIC MICROSCOPE' which was 'invented by him for the Examination of Water Animals' [The Cuff-type microscope was] 'a very popular design, being easy to focus and, with its box-shaped stand, more stable than many other microscopes, which tended to have tripod stands'.[175]

Stukeley described it being exhibited in a Royal Society meeting of 20 November 1741:

> Mr Cuff brought his solar reflecting microscope; a late, and vast improvement. Which being plac'd in a hole in a window shutter of a darkened room, casts a magnify'd spectrum of huge dimensions, on a white sheet;…that is done by the solar rays. It shows exceeding large, and distinct, the vessels in a frogs foot, or the like; a fishes fin. Where even the blood vessels shall appear an inch in diameter; and the globules of blood running thro them, as big as pepper corns.[176]

On 16 June 1743, Folkes again urged Trembley to publish, fearing someone would make his discoveries 'public without rendering you all the justice that you deserve. What I have given in the Transactions and that I have sent to you in manuscript has credited you with your discovery; but we cannot be responsible for someone else who could gain in printing this somewhere else'.[177] Folkes had the right to be worried. As Ratcliff has remarked, Trembley's 'openness in distributing polyps was strategic because he wanted a new scientific field to grow from his work. But his generosity also had a price: the risk that other scholars would take

---

[173] RS/MS/250, 7 June 1743, Trembley to Folkes, Royal Society Library, London.

[174] RS/MS/250, 31 May 1743, Trembley to Folkes, Royal Society Library, London.

[175] 'Cuff microscope', Science Museum, http://www.sciencemuseum.org.uk/broughttolife/objects/ display?id=93111 [Accessed 15 October 2019]. The link is now broken, as the Science Museum is no longer maintaining the site. The Museum explains 'access to these collections [is] available elsewhere and the technology behind the multimedia [is] no longer supported. See https://www.sciencemuseum. org.uk/brought-life [Accessed 21 November 2020].

[176] William Stukeley, *Memoirs of the Royal Society*, vol. 1, p. 59, Spalding Gentlemen's Society, Spalding, Lincolnshire.

[177] Letter of 16 June 1743 from Folkes to Trembley, Trembley Family Archives, Geneva.

unwarranted credit for his experiments'.[178] Unfortunately for Trembley, in 1743, Henry Baker FRS (1698–1774) did just that, opportunistically publishing *An Attempt towards a Natural History of the Polype: In A Letter to Martin Folkes*, describing his replication of Trembley's discoveries. This was one year before Trembley's own *Mémoires*.[179]

Although Baker acknowledged Trembley in his book, it was an unfortunate situation for Folkes, who criticized Baker's behaviour strenuously.[180] On 28 September 1743, Folkes admitted to Trembley, 'it had been reported that Mr Baker was writing something; I would better prefer that no one had done it, although I am convincing myself that he will act as an honest man in recognising you as author of these beautiful discoveries'.[181] Trembley was not convinced, and in a series of curt letters to Baker, told him what he thought of his behaviour, which was not much.

To set things right, Folkes nominated Trembley for the Copley Medal, telling him 'if God conserves me, I will have the honour of designating you [as the winner] on November 30 and give you immediately this small token of respect from the Society'.[182] Although Folkes described the medal as a trifle, so small that Trembley's full name was too large to fit on its entirety, he emphasized that it was 'still a mark of respect from the Society'.[183] He also reassured Trembley that Baker's book had plates 'no more than simple woodcuts' in contrast to the planned engravings by Pierre Lyonnet for Trembley's own work. Lyonnet was a Dutch naturalist and polymath who completed the engravings for Trembley's studies when the 'commissioned professional artist, Jan Wandelaar, fell behind schedule'.[184] His rapid mastery of drawing, engraving, and insect dissection led to his illustration of Friedrich Christian Lesser's *Théologie des Insectes* (1742). In this way, Folkes preserved his friendship with the Swiss naturalist, showing his esteem, and their correspondence continued for several years until 1747.

As there was reduced postal service due to the imminence of the War of the Austrian Succession, when Trembley's *Mémoires* about hydra were published, Folkes arranged for their shipment, suggesting 'two copies could be sent with freedom through the post under cover in Mr Sheloche's name, who is the Secretary of the Post, when Milord Lovel, who presently is Count of Leicester, is in the countryside'.[185] Folkes was right to take such precautions; in 1747, Réaumur

[178] Ratcliff, 'Abraham Trembley's Strategy of Generosity', pp. 573–4.
[179] Lenhoff and Lenhoff, 'Trembley', p. 264.
[180] Susannah Gibson, *Animal, Mineral, Vegetable: How Eighteenth-Century Science Disrupted the Natural World* (Oxford: Oxford University Press, 2015), p. 52. A point also made by Ratcliff, *Quest for the Invisible*, p. 109.
[181] Letter of 28 September 1743 from Folkes to Trembley, Trembley Family Archives, Geneva.
[182] Letter of 28 September 1743 from Folkes to Trembley, Trembley Family Archives, Geneva.
[183] Letter of 14 November 1743 from Folkes to Trembley, Trembley Family Archives, Geneva.
[184] Cope, 'Notes from Many Hands', p. 5.
[185] Letter of 4 May 1744 from Folkes to Trembley, Trembley Family Archives, Geneva.

went so far as to ask Jacques François Artur, a physician and collector in Cayenne, to put Folkes's name and the address of the Royal Society on packages of *naturalia* being shipped to him in Paris, in case they were seized by English corsairs and officials.[186] Réaumur mentioned that Folkes 'would not have neglected to send them to me, being a very good friend of mine'.[187] Folkes also wrote the abstract of Trembley's *Mémoires* for publication in the *Philosophical Transactions*, and begged that 'everything should be inserted in the same order that they were read', so Trembley's discoveries would precede the work of Henry Baker.[188]

Folkes was also correct to take these precautions, as several members of the Royal Society, like Henry Miles, Arderon, and members of the nobility, such as Hugh Hume-Campbell, 3rd Earl of Marchmont, wrote to Baker about their own experiments with the hydra, as well as providing reports about finding them in ditches in places like Ealing and Tooting, leaving out mention of Trembley almost entirely.[189] Their actions indicate that these gentlemen-naturalists (wrongly) perceived Baker, rather than Trembley, as the authority about the creatures, most likely due to Baker's several works popularizing microscopy.

7.5 Henry Baker: Gentlemen's Microscopist, Antiquary, and Opportunist

Baker's life and career, in fact, seemed to be built upon repurposing the work of others for commercial gain.[190] Baker was born in London on 8 May 1698. After an apprenticeship with a bookseller, he made a fortune by refining a method of instructing the deaf and dumb invented in the previous century by William Holder and John Wallis, whose treatise of speech, *De loquela*, Baker paraphrased into English as *A Short Essay on Speech* (1723)'.[191] Baker's success brought him to the notice of Daniel Defoe, whose youngest daughter Sophia he married in 1729. A year before, under the name of Henry Stonecastle, Baker was associated with Defoe in starting the *Universal Spectator and Weekly Journal*. In 1740 he was elected a Fellow of the Society of Antiquaries and of the Royal Society. He

---

[186] Mary Terrall, *Catching Nature in the Act: Réaumur and the Practice of Natural History in the Eighteenth Century* (Chicago and London: University of Chicago Press, 2015), p. 158.

[187] Réaumur to Folkes, 9 February 1747, in Jean Chaïa, 'Sur une correspondance inédite de Réaumur avec Artur, premier médecin du Roy à Cayenne', *Episteme* 2 (1968), p. 123, in Terrall, *Catching Nature in the Act*, p. 158.

[188] Letter of 19 May 1744 from Folkes to Trembley, Trembley Family Archives, Geneva.

[189] A. Hume-Campbell to Henry Baker, 19 September 1743, John Rylands Library, Manchester, Henry Baker Correspondence, vol. 1, ff. 240r–241v.; Henry Miles to Henry Baker, 19 September 1743, John Rylands Library, Manchester, vol. 1, ff. 242r–243v.

[190] Also a point made by Ratcliff, *Quest for the Invisible*, p. 80.

[191] G. L'E Turner, 'Henry Baker (1698–1774), Natural Philosopher and Teacher of Deaf People', *Oxford Dictionary of National Biography*. 23 Sep. 2004. https://doi.org/10.1093/ref:odnb/1120 [Accessed 20 October 2019].

contributed over eighty papers to the *Philosophical Transactions*, and due to Folkes's patronage received the Copley Medal in 1744 for microscopic observations on the crystallization of salts. The Bakerian Lecture at the Royal Society was founded as a result of his bequest of £100 to the Society. He was one of the founders of the Society of Arts (1754) and acted as its secretary, encouraging ties between it and the Royal Society and Society of Antiquaries.

Baker's influence on the popularization and development of the microscope was considerable. He wrote *The Microscope Made Easy* (1742), the first microscopy laboratory manual for amateur naturalists. One thousand copies of this work were printed, and it was reprinted the following year. Baker followed this volume with the *Employment for the Microscope* (1753), which was equally popular, appealing to Enlightenment interest in education and improvement. Baker's book, which appealed to gentlemen naturalists, was thus seen as authoritative by this particular audience.

In the second half of the seventeenth century, Royal Society secretary Henry Oldenburg cultivated provincial naturalists like Martin Lister to encourage their contributions to the *Philosophical Transactions*, giving them books, information, or materials to further their work. In a similar manner, Baker used his publications to promote talented naturalists outside of London such as Norwich excise officer and naturalist William Arderon FRS and, by doing so, promoted himself. Arderon's correspondence and diaries were purchased in the early nineteenth century by Dawson Turner, who, after examining them, said that, in his opinion, Arderon 'was not at all inferior to Gilbert White the Historian and Naturalist of Selborne'.[192] But we know little about Arderon, whereas we venerate Gilbert White. Arderon invented the board hygrometer for measuring atmospheric humidity, made some of the earliest references to biological control concerning the reduction of the number or predators from the ecosystem, and provided the earliest description of the fossils in the East Anglian Boulder Clay. Inspired by Baker's works on the microscope, he identified the water mould *Saprolegnia* as a fish pathogen, which is visually spectacular under the microscope: the mould has branched filaments which exude sugars, encouraging other microbes like rotifers and Paramecia to dart in and out.[193] In turn, Arderon's discoveries of these and other newly discovered rotifers appeared not only in the *Philosophical Transactions* but also in Baker's publications about popular microscopy. Arderon is representative of how eighteenth-century naturalists maintained and expanded the role of natural history by combining it with local chorography. His knowledge of the local ecosystem was internationalized in his works of natural philosophy in

---

[192] Dawson Turner, *Catalogue of the Manuscript Library of the Late Dawson Turner* (London: Puttick and Simpson, 1859), p. 5.
[193] Laura Bowater and Kay Yeoman, 'Case Study 5.2: Evaluating an Activity for the "Norfolk Science Past, Present and Future" Event', in *Science Communication: A Practical Guide for Scientists* (New York: John Wiley and Sons, 2012), pp. 110–13.

the *Philosophical Transactions* and spread to gentlemen naturalists in Baker's works, who could recreate his observations, or armchair-travel through Arderon's macroscopic and microscopic world.

In his works, Baker covered the preparation of specimens and making of infusions to create cultures of animalcules, discussed and dismissed spontaneous generation, and in prose reminiscent of the pioneer microscopist Robert Hooke described the inherent beauty of nature: 'Our finest miniature paintings appear before this instrument as meer dawbings, plaistered on with a trowel, and devoid of beauty'.[194] Baker, in fact, published a new edition of Robert Hooke's ground-breaking *Micrographia* in March 1745, having some of its original plates re-engraved and rewriting and updating Hooke's work as *Micrographia Restaurata* to stimulate interest in amateur microscopy and, in his words, 'to make this work more valuable'.[195] 'For Baker, this work substituted for the lack of illustrations of microscopic objects in his own text *The Microscope Made Easy*'.[196] In a letter to William Arderon of 21 November 1744, Baker wrote:

For two months past I have also had upon my Hands, not only the Preparation of Experiments for the Royal Society, but the Daily Care likewise of overlooking the press, which a little work of mine has just now passed through. I call it Micrographia Restaurata of the Copper Plates of Dr Hookes wonderful Discoveries by the Microscope reprinted and fully explain whereby the most valuable Particulars in that celebrated Authors Micrographia are brought together in a narrow Compass; many Discoveries and Observations made since the Doctor's Time on the same Subjects are also occasionally intermixed. The Plates to be republished are what were originally engraven for the Micrographia, and have lately been discovered in an exceedingly good Condition; there are 33 of them, finely done, and as the Book they belong to is extremely scarce and worth above two guineas, I hope my Undertaking will be acceptable to the Curious; as it will furnish them with these Plates, and much more in Reality that Dr Hooke wrote, at about the Quarter of the expense; for though I have nothing at all to do with the Sales, I have engaged the Booksellers to set the Price as low as possible.[197]

[194] Henry Baker, *The Microscope Made Easy* (London: R. Dodsley, 1743), p. 297.
[195] Henry Baker, *Micrographia Restaurata* (London: John Bowles, 1745), p. 65.
[196] Brian Gee, Anita McConnell, and A. D. Morrison-Low, *Francis Watkins and the Dolland Telescope Patent Controversy* (Abingdon: Ashgate, 2004), p. 69.
[197] Autobiographical memoranda, correspondence with William Arderon in 4 vols, GB 072 Forster Collection, Victoria & Albert Museum, National Art Library, London; See also Paul E. S. Whalley, 'The authorship and date of the *Micrographia restaurata*, with a note on the scientific name of the silverfish (Insecta, *Thysanura*)', *Journal of the Society of the Bibliography of Natural History* 6, 3 (1972), pp. 171–3.

Baker was correct that the first edition was scarce and expensive, Arderon replying, 'Your undertaking, to reprint Dr Hook[e]s Micrographia, is highly commendable...had I but known 3 months sooner it might have saved me two Guineas...and what's still Worse, the Cuts are but in a very bad Condition'.[198] Baker also arranged for Cuff to sell *Micrographia Restaurata* in his shop on Fleet Street, Cuff's instruments and the publication mutually stimulating consumer interest. It was a competitive market; James Ayscough with James Mann from 1745–7 produced a universal instrument combining single, compound, and solar microscopes, and in 1751 George Sterrop announced his own newly invented single and double instrument for use with a solar apparatus.[199]

As a natural philosopher, antiquarian-archivist, and commercial author, Baker found ways of authenticating their documents and thus rendering them, if not authoritative sources, at least starting points for further discussion and experimentation. In repurposing Hooke and bringing a connoisseur's eye to scientific instrumentation, Baker, like Folkes, represented the antiquarian impulse in natural philosophy at this time to preserve the history and heritage of scholarly organizations. In 1736, Folkes had proposed that:

> some Impressions might be taken of the Title Page to Sprat's History of the Royal Society from the Plate in the Society's Custody; that Cut being now very scarce and seldom to be met with. Which was accordingly agreed to and ordered to be done: and Dr Mortimer was desired to look out that as well as other Plates that might be cleaned and fit to take off Impressions of any such Cutts as are become scarce.[200]

This frontispiece to Thomas Sprat's *The History of the Royal-Society of London, for the Improving of Natural Knowledge,* was a key piece of iconography for the early Royal Society. The engraving, designed by John Evelyn and executed by Wenceslaus Hollar, featured the Society's early instruments and included portraits of its philosophical founder Francis Bacon, the first President Lord Brouncker, and the royal patron Charles II. Cromwell Mortimer, the Royal Society secretary, indeed edited an eighteenth-century edition of another iconic natural history work of the early Royal Society, Francis Willughby's and John Ray's *Historia Piscium* (1743), again using the original ichthyological plates and including an updated taxonomic index.

Seven of the original copperplates for the *Micrographia* which Baker had intended to re-engrave were lost, but for the other thirty-three, 'well-preserved...excepting a little Rust', he had the engravers use a technique akin to

[198] Correspondence with William Arderon; Whalley, '*Micrographia restaurata*', p. 172.
[199] Gee, McConnell, and Morrison-Low, *Francis Watkins*, p. 67.
[200] CMO/3/70, 7 December 1736 Council Minutes, Royal Society Library, London.

repoussage, where the copperplate was beaten out from the back with a punch or small hammer to knock out the old engraving and achieve a smooth surface that could be cut again with a burin.[201] Baker labelled each figure with dimensions for the objects using a micrometer, providing more detailed descriptions of what they were supposed to depict, provided objects in actual sizes to indicate magnification, replaced Hooke's text with 'short and planned descriptions of the Pictures' for a more general eighteenth-century audience, and gave other examples of similar specimens to those Hooke illustrated. For instance, Hooke portrayed the middle 'part of the Hair of an Indian deer' and Baker recommended his readers use 'hairs taken from a Mouse's body', as they were 'least opake and fittest for examination' of their structure.[202] He gave references to what Hooke, Leeuwenhoek, Henry Power, and Malpighi stated about the hollow structure of hair. Ever opportunistic, Baker then included a page reference to his own examinations of mouse hair in his *Microscope Made Easy*.

Baker also published a paper about Folkes's account of Leeuwenhoek's bequest of microscopes to the Royal Society in the *Philosophical Transactions* and assessed their designs and effectiveness.[203] Leeuwenhoek's microscopes were quite difficult to use, and Leeuwenhoek had faced much scepticism about the veracity of his observations due to his relative lack of social capital as a draper and lens grinder. Folkes, in his 1723 assessment of his instruments (see chapter four), preserved the scientific reputation of the Dutch microscopist and affirmed his importance, stating Leeuwenhoek's discoveries: were 'so numerous as to make up a considerable Part of the *Philosophical Transactions*, and when collected together, to fill four pretty large Volumes in Quarto.... And of such Consequence, as to have opened entirely new Scenes in some Parts of Natural Philosophy'.[204] Baker, however, had a different purpose. Although Baker praised Folkes's previous 'exact and full Descriptions of their Structure and Use', he used his own analysis of the instruments for two reasons: to sing the praises of the latest Cuff microscopes which he featured in his own books and, as a philosophical vitalist, to question past claims about observations of male semina, particularly by preformationists. As Ratcliff stated, Baker's critique of Leeuwenhoek's microscope was accomplished to advertise 'the new models of microscopes made by Cuff', to praise them as the 'very best available in England', while at the same time to claim the 'credit for requesting Cuff to improve the compound microscope'.[205]

To the first end, Baker provided a table of the focal distances of the twenty-six microscopes that Leeuwenhoek left to the Royal Society, along with their

---

[201] Baker, 'Preface', *Micrographia Restaurata*, p. 1.   [202] Baker, *Micrographia Restaurata*, p. 7.
[203] Henry Baker, 'An Account of Mr Leeuwenhoek's Microscopes', *Philosophical Transactions* 41, 458 (1739–41), pp. 503–19.
[204] Martin Folkes, 'Some Account of Mr Leeuwenhoek's curious Microscopes, lately presented to the Royal Society', *Philosophical Transactions* 32, 380 (November–December 1723), pp. 446–54, on p. 449.
[205] Ratcliff, *Quest for the Invisible*, p. 80.

magnifying powers; sixteen or seventeen of the specimens affixed to the mount-
ing pin were 'destroyed by time, or Struck off by Accident', evidence of the depre-
dations of the museum which Folkes had previously addressed as a member of
the Repository Committee (see chapter five).[206] Baker also evinced surprise at the
relatively low magnification power of the instruments. He stated, 'it appears, by
the foregoing Table, that One only of these 26 microscopes is able to magnify the
Diameter of an Object 160, and its superficies 25600 times; all the rest falling
short of that Degree'. Although Baker acknowledged that Leeuwenhoek provided
the Society with a cabinet of microscopes with fixed specimens, and there 'must
certainly have been much greater Magnifiers than any in our possession', Baker
concluded it may have been nearly impossible for Leeuwenhoek to have observed
semen using the instruments he had given to the Society. Baker continued,

> It may perhaps be objected, that Mr Leeuwenhoek declares, he did not use such
> small Glasses as some People boasted of; and that, although for 40 years together
> he had been possessed of Glasses exceedingly minute, he had employed them
> very seldom; since, in his Opinion, they could not so well serve to make the first
> Discoveries of Things, as those of a larger diameter.[207]

Baker then compared Leeuwenhoek's microscopes with a 'curious Apparatus of
silver with Six different Magnifiers, belonging to Mr Folkes, and then newly made
for him by Mr Cuff in Fleetstreet'.[208] These were part of a pocket microscope also
designed to be used with a solar microscope; in 1744, Folkes had mentioned to
Trembley 'we have enjoyed ourselves over here for some time with the microscope.
It makes the workers [instrument makers] dream of perfecting this instrument;
the enlightenment that you could give us will perhaps allow us to add some more
commodities we have not thought of yet'.[209] Folkes was correct about the prolif-
eration of new instruments: he himself owned no fewer than nine microscopes by
makers such as Lindsey, Cuff (three), Baker, Scarlet, Culpeper, and Filason.[210]

Baker measured the magnification powers and focal distance of Folkes's apparatus, concluding that 'Mr Folkes First Glass magnifies the Superficies of an Object
six times as much as the greatest magnifier of Mr Leeuwenhoek', and went on to
extol the virtues of ground convex lenses versus the bead lenses of the Dutch instrument maker. He then followed with a thinly disguised commercial description of
Cuff's instruments, concluding by speculating that Leeuwenhoek may have been
the inventor of the double reflecting microscope subsequently improved by Cuff!

Due to his social status, Leeuwenhoek was not elected FRS until 1680, well
after secretary Henry Oldenburg's death. Despite the excellence of his

---

[206] Baker, 'An Account', p. 504.   [207] Baker, 'An Account', p. 510.
[208] Baker, 'An Account', p. 512.
[209] Folkes to Trembley, 26 October 1744, Trembley Family Archives, Geneva.
[210] A Catalogue of the Genuine and Curious Collection, pp. 3–5.

instruments and Baker's puff pieces, Cuff would not be so lucky. Although Folkes and Baker nominated him on 19 April 1744 for his 'inclination and Ability to promote those Parts of Natural knowledge and Philosophy that he in the way of his Profession, and of his Zeal and readiness to be Serviceable to the Society', Cuff was balloted and rejected.[211] Cuff may have been refused because he was more largely perceived as an imitator and technician rather than an innovator. Naturalist and antiquary Henry Miles, for example, complained bitterly to Henry Baker that Cuff had wrongly claimed he invented a micrometer even when he had done so; the account of Cuff's instrument was appended to Baker's book, and Miles mistakenly thought that Baker was the original creator! The rejection of Cuff for the Fellowship was especially unfortunate because, as we have seen, Folkes himself was generally supportive of technicians and instrumentalists with high degrees of expertise in their bids to become Fellows of the Society.

7.6 Baker, Folkes, and Preformation

Although Baker questioned the efficacy of Leeuwenhoek's instruments in an attempt to praise the ill-fated Cuff, Baker did give Leeuwenhoek credit for dismissing preformationist claims about sperm by Nicolaas Hartsoeker (1656–1725). Hartsoeker invented the screw barrel simple microscope, and was involved in a priority dispute with Leeuwenhoek over the discovery of sperm. In Baker's eyes, Hartsoeker 'pretended to discover the Animalcules in Semine virile to be exactly of a human Shape'. Baker's dismissal of preformationist claims reflected his larger beliefs which he shared with Folkes about vitalism, albeit with Baker presenting a more conventionally theological view.

In his *Employment for the Microscope*, Baker recalled that in August 1743, naturalist Joseph Turberville Needham, the first member of the English Catholic Clergy to be elected FRS, had sent a parcel of blighted wheat to Folkes, who, in turn, gave it to Baker to examine.[212] In the mid-1740s Needham was in Paris, where he was recommended by Folkes to naturalist Georges-Louis Leclerc, Comte de Buffon, whereupon 'there emerged a brief but significant scientific collaboration that was to form the foundation for Needhams's biological theory of generation' that challenged preformationism.[213] Needham concluded that there was a 'vegetative force...in every microscopical point of Matter'; as Shirley Roe explained, 'he began to view the production of microorganisms in infusions as a process of vegetation, resulting from the actions of a vegetative force released...by the

---

[211] EC/1744/02, Royal Society Library, London.
[212] Baker, *Employment for the Microscope*, pp. 252–3.
[213] Roe, 'John Turberville Needham', p. 161.

decomposition of animal or plant material.[214] In 1743 and 1745, Needham presented his observations on eels found in the grain who lived in a senescent state when desiccated, but 'needed only to be soaked in water to exhibit characteristics of life'.[215] To explain the origins of these eels, Needham used his theory of vegetative force, explaining that the eels resulted from the exaltation of the grain contents.

Baker came to a similar conclusion when repeating Needham's experiments, convinced the eels could endure three or four years of senescence without being deprived of their 'living Power'. Baker went on,

> the cutting of the Polyp…into Pieces, the Continuance of Life in those Pieces, and their producing all the Parts necessary to make each of them a perfect Polyp…prove beyond all Contradiction, I will not say that Life itself may be divided (lest I should give Offence,), but that an Animal possessed of Life may cut asunder…whatever the Essence of Life, it is perhaps not to be destroyed, or really injured, by any Accidents that may befall the Organs wherein it acts, or the body it inhabits. Dr Butler, the late Bishop of Durham…gives it as his Opinion 'We have no more Reason to think a Being endued with living Powers, ever loses them…than to Believe that a Stone ever acquires them—The Capacity of exercising them for the present…may be suspended, and yet the Powers themselves remain undestroyed'.[216]

Although Baker remained safely pious in his assertions, French intellectuals such as Maupertuis were convinced that the senescent eels were evidence for materialism, or implied support for it, as did Royal Society critic John Hill: 'it cannot but be observed, that his doctrine tends, tho' without his intending it, to that monstrous and absurd system of materialism…We know him well enough…that perhaps there is not a man living who has greater and more reverend notions of a God than Mr Needham'.[217]

7.7 The Plant–Animal Continuum

Unlike the pious Needham, Folkes expressed heterodox and materialist ideas to Trembley about vitalism and the relationship between humans, animals, and plants. The nature of animals and whether they had feeling, understanding, and souls was a common item of discussion in the early modern salon.[218] In his

---

[214] Roe, 'John Turberville Needham', p. 162.   [215] Roe, 'John Turberville Needham', p. 175.

[216] Baker, *Employment for the Microscope*, p. 255.

[217] John Hill, 'Some Late Observations', in *The Monthly Review, or New Literary Journal* 4 (Nov. 1750), pp. 58–62 on p. 58, as quoted in Roe, 'John Turberville Needham', p. 178.

[218] See Anita Guerrini, *Experimenting with Humans and Animals: From Galen to Animal Rights* (Baltimore: Johns Hopkins University Press, 2003).

Summa Theologica, Thomas Aquinas used the medieval idea of the Great Chain of Being to establish that humanity's superiority was based on possession of reason, implying the possession of an immortal soul. As animals lacked reason, they thus lacked souls, and the Cartesian animal machine reinforced such opinions.[219]

Folkes had different ideas about the matter, and he was not alone. In the 1730s, George Edwards painted a watercolour of a stuffed orangutan in Sloane's collection; the ape had been dissected in 1698 and was posed anthropomorphically sitting jauntily on a stool and holding a staff; the illustration was later published in 1758 by Edwards accompanying an essay entitled 'The Man of the Woods', although he 'mixed up the characteristics of chimpanzees, orang-utans and gibbons'.[220] James Parsons also sent Folkes an account of a 'very small monkey', noting the great similarity of its thumbs and genitalia to human beings, and describing his food preferences—'he will not touch currants, because of their acidity'—and his fondness for gum senegal (gum Arabic) for curing digestive ills.[221]

In the art world, there was also a fashion for anthropomorphized *singeries*, or decorations with monkey themes, particularly strong in France. 'These had first appeared in the late 17th century in the designs of Jean Berain, and had grown in popularity in the new century'.[222] *Singe Antiquaire* (*The Monkey Antiquarian* (1726)) by Jean-Siméon Chardin was exhibited in paintings in the Salon of 1740, where it was described as '*Le singe de la philosophie*'; 'it could also have been seen by Hogarth, who was well aware of the composition' (see figure 7.1).[223] Folkes was a patron of Hogarth's, and by this means may have become aware of Chardin's image featuring a monkey aesthete examining his prints and coin collection with a magnifying lens. Three years later, '*Le singe de la philosophie*' was engraved by Pierre-Louis Surugue, with the 'following verses appended by the writer Charles-Etienne Pesselier, an intimate of Chardin's':

Dans le Dédale obscur de monuments Antiques
Homme Docte, à grands frais, pourquoi t'embarrasser?
Notre siècle à deux yeux vrayment philosophiques,
Offre assez de quoi s'exercer.

---

[219] Guerrini, *Experimenting with Humans and Animals*, p. 57.

[220] Chris Herzfeld, *The Great Apes: A Short History* (New Haven: Yale University Press, 2017), p. 24.

[221] James Parsons, 'An Account of a Very Small Monkey, Communicated to Martin Folkes Esq; LL. D. and President of the Royal and Antiquarian Societies, London; By James Parsons M. D. F. R. S', *Philosophical Transactions* 47 (1751–2), pp. 146–50. My thanks to Keith Moore for pointing this article out to me.

[222] 'Jean-Siméon Chardin, 'Le singe peintre', Lot 262, Important Old Master Paintings Including European Works of Art, 24–5 January 2008, Sotheby's. http://www.sothebys.com/en/auctions/ecatalogue/2008/important-old-master-paintings-including-european-works-of-art-n08404/lot.262.html [Accessed 1 November 2019].

[223] Simon, *Hogarth, France and British Art*, p. 33. The painting is now in the Louvre.

Fig. 7.1 Pierre Louis Surugue, *L'Antiquaire*, etching and engraving, 1743, 305 mm × 230 mm, after painting by Jean Siméon Chardin, *Le Singe de la philosophie*, shown at the Louvre in 1740 and now in the Musée du Louvre, Paris, Metropolitan Museum of Art, New York, Public Domain.

> In the shadowy maze of ancient monuments
> Learned man, why bother yourself at such great expense?
> Our century, to truly philosophical eyes
> offers enough to learn about.[224]

Folkes also owned several prints engraved of Chardin paintings, so he may have been familiar with this piece or had it in his collection.[225] The topic matter was certainly self-referential to the 'godfather of all monkeys', as Stukeley had called him.

Not only were there debates about the relative distinctions between animals, but also between animals and plants. In the late seventeenth century, naturalists like Martin Lister considered them to be on an anatomical continuum, with similar circulatory systems. Lister argued that plants had vessels analogous to the veins of mammals, carrying a nutritive sap that was usually milky in colour. Other theorists like John Wallis (1616–1703), Savilian Professor of Geometry at Oxford, thought the veins of leaves were analogous to animal nerves and contained a

---

[224] Jean-Siméon Chardin, 'Le singe peintre'.
[225] *A catalogue of the genuine, entire and curious collection of prints and Drawings . . . of Martin Folkes*, pp. 3, 5, 15.

nervous fluid.[226] Wallis noted that, while veins in leaves were branched, the veins in plant stems were not 'ramified', as were mammalian circulatory vessels, but rather smaller and bundled, resembling animal nerves. The debate was settled by Nehemiah Grew, Grew demonstrating that while the bleeding of plants could be possible due to the 'internal pressure forcing the plant to yield its sap when cut', the circulation of the sap in plants was not the same as it was in animals.[227]

By the eighteenth century, discussions had moved from anatomical comparison between animals and plants to their relative possession of reason, as seen by a letter to Trembley of 28 June 1744 from Folkes:

> I have been convinced for a long time that animals and plants are not essentially different as one would ordinarily think...seeds or productive parts of certain plants are inclined to seek for each other...having the propensity to direct themselves, like a weak vine...trying to intertwine itself around stronger trees...I remember that Cardinal Polignac showed me some beautiful verses from his Anti-Lucretius[228] where he represented the vine 'reasoning' on its means of supporting itself on a stronger elm. It is true that he only used this verse to argue with Descartes who was not willing to grant feelings to the animals. But I did not dare tell him that he should rather convince me that the vine has some analogous aspects to the animal instincts if he had not tried to make me believe we should remove the degree of reason that we would give to the animals. During dinner, on that very day, as he petted a small dog he was very fond of, I asked him if...[others]...did not caress their watches in the same manner. He agreed with me, while smiling, that whatever philosophy we believed in, we could not help acting towards animals as if they had reason.[229]

Polignac's *Anti-Lucretius*, to which Folkes referred, was a 12,000-line poem in Latin that was a sustained attack on Epicurean atomism, 'the high priest of

---

[226] John Wallis, 'A Note of Dr Wallis, Sent in a Letter of Febr 17 1672/3 Upon Mr Lister's Observation Concerning the Veins in Plants, Published in Number 90 of these Tracts', *Philosophical Transactions* 8, 95 (1672), p. 6060.

[227] Brian Garrett, 'Vitalism and Teleology in the Natural Philosophy of Nehemiah Grew (1641–1712)', *British Journal of the History of Science* 36, 1 (March 2003), pp. 63–81, on p. 72.

[228] Cardinal Melchior de Polignac, *Anti-Lucretius sive de Deo et Natura* (Paris: H.-L. and J. Guérin, 1747; Amsterdam: M. M. Rey, 1748).

[229] Letter of 28 June 1744 from Folkes to Abraham Trembley, Trembley Family Archives, Geneva. [Je me souviens que Monsieur le Cardinal Polignac ne fit voir de beaux vers de son Anti Lucrèce, où il représentait la vigne comme raisonnant sur les moyens de s'aider par la force de l'ormeau. Mais il est vrai qu'il n'sen servit que pour l'argument de Des Cartes qui ne voulait point accorder de sentiment aux bêtes; mais je me hasardai de lui dire qu'il m'avait bien plutôt convaincu que la vigne avait quelque chose d'analogue à l'instinct des bêtes, qu'il ne m'avait fait croire qu'il fallait ôter aux bêtes ce degré de connaissance que nous étions portés à leur donner. A dîner, le même jour, comme ill caressait un petit chien qu'il aimait beaucoup, je lui dis que S.E. ne caressait pas de même sa montre, et il m'accorde bein en souriant qu quelque philosophie qu'on embrassât, on ne pouvait s'empêcher d'agir envers les animaux comme s'ils avaient de la connaissance]. Folkes had Polignac's *Anti-Lucretius* in his library (*A Catalogue of the Library of Martin Folkes*, p. 6).

atheists'; 'For him, Epicureanism under any guise was a materialistic philosophy diametrically opposed to the dualistic Mind-ordered universe of the Cartesians'.[230] Polignac also doubted the possibility animals had souls with which to feel. In his view, 'to equate, if necessary, men and animals in the matter of soul would not be to lessen the nobility of man nor the glory of God, nor would it be to make men die forever, as Lucretius would have it, but to make animals immortal'.[231] Although Polignac and Folkes had philosophical differences, Folkes still had a copy of *Anti-Lucretius* in his library, for the aesthetics of its poetic expression and the memory of the conversation he had with his friend.

Folkes subsequently went on to discuss with Henry Baker the same sorts of philosophical questions with regard to parasitic twins; parasitic twins are a type of conjoined twin, where one twin ceases development during gestation and is vestigial to the dominant twin. On 7 January 1747/8, Folkes wrote to Baker, recalling that on an excursion with George Graham to the Marlborough's Head in Fleet Street he saw a parasitic twin in the chest of Jacomo Poro, a 'middle siz'd black man' of '28 years of age' with a 'singularly melancholy Aspect'[232] (see figure 7.2). As is well known from the work of Stewart and Guerrini, taverns and coffee-houses were sites both of natural philosophy and racist hierarchies, displays of 'monstrosities' of nature. Guerrini stated,

The eighteenth century was a great era of classification, of putting nature's productions into a logical order that would tell us something about its purposes. Where monsters fit this order, or by definition didn't fit, was an important part of this enterprise. Popular display and natural philosophy therefore quite often were about the same things, and this is particularly true when we enter the realm of defining what is human.[233]

Parasitic twins mediated between natural philosophy, taxonomy, moral reflection on the creativity of the divinity, as well as provoking curiosity and unpalatable gawping.

Folkes continued,

The small Head grew out just below the whole Ribs, and was about the Size of a Child's Head of a quarter old, with very long black Hair kept braided; it had no Eye, but the Appearance of one crush'd out, it had, I think, an imperfect Ear and

[230] Ernest J. Ament, 'The Anti-Lucretius of Cardinal Polignac', *Transactions and Proceedings of the American Philological Association* 101 (1970), pp. 29–49, on. p. 36.
[231] Cardinal Polignac, *Anti-Lucretius* (1748), 6.310–6.314; Ament, 'The Anti-Lucretius', p. 39.
[232] Letter of 7 January 1747/8 from Folkes to Baker, John Rylands Library, Manchester, Henry Baker Correspondence, vol. 3, ff. 218–19.
[233] Anita Guerrini, 'Advertising Monstrosity: Broadsides and Human Exhibition in Early Eighteenth-Century London', in: *Ballads and Broadsides in Britain, 1500–1800* (Aldershot: Ashgate, 2010), pp. 109–30, on p. 113.

Fig. 7.2 Giacomo [James] Poro. 1714, engraving, Wellcome Collection, Attribution 4.0 International (CC BY 4.0).

Nose, and a Mouth with either three or five Teeth in it: it slobber'd and voided, as I remember, Mucus at the Nose, and was a shocking and a nasty sight. The Man affirm'd he felt nothing that was done to it, but was extreamly tender of it, saying to us in a mournful Manner, that his own Life must be the Price of any Hurt that should come to it. He called it his Sister, and said it had been baptiz'd by the Name of Martha.[234]

Martha, a type of twin called an *acardius acormus*, where there is a well-developed head but rudimentary heart and trunk, had a small involuntary motion to the neck, which put Folkes 'in the mind of a motion of small kittens or puppies just litter'd', his expression of what he saw as the boundaries between animals and humans at once frank, empirical, and rather cruel.[235] Poro was born in Genoa in 1686 and brought to London in 1714, where he 'attracted the attention of Sir Hans Sloane, who commissioned his portrait to be painted'.[236] The illustration conveys Poro's protective and tender feelings towards Martha, her hair braided and beribboned, her mouth agape with two small teeth. Folkes

---

[234] Letter of 7 January 1747/8 from Folkes to Baker.
[235] Letter of 7 January 1747/8 from Folkes to Baker. For a recent example of such a parasitic twin see Tatsuma Sakaguchi, Yoshinori Hamada, Yusuke Nakamura, Yuki Hashimoto, Hiroshi Hamada A-Hon Kwon, 'Epigastric heteropagus associated with an omphalocele and double outlet right ventricle', *Journal of Paediatric Surgery Case Reports* 3, 10 (October 2015), pp. 469–72.
[236] C. J. S. Thompson, *The Mystery and Lore of Monsters: With Accounts of Some Giants, Dwarfs, and Prodigies* (London: Williams and Norgate Ltd, 1930), p. 56.

concluded his letter to Baker by appending an excerpt in Latin describing Poro from a work by Bartholin.

In his letter to Baker, Folkes then went on to discuss the Hungarian twins, Helena and Judith, born in 1701 'every way compleat, but…conjoynd at their Haunches', who came to England in 1708 after being paraded in The Hague for elite enjoyment. Although Folkes did not see them himself, on 12 May 1708, they were the subject of a letter written by William Burnet to Hans Sloane and read to the Royal Society. After the scandal of the Mary Toft rabbit case, Burnet declared 'it was no cheat' and claimed their urinary and foecal vessels and passages were united, which prevented separation—thus, 'when one stopped to take up any thing, she carried the other quite from the ground'. Folkes indicated to Baker he would 'make enquiry when they died, and what Circumstances attended their Sickness and Death'. The twins were indeed dissected in 1723, although a posthumous tract on their anatomy was not published in the *Philosophical Transactions* until 1757.

The journal reported that the interval between the reading of the paper before the Royal Society and its publication was due to Folkes himself, who 'having taken…[the paper] to his house with a view of collecting and adding to it some further particulars, it could not be found after his decease'.[237] When the report was finally published, it included all accounts of the twins to supply in 'some measure the want of what Mr Folkes's extensive reading and industry might have furnished the public with'. Among these excerpts was the account by Burnet about their display, passages about the twins from a work by anatomist James Paris du Plessis, as well as Latin treatises recounting their dissection and anatomy.[238]

Folkes's desire to assemble and publish a complete account of parasitic and conjoined twins at once displayed the dark side of his curiosity, and his antiquarian and historical interests. As an antiquary, he collected examples of human deformities like rare coins; as a natural philosopher, he used these examples to further his understanding of Trembley's experiments and to support his freethinking belief that plants, animals, and humans were all rational, existing on a hierarchical continuum. At the end of his letter to Baker, Folkes stated that he could not help but mention what 'Mr Trembley lately inform'd me of, that a small Bird, I think a Chaffinch, was accidentally shot, a few Months since near the Hague, which had two complete Heads, only the one somewhat

---

[237] Justo Johanne Torkos and William Burnett, 'Observations anatomico-medicœ, de monstro bicorporeo virgineo a. 1701. Die 26 Oct. In Pannonia, infra comaromium, in possessione szony, quondam quiritum bregetione, in lucem edito, atque A. 1723. Die 23 Febr. Posonii in Cœnobio Monialium S. Ursulæ morte functo ibidemque sepulto. Authore Justo Johanne Torkos, M. D. Soc. Regalis Socio', *Philosophical Transactions* 50 (1757), pp. 311–22, on pp. 314–15.
[238] Torkos and Burnett, 'Observations anatomico-medicœ', p. 315; Guerrini, 'Advertising Monstrosity', p. 121.

less that the other'—just like the hydra snipped in two at Queen Square and Sorghvliet. The perceived boundaries between life, ensoulment, humans and the rest of the animal and plant kingdom were growing ever more tenuous as the eighteenth century progressed, with the Royal Society at the forefront of such debates.

8

Martin Folkes and the Royal Society Presidency

The Electric Imagination

8.1 Keeping Current: Folkes and Electrical Research in the Royal Society

The Royal Society under Folkes's presidency not only promoted pioneering experiments in biological vitalism, but also embraced new investigations in static electricity, primarily with the use of insulators. These experiments were usually in a Newtonian cast. Newton was convinced that the future of natural philosophy was in the investigation of electrical forces. Louis Elisabeth de la Vergne, Comte de Tressan, noted in his work *Essai sur le fluide électrique* (begun in 1747 and published in 1786) that during one of Folkes's stays in Paris they shared lodgings. The Comte heard Folkes mention a conversation with Newton about electricity. According to Folkes, Newton said 'Mes yeux s'éteignent…mon esprit est las de travailler, c'est à vous à faire les plus grands efforts pour ne pas lasser échapper un fil qui peut vous conduire' [My eyes extinguish themselves, my mind is weary of working, it is to you to make the greatest efforts you can, not to let slip a thread which may lead you on].[1] The Comte then noted of the conversation, 'Electricity offers to our research an inexhaustible source of new facts: Phenomena as varied, so marvellous that can appear to us in everything connected with movement, light, and violent explosions, can be presumed to arise from a very general cause…Electricity is the greatest discovery we can make to illuminate the mechanical power of the great movements of the Universe'.[2]

Due to the experiments of Stephen Gray (1666–1736) earlier in the century, it was shown that substances such as amber, glass, or gemstones could remain charged. These substances called 'electrics' were objects of fascinating 'epistemological dramas', with Desaguliers doing a series of experiments in 1741 at the Royal

---

[1] Comte de Tressan, *Essai sur le fluide électrique*, 2 vols (Paris: Chez Buisson, 1786), vol. 1, pp. xliv–xlv; Peter Rowlands, *Newton and Modern Physics* (London: World Scientific, 2018), p. 224.
[2] Comte de Tressan, *Essai sur le fluide électrique*, p. xlv. The Comte de Tressan also noted that in 1748, Folkes put Tressan in touch with William Watson FRS to verify Tressan's own experimental work, them both concluding that electricity must be of the same nature as elemental fire present in all bodies.

Martin Folkes (1690–1754): Newtonian, Antiquary, Connoisseur. Anna Marie Roos, Oxford University Press (2021).
© Anna Marie Roos. DOI: 10.1093/oso/9780198830061.003.0008

Society attempting to distinguish between electricity generated from resinous and vitreous substances after the work of Charles Du Fay.[3] As Fara noted, in contrast,

> since metals and other conductors transmit any charge—or 'electrical virtue'... they were labelled 'non electrics'. Iron gun barrels, lead-lined flasks and gold leaves became essential components of electrical apparatus. Water was another non-electric that came to play a vital role in storing and transporting large electric charges. For many experimenters, the most interesting non electrics were people.[4]

Machines made by experimenters like those by William Watson FRS (1715–87) literally 'electrified' Fellows of the Royal Society. Watson basically modified an earlier machine by Royal Society demonstrator Francis Hauksbee in which an experimental assistant turned the handle, whereupon the horizontal metal rod, called the 'prime conductor', transmitted the electricity generated by a glass globe rubbed with leather or wool pads.[5] Even Henry Baker kept current in investigations of electricity. The Duchess of Bedford wrote, 'I supped at the Duchess of Montagu's on Tuesday night, where was Mr Baker of the Royal Society, who electrified: it really is the most extraordinary thing one can imagine'.[6]

Folkes even had his own electrical apparatus created by Royal optician and instrument maker Francis Watkins for gentleman natural philosophers, advertised as 'a most compleat Electrical Machine, which, with its whole Apparatus, is contained in a box exceeding the Size of an ordinary Tea-chest'.[7] To attract customers, Watkins first demonstrated his machines at the Sir Isaac Newton's Head Tavern in Charing Cross. In his handbooks accompanying his inventions, Watkins included experiments with down, and a demonstration that made use of the loadstones, handy because Folkes had an experimental set of magnets in his collection of instruments at Queen Square to create his own spectacles of industry and nature to his fellow natural philosophers. Watkins noted that a 'loadstone being hung to the prime conductor, and a key or other piece of iron hung to the armature of the loadstone; if the finger or other non-electric be brought near the iron, it will snap and emit fire'.[8] He then concluded, somewhat

---

[3] William Stukeley, Memoirs of the Royal Society, vol. 2, f. 21, Spalding Gentlemen's Society, Spalding, Lincolnshire. See Larry Stewart, 'The Laboratory, the Workshop, and the Theatre of Experiment', in Bernadette Bensaude-Vincent and Christine Blondel, eds, *Science and Spectacle in the European Enlightenment* (Aldershot: Ashgate, 2008), pp. 11–24, on p. 11.

[4] Patricia Fara, *An Entertainment for Angels: Electricity in the Enlightenment* (London: Icon Books, 2017), p. 27. See also J. L. Heilbron, *Electricity in the 17th and 18th centuries: A Study of Early Modern Physics* (Berkeley: University of California Press, 1979).

[5] Fara, *An Entertainment for Angels*, p. 39.

[6] Brian Gee, Anita McConnell and A. D. Morrison-Low, *Francis Watkins and the Dolland Telescope Patent Controversy* (Abingdon: Ashgate, 2004), p. 52.

[7] *The Remembrancer*, 14 May 1748, as quoted in Gee and McConnell, *Francis Watkins*, p. 54.

[8] Francis Watkins, *A Particular Account of the Electrical Experiments Hitherto made publick* (London: the author, 1747), p. 25.

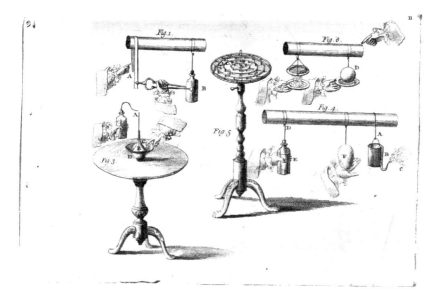

Fig. 8.1 Francis Watkins, *A Particular Account of the Electrical Experiments Hitherto made publick* (London: By the author, 1747), Wellcome Collection, CC-BY.

presciently, that this was 'an unquestionable proof that both the magnetic and the electric virtues can act at the same time on the same subject, without destroying or interfering with one another'. Watkins also provided instructions to make an electric orrery, noting the circular motion of the model planets caused by electricity 'has been thought by some to have a kind of analogy with that of planets round the sun'.[9]

Watkins was, of course, referring to Newton, who made similar analogies. Just as we have seen, Folkes's application of Newton's natural philosophy to geodesy, the ultimate form of metrology, and most of the electrical theories and theorists Folkes promoted in the Royal Society were essentially Newtonian in their basis.

Newton explained static electricity by 'an aetherial wind'. As John Henry noted, the movements of small pieces of chaff, or feathers, which are observed when glass is rubbed and held over them gave rise to Newton's speculation that 'some kind of subtil matter lying condensed in the glass, and rarefied by rubbing, as water is rarefied into vapour by heat, and in that rarefaction diffused through the space round the glass to a great distance, and made to move and circulate variously, and accordingly to actuate the papers till it return into the glass again, and be recondensed there'.[10] Similarly, 'the gravitating attraction of the earth', Newton

---

[9] Watkins, *A Particular Account*, p. 22.
[10] Bernard Cohen, ed. *Isaac Newton's Papers and letters in natural philosophy*, 2nd ed. (Cambridge: Harvard University Press, 1978), p. 180, as quoted in John Henry, 'Gravity and *De gravitatione*: the development of Newton's ideas on action at a distance', *Studies in History and Philosophy of Science Part A* 42, 1 (March 2011), pp. 11–27, on p. 20.

suggested, 'may be caused by the continual condensation of some other such like aetherial spirit, not of the main body of phlegmatic aether, but of something very thinly and subtilly diffused through it, perhaps of an unctuous or gummy, tenacious, and springy nature.[11] As Home has indicated, Newton at one point may have thought that static electricity could displace his belief in actions at a distance.[12]

8.2 Patronage of Benjamin Wilson

Other investigators such as Benjamin Wilson (1721–88) also attempted to find a relationship between the aether and electricity. Wilson, as well as being a natural philosopher, was a portrait painter for whom Folkes sat early in his career; Wilson's diary records several encounters, artistic and scientific, with the President of the Royal Society. Wilson developed an early interest in painting from the French artist Jacques Parmentier, whom his father employed to decorate his Leeds mansion, and he subsequently received instruction from Longueville, another French artist. His father's reduced circumstances, however, forced Wilson to seek a livelihood in London, where he became employed as a clerk in the registry of the prerogative court in Doctors' Commons.[13] 'When he had amassed £50 he obtained a more lucrative post as clerk to the registrar of the Charterhouse, which enabled him to resume his artistic studies under the guidance of the master of the Charterhouse, Samuel Berdmore, and the painter Thomas Hudson. He became acquainted with William Hogarth, George Lambert, and other leading painters.'[14] Wilson also knew Roubiliac and Francis Hayman; all had ties to the St Martin's Lane Academy founded by Hogarth, a successor to James Thornhill's drawing school.[15] Wilson recorded, 'I entered myself a member of the painting academy in St. Martins lane, in order to draw naked figures from the life: the hours of attendance were from six to 9 in the evening, and therefore the more agreeable, because those hours did not interfere with my other avocations.'[16]

---

[11] Cohen, *Papers and letters in natural philosophy*, p. 180, as quoted in Henry, 'Gravity', p. 20.

[12] R. Home, 'Force, Electricity, and the Powers of Living Matter in Newton's Mature Philosophy of Nature', in Margaret J. Osler and Paul L. Farber, eds, *Religion, Science, and Worldview: Essays in Honor of Richard S. Westfall* (Cambridge: Cambridge University Press, 1985), pp. 95–117.

[13] E. I. Carlyle and John A. Hargreaves, 'Wilson, Benjamin (bap. 1721, d. 1788), Portrait Painter and Scientist', *Oxford Dictionary of National Biography*. 23 Sep. 2004. https://doi.org/10.1093/ref:odnb/29641 [Accessed 7 November 2019].

[14] Carlyle and Hargreaves, 'Wilson, Benjamin (bap. 1721, d. 1788)'.

[15] Andrew Graciano, 'The Memoir of Benjamin Wilson FRS (1721–88): Painter and Electrical Scientist', *Walpole Society* 74 (2012), pp. 165–243, on p. 166. The memoir can also be seen on The Royal Society's *Turning the Pages* website: https://royalsociety.org/collections/turning-pages/ [Accessed 9 November 2019].

[16] Graciano, 'The Memoir of Benjamin Wilson FRS', p. 187.

These 'other advocations' were electrical experiments that Wilson performed during his spare hours at The Charterhouse. Wilson modified William Watson's machine which used a glass tube rubbed to generate sparks, substituting a glass globe instead: Wilson noted that he 'turned this globe with a Coach wheel; and excited the electric fluid, by rubbing the glass with [his] hand', having good success. Michael Broughton—one of the Duke of Montagu's clerical friends—mentioned that Folkes, Charles Lennox, and others went shortly afterwards to be 'electrified by the new method by the wheel' which probably was via Wilson's machine. Wilson's diary then noted:

In philosophy, I made so many experiments, about this time, that I began to give some account of them in writing, as I apprehended that all the electrical phenomena might be easily explained upon a very simple principle: because many experiments tended to shew, that the electric fluid did not proceed from vitrious and resinous substances only (as philosophers had all along imagined) but from the Earth itself, and all bodies surrounding the apparatus. This new Theory I communicated to Francis Wolleston Esq., Mr. [John] Hyde, Martin ffolkes the president of the Royal Society, *and Mr. Watson* [emphasis mine]. The President from that time shewed me always great respect, and was very much my friend. He pressed me to pursue my Theory, and make the experiments I proposed, to confirm it. He also advised me, as I intended to practice painting of portraits, to visit Ireland for a couple of years and paint there: because the works I did in that Kingdom would not appear against me in this: for which reason he apprehended I should succeed better on my return. This advice was very agreeable, not only to myself; but to most of my friends: and it was thought prudent, that I should first go there and make a short visit, in order to pave the way for a longer visit.[17]

The anecdote was relayed many years later by Wilson to King George II when Wilson was exhibiting a history painting of the 'History of Belshazzar'; Wilson noted in his diary:

The King…enquired, what I did at the Charterhouse? I told him that I had a place there as a Clerk: and having a great deal of leisure, I studied painting. That Martin Folkes the President of the Royal Society, who was a particular friend of mine, recommended it to me to go over to Ireland, for a couple of years and paint there; that my first beginnings might not appear against me in this country. Their Majesties laughed heartily at this, and said, it was excellent advice.[18]

[17] Graciano, 'The Memoir of Benjamin Wilson FRS', p. 187.
[18] Graciano, 'The Memoir of Benjamin Wilson FRS', p. 207.

Folkes's advice indeed proved very wise, leading to more patrons for Wilson's portraiture and to Wilson's first discoveries in electricity. In 1745, Wilson reached Dublin where he stayed 'not exceeding three weeks'. After waiting upon the Surgeon General of Ireland, he noted in his diary that he became acquainted with, and greatly influenced by, Dr Bryan Robinson (1680–1754) 'who had written a learned Treatise upon the Aether of Sir Isaac Newton', which gathered all of Newton's queries about the subject.[19] In his work, Robinson also published a letter from Newton to Robert Boyle for the first time, which described Newton's conception of an active and springy aether in all bodies in amounts that were inversely proportional to their densities. Bryan Robinson was not alone in his efforts; in the 1740s there were 'at least half a dozen major efforts to explain the behaviour of observable bodies by postulating a variety of invisible (and otherwise imperceptible) elastic fluids'.[20] Robinson also wrote a *Treatise of the Animal Oeconomy*, a work on physiological iatromechanism, a discipline that regarded the body as a machine, conforming in its functions to mechanical laws, and in its physiological phenomena to the laws of physics.[21]

Past scholarly analysis has portrayed Robinson's work in the context of two intellectual influences: first, Leiden physician Herman Boerhaave's (1668–1738) emphasis on the hydraulics of bodily fluids flowing through the veins and arteries, as well as his study of solids and their fibres, whose faults caused distempers; second, a 'Newtonian physiology' based on the queries on aether in Newton's *Opticks*.[22] Indeed, it is true that Robinson 'was an ardent admirer of Newton, and

[19] Graciano, 'The Memoir of Benjamin Wilson FRS', p. 187. This was *Isaac Newton's Account of the Aether* (Dublin: G. and A. Ewing, 1745).

[20] Larry Laudan, *Science and Hypothesis: Historical Essays on Scientific Methodology* (Dordrecht: Springer, 1981), p. 112.

[21] Bryan Robinson, *A Treatise of the Animal Oeconomy* (Dublin: George Grierson, 1732). A second edition appeared in Dublin, printed by S. Powell, for George Ewing and William Smith in 1737 (with the date 1734 in its first part)—it is this second edition of the work which I will refer to throughout the paper. A subsequent edition was printed in London for W. Innys and R. Manby in 1738. The 1734–7 edition includes 'A Continuation of a Treatise of the Animal Oeconomy' which has a separate title page ([chi]1r) dated 1737, and 'A Letter to Dr Cheyne containing An Account of the Motion of Water through Orifices and Pipes; And an Answer to Dr Morgan's Remarks on Dr Robinson's Treatise of the Animal Oeconomy', printed by the same printer and dated 1735, the year it had initially been printed. Some of the material on Robinson is taken from my own 'Irish Newtonian Physicians and their Arguments: The Case of Bryan Robinson', in Elizabethanne Boran and Mordechai Feingold, eds, *Reading Newton in Early Modern Europe* (Brill, Leiden, 2017), pp. 116–43.

[22] For Boerhaave's influence, see Theodore Browne, 'The Mechanical Philosophy and the "Animal Oeconomy"—A Study in the Development of English Physiology in the Seventeenth and Early Eighteenth Century', PhD Dissertation (Princeton University, 1968), pp. 351–3. For the aether's influence on Robinson, see Arnold Thackray, *Atoms and Powers: An Essay on Newtonian Matter Theory and the Development of Chemistry* (Cambridge, MA: Harvard University Press, 1970), pp. 135–41; J. R. R. Christie, 'Ether and the Science of Chemistry: 1740–1790', in G. N. Cantor and M. J. S. Hodge, eds, *Conceptions of Ether: Studies in the history of ether theories, 1740–1900* (Cambridge: Cambridge University Press, 1981), pp. 86–110, on pp. 96–8. Christie sees Robinson primarily as an influence on the Scottish chemist William Cullum.

tried to account for animal motions by his principles'.[23] Robinson attributed the 'motion of muscles to the vibration of an ethereal fluid pervading the animal body', grounding his theory of the aether on his mentor Richard Helsham's analysis of pneumatics.[24]

As Heilbron has noted, all of Robinson's publications had an effect: 'beginning in 1745, all significant British electricians postulated a special electrical matter identical with, or similar to, the spring, subtle, universal Newtonian aether. At least one of these electricians, Benjamin Wilson, drew his inspiration directly from Robinson'.[25] Wilson did an etching of the elderly Robinson, portraying him seated next to a table on which was placed a Roubiliac bust of Newton and two books: Newton's *Opticks* and Hippocrates' *Aphorisms*[26] (see figure 8.2). Looking at the portrait, we imagine that the artist and his subject were probably engaged

Fig. 8.2   Benjamin Wilson, *Bryan Robinson, M.D. aetatis suae 70*, engraving, 1750, 32.7 × 26.6 cm, Wellcome Collection, CC-BY.

[23] G. Le G. Norgate, 'Robinson, Bryan (1680–1754)', Rev. Jean Loudon, *Oxford Dictionary of National Biography* (Oxford: Oxford University Press, 2004).
[24] Turlough O'Riordan, 'Bryan Robinson', *Dictionary of Irish Biography Online*. https://dib.cambridge.org/ [Accessed 14 January 2019].
[25] John Heilbron, *Electricity in the 17th and 18th Centuries: A Study of Early Modern Physics* (Berkeley: University of California Press, 1997), p. 69.
[26] Benjamin Wilson, 'Bryan Robinson, M.D. aetatis suae 70', Wellcome Collection.

in lively conversation, as Robinson leans forward, ready to speak. The portrait also indicated the origin of Robinson's intellectual sympathies, and the Newtonian Folkes had a copy of it in his art collection, 'framed and glazed'.[27] Not surprisingly, Robinson even organized his works much as Newton did in his *Principia*, with Propositions and General Scholia. In his youth, Robinson was best known as a mathematician, having published a translated edition of Pierre de la Hire's *Conic Sections* (1704), so organizing his medical treatises in the manner of a geometrical argument (as Newton did) would have held great appeal for him.[28] Robinson's name was also prominent in the lists of subscribers to works popularizing Newtonian theory; for instance, he was a subscriber to *An Account of Sir Isaac Newton's Philosophical Discoveries in Four Books* (1748) compiled from the notes of Colin Maclaurin, Professor of Mathematics at Edinburgh.[29]

Robinson, in turn, introduced Wilson to 'many other learned men of the College &c. Dr. Cartwright — professor of experimental philosophy, complimented me with the use of the College experimental room, and all the apparatus's belonging to it; together with his operator to assist me in my experiments: his name was Maple, a very worthy and ingenious man'.[30] This was William Maple, (*c*.1661–1762) a Dublin chemist, who also worked at the Rotunda Hospital and as keeper of the Parliament House at College Green. A founding member of the Royal Dublin Society, Maple was later involved in the assay of Irish coins produced under Newton's tenure as warden of the Royal Mint, and was known to Newton.[31]

When in Dublin, Wilson did a series of experiments with Robinson to determine if electricity really came from the Earth. He ultimately rejected this hypothesis and concluded instead that 'the electric matter is an original Element and diffused through all Bodys in the Universe; and not produced but only collected and retained by these Bodys which are called electricks per se'. Wilson also concluded that because a 'luminous matter could be produced infinitely from rubbing a glass globe', it was 'impossible that the electric matter should issue from the glass'. He then did a series of trials to test if the 'electric matter came out of the glass or not, for I was greatly persuaded from what I had done before, that the electric matter did not come from the glass but the bodies contiguous'.[32] Wilson claimed that this amount of electric matter, just like the aether, was in reciprocal

---

[27] *A catalogue of the genuine, entire and curious collection of prints and drawings...of Martin Folkes*, p. 14.

[28] [Philippe de la Hire], *New Elements of Conic Sections*, trans. Bryan Robinson (London: Dan Midwinter, 1704).

[29] Patrick Murdoch, *An Account of Sir Isaac Newton's Philosophical Discoveries in Four Books by Colin Maclaurin* (London: printed for the author's children, 1748), List of Subscribers.

[30] Graciano, 'The Memoir of Benjamin Wilson FRS', p, 187.

[31] Gregory Lynall, *Swift and Science: The Satire, Politics, and Theology of Natural Knowledge, 1690–1730* (New York: Palgrave, 2012), p. 96.

[32] Benjamin Wilson, *An Essay Towards an Explication of the Phaenomena of Electricity, Deduced from the Aether of Sir Isaac Newton* (London: C. Davis, 1746), p. xii.

proportion to the bodies' density, with the exception of oily or sulphurous sub-stances that were more electric than others of the same density. As we will see in section 8.4, sulphur was often associated by investigators in the Royal Society such as Martin Lister and Stephen Hales with electrical charge; lightning does smell of ozone which is similar in smell to sulphur, sulphur is in fact an excellent insulator, and static electricity accumulated on it discharges in electrical sparks towards proximate objects. Further, in early modern German mining literature, ore exhalations from sulphurous minerals such as iron pyrites were implicated in meteorological effects such as lightning. Wilson finally argued, just as Newton had speculated, that electricity was the ultimate cause of gravity.[33]

Wilson then noted in his diary, 'I wrote by that days post, an account of my discovery to Martin Folkes' and upon his return to London on 23 October 1745, he immediately went to the Royal Society to hear his work read that evening, to 'know what was said upon it.'[34] It transpired that Watson at the preceding meeting had a paper read on the very same discovery, which we recall Wilson had relayed to him in some detail before he went to Dublin. Wilson wrote in his journal:

> This made me suspect, that he had availed himself of the manuscript I had put into his hands, before I started for Ireland: and which he kept in his hands twelve days at least; for the theory upon which that discovery was made, I fully explained in that manuscript. I could not forbear mentioning this fact to several of the members, and an altercation immediately insued upon it. But being very young, and not having so many friends, as Mr. Watson, I had not the advantage in that dispute.[35]

What Wilson also did not know is that Watson had subsequently written a letter to Folkes and the Royal Society Council in which he claimed he 'never received any hints' from Wilson. When Wilson was in Dublin, Watson had also been busily demonstrating 'his discovery' to a variety of Royal Society Fellows, includ-ing Daniel Wray, William Stanhope, George Graham, and Folkes himself, on 3 October.[36] Rumours—or, as Watson termed it, 'very industrious reports about town'—however, had subsequently spread about his misdeed, so he also strenu-ously protested to Folkes that,

> the date of my papers ought not to be fixed on Octob. 28, when I put them in the presidents hands in order for their being laid before you; but on the third of that month, at which time to himself and the other gentlemen before mentioned they

---

[33] Graciano, 'The Memoir of Benjamin Wilson FRS', p. 187.
[34] Graciano, 'The Memoir of Benjamin Wilson FRS', p. 187.
[35] Graciano, 'The Memoir of Benjamin Wilson FRS', p. 189.
[36] MS 2392/7, Wellcome Library, London.

were communicated. As my papers and Mr Wilson's have no connection with each other…I cannot but regard the report insinuating my being a plagiary as a gross calumny.[37]

Although not as famous as Newton's dispute with Leibniz or the Cavendish–Watt–Lavoisier controversy over the discovery of the compound nature of water, the 'calendar of disputes' was thus full during Folkes's tenure as President.[38]

Watson's reputation was not unduly affected. The next year in 1746, Watson would go on to demonstrate that electrical aether could not be created or destroyed, but only transferred (conservation of charge), something that Benjamin Franklin announced only a few months later. Unlike Wilson, with whom Watson's relationship was irrevocably damaged, Franklin would become a close colleague. Watson would become famous for his experiment of 14 August 1747 to conduct electricity through a 6,732-foot-long wire at Shooter's Hill near Greenwich, designed to determine electricity's speed and whether it was faster than sound.[39] In the previous month, Folkes, along with the Earl of Stanhope and other members of the Royal Society, had participated in several smaller preparatory experimental trials at Westminster to determine the conductivity of electricity through different substances: 'to wit, whether the circuit was completed by the Ground, or by the Water of the [Thames] River'?[40]

Participants in these trials at Westminster Bridge and Shooter's Hill included James Bradley, Peter Davall, George Graham, Richard Graham Esq., George Lewis Scott, and Charles Stanhope, all FRS. George Graham and Charles Stanhope were among those who served as observers, Graham employing one of his precision timepieces, with another observing him to note the time lapse.[41] Each person that was part of the circuit stood on a wax insulator and held an earthed iron bar in one hand and the end of the transmission wire, which was wound around a series of dry sticks, in the other. At Shooter's Hill, the transmission wire was connected to a Leyden Jar that was placed upstairs in a house on the western side of the hill [visible at one of the observers' stations].[42] A musket was fired when the transmission started and the observer timed the difference between when they heard the gun and when they received an electric shock convulsing their arms on discharge of the Leyden Jar. It did not always work, usually

[37] MS 2392/7, Wellcome Library, London.
[38] The phrase, of course, was taken from Robert K. Merton's classic essay, 'Priorities in Scientific Discovery: A Chapter in the Sociology of Science', *American Sociological Review* 22, 6 (December 1957), pp. 635–59, on p. 636.
[39] Paola Bertucci, 'Domestic Spectacles: Electrical Instruments between Business and Conversation', in Bensaude-Vincent and Blondel, eds, *Science and Spectacle in the European Enlightenment*, p. 77.
[40] William Watson, 'A Collection of the Electrical Experiments Communicated to the Royal Society', *Philosophical Transactions* 45, 485 (1748), pp. 49–120, on p. 52 and p. 65.
[41] Deborah O'Boyle, 'Shocking Experiments in Shooters Hill', 4 August 2004, Royal Greenwich Time. https://royalgreenwichtime.com/2014/08/14/shocking-experiments-in-shooters-hill/ [Accessed 16 January 2019].
[42] O'Boyle, 'Shocking Experiments in Shooters Hill' [Accessed 16 January 2019].

because a member of the audience 'broke the connecting wire and otherwise greatly incommoded them'.[43] Folkes himself was left waiting for the signal of the electrical discharge on Westminster Bridge due to a broken wire.[44] Transmission of the electrical charge among the 'circuit', a term Watson invented, was 'nearly instantaneous' with the sound of the musket shot, but Watson wanted to further ascertain the 'absolute Velocity of Electricity at a certain Distance'.[45] The next year on 5 August 1748 he tried a similar experiment again, using an electrical circuit of two miles. Folkes was present with (*inter alia*) Thomas Birch, James Bradley, George Graham, and the Earl of Stanhope, as well as 'Mr Grischow', a representative of the Royal Academy of Sciences at Berlin, but the results were inconclusive, merely stating that 'through the whole Length of this Wire, being...twelve thousand two hundred and seventy-five Feet, the Velocity of Electricity was instantaneous'.[46]

As for Wilson, he continued to have Folkes's patronage and encouragement, so it is fairly clear that Folkes ascertained what had transpired between him and Watson.[47] Wilson went on to publish his discovery as *An essay towards an explication of the phænomena of electricity deduced from the æther of Sir Isaac Newton* (1746), dedicating it to Folkes. The work, imitating both the layout of Newton's *Principia* and Robinson's iatrochemical works, was organized similarly to a mathematical treatise, with propositions followed by experiments, and it was a summary of three papers that he subsequently read to the Royal Society. Wilson proposed in a paper of 6 November 1746 that 'if a fluid of a uniform density surround a globe, and that be electrified and turn'd round on an axis, passing through its center, and there be held a non-electric near the equator of that body, the fluid will rise successively towards the non-electric, as it turns round, in *like manner as the seas is affected by the moon*' [italics original].[48] To prove his assertion, Wilson perforated a pumice stone through the centre for the poles by which it was to be turned, and dipped it into water. A 'non' electric was held near the ball which was electrified and rotated. Wilson noted that 'the water on the surface so electrified will be seen to rise, and always towards the non-electric, so that the diameter at the equatorial parts was greater than at the poles', just as the seas were affected by the moon's gravity.[49] Using analogical thinking, he then concluded that the electric or subtle manner 'in all bodies more or less...seems to be the great *Desideratum* to account for...an universal attraction and gravity'.[50] He followed this publication in 1750 with *A Treatise on Electricity* (2nd ed., 1752), inventing

[43] Watson, 'A collection of the Electrical Experiments', p. 53.

[44] Watson, 'A collection of the Electrical Experiments', p. 56.

[45] Watson, 'A collection of the Electrical Experiments', pp. 85–6.

[46] William Watson, 'An Account of the Experiments made by some Gentlemen in the Royal Society, in order to measure the absolute Velocity of Electricity', *Philosophical Transactions* 45, 489 (1748), pp. 491–6, on p. 491 and p. 496.

[47] Graciano, 'The Memoir of Benjamin Wilson FRS', p. 191.

[48] Wilson, *An Essay Towards an Explication of the Phaenomena of Electricity*, p. 22. These results were probably due to a combination of surface cohesion and centripetal force.

[49] Wilson, *An Essay Towards an Explication of the Phaenomena of Electricity*, p. 23.

[50] Wilson, *An Essay Towards an Explication of the Phaenomena of Electricity*, p. 28.

and exhibiting a large electrical apparatus; on 5 December 1751 Wilson was elected a Fellow of the Royal Society.[51]

8.3 Illusion and Reality: The Rembrandt Craze

Wilson subsequently established a studio on Great Queen Street, where the artist Godfrey Kneller had previously lived. Here, Wilson painted a portrait of Folkes, as well as Lord Orrery and Lord Chesterfield, whereby, as he stated, 'by my succeeding with these tolerably well, I began to creep a little into esteem'.[52] Wilson ended up as a very successful artist, his fame arising from a clever joke. Thomas Hudson, a fellow artist and collector, had treated Wilson rather badly, so Wilson duped him into buying Rembrandt etchings that he himself faked; after the joke was revealed, with the encouragement of Hogarth Wilson was able to sell the counterfeit prints to such huge demands 'that both plates were almost entirely wore away'.[53]

One wonders if Folkes bought a counterfeit print from Wilson because of his interests in optical illusions, perceptions, and even having portraits made of himself to look like Newton (see chapter five).[54]

Certainly, Folkes was more generally a connoisseur of etchings, engraving, and mezzotints, often a customer of the artist, publisher, and dealer Arthur Pond (1705–58); Pond's account books reveal that in 1748 alone, Folkes bought engravings of *Storm: Jonah and the Whale* (Vivares after Nicolas Poussin and Gaspard Dughet); two engravings after works by Claude (Lorrain) Gellée; and three sets of engravings of Roman Antiquities by J. S. Muller after Giovanni Paolo Panini.[55] But Folkes was a particularly avid collector of Rembrandt etchings and

---

[51] E. I. Carlyle and John A. Hargreaves. 'Wilson, Benjamin (bap. 1721, d. 1788), portrait painter and scientist', *Oxford Dictionary of National Biography*. https://doi.org/10.1093/ref:odnb/29641 [Accessed 23 November 2020].

[52] Graciano, 'The Memoir of Benjamin Wilson FRS', p. 191.

[53] Graciano, 'The Memoir of Benjamin Wilson FRS', p. 193.

[54] See *A Catalogue of the genuine and curious collection of mathematical instruments, gems*, pp. 4, 6 8, 14, 18, 19. Many Rembrandts from Folkes's collection are in major galleries. *The raising of Lazarus* by Rembrandt from Folkes's collection is in the British Museum, museum number F, 4. 152. The National Gallery, Washington DC has *Canal with an Angler and Two Swans*, number 1943.3.7227, https://www.nga.gov/collection/art-object-page.10023.html; A Rembrandt etching of *David slaying Goliath* with Folkes's collector's mark is in the Royal Collection, RCIN 808196; *The raising of Lazarus* from Folkes' collection is in the British Museum, F, 4. 152. They also frequently appear at auction sales, *The Flight into Egypt* selling at Christie's on 5 July 2016, Fifty Prints by Rembrandt van Rijn: A Private English Collection. https://www.christies.com/lot/lot-rembrandt-harmensz-van-rijn-the-flight-into-5990634/ [All URLS accessed 9 November 2019].

[55] On 8 August 1748, Folkes purchased two works by Claude (Lorrain) Gelleé, as well as three sets of works by Panini of Roman Antiquities, including views of the mausoleum of Hadrian and the Arch of Constantine (p. 299); On 7 October 1748, Folkes bought five engravings of *Storm: Jonah and the Whale* (Vivares after Gaspard Dughet and Nicolas Poussin), (p. 300). See Louise Lippincott, 'Arthur Pond's Journal of Receipts and Expenses, 1734–50', *The Volume of the Walpole Society* 54 (1988), pp. 220–333, on pp. 299–300.

engravings, including the famous *Christ Healing the Sick* (known as the Hundred Guilder Print), the *Flight into Egypt, Crossing a Brook* (1654), and *St Jerome Reading in an Italian Landscape* (see figure 8.3).[56] In the eighteenth century, there was a 'craze' for Rembrandt's works among connoisseurs, spurred on by the publication of Roger de Piles's *The Art of Painting* (1706), which described the merits of his chiaroscuro, drawings, and etchings, noting the artist's preference for 'nothing more than to imitate *Living Nature*'.[57] Pile enthused, 'His manner in Etching was very like that in Painting. 'Twas expressive and lively, especially in his Portraits, the Touches of which are so a propos, that they express both the Flesh and the Life'.[58]

Perhaps influenced by Pile's art criticism, Folkes possessed several of Rembrandt's portrait etchings that featured immediate and evocative gazes, including some rare prints such as *Old Man Shading His Eyes with His Hand* (1638); its being left unfinished by the artist added to its novelty, making it desirable for connoisseurs.[59] All of the etchings in Folkes's collection were stamped with his distinctive owner's initials, creating a pedigree that confirmed 'authenticity and importance, as well as increasing their value'.[60]

Folkes's penchant for Rembrandt was more definitively influenced by his friend, artist and neighbour Jonathan Richardson the Elder, the 'towering figure of Rembrandt's discovery in Britain'.[61] Like Rembrandt, Richardson was a devoted self-portraitist, considering the genre as important as history painting; for him, a portrait was 'a sort of General History of the Life of the Person it represents, not only to Him who is acquainted with it, but to Many Others, who upon Occasion of seeing it are frequently told, of what is most Material concerning Them, or their General Character at least'.[62] And, as Rembrandt, Richardson used self-portraiture to explore his own soul, his habit of self-portraiture, 'charting his declining physical appearance...married over a decade to a discipline of almost

---

[56] *The Flight into Egypt*, Christie's, Fifty Prints by Rembrandt van Rijn: A Private English Collection, 5 July 2016, https://www.christies.com/lotfinder/Lot/rembrandt-harmensz-van-rijn-the-flight-into-5990634-details.aspx [Accessed 31 July 2020]. Martin Folkes (1690–1754), London, his stamp *recto* (Lugt 1034); presumably his sale, Langford, London, 17 January 1756 and seven following days.

[57] Robert de Piles, *The Art of Painting: and the Lives of Painters: Containing, a Compleat Treatise of Painting, Designing, and the Use of Prints* (London: J. Nutt, 1706), p. 317. Christian Tico Seifert, 'Rembrandt's Fame in Britain, 1630–1900', in Christian Tico Seifert, ed., *Rembrandt: Britain's Discovery of the Master* (Edinburgh: National Galleries of Scotland, 2018), pp. 11–50, on p. 22.

[58] De Piles, *The Art of Painting*, p. 317.

[59] *Old Man Shading his Eyes with his Hand*, etching with drypoint. 5–3/8 w. 4–1/4 in. circa 1638, [Bartsch, Holl, 259; New Holl. 175; H. 169], with collector's mark of Martin Folkes (British, 1690–1754), 22 July 2018, Lot 48, Butterscotch Auction House, Bedford Village, New York. The previous owner had bought the print from Associate American Artists, 663 Fifth Avenue, New York, New York, 10022 in 1972. See https://www.invaluable.com/auction-lot/rembrandt-van-rijn-dutch-1606-1669-48-c-f414e75bda [Accessed 21 June 2019].

[60] Seifert, 'Rembrandt's Fame in Britain', p. 23.

[61] Seifert, 'Rembrandt's Fame in Britain', p. 23.

[62] Jonathan Richardson, *An essay on the whole art of criticism as it relates to painting and an argument in behalf of the science of the connoisseur* (London: W. Churchill, 1719), p. 45.

Fig. 8.3 Rembrandt van Rijn, *St. Jerome Reading in an Italian Landscape*, Etching, Engraving and Drypoint, *c.* 1653, 26.2 × 21.4 cm, Bequest of Mrs. Severance A. Millikin 1989.233, Cleveland Museum of Art, CC Open Access Licence.

daily poems, where he examined his state of mind'.[63] Richardson's drawings of himself show his consciousness stripped bare, with startling intensity and vivid presence.

---

[63] Emma Crichton-Miller, 'Jonathan Richardson by himself', *Apollo*, 27 July 2015, https://www.apollo-magazine.com/jonathan-richardson-by-himself [Accessed 20 March 2020].

Richardson's self-portraiture could also be playfully illusory. Just as Folkes had himself painted by Vanderbank to imitate Newton's appearance, Richardson's 'graphite on vellum self-portrait, pretty much as Rembrandt, in fur hat, is evidence of his adulation of this Dutch master'.[64] Richardson was like his hero in more ways than one. While Rembrandt famously featured himself dressed up alternatively as a burgher, gentleman, or an artist in a variety of headdresses and poses, his self-portraits were often in historical dress to emulate his celebrated predecessors, such as Albrecht Dürer.[65] In this manner, Rembrandt 'wanted to present himself as an artist in the tradition of the Dutch and German masters of the past', much like Folkes and Richardson wished to model themselves on their own heroes.[66]

Like Folkes, Richardson had a renowned collection of Rembrandt's works on paper, and he also owned the Hundred Guilder Print of *Christ Healing the Sick*, analysing the print in some detail in his *Essay on the Theory of Painting*.[67] Richardson particularly praised Rembrandt's inclusion of a few choice figures to suggest the magnitude of the audience gathered around the Messiah to be healed without unnecessarily crowding the composition.[68] One recalls Folkes's discussion with Algarotti on similar methods employed by the sculptors of Trajan's columns of creating illusions of Roman crowds with aesthetic sensitivity (see chapter five). Both Richardson and Folkes clearly delighted in Rembrandt's ability to create works that combined startling intensity and directness with subtle illusion.

8.4 Public Upheavals

The 1740s and 1750s in the Royal Society were not only electric and artistic, but tumultuous for Folkes publicly and privately. On 8 February, and then again on 8 March 1750, an earthquake shook London, an event which saw the birth of seismology. Folkes was home in bed at Queen Square, realizing that the noise and aftershocks could not have been due to a horse cart because of the early hour. The shock was felt outside of London, and as he reported to the Royal Society:

> I sent a servant out about 7 o'clock, and he met a countryman, who was bringing
> a load of hay from beyond Highgate, and who was on the other side of the town

---

[64] Emma Crichton-Miller, 'Jonathan Richardson by himself'.

[65] Marieke de Winkel, *Fashion and Fancy: Dress and Meaning in Rembrandt's Paintings* (Amsterdam: Amsterdam University Press, 2006), p. 189.

[66] De Winkel, *Fashion and Fancy*, p. 189.

[67] Seifert, 'Rembrandt's Fame in Britain', p. 23; Carol Gibson-Wood, *Jonathan Richardson, Art Theorist of the English Enlightenment* (New Haven and London: Yale University Press, 2000), pp. 156–7. In 1747, Richardson's collection was dispersed by his son at a sale, many of the Rembrandt drawings going to his son-in-law Thomas Hudson (1701–79).

[68] Jonathan Richardson, *An Essay on the Theory of Painting* (London: A.C., 1725), p. 66.

when the shock happened; he did not, he said, feel it, as he was driving his waggon; but that the people he saw in the town of Highgate were all greatly surprised, saying they had had their houses very much shocked, and that the chairs in some were thrown about in their rooms.

The following week on 15 February, there was another tremor. Trembley was visiting Folkes from Holland, and he, along with the Earl of Macclesfield and Charles Bentinck, were in Folkes's study in Queen Square. As the *Philosophical Transactions* reported,

> they all felt themselves at the same instant strongly lifted up, and presently set down again: they also heard a noise over their heads as of some heavy piece of furniture being thrown down, whilst those who were in the room over them were frighted, and apprehended the like accident had happened below stairs. The coachmen on the boxes of 2 coaches then standing at the door, were extremely sensible of the shock, and apprehended the house was going to fall upon them.[69]

Accounts poured into the Royal Society. Cromwell Mortimer, the Royal Society secretary, reported from his house on Devonshire Street around the corner from Queen Square:

> On Thursday morning March 8. 1749 I awaked a little after five, I opened my curtains and observed the sky hazy, I drank a draught of water…and therefore was going to lie down again in my bed, when leaning on my right elbow I first felt a shock as if the whole house was violently pushed from the NW to the SE and then with equal force pushed back again…I saw the cornice and ceiling of the Room sensibly move.[70]

James Parsons, physician and Assistant Foreign Corresponding Secretary of the Society, reported that after the shock and aftershock, 'all Newgate-market was in the greatest confusion imaginable no one thinking himself safe', and a maidservant at a house in Marylebone, where he was visiting a patient, said 'she was thrown first to one side and then back to the other, and many compared to the rocking of a cradle'.[71] Folkes wrote down a particularly interesting

---

[69] John Martyn, *The Philosophical Transactions from the Year 1743, to the Year 1750, Abridged and Disposed under General Heads* (London: Lockyer Dvais and Charles Reymers, 1750), vol. 10, p. 494. As this book was going to press, Kerrewin Van Blanken's work was published ahead of print: 'Earthquake observations in the age before Lisbon: eyewitness observation and earthquake philosophy in the Royal Society, 1665–1755', *Notes and Records: The Royal Society Journal of the History of Science*, https://doi.org/10.1098/rsnr.2020.0005.

[70] RS/L&P/2/10D/63, Royal Society Library, London.

[71] RS/L&P/2/10D/60, Royal Society Library, London.

account from Mr Josiah Bayfield of Gravel Lane near St George's Fields of roach fish leaping from the canals, then shooting away 'in all sorts of directions and seem to shift for themselves, as if they were frighted and alarmed at what had happened'.[72]

Early modern natural philosophers postulated that earthquakes were the result of vapours arising from minerals in passages underground which would ignite and cause explosions. In the seventeenth century, the opinion of physician and FRS Martin Lister, who attributed earthquakes to the vapours emitted by pyrites or iron sulphides, was fairly influential, as fifty years later, Benjamin Franklin wrote about it in the *Pennsylvania Gazette* in December 1737: 'Dr Lister is of the opinion, that the material cause of thunder, lightning, and earthquakes, is one and the same. Viz. the inflammable breath of the pyrites, which is a substantial sulphur, and takes fire in itself'.[73] In his work on mineral waters, the *De Fontibus*, Lister also briefly mentioned that earthquakes were primarily due to the firing of the 'inflammable breath of Pyrites...underground', and earthquakes could occur 'if by chance the fire [from pyrites] is contained in subterranean hollows, and hot springs' or 'if it is transported in abundance among water channels, even if it is not set on fire'.[74]

The possible involvement of pyrites in earthquakes also continued to provide fodder for Royal Society experiments by Stephen Hales (1677–1761) into the 1750s. Stukeley took notes on some papers presented in the Royal Society in 1752, in which Hales:

gives an experiment of putting some pyrites stone, with some aqua fortis [concentrated nitric acid (HNO_3)], into a vessel set in water, and covered with a large glass, whose mouth must be immersed in the water. A brisk fermentation arises, a black cloud, and a destruction of some quantity of air...Then suddenly taking up the glass out of the water, and letting in fresh air, a new ebullition arises...From this experiment the doctor apprehends that the cause of earthquakes is much illustrated. He says sulphureous vapour arises out of the earth

---

[72] RS/L&P/2/10D/62, Royal Society Library, London.
[73] Martin Lyster [Lister], 'The Third Paper of the Same Person, Concerning Thunder and Lightning being from the Pyrites', *Philosophical Transactions* 14, 157 (1684), pp. 517–19; see Alfred Owen Aldridge, 'Benjamin Franklin and Jonathan Edwards on Lightning and Earthquakes', *Isis* 41 (1950), pp. 162–4, on p. 162. Franklin may also have seen Lister's 'Three Papers of Martyn Lister, the first of the Nature of Earth-quakes; more particularly of the Origine of the matter of them, from the Pyrites alone', *Philosophical Transactions* 14, 157 (1684), pp. 517–19, in which Lister states in the first paragraph on p. 517: 'The material cause of Thunder and Lightning, and of Earthquakes is one and the same; viz, The inflammable breath of the Pyrites, the difference is, that one is fired in the Air; the other underground'. See Anna Marie Roos, 'Martin Lister (1639-1712) and fools' gold', *Ambix* 51, 1 (2004), pp. 23–41.
[74] Martin Lister, *De Fontibus medicates Angliae* (London: Walter Kettilby, 1684), pp. 78–9. [Pyritae autem halitus effectus sunt Fulmina et Fulgura, si in caelo accendatur; Terrae motus, si forte cavis subterraneis accensus contineatur: Thermae si per aquarum ductus subterraneos copiose feratur, etiamsi non accendatur.]

generated probably by the pyrites abounding therein...through cracks and chinks of the gaping earth. The vapours fly into the upper regions of the air, where they meet with pure and uncorrupted air, clouds intervening like as in the glass receiver, they ingage with violence...through the clouds, and cause a prodigious tumult above...These concussions in the air act upon the surface of the earth and cause earthquakes.[75]

Hales indeed would have observed an exothermic reaction (brisk fermentation). The sulphurous vapours he reported were most likely sulphur dioxide produced from the reaction of pyrites with oxygen: $4\,FeS_2 + 11O_2 \rightarrow 2Fe_2O_3 + 8\,SO_2$, or just the production of free sulphur. The reaction that Hale described was likely (assuming some pyrites were iron-copper pyrites, as he does not distinguish): $CuFeS_2 + 18HNO_3 = Cu(NO_3)_2 + Fe(NO_3)_3 + 2SO_2 + 2SO_3 + 13NO_2 + 9H_2O$. Stukeley was particularly interested in Hales's account of earthquakes, because he offered an alternative idea inspired by the work of Benjamin Franklin, namely that earthquakes were created by electricity.[76]

The Royal Society's first account of Franklin's experiments on electricity were sent to Peter Collinson on 25 May 1747 and shown to William Watson, who approved of Franklin's idea.[77] Franklin then sent letters on 29 April 1749 on electricity and thunder, addressed respectively to Collinson and Dr John Mitchell, which were read in the Society in the fall of 1749.[78] 'Franklin's conjectures that earthquakes [and lightning] were caused by electricity were studied when earthquakes shook London in the spring of 1750, and William Stukeley explained the phenomenon entirely in terms of Franklin's hypothesis'.[79] Franklin went on to publish his own account in his *Experiments and Observations on Electricity* (1751), in which he stated:

From the similar effects of lightning and electricity our author has been led to make some probable conjectures on the cause of the former; and at the same time, to propose some rational experiments in order to secure ourselves, and those things on which its force is often directed, from its pernicious effects; a

---

[75] William Stukeley, *The Family Memoirs of the Rev. William Stukeley* (London: Surtees Society, 1882–7) vol. 2, pp. 378–9, as quoted in I. Bernard Cohen, 'Neglected Sources for the Life of Stephen Gray (1666 of 1667–1736)', *Isis* 45, 1 (May 1954), pp. 41–50, on pp. 44–5.

[76] William Stukeley, *The Philosophy of Earthquakes, Natural and Religious* (London: C. Corbet, 1750).

[77] 'Experiments and Observations, [April 1751]', *Founders Online*, National Archives. https://founders.archives.gov/documents/Franklin/01-04-02-0039 [Accessed 29 September 2019]. [Original source: *The Papers of Benjamin Franklin*, vol. 4, *July 1, 1750, through June 30, 1753*, ed. Leonard W. Labaree (New Haven: Yale University Press, 1961), pp. 125–30.]

[78] 'Experiments and Observations, [April 1751]', *Founders Online*.

[79] 'Experiments and Observations, [April 1751]', *Founders Online*; William Stukeley, 'On the Causes of Earthquakes', *Philosophical Transactions* 46, 497 (1749–50), p. 641–6, on p. 643.

circumstance of no small importance to the publick, and therefore worthy of the utmost attention.[80]

Although the theory of earthquakes was of course abandoned, Franklin's work was widely read. Joseph Priestley, who reviewed the history of electricity some fifteen years later, stated:

Nothing was ever written upon the subject of electricity which was more generally read, and admired in all parts of Europe than these letters. There is hardly any European language into which they have not been translated; and, as if this were not sufficient to make them properly known, a translation of them has lately been made into Latin.[81]

As for Stukeley, in his work, *The Philosophy of Earthquakes* (1750), he first discounted the vapours theory, writing:

We never hear, from the many hundreds of thousands of workmen in this kind, at *Newcastle, Nottinghamshire, Yorkshire, Derbyshire, Staffordshire, Somersetshire,* and *Wales:* from the infinite numbers of workmen in the mines of lead, tin, and the like, of the cavernous state of the earth, so as to give any colour for this *hypothesis* of earthquakes. The earth is generally of solid rocks in which there must be now, and then, some clefts, and vacuities, small in compass, as naturally so many heterogeneous *strata* of the earth consolidate together. But there can be no imagination of vapours breaking through, uniting, traversing so suddenly, a large space of earth, so as to produce those earthquakes, we have seen, and felt; much less such as we read of. The workmen in all sorts of mines confess by their hard labour, that the earth is not cavernous; nor are there mines of sulphur, nitre, and the like inflammable materials in *England*.[82]

He continued,

Electricity may be call'd a sort of soul to matter, thought to be an ethereal fire pervading all things, and acting instantaneously, where, and as far as it is excited. 'Tis every body's observation, that there never was a winter, like the last past, in any one's memory, so extremely remarkable for warmth and driness, abounding with thunder and lightning, very uncommon in winter; coruscations in the air frequent, justly thought electrical by all philosophers; particularly, twice we had

---

[80] Benjamin Franklin, 'The Preface', *Experiments and Observations on Electricity, Made at Philadelphia in America, by Mr Benjamin Franklin, and Communicated in several Letters to Mr P. Collinson, of London, F.R.S.* (London: Printed and sold by E. Cave, at St. John's Gate, 1751).
[81] Quoted by 'Experiments and Observations, [April 1751]', *Founders Online.*
[82] Stukeley, *The Philosophy of Earthquakes,* pp. 11–12.

the extraordinary appearance of that called *aurora australis* with colours altogether unusual and this just before the first earthquake: All the while the wind constantly south and south-west, and that without rain, which is unusual with these winds.

As these meteorological conditions prevailed five months before the earthquakes, Stukeley thought it was 'reasonable' to conclude that the earth was brought 'into an unusual state of electricity; into that vibratory condition wherein electricity consists; and, consequently, nothing was wanting but the approach of a non-electric body, to produce that snap, and that shock, which we call an earthquake; a vibration of the superficies of the earth.'[83] He was supported in his ideas by Stephen Hales, whose publication in the *Philosophical Transactions* immediately followed Stukeley's work.[84]

In his presentation of the paper to the Royal Society on 5 April 1750 Stukeley also spoke about the divine causation of the earthquake and subsequently gave a sermon about it at St George the Martyr, though the theological part of his discourse was truncated in the final article published in the *Philosophical Transactions*.[85] Stukeley's comment, however, 'yet it is notorious, that London was the Centre; the Place to which the Finger of God was pointed', remained. It would not have been surprising that when Stukeley read his work, he faced objections from those who believed in subterraneous vapours as well as a more secular form of explanation. Stukeley recorded in a letter to Maurice Johnson that 'no less than 5 rose up against it. the president [Folkes], Lord Macclesfield, Mr Burroughs, Dr Squire & de la costa with a stale joke. I laught at em: my friends were excessively angry. The next week I gave in a much larger paper, with a gentle reproof, and full confirmation of my opinion, all was hush.'[86] Nonetheless, Stukeley's theory was not supported by Benjamin Franklin and, of course, eventually discounted.

The aftershocks continued through the year, Folkes's brother William feeling the tremors at Newton in Northamptonshire while he and his family were at Church, as well as when he 'went thro' Stamford to Grantham' in his way to Yorkshire...none of them attended with any ill Consequences, any further than furnishing Room for melancholy Reflections upon such a disagreeable Alteration

[83] Stukeley, *The Philosophy of Earthquakes*, pp. 21–2.
[84] Stephen Hales, 'Some Considerations on the Causes of Earthquakes', *Philosophical Transactions* 46, 497 (31 December 1750), pp. 669–81.
[85] Stukeley, 'Causes of Earthquakes', pp. 657–69.
[86] Stukeley to Johnson, 15 May 1750, SGS/Stukeley/25, reprinted in Honeybone and Honeybone, *Correspondence*, p. 150. The other gentlemen were George Parker, 2nd Earl of Macclesfield (1698–1764), who succeeded Folkes as Royal Society President, James Burrow (1701–82), who later became President of the Society in 1768 and in 1772, Dr Samuel Squire (1713–66), later Bishop of St Davids, and Emanuel Mendes da Costa (1717–91), naturalist and the first Jewish clerk of the Royal Society (see section 8.6).

in our climate, which had been generally thought before tolerably free from this Calamity'.[87]

8.5 Private Upheavals

The decade of the 1740s also saw seismic upheavals for Folkes, particularly with his involvement as one of the administrators of the Coram Foundling Hospital, the world's first incorporated charity for abandoned children. Via the patronage of aristocratic women to confer social legitimacy, the acquisition of a Royal Charter, and a personality that combined 'strong social aims and weak social ambitions', Thomas 'Captain' Coram (1668–1751), a retired shipwright and ship-master, was able to realize his vision after two decades of effort.[88] As Helen Berry has remarked, 'the willingness of elite women to sign up to Coram's petition reflected the tenets of Christian charity deemed appropriate for their own sex'.[89] Queen Caroline and Sarah, Duchess of Richmond, her lady of the bedchamber, were signatories to the Foundling Hospital petition.[90] On 17 October 1739, King George II signed and sealed the charter of incorporation for the 'Hospital for the Maintenance and Education of Exposed and Deserted Young Children', giving the corporation the right to receive benefactions and hold property in perpetuity.[91] Charles Lennox and the Duke of Montagu, 'right trusty, and right entirely-beloved Cousins', were listed at the front of the charter due to their protracted involve-ment with planning the hospital, and it is likely via their influence that Folkes was given office. Folkes himself was listed as signatory to the original petition for incorporation. *The Gentleman's Magazine* of 25 November 1739 reported the passing of the charter and seal:

> To incorporate Charles, Duke of Richmond and several other great officers and ministers of state and their successors into one body politic and corporate, and by the names of governors and guardians of the hospital for the maintenance and education of deserted and exposed young children with powers to purchase lands and mortmain not exceeding yearly value of £4000.[92]

[87] 'Part of a letter from William Folkes, Esq; F.R.S. to his brother the President, concerning a shock of an earthquake felt at Newton in Northamptonshire, on Sunday. September 30, 1750', *Philosophical Transactions* 46, 497 (31 December 1750), pp. 701–2.
[88] Douglas Fordham, *British Art and the Seven Years' War: Allegiance and Autonomy* (Philadelphia: University of Pennsylvania Press, 2010), p. 28. See also Gillian Wagner, *Thomas Coram, Gent (1668–1751)* (Woodbridge: The Boydell Press, 2004).
[89] Helen Berry, *Orphans of Empire, the Fate of London's Foundlings* (Oxford: Oxford University Press, 2019), e-book.
[90] Berry, *Orphans of Empire*.
[91] *A copy of the Royal Charter, Establishing an Hospital for the Maintenance and Education of Exposed and Deserted Young Children* (London: J. Osborn, 1739). See also Wagner, *Thomas Coram*.
[92] *Gentleman's Magazine* 9 (25 January 1739), p. 552, as quoted in Wagner, *Coram*, p. 136.

Folkes was appointed Vice President, along with five others, his name announced at the first executive meeting of the governors. As Gillian Wagner has related, the other five Vice Presidents were: 'Micajah Perry, the ex Lord Mayor, [who] had been a supporter, Lord Vere Beauclerk…an old friend of Coram's, and Sir Joseph Eyles, a rich merchant'; the other two were Peter Burrell and James Cook, also merchants.[93]

The committee got to work. Immediately after Christmas Day 1739, Folkes acquainted the governors with a proposal 'for taking a 21 years lease of Montagu-house, for the use of the charity' for £400 per year to serve as temporary premises.[94] The building was not altogether suitable for use as a hospital, but as the Duke of Montagu was Folkes's friend, there was an obligation to take the offer seriously; the governors 'engaged a surveyor, Mr Sanderson, to inspect the house and prepare plans'.[95] There were, however, extensive renovations required, as well as legal difficulties with the lease for the use of the charity. Montagu House was eventually bought from Montagu's two daughters to house the collections left by Sir Hans Sloane and to create the British Museum. Coram was deeply frustrated by the delay in finding proper premises for his hospital, which was also hindering further bequests, but eventually the charity found temporary premises in Hatton Garden for a more sensible £48 per annum, while the executive committee negotiated with the Earl of Salisbury's agent, Mr Lamb, to buy fifty-six acres near Queen Square.

Gillian Wagner has portrayed Folkes's role in these negotiations in a fairly pejorative light. She wrote, 'Coram, impatient at not having been able to arrange a meeting with the other two governors…went ahead and met Lamb and reported back to the general committee that Salisbury would only sell all four fields, and not the two the governors wanted to buy. Folkes was in the chair and snubbed Coram by resolving that the matter "would be referred to [William] Fawkener [one of the governors] and the Treasurer so that they could join with Mr Coram"'.[96] This may have been not so much of a snub as an enforcement of proper procedures; Folkes had spent a lifetime being a broker for the requests of members of the aristocracy, and as a successful committee man of genial disposition, he may have made his request so as not to jeopardize the negotiations. Eventually the governors offered £6,500 for all five fields; although Salisbury demanded £7,000, he donated the remaining £500, and the purchase was made on 17 October 1740.[97]

[93] Wagner, *Thomas Coram*, p. 133.

[94] *An Account of the Foundling Hospital in London for the Maintenance and Education of Exposed and Deserted Young Children* (London: Luke Hansard, 1817), p. 17.

[95] Wagner, *Thomas Coram*, p. 137.

[96] Wagner, *Thomas Coram*, p. 143; LMA:A/FH/K02/001–023, General Committee Minutes, Foundling Hospital, 25 June 1740, London Metropolitan Archives.

[97] Wagner, *Thomas Coram*, p. 143.

The following March, the governors made a public announcement that the first infants would be admitted:

On Wednesday 25th of this instant March at eight at night and from that time until the house is full, their house over against the charity school in Hatton Garden will be open for children under the following regulations: that no child exceed in age two months nor shall have the French Pox [syphilis] or disease of like nature; all children to be inspected and the person who brings it to come in at the outer door and not to go away until the child is returned or notice given of its reception. No question asked whatsoever of any person who brings a child, nor shall any servant of the Hospital presume to enquire on pain of being dismissed.[98]

The governors met regularly and recorded all their admission decisions in meeting minutes and in Books of Regulation. They arranged for registers or billet books to record the admission of each child, any tokens it brought in, its placement at nurse, the inspector supervising the nurse in her home parish, any illness, death or survival of the child to apprenticeship.[99]

One of the children admitted on 17 April 1741, baby number 53, called George Hanover, was unusual in its dress and state of cleanliness, 'dressed in white satin sleeves bound up with a blue mantua, a fine double cambric cap with double cambric border, and three fine diaper clouts'.[100] Wagner speculated if the child with the King's name was a royal bastard, as the copy of the letter accompanying the baby noted 'begs if asked by that name that he may be seen if agreeable to the hospital'.[101] The baby was eventually christened John Montagu in honour of the Duke of Montagu and died on 2 May 1741. What Wagner did not realize is that the copy of the entry record was written by Folkes himself and subsequently sealed (see figure 8.4).

Folkes had spent much of his life acting as a broker for his aristocratic friends' needs, buying artwork, procuring instruments, serving as a witness to their wills. Did he intervene in this case to save the reputation of a member of the nobility? Helen Berry noted that one of the reasons female aristocrats supported Coram's petition was that some had husbands 'who were well-known womanizers with many illegitimate children by different mistresses, while others had first-hand experience of unorthodox family arrangements... Aristocrats had their own

[98] LMA:A/FH/K02/001–023, General Committee Minutes, Foundling Hospital, 4 March 1752, London Metropolitan Archives, quoted by Wagner, *Thomas Coram*, p. 145.

[99] 'The Foundling Hospital Records', London Metropolitan Archives. https://www.cityoflondon. gov.uk/things-to-do/london-metropolitan-archives/the-collections/Pages/foundling-hospital-records.aspx [Accessed 23 November 2019].

[100] A/FH/A09/001, Billet Book, 1741, March–May, London Metropolitan Archives, as quoted by Wagner, *Thomas Coram*, p. 150.

[101] Wagner, *Thomas Coram*, p. 151.

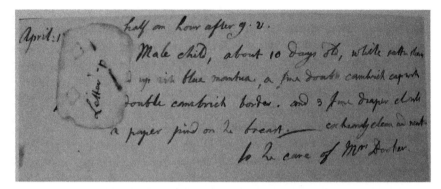

Fig. 8.4 Copy of Entry Record about baby 53, written by Martin Folkes, with an indication where the letter was pinned, A/FH/A09/001, Billet Book, 1741, March-May. © Coram.

moral codes that were different from those of the gentry and middling sorts'.[102] Lennox himself was the child of an illegitimate son of King Charles II.

After all, Folkes himself was not unfamiliar with the consequences of illegitimacy. As we have seen, Folkes provided Montesquieu (as well as Maupertuis) with scientific books and 'capotes anglaises', or condoms, via the post, as condoms were available in England and banned in France.[103] Folkes also received a letter from Louisa Edwards, Charles Street, Westminster, on 12 November 1743 that indicated he may have had an illegitimate child or two himself. Edwards wrote:

> Being at Bristol Last week, and knowing my cousin, from her infancy I made her a visit, I beg you Dr Sr not to think me inpertnent, or pretending not to advise, I shall only say what I observe. She is big with Child. And has one about seven Months very much like yourself, the sight of me put her spirits in a great hurry, and upon a little talk I found she is vastly desirous of a kind letter from you, I doe assure you if I did not think, her Condition required som tenderness and that it is absolutely necessary for her Health I should not have intermeded because I know tis a tender point...respects to Miss Foulks, pardon this trouble.[104]

Edwards mentioned a 'Mr [William] Rishton', who Dorothy, Folkes's youngest daughter, would eventually marry, so Louisa was known to his family. The identity of Edwards's cousin or her child remains unknown.

As Wagner has related, Coram also lost his position of governor of the organization as he made 'several scandalous insinuations about Martin Folkes', one of which was a charge that Folkes had an inappropriate sexual relationship with a

---

[102] Berry, *Orphans of Empire*.

[103] Battin, 'Montesquieu, les sciences et la médecine en Europe', p. 251.

[104] RS/MS/250/58, Letter of 12 November 1743 from Louisa Reynolds to Folkes, Royal Society Library, London.

Foundling Hospital laundry maid, Sarah Wood, who subsequently was inexplic-ably promoted to the rank of Chief Nurse.[105] As all the relevant evidence was released to Folkes and destroyed, and Folkes was subsequently cleared of all charges to his character and retained his position on the board of governors, it is difficult to ascertain exactly what transpired.

Although it is not to excuse any bad behaviour, Folkes's actions may have been explained by two personal events in his life: his wife's growing mental ill-ness, and the death of his son Martin in a riding accident in 1740. Folkes had clearly been agonizing over his wife's mental illness for some years, and in July 1735 even privately entertained moving to Florence permanently and leaving Lucretia in England, writing to his friend Antonio Cocchi about settling in Florence.[106]

Cocchi replied the following month:

> We are glad to hear Shee [Lucretia] is better and I believe if she could be brought to a proper method of cure She might entirely recover. As for living with her, I well conceive it may be very troublesome to you. I have some experience of People who Suffered any disorder in their mind, and I think the Surest way for quiet and ease is to break all commerce with them.[107]

Cocchi then stated, 'after having settled what justice and humanity demands of us, so that I approve mightily of your design of coming abroad which will be a good pretext to separate your dwelling'.[108] In his account book of his travels, Folkes subsequently recalculated the costs of going to Italy without his wife and children, but he ultimately decided against settling abroad.

Some of his decision to stay in England may also have been because of his pre-vious status as Deputy Grand Master of the English Lodge and of the growing suppression of Masonry in Rome and in Florence. As Eisler has noted, 'the Roman lodge, whose initial documented session took place on 16 August 1735, soon ran foul of the Inquisition', and 'overt Masonic activities in Florence came to a halt with the arrest of an Italian brother, the poet Tommaso Crudeli, on 11 May 1739'.[109] Folkes wrote to Lennox about the activities of the Italian Inquisition:

> Our brethrens virtue will sure be tryd and if any should happily be crownd with Martyrdom in so good a cause, we may overspread the earth, and unborn

---

[105] As quoted by Wagner, *Thomas Coram*, p. 156.
[106] Leghorn MS Facs 589, 584B, Martin Folkes to Antonio Cocchi, 6 November 1735, British Library, London.
[107] RS/MS/790/25, f. 1r. Antonio Cocchi to Folkes, Royal Society Library, London.
[108] RS/MS/790/25, f. 1r. Antonio Cocchi to Folkes, Royal Society Library, London.
[109] William Eisler, 'The Construction of the Image of Martin Folkes (1690–1754), Part I', *The Medal*, 58 (2011), pp. 4–29, on pp. 20–1.

masons may happen to use the intercession of Saint Dessy, and read the legends of the faithfull that were tyed up in sacks on the other side the water, the lodge will doubtless be strengthened every day by these persecutions and become more glorious. There is now however a Mason Prince in Tuscany, tho a friend of ours you will hardly allow him so true a mason as his predecessor.[110]

Folkes's remarks were prescient. Crudeli died soon after his arrest due to the torture and the hardship of his imprisonment. In April 1738, Pope Clement XII issued the first papal bull against Freemasonry, the political crisis of the Grand Duchy of Tuscany the specific event that triggered the repression:

> Franz Stephan of Lorraine, the new ruler, was himself a well-known freemason and more than keen on the anti-Jesuit cultural policy of the radicals affiliated to the Florentine lodge. The clash for the cultural hegemony over the Grand Duchy ended with a political and cultural compromise, following which the activity of the Florentine lodge was suspended.[111]

Furthermore, one of Folkes's personal medals, created to honour his contributions to Italian science but also containing symbols of Freemasonry, was also provocatively struck in Rome between 1739 and 1742, at the height of the Catholic Inquisition's suppression of Masonic Lodges.[112] Italy for Folkes thus became an area to avoid.

With his plans to relocate to Italy defeated, Folkes ultimately decided to confine Lucretia to a lunatic asylum in Chelsea.[113] Although we do not know when Lucretia entered a private asylum, Folkes's will, written in 1751, noted, 'And Whereas the unfortunate State of the health of my dear Wife Mrs Lucretia ffolkes render it absolutely necessary several years since according to the joint advice of several Eminent physicians and of all my most particularly ffriends, whom I consulted on that occasion to confine her where she now remains at Chelsea in the County of Middlesex'. With her income, Lucretia would not have been in Bethlem Royal Hospital (Bedlam) or its sister asylum, Saint Luke's Hospital for Incurable Lunatics, which opened in 1750, but would have been privately treated. Although there were a cluster of private madhouses in Great and Little Chelsea run by

---

[110] Goodwood MS 110, Folkes to Lennox, 23 July 1737, West Sussex Record Office.

[111] Massimo Mazzotti, 'Newton for Ladies: Gentility, Gender and Radical Culture', *British Journal of the History of Science* (June 2004), pp. 119–46, on p. 145.

[112] William Eisler, 'The Construction of the Image of Martin Folkes (1690–1754) Part II: Art, Science, and Masonic Sociability in the Age of the Grand Tour', *The Medal* 59 (2011), pp. 4–16.

[113] In his library, Folkes had 'Bunny against Divorce for Adultery and marrying again (1610)' Item 205, Second day's sale, 3 February 1756, *A Catalogue of the Entire and Valuable Library of Martin Folkes*. This work by Edward Bunny argued against dissolution of marriage for any reason; as Folkes was heterodox, the inclusion of this work in his library is interesting.

various operators such as Michael Duffield (d. 1798) and his nephew, it is unknown exactly in which asylum Lucretia was confined until she died.[114]

While it has been suggested (and contested) that in the Georgian period 'confinement, as both a cause and a symptom of marriage breakdown, was exercised in similar ways to wife beating, to demonstrate male power and limit women's agency over both their personal space and their property', this does not seem at all to be the case with Lucretia.[115] Andrews and Scull have in fact demonstrated that the more prominent literary depictions of individuals that were wrongly incarcerated were created rather to expose injustice than to act as authentic social commentary.[116] Folkes was indeed careful to make sure she had financial provision after he died. In his will, Folkes wrote that he had 'constantly been informed by those under whose care and direction she is, that there is but small prospect or probability of her recovery from her present state and condition', and thus he set in trust investments that would pay £400 per year for her 'support and maintenance'. He charged his daughters Lucretia and Dorothy to administer the payment for his 'said dear Wife their Mother who was ever whilst God gave her the ability a fond Carefull and most Indulgent Parent to them and whom I have ever most dearly and affectionately loved'.[117]

Richard Mead (1673–1754), physician to St Thomas's Hospital and Folkes's colleague and friend, wrote about the effects of the Sun and the Moon upon occurrences of 'lunacy', and served as a consultant physician for John Munro (1715–91), visiting physician to Bethlem Hospital (Bedlam).[118] It would be interesting to speculate as to whether Mead were one of the 'eminent physicians' mentioned in Folkes's will who visited and diagnosed Lucretia's illness. It was a horrific end for a brilliant actress, from star of Drury Lane to a melancholy object who, in Folkes's words, had 'in her power to make me and all my ffriends uneasie'.[119] Folkes kept a plaster figure of the heroine Roman figure Lucretia in his study, perhaps in commemoration of his wife.[120] It would have been appropriate. The Roman heroine

---

[114] Jonathan Andrews and Andrew Scull, *Customers and Patrons of the Mad-Trade: The Management of Lunacy in Eighteenth-Century England* (Berkeley: University of California Press, 2003), p. 9.

[115] Elizabeth Foyster, 'At the Limits of Liberty: Married Women and Confinement in Eighteenth-century England', *Continuity and Change* 17, 1 (2002), pp. 39–62, on p. 42. But for an opposing view, see R. A. Houston, 'Madness and Gender in the Long Eighteenth Century', *Social History* 27, 3 (2002), pp. 309–26.

[116] Jonathan Andrews and Andrew Scull, *Undertaker of the Mind: John Monro and Mad-doctoring in Eighteenth-century England* (Berkeley: University of California Press, 2001), pp. 149–53; Houston, 'Madness and Gender', p. 315, note 25.

[117] PRO/11/809/301, Will of Martin Folkes, 3 July 1754, The National Archives, Kew.

[118] Anna Marie Roos, 'Luminaries in Medicine: Richard Mead, James Gibbs, and Solar and Lunar Effects on the Human Body in Early Modern England', *Bulletin of the History of Medicine* 74, 3 (Fall 2000), pp. 433–57; Andrews and Scull, *Customers and Patrons of the Mad-Trade*, pp. 32 and 52.

[119] Leghorn MS Facs 589, 584B, Martin Folkes to Antonio Cocchi, 6 November 1735, British Library, London.

[120] *A Catalogue of the Genuine and Curious Collection of Mathematical Instruments...of Martin Folkes, Esq*, p. 3.

Lucretia had an equally tragic and dramatic life; beautiful and virtuous, she was raped by Sextus Tarquinus, son of the tyrannical Etruscan King at Rome. Her honour defiled, she stabbed herself to death. So too did Lucretia Folkes metaphorically disappear, mental illness causing her segregation and ultimate disappearance from larger society. With his wife in an asylum, it thus would not necessarily be surprising (if disappointing) that Folkes would have sought companionship with a Coram Hospital nurse, but the archives fall silent on this matter.

In 1740, Folkes faced another tragedy: his son Martin was killed in a riding accident at the age of twenty. His son's short life means that there is not a large amount of archival evidence about him. We do know that Folkes took particular care over his son's education. Just as his father had been, Martin was admitted as Fellow Commoner at Clare Hall, Cambridge on 29 October 1737. Over the years, Folkes had carefully nurtured his son's 'extraordinary taste for medallic [numismatic] knowledge', and frequently had him attend as a visitor to Royal Society meetings and gatherings with John Byrom at the Mitre Tavern afterwards to learn to debate and to drink.[121] In 1739, Folkes travelled with his son to France, 'chiefly with a view of seeing the Academies there, and conversing with the learned men'.[122] Folkes wished for Martin to practise his French and attain the discipline and accomplishments belonging to a gentleman of quality, a cosmopolitanism to add polish to his English education. Sir Thomas Browne wrote to his son in 1661, advising him to lose his *pudor rusticus* (rustic bashfulness) abroad by practising a 'handsome guard and Civil boldness which he that learneth not in France travaileth in vain'.[123] It seemed Folkes was taking much the same advice in educating his son.

Folkes ultimately and fatefully decided to place young Martin at the Caen Military Academy in Normandy, writing to his brother William on 23 April 1739 (NS):

> the Gentleman I like very well at the Academy...a plain man perfectly I believe understands his own business, and keeps the young gentlemen in exceeding good order...tho' it is quite in a genteel way, but I hope seeing others and some of much superior quality...will accustom [h]im to it. Martin makes the 36th English Gentleman now in the house and here are 18 ffrench besides the day scholars, and for their exercise here are about 40 horses some very good; the eating I believe will do...Drinking I think here is hardly any...and as there are so many other young people that do without it hope we shall, and he assures me there is no other play among them, but quadrille chiefly with the women in familys and that exceedingly low. I had rather there had been less of that but if there is no hazard...I should hope there can not be much harm done...the

[121] Stukeley, *Family Memoirs*, vol. 1, p. 99.
[122] BL Add MS/4222, ff. 25–6, British Library, London.
[123] Sara Warneke, *Images of the Educational Traveller in Early Modern England* (Leiden: Brill, 1995), p. 49.

Education is stricter than our University at least with fellow Commoners, and much more attention had to the regularity of their conduct...I only fear that Martin will think the allowance I intended to himself too large for the place as things are very cheap...the young Gentlemen who have been longest seem to like best which is I think a good sign. Martin was pleased I neither brought a governor nor an English servant and he has helpt us to a civil young fellow for a footman who has already servd one here that is just gone.[124]

Folkes's fears about adolescent mischief were well founded; later in the eighteenth century, while studying at the Caen Academy, 'Henry Pelham ran up a debt of £50 as a result of gambling on backgammon games at Harcourt'.[125] On the other hand, the Caen Academy had a good reputation for its training in geography, ballistics, and mathematics, as well as its riding school. In 1738, Montesquieu wrote to Folkes that he intended to take his son Jean-Baptiste there as he had made 'some progress' in natural philosophy, and hoped to see him in Caen when he visited, adding 'Je crois pouvoir vous dire cela, car quand on parle à son ami on parle à soi-même' [I think I can tell you that because when you talk to your friend, you talk to yourself].[126]

Folkes's close bond with his only son and heir, and his feeling that he was in generally safe hands at the Academy must have made the news of his son's fatal accident there—a fall from a horse—particularly heart-wrenching. His grief over the loss of his son was no doubt complicated and exacerbated by feelings of injustice—that this loss should simply never have occurred. Lennox was apparently the first to hear, as he had diplomatic contacts in France. On 2 August 1740, he wrote to Folkes's brother William:

I am so sho[c]k'd at this dreadfull piece of news that I find my self incapable of breaking it to your brother. besides my Wife is now in bed, & in danger of miscarrying, so I cannot take upon me tell it him, not knowing what the consequences may be. So I have told him you are very ill, & beg to see him. & you are certainly the only person that can properly break it to him. I am half distracted myself.[127]

[124] MS/MC/50-3-5, Martin Folkes, Caen to William Folkes, Chancery Lane, London, 23 April 1739, Norfolk Record Office, Norwich.

[125] BL Add MS/33127, Henry to Thomas Pelham, 6 November 1776, f. 102, British Library, London, as quoted in Jeremy Black, *France and the Grand Tour* (Houndmills: Palgrave, 2003), p. 136.

[126] Charles-Louis de Secondat, Baron de La Brède de Montesquieu to Folkes, 19 August 1738, in *Correspondance de Montesquieu*, published by François Gébelin with M. André Morize (Bordeaux: Imprimeries Gounouilhou, 1914), vol. 1, pp. 332–3. [J'espère de venir vous rendre visite l'année prochaine; je compte y mener mon fils qui s'applique aux sciences, et qui y fait même quelque progrès. Je crois pouvoir vous dire cela, car quand on parle à son ami on parle à soi-même].

[127] MS/MC/50-2-114, Letter from Charles Lennox to William Folkes, 2 August 1740, Norfolk Record Office, Norwich.

Folkes was clearly stunned, Lennox subsequently writing a concerned letter to Folkes's brother William about his friend's emotional state, "let me beg of you to write if it is butt two lines to let me know how he does', continuing that he loved and valued Folkes 'equal to any man living'.[128]

> I can't butt still be in very great pain for him. I don't like the account he gives me of his starts, and talking of the thing as if it had been a dream. I thinke those are bad symptoms. I hope in God you will be able to govern him and keep him low. for I can not help being in pain for his senses which strong as they are may not be proof against such a terrible shock.[129]

Folkes's friend the clergyman and writer Montagu Bacon (1688–1749) wrote to him soon after on 16 August 1740:

> It was with incredible sorrow that I heard of the sad accident happen'd in your family: at first it being only in one newspaper and not in the rest, we were in hopes that it might be a mistake...But seeing it confirmed...I can no longer defer troubling you with this. I cannot call it a consolatory epistle, because I want to be consoled by self, for the loss of so amiable a young gentleman: and I shall never forget how we embrac'd at parting. I won't enter into the topicks, which I might easily steal out of Seneca, Cicero, or Bunyan. You have seen enough of the world, I dare say, to know the fragility of human life; and to question your Philosophy would be contradicting all Mankind. Neither would it become me, sinking under so many wrongs, and calamities my self, to think of administring comfort to other People. The Lord hath given and the Lord hath taken away—You know the rest. I won't preach to You, but leave You to guess whither my thoughts would direct you for comfort. But I cannot but take notice, that You have still left two, as agreeable, as deserving, as virtuous young Ladies, I may say, whether 'tis my fondness or not, but I will venture to say, as the world can afford. In them, in your Philosophy, in the esteem You meet with every where, in your Brothers, in your books and your fortune, but above all in the God of all mercies be your comfort.[130]

Four years later, on 4 April 1744, Folkes also suffered a threat to his life, as he was mugged. Folkes wrote to Trembley that he had been assaulted at 7:30 p.m.,

---

[128] MS/MC/50-2-115, Letter from Charles Lennox to William Folkes, 3 August 1740, Norfolk Record Office, Norwich.
[129] MS/MC/50-2-116, Letter from Charles Lennox to William Folkes, 8 August 1740, Norfolk Record Office, Norwich.
[130] RS/MS/790/5, Royal Society Library, London.

only 20 steps, from my house where I was coming out. The rascals [coquins] were three people who threw me down on the ground, took away my golden watch with the chain, some trifle objects and the purse which contained about 10 guineas. They gagged my mouth and, when they spoke I heard one of them say: *Quiet him* and at that very moment I got hit with a stick which had some sort of lead in it…My daughter whom I had just left, heard the noise and screamed to my people to run out and help me when I came back to my door.[131]

Folkes suffered a puncture of the inner cartilage of the ear, and was seen by the surgeon.

Robbery was a fact of life in Georgian England.[132] In 1730, John Byrom turned a highway robbery outside of Cambridge into a comedic poem, dedicating it to Folkes:

> Dear Martin Folkes, dear Scholar,
>   Brother, Friend;
> And Words of like Importance without End;
> This comes to tell you, how in *Epping* Hundred,
> Last *Wednesday* Morning I was robb'd and plunder'd.[133]

In the poem, mischievously subtitled *Arma Virumque Cano* [I sing of arms and of the man], the opening words of Virgil's *Aeneid*, Byrom was saved when the Goddess Shorthand presented a note to the robbers, who mistook it for a magical incantation and fled. Byrom, as we recall, invented a system of shorthand he published and discussed with Folkes (see chapter four). In 1733, Lennox even went so far as to stage a fake highway robbery on the road from Portsmouth to London as a practical joke to tease Parson Sherwin, the canon of Chichester. 'For the upper class men who controlled clerical livings, teasing a clergyman…was a routine form of fun', and Lennox considered Sherwin especially tiresome.[134] Lennox wrote to Folkes:

> the Robbed were the Duchess of Richmond, Lady Tankerville, Lady Hervey, Mr. Fox, & Dr. Sherwin; butt the Robbers, were my self & Liegois, the whole scheme as you may imagine, for the Doctor. the three Ladys, & Fox, being in the secret.

---

[131] Letter of 2 April 1744 from Folkes to Trembley, Trembley Family Archives.

[132] See for instance, SP/36/19/54–58 'An account of a highway robbery', 19 June 1730, The National Archives, Kew.

[133] [John Byrom], *A Full and True Account of an Horrid and Barbarous Robbery, Committed on Epping-Forest, upon the Body of the Cambridge Coach, in a letter to M.F. Esq* (London: J. Roberts, 1728).

[134] Simon Dickie, *Cruelty and Laughter: Forgotten Comic Literature and the Unsentimental Eighteenth Century* (Chicago: University of Chicago Press, 2011), p. 180.

Every mortal now in England butt the Doctor know the joke, butt the Doctor to this minute knows nothing of it, nor does he smoke the least trick in it.[135]

Richmond related that Sherwin went on after the experience to describe his bravery, 'tho he was ready to beshit himself the whole time'. The Duchess recognized her husband and kept a 'straight face, and Richmond and his chums eventually made their way to London, where they divide[d] the spoils'.[136]

8.6 The Constancy of Friendship: Charles Lennox, the 2nd Duke of Richmond and Goodwood

Through all these ups and downs in Folkes's life, his friendship with Lennox, and their time together at Lennox's estate at Goodwood, was a constant. Lennox wrote to him repeatedly when he was on diplomatic missions, and reported family news, such as his wife Caroline's frequent pregnancies and miscarriages, and his children's illness. On one occasion, with some exasperation Lennox wrote to Folkes:

I have been almost always with a sick family for as soon as the Dutchess of Richmond, & the children met me at Brusells, Carolina fell ill of a fever, which had no intermission for three weeks, which confined us very disagreeably there, then as soon as we arrived here, Louisa fell ill, & this is also the first day since three weeks that the feaver has quitted her; besides all this, my wife being with Child & she is oblig'd to keep her room, so that if I had not pos'd my Doctors degrees at Cambridge, I should have thought my life for these last seven weeks, about as agreeable as that of a nurses.[137]

Although Lennox was one of the first aristocrats to become involved in formulating the rules of cricket, his own life in sport prematurely ended in 1732, as his leg had been broken 'by an Iron Back, which they burn wood against, falling on him'.[138] Lennox described his treatment to Folkes. William Cheselden (1688–1752); Claude Amyand (d.1740), Sergeant Surgeon to Kings George I and II; and William Harris (1647–1732), a pioneer in smallpox inoculation, and discussed their opinions on the case.

Cheselden was about to publish his landmark *Osteographia or the Anatomy of the Bones* in 1733, which used a *camera obscura* to provide an accurate, aesthetic,

---

[135] RS/MS/865/14, 11 October 1733, Richmond to Folkes, Royal Society Library, London.
[136] Dickie, *Cruelty and Laughter*, p. 180.
[137] RS/MS/865/5, Lennox to Folkes, Paris. 4 May N.S. 1729, Royal Society Library, London.
[138] *Daily Journal*, 22 May 1732.

and three-dimensional rendering of anatomy. Allister Neher has argued that Cheselden, considered the finest surgeon in Britain, used the camera obscura to portray anatomy as a means of 'drawing the images that lay on [Locke's] "tabula rasa" of unmediated human consciousness, as if produced by the "Crystalline Humour and Retina in the Eye" itself'.[139] His impulse was similar to that of Folkes, who as we recall discussed depth perception, optical illusion, and perspective with Algarotti, and who prized portrait busts by Roubiliac that promoted a Lockean mode of empiricist viewing (see chapter five). Apart from concern from Lennox, Folkes would have been interested to hear of Cheselden's professional activities.

On 26 September 1732, Lennox wrote to Folkes:

I had Chiselden upon the opening of my leg on Sunday, & upon examination, he & all the other Surgeons agreed the Callous was strong & confirm'd, so I have left of my splints, & have taken again to my crutches, & now use them every day; I can lift up my legg, turn it any way, & sett it to the ground, butt not to lay my whole weight upon it yet, however I am sensibly stronger & feel that I grow more so every day, & want nothing now butt to recover the tone of my muskles, which motion will I dare swear soon procure me. butt I have had some pretty acute pains, only within these four days, in the shin bone, about an inch below the fracture; which I do not at all like, for fear it should be a little detatch'd splinter of the bone, which if it is, must come out, & that would be both a troublesome & painfull job; some of the surgeons say it may be so, butt they all say they believe it to be only a little inflammation in the Periostium; butt Chiselden thinks the bone of the Tibia was split down thither in short it is a little bump, & very sore, & by fitts pretty painfull; butt in a few days we shall know what it is. Amyand says he is absolutely sure it is nothing butt a little inflamation in the Periostium, & that it will be quite gon in two or three days. the Legg is perfectly straite. Adieu Dear Martin; I hope you will soon be in town, & that youl favour me with your company at Goodwood, where I now really hope to be going to, in about a fortnight.[140]

Indeed, Cheselden was correct, and Lennox quickly resumed his duties, the *London Evening Post* reporting that 'yesterday his Grace took the Air for the first Time, before Dinner, in his Coach in Hyde Park. On Monday next his Grace will go to Court to pay his Duty to his Majesty and the Royal Family'.[141] Lennox spent the next five months convalescing at Goodwood, inviting Folkes to visit.

---

[139] Alistair Neher, 'The Truth about Our Bones: William Cheselden's Osteographia', *Medical History* 54 (2010), pp. 517–28, on p. 524.

[140] RS/MS/865/10, Lennox to Folkes, Whitehall, 26 September 1732.

[141] *London Evening Post*, 5 October 1732, as quoted in Timothy J. McCann, 'Two of the Stoutest Legs in England: The 2nd Duke of Richmond's Leg Break in 1732', *Sussex Archaeological Collections* 150 (2012), pp. 139–41, on p. 141.

Goodwood was a frequent gathering place for Folkes and his circle of natural philosophers. From the 1720s, the estate provided an informal salon and place of entertainment, experiment, and relaxation. Lennox asked Folkes on one occasion to 'pray bring a small reflecting Telescope with you, that I would have you buye for me; & some prisms, & any other things to Conjure with; & also the transaction in which there is an account of Santorini'.[142] There were activities for Folkes's children as well, and arrangements for his daughter 'Miss Lucrece' to go riding at Goodwood.[143] Lennox wrote to Folkes on 4 June 1742:

> Miss Lucrece has found so much benefit by rideing, I would have her by all means continue it, so the Duchess of Richmond, my daughter Caroline, my self & all of us here, beg you would come, & bring her with you, where she may ride her old friend punch every day; I am sure it will do her a great deal of good.[144]

When Folkes could not visit, Lennox sent him game from the Estate, in one case stating he took liberty to send them 'a warrant for a doe, they are never at Goodwood so fatt as the bucks, but they are very sweet venison'.[145]

Lennox also frequently invited a select group of their friends to his home, often using Folkes as an intermediary. On 9 August 1747, Folkes sent a letter of invitation to Emanuel Mendes da Costa (1717–91) to visit Goodwood[146] (see figure 8.5). Da Costa was an émigré Jewish conchologist, geologist, and antiquary born into the nomadic population of Portuguese Sephardim that fled the

Fig. 8.5 Portrait of Emanuel Mendes da Costa, LMA/4553/01/06/001, London Metropolitan Archives, City of London Corporation.

[142] RS/MS/865/11, Lennox to Folkes Goodwood 12 November 1732, Royal Society Library, London.
[143] RS/MS/250/133, Royal Society Library, London.
[144] RS/MS/865/22; RS/MS/865/3, Royal Society Library, London.
[145] RS/MS/250/133, Royal Society Library, London.
[146] RS/MS/849, vol. 4., Royal Society Library, London. Also in: Folkes to Mendes da Costa, 9 and 28 August 1747, in Nichols, *Illustrations of Literary History*, vol. 4, pp. 635–7.

Inquisition.[147] He was a leading collector in the crux of a transition in natural history, which was moving from the baroque tradition of constructing natural history collections as cabinets of curiosities in the country house or gentleman's collection to the Enlightenment passion for order and taxonomy that flowered in the work of Linnaeus. Da Costa formulated his own system of taxonomy in his publications *Elements of Conchology* and *British Conchology*, and took Linnaeus to task for changing the names of species and genera. For instance, Da Costa thought it was incorrect that the shells known as volutes 'from à volvendo, or rolled up' were changed by Linnaeus to Conus [cone].[148] In his popular *Natural History of Fossils* (1757), Da Costa also 'took issue with Linnaeus's quintuple nomenclature (kingdom, class, order, genus, species) and had created his own classification which ran in the following order: series, chapter, genus, section and number'.[149]

When Folkes contacted him, Da Costa had just spent time going mineral prospecting in Derbyshire. Da Costa's cousin, Joseph Salvador (1716–86), was the first and only Jewish Director of the East India Company and head of a trading empire in diamonds, coral, and ostrich feathers.[150] Da Costa utilized these same networks for the trading of flora and fauna to advance knowledge in natural history. To pique Da Costa's interest, Folkes first mentioned that Lennox, 'had just founded a wild receptacle for fossils in his garden: he has transplanted from the coast large fragments of Rock rich in all sorts of fossil shells, he has interspersed corals and other marine productions, and a spacious grotto where nature has furnished a vast variety of shells…an entertainment much to your taste'. As Lennox was 'desirous that you might impart to him your observations', Folkes proceeded to give him directions to Goodwood, 'one of the pleasantest places, and the best company you can possibly meet…The Duke being the most honourable and the best man living'. Folkes, ever the tolerant freethinker, then assured Da Costa there was food 'without breach of the Laws of Moses, unless the lobster of Chichester should be a temptation, by which a weaker man might be seduced'. Folkes also mentioned 'there is also a Chaplain I should suspect originally of your nation, for he talks hebrew almost naturally, and will not wish to turn you any more than myself'. Folkes later that month wrote to Da Costa, that 'we are all citizens of the world, and see different customs and tastes without dislike or prejudice, as we do different names and colours'.[151]

---

[147] My PhD student Aron Sterk at the University of Lincoln is currently finishing his second doctorate, a Dissertation on Da Costa. See our jointly curated slideshow with Google Arts and Culture and The Royal Society: https://artsandculture.google.com/exhibit/emanuel-mendes-da-costa-1717–1791/gAKCb0daZKtiIA

[148] My thanks to Dr Aron Sterk for this information.

[149] Matthew Eddy, *The Language of Mineralogy: John Walker, Chemistry and the Edinburgh Medical School, 1750–1800* (Aldershot: Ashgate, 2008), pp. 98–9.

[150] Tijl Vanneste, 'The Eurasian Diamond Trade in the Eighteenth Century: A Balanced Model of Complementary Markets', in M. Berg, F. Gottman, H. Hodacs, C Nierstrasz, eds, *Goods from the East, 1600–1800: Trading Eurasia* (New York: Palgrave Macmillan, 2015), pp. 139–53.

[151] Folkes to Da Costa, 28 August 1747, in Nichols, *Literary Anecdotes*, vol. 4, p. 635.

Da Costa's reputation as an expert in 'the mineral and fossil part of the Creation' led to his election as Fellow of the Royal Society on 26 November 1747, sponsored by Folkes and the Duke of Montagu.[152] Da Costa would in 1763 become the first Jewish clerk of the Royal Society, although he was not the first or only Jewish member. Unlike many British institutions, the 'Royal Society did not require a new member to take a religious oath (although Council members were not exempt from this). By the mid-eighteenth century its membership included Anglicans, Catholics, Dissenters, Jews and atheists'.[153] His cousin Joseph Salvador was elected FRS in 1759, as was Naphtali Franks (1715–96), an emigrant from New York working as a diamond trader, who also did significant work in botany. Dr Jacob de Castro Sarmento (1692–1762) was also FRS, a physician and *converso* who left Portugal 'with the onset of a new wave of persecution of New Christians in 1720'.[154] Sarmento translated Newton's ideas into Portuguese for the benefit of his countrymen; he also sent Portugal's first microscope to the University of Coimbra from London.[155] He later renounced his Sephardic beliefs in the late 1740s and became a Deist, perhaps influenced by the freethinking atmosphere at the Royal Society encouraged by Folkes.[156]

8.7 Friendship Beyond the Grave: Folkes and John, 2nd Duke of Montagu

Folkes was also a frequent visitor to Ditton Park in Buckinghamshire and Boughton House, Northamptonshire, the properties of John, 2nd Duke of Montagu. When Folkes had just left for his Grand Tour, John wrote 'you cant wish so much to be att home as all your friends desire to have you…everybody [at Ditton] wishes for you and all think you would not deslyke [dislike] to be one of the Company,' before passing his regards to Folkes's son, 'the young philosopher'.[157] Boughton was modelled after a French chateau, its gardens laid out in the style of Versailles, which would have appealed to Folkes's Francophilia. The cosmopolitan Montagu was a participant in Folkes and Trembley's polyp experiments, expert in ordnance, and an antiquary with expertise in heraldry, having a prominent role in the revived Order of the Bath. Montagu was also fascinated by Chinese culture, participating in the Georgian penchant for Chinoiserie just as Lennox did with his collection of Chinese goldfish. Montagu went so far as to have a Chinese pavilion of oilcloth and timber constructed as a

---

[152] RS/EC/1747/11, Royal Society Library, London.

[153] Geoffrey Cantor, *Quakers, Jews, and Science: Religious Responses to Modernity and the Sciences in Britain, 1650–1900* (Oxford: Oxford University Press, 2005), p. 104.

[154] Cantor, *Quakers, Jews, and Science*, p. 106.

[155] My thanks to Dr Aron Sterk for this information, as well as for his rediscovery of Mendes Da Costa's portrait in the London Metropolitan Archives.

[156] Cantor, *Quakers, Jews, and Science*, p. 107.

[157] RS/MS/790, letter from Montagu to Folkes, 20 November 1733, Montagu Folder, Royal Society Library, London.

'tea pavilion' in 1745 for his riverside terrace at Montagu House, Whitehall (later the site of the British Museum); a small portion of it was portrayed by Canaletto in his contemporary portrait of the Thames, a subtle nod to the aristocratic presence of one of his patrons.[158] Montagu's interests were not confined to Chinese architecture. He subsequently wrote Folkes this missive only dated 'Fryday night':

Tomorrow morning is not a bad day the Experiment of rowing a ship with Chinese Sculs [oars] will be tried on the River over against my house at about 11 a clock, where you shall be welcome to see it, and if you please dine with me afterwards. Montagu.[159]

Like Lennox, Montagu was Folkes's close friend. When Folkes's daughter Lucretia was ill, Montagu sent along a medical powder to him remarking 'the Dutches of Montagu says he has given it to above twenty people and never knew it faile…or do any hurt, if you should have any reason to think your daughter is quite well before she has taken the whole quantity, then she may leeve its use'.[160] Montagu also arranged for Folkes's younger brother William to be appointed agent to pay the garrisons in the Isle of Wight, and William served as Montagu's agent.[161]

After the Duke's sudden decease of a violent fever on 5 July 1749, Folkes subsequently served as a witness to the Duke's will and executor to his estates, arranging with Roubiliac for the construction of a monument at St Edmund's Church, Warkton[162] (see figures 8.6 and 8.7). Stukeley wrote to Maurice Johnson of the Spalding Gentlemen's Society in 1749, 'Mr Folkes and I walked to Kentish Town to take our last leave of my great Patron'.[163] With the letter, Stukeley enclosed a 'meditation…on seeing the Duke of Mountague's herse setting forward from London for his Interrment at Warkton in Northamptonshire 18 July 1749'. Stukeley wrote in his autobiography: '5th July the year following the Duke Dyed, the most regretted of any subject in England. The Dr. went to pay his last respects to his great patrons remains passing thro' Kentish town and sprinkled his herse with woodbine flowers'.[164]

[158] 'Canaletto, London: The Thames and the City of London from Richmond House, 1747', in The Duke of Buccleuch, *Vistas of Vast Extension*, Catalogue, Boughton House, 2017.
[159] RS/MS/790, Montagu Folder, Royal Society Library, London.
[160] RS/MS/790, Montagu Folder, Royal Society Library, London.
[161] RS/MS/790, Letter from Montagu to Folkes, 20 November 1733, Montagu Folder, Royal Society Library, London; papers of William Folks as agent of the Duke of Montagu, MC 50, Norfolk Record Office.
[162] RS/MS/790, Montagu Folder. The Duke wrote Folkes an undated letter indicating 'You may remember you was last year witness to my will, and as sum small purchases I have made lately makes it necessary for me to execute a new one I must by the same favour of you, and that you would be with me tomorrow morning by nyne a clock for that purpose. Montagu'. More about Folkes's role in the Montagu monument can be seen in BL Add MS 35397, Thomas Birch to Lord Hardwick, 8 September 1750. See also Tessa Murdoch, 'Roubiliac as an Architect? The Bill for the Warkton Monuments', *Burlington Magazine* 122, 922 (January 1980), pp. 40–6, on p. 40.
[163] Honeybone, *The Correspondence of William*, p. 136.
[164] William Stukeley, 'Autobiography', *The Family Memoirs of the Rev. William Stukeley*, p. 57. Stukeley's paean to Montagu was in book two of his *Medallic History*, pp. xix–xxii.

Fig. 8.6 Monument of John, Second Duke of Montagu by Louis-François Roubiliac, St Edmund's Church, Warkton, Northamptonshire, 1752. Photograph by Ian Benton.

Folkes and his brother William acted on the behalf of the Duke's widow, Lady Mary Churchill, who entrusted the details of the monument to him. Thomas Birch reported in September 1750 that:

> The Duchess of Montagu has impos'd upon Mr Folkes the whole care of a monument for the Duke her Husband in a country Church about two miles from Boughton. But her Grace would neither give him Instructions herself not allow him to consult any other person about it nor see any of the models which he had prepared. However, she has already sent him [Mr Folkes] £1000 for the Expense of it. He [presumably Folkes] intends to introduce a figure of the Duchess in the Groupe. Roubilliac is the Scu[l]ptor, who has gain'd new Reputation by the Mo[nu]ment of General Wade.[165]

---

[165] BL Add MS 35397, Thomas Birch to Lord Hardwich, 8 September 1750, British Library, London; Murdoch, 'Roubiliac as an Architect?', p. 41. See also David Bindman and Malcolm Baker, *Roubiliac and the Eighteenth-Century Monument: Sculpture as Theatre* (New Haven and London: Yale University Press, 1995), p. 298. Philip Lindley, 'Roubiliac's Monuments for the Second Duchess of Montagu and the Building of the New Chancel at Warkton in Northamptonshire', *The Volume of the Walpole Society* 76 (2014), pp. 237–288.

Fig. 8.7 Louis-François Roubiliac, Monument of John, Second Duke of Montagu by Louis-François Roubiliac, St. Edmund's Church, Warkton, Northamptonshire, 1752. Detail of Lady Mary. Photograph by Ian Benton.

The Duke had left £1,000 to Lady Mary for mourning in his will, and it seems this money was transferred to Folkes, something collaborated by the Account Book at the Duke's estate at Boughton, which records £500 being paid on 8 November 1751, 'By part of the £1000 deposited with my Brother Folkes by the late Ds of Montagu to be laid out in a monument of the Duke'.[166] William Folkes, Martin's younger brother, was John, Duke of Montagu's factor for his estates. The *Dublin Courant* reported in December 1749 that 'Roubillac [sic] is preparing a noble monument to his grace the Duke of Montagu, the Device of which is Charity erecting a Shrine to his memory and Fame applauding her'.[167] Fame would eventually be replaced by the figure of Lady Mary, perhaps due to Folkes's intervention.[168] The widow's constant vigil in sculpture also commemorates her own life and transcends it.

[166] Bindman and Baker, *Roubiliac*, p. 298.
[167] As quoted by Louise Allen, *St. Edmund's Church and the Montagu Monuments* (Oxford: Shire Publications, 2016), p. 23.
[168] Allen, *St. Edmund's Church*, p. 23.

Fig. 8.8 Louis-François Roubiliac, Monument to Lady Mary, Duchess of Montagu (detail), St. Edmund's Church, Warkton, Northamptonshire, 1752. Photograph by Ian Benton.

Roubiliac's use of gesture and physiognomy in Lady Mary's figure to express her grief is set within a 'sequence of actions, either shown or implied' in a monument that would have appealed to Folkes's familiarity with the theatre (see figure 8.8).[169] Although we do not know how much Folkes attended the theatre due to any possible discomfort of his wife revisiting the scene of her formal career, or her illness, he did retain a substantial number of books about the theatre in his library. These included Luigi Riccoboni's *Histoire du theatre italien* (1728), François-Hédelin D'Aubignac's *Térence justifié* (1656), which also examined the history of the ancient theatre, Shakespeare's first folio, Theobald's edition of *Shakespeare restor'd* (1726), five separate editions of the plays of Corneille, the complete works of de la Motte (1730), and some operatic libretti, including the *opera seria P.C. Scipione* (1726) composed for the Royal Academy of Music by Handel, its dedication to Lennox signed by the librettist Paolo Rolli.[170]

---

[169] Bindman and Baker, *Roubiliac*, p. 37.
[170] *A Catalogue of the Entire and Valuable Library of Martin Folkes* (1756).

Roubiliac's stunning rococo design of the monument as well as its installation meant that the medieval chancel in Early English Style at St Edmund's had to be rebuilt in Georgian style to accommodate the tomb sculptures. Ever thrifty, Roubiliac used fragments of the medieval chancel in the monument's core structure.[171] As the Duke, along with Folkes, was a forefather of the Foundling Hospital to the poor in London, Roubiliac included a figure of Charity which stands to the right of the monument, holding the Duke's sculpted portrait as a weeping child snuffs out the candle of life. The work is thus a sculpture within a sculpture. Lady Mary turns plaintively to the portrait of her husband, holding the Lesser George of the Order of the Garter Star which he received in the ceremony to which Folkes was squire in 1725. The Duke's service as Master-General of the Ordnance was reflected in the cannon, cannonballs, and gunpowder barrel that Roubiliac sculpted, along with Fame's trumpet and a military standard, but these play a muted role. In the eyes of his widow, his private charitable works were more important than his public office.[172]

Lady Mary died in May 1751, and Roubiliac was also commissioned to sculpt her monument under William Folkes's administrative direction and perhaps Martin Folkes' aesthetic guidance. (see figure 8.8). Malcolm Baker speculated that Martin Folkes may have been influenced by a book in his library collection, Tommaso Porcacchi's illustrated *Funerali antichi* (1591), to direct the sculptor to include iconography that re-enacted 'in their imagery the rites of homage associated with such temples... Both the hanging of the Duke's medallion on the 'temple' and the adorning of the Duchess's urn with flowers may have to be read in this way'.[173] In Lady Mary's monument, Roubiliac portrayed the Three Sisters of Fate cutting the Duchess's life short, which was overtly pagan iconography— something the freethinking Folkes would have accepted, but a choice that was not without its critics. A writer (likely Connell Thornton, a man of moralistic Low Church opinions) for *Connoisseur* in 1755 complained, stating:

> Our pious forebears were content with exhibiting to us the usual emblems of death, the hourglass, the skull, and the cross-marrow-bones, but these are not sufficient for our present more refined age. The Three Fatal Sisters, mentioned in Heathen Mythology, must be introduced spinning, drawing and cutting the thread of Life'.[174]

The use of the Fates may have appealed to the antiquarian Folkes because of their pagan antiquity and because the trope of the Three Fates was a common one in eighteenth-century French clocks. As Tessa Murdoch has indicated, the Duke of

---

[171] Allen, *St. Edmund's Church*, p. 20.   [172] Craske, *The Silent Rhetoric of the Body*, p. 289.
[173] Bindman and Baker, *Roubiliac*, p. 298. For William Folkes's role, see Lindley, 'Roubiliac's Monuments', p. 247.
[174] As quoted by Bindman and Baker, *Roubiliac*, p. 74.

Montagu had a longcase clock attributed to André-Charles Boulle in his house in Whitehall, where the Three Fates appeared in ormolu in front of the dial, and it was crowned with Father Time.[175] Both Folkes, his brother William, and Roubiliac could have seen it there, as Martin 'would have known the Second Duke through his membership of the Freemasons', and William was the Duke's financial steward.[176]

Memorializing the dead was also part of a tradition in Folkes's family, but in a more unique manner. Folkes was in frequent contact with his 'cousin' Mr Martin Challis, his rector at his Hillington estate from 1728 until 1757.[177] Challis wrote to Folkes on Christmas Eve 1735, relating a ghost story that was apparently told during Christmas, and it was also bound with the theme of precision clocks.[178] William Folkes appended a note to the letter attesting to the validity of the account. The premise was that Martin Challis's brother Thomas, who was sleeping at an inn at Bury St Edmunds, was awakened by the apparition of a shrouded woman, who predicted the hour of his death: 'Three times Nine & Twice Three your Age shall be & then you shall be Dust like me'. The next morning Thomas was informed that the parish bell tolled at the exact moment signifying the woman's death. As Moore relates:

> Naturally the augury gradually preys on his mind until the appointed hour of his thirty-third birthday when in the midst of a party he expires, fulfilling the weird woman's prophecy to the minute. And it is that last detail which makes the story fascinating. On the surface it is a tale of the uncanny: but beneath that is a text about the ability to measure time. The manuscript dates from 1735 and it describes events supposed to have taken place in the narrator's childhood, perhaps at the turn of the eighteenth century. Our first measure of time is the parish bell tolling in the fashion of Shakespearean chimes at midnight. But years later, the traveller's death is in a private house: here, the gentlemen present check their pocket-watches and a house-clock '& 'twas agreed by all these present that he expired within a minute or two of the time of his Nativity which if I mistake not was between the Hour of Eleven & Twelve'.[179]

---

[175] Tessa Murdoch, 'Spinning the Thread of Life: The Three Fates, Time and Eternity', in Diana Dethloff, Tessa Murdoch, Kim Sloan and Caroline Elam, eds, *Burning Bright: Essays in Honour of David Bindman* (London: UCL Press, 2015), pp. 47–54, on pp. 52–3. André-Charles Boulle, *Clock case with the Three Fates, c.1690–1700*. Oak, ebony, brass, and tortoiseshell with gilt bronze figures, 22.6 cm × 37.7 cm × 20.7 cm, dimensions of upper clock case (Buccleuch Collection, Boughton House).

[176] Murdoch, 'Spinning the Thread of Life', p. 52.

[177] Challis was the son of a Warrington mercer and a graduate of Gonville and Caius (BA 1703–1704; MA 1707), having previously served as a vicar at Elm, Cambridgeshire and Emneth, Norfolk. See John Venn, *Biographical History of Gonville and Caius College, 1349–1713* (Cambridge: Cambridge University Press, 1897), p. 504.

[178] This manuscript was first noted by Keith Moore. See Keith Moore, 'Chimes at Midnight', Repository Blog, The Royal Society, 22 December 2014. https://blogs.royalsociety.org/history-of-science/2014/12/22/chimes-at-midnight/ [Accessed 19 November 2019] The ghost story is in RS/MS/790/23/1.

[179] Moore, 'Chimes at Midnight'.

As we have seen, Folkes had a penchant for George Graham's watches and other accurate timepieces; the auction inventory of his possessions showed he left a silver stop watch by Graham, another by William Garfoot, and an eight-day clock by Daniel Quare, as well as a 'most curious sidereal clock by Graham, with a compound pendulum, that goes a month, the banks in the arch go by equal time, and those below by sidereal time'.[180] George Graham's clocks achieved an accuracy of a second each day utilizing temperature compensation, and the ghost story related by Challis only works because of the development of accurate timepieces. The story would have been of great appeal to Folkes because of his love of instrumental precision. But with the close of the decade of the 1740s, although he did not know it, Folkes himself was running out of time.

---

[180] *A Catalogue of the genuine and curious collection of mathematical instruments, gems*, p. 6 and p. 8.

9

'Charting' a Personal and Institutional Life

9.1 President of the Society of Antiquaries of London

In August 1750, Charles Lennox died, interred in full panoply at Chichester Cathedral, Folkes in the procession. Folkes perhaps recalled a happier ceremony twenty-five years earlier, when he served as Lennox's squire in the installation of the Order of the Bath. A contemporary publication eulogized Lennox:

> He was polite, affable, and generous; a man of strict honour, and was greatly admired at the Courts of Europe which he visited for the eminent qualities of mind which he possessed. He was an amiable father, and so worthy a nobleman that he never lost a friend nor created an enemy, even when political rage seemed to animate every breast; and he was a patron and admirer of The Fine Arts.[1]

That year, the artist John Faber the Younger dedicated to Folkes his mezzotint of a painting of Charles II by Sir Peter Lely portraying the sovereign wearing the robes of the Order of the Garter; not only was Charles II patron of the Royal Society, but he was Lennox's grandfather.[2] It is likely Folkes had commissioned the piece in commemoration of his friend and their work together in the Royal Society.

On 25 August 1751, the grief-stricken Lady Sarah, Lennox's beloved 'Taw', suddenly passed away—death by a broken heart. As we recall, the Duke of Montagu had also died in 1749, and Folkes had had the unhappy task of commissioning his and his wife's monument from Roubiliac. The amiable gatherings at Goodwood and Ditton and the succour of genial friendship and stimulating conversation was no more.

---

[1] The Renaissance Duke. Catalogue to 2018 exhibition, Goodwood House, Sussex, p. 16. https://www.goodwood.com/globalassets/venues/goodwood-house/summer-exhibition-2017.pdf [Accessed 9 October 2019].

[2] *Charles II* (John Faber the Younger after Peter Lely, 1750). Lettered below the image with the title and a dedication to Martin Folkes in Latin: 'Carolus II. Magnae Britanniae, Franciae et Hiberniae Rex/ Regalis Societatis Fundator et Patronus / Viro Doctissimo Martino Folkes Praesidi, Concilio et Sodalibus Regalis Societatis Londini, Hanc Regis Caroli Effigiem, Accurate Expressam, Eá, quá par est, Observantiá/ D.D./ Johannes Faber 1750 / Ex Tabulá Archetypá Petri Lely, Equitis Aurati in Aedibus Illustrissimi Principis, de Richmond, Lennox et Aubigny Ducis, Conservatá'. British Museum, London, 1902, 1011.1300.

Martin Folkes (1690–1754): Newtonian, Antiquary, Connoisseur. Anna Marie Roos, Oxford University Press (2021).
© Anna Marie Roos. DOI: 10.1093/oso/9780198830061.003.0009

William Folkes, who served as an executor for Richmond's estates, quickly arranged for the guardianship of their children, including their fifteen-year-old son and heir. William wrote to his brother Martin worriedly,

the Death of the Dutch[es]s during the Extreme Infancy of the Children is an Event I have dreaded tho' very little expected, looking upon her life as a very good one, her Capacity for the Management of the late Duke's Affairs, and the Interest she had therein was a great Satisfaction and Ease both to Mr Hill and Myself, in so much that tho we are both Executors with her to the Duke, yet we have neither of us acted any further therein than in concurring with her, in what she desired to have done, and I doubt not but she has kept a very faithful and just Account and made the most of everything for herself and the Children...how this can be carried on now I am at a Loss to know, unless with the Concurrence of the family.[3]

In autumn 1750, the engraver George Vertue noticed Martin Folkes's subsequent depression, 'occasiond by reflection of the loss of his good & noble Friends lately. Earl of Pembroke[,] the Duke of Montague & lastly the Duke of Richmond—these were noble worthy Friends...such as gave life and spirit to his Studies his amusements and conversation.'[4] Even William Stukeley commented on the spate of deaths of those in Royal Society circles: 'the few old acquaintances I had left here are dropping off, every day. Lord Pembroke was well on Sunday, our president Folkes din'd with him, on Tuesday 6 a clock, he [Pembroke] was dead'.[5]

It was some small comfort that on 22 November 1750, Folkes was elected to succeed Lennox as President of the Society of Antiquaries of London, a position he held until his own death in 1754. He now had the historical distinction of being the only sitting President of both the Antiquaries and the Royal Society, and he began his term of office in the Antiquaries by attempting to get them a Royal Charter, which was granted in November 1751. Having a Royal Charter for the Society of Antiquaries had been proposed long before: 'indeed, [Humphrey] Wanley appears to have anticipated petitioning for incorporation in 1708, only to be deterred by the dismissal of his patron, Edward Harley, from office'.[6] On 26 April 1750, when Lennox was still President, a committee had considered incorporation again, and 'the duke counselled caution on the grounds of cost'.[7] As Pugh indicated,

---

[3] Goodwood MS 112, William Folkes to Martin Folkes, 8 September 1751, West Sussex Record Office.

[4] BL Add MS/23096, f. 23v, British Library, London; George Vertue's copybook as quoted by George S. Rousseau and David Haycock, 'Voices Calling for Reform: The Royal Society in the Mid-Eighteenth Century: Martin Folkes, John Hill and William Stukeley', *History of Science* 37, 118 (1999), pp. 377–406, on p. 393.

[5] Stukeley to Maurice Johnson, 16 February 1750, SGS/Stukeley/23, in Honeybone, Diana and Michael, eds, *The Correspondence of William Stukeley and Maurice Johnson, 1714–54* (Woodbridge: Boydell Press, 2014), p. 141.

[6] Rosemary Sweet, *Antiquaries: the Discovery of the Past in Eighteenth-Century Britain* (London: Hambleton, 2004), p. 87.

[7] R. B. Pugh, 'Our First Charter', *The Antiquaries Journal* 62 (March 1982), pp. 347–55, on p. 347.

Nevertheless when the Society met next, in seven days' time, it was there and then decided that Henry Rooke, keeper of the records in the Rolls Chapel and himself an antiquary, should consult one Grub (or Grubb), clerk of the patents to the law officers. Only one week passed before he had done so and had secured an estimate of cost. A committee of eleven was thereupon chosen and told to meet on 16th May, that is in less than a week therefrom.[8]

Richmond subsequently died, and 'nothing further occurred until 28th February 1751. [Attorney] Philip Carteret Webb then read the prepared text of a charter to all the members'. Folkes then convened a meeting on 14 March to vote on the need for incorporation. 'In all these activities and posts [Folkes] displayed commitment, energy, and goodwill'. And, at first 'his antiquarian colleagues had complete confidence when electing him that he would, like Richmond before him, represent their best interests'.[9]

Unfortunately for Folkes, it proved not to be such a simple process. Not all of his colleagues agreed with his decision to incorporate, as some of the older Fellows of Antiquaries feared being amalgamated into the Royal Society. They remembered that in 1728, after the Society of Antiquaries were at a low ebb in membership and funds, they 'renewed their meetings at the Mitre Tavern in Fleet-street, having so far complied with the desire of those gentlemen who were also Members of the other [Royal] Society, as to fix them to Thursday evening, after the Royal Society had broke up'.[10] The Mitre was conveniently proximal to the Royal Society headquarters in Crane Court, also off the Fleet, and became the site of the Royal Society Dining Club. The resulting revitalization of the Society of Antiquaries led some members to consider that perhaps the two organizations should be united. The consolidation of resources made some financial sense, and at the time it was thought amalgamation was far cheaper than the legal expenses involved in getting the Society of Antiquaries its own Charter.[11] Indeed, after a meeting of the Antiquaries on 27 November 1729, George Vertue recorded that such 'an attempt was made to unite 'em with the RS, but fail'd'.[12]

Although Folkes was at this date not part of the Royal Society Council due to his defeat against Sloane in the presidential election, he was still a frequent attendant at Antiquaries meetings, having been elected a Fellow in 1720. It is not entirely clear from archival evidence that Folkes was the moving force behind such an amalgamation in 1729, but it would have served his combined interests in natural

[8] Pugh, 'Our First Charter', p. 347.
[9] Rousseau and Haycock, 'Voices Calling for Reform', p. 382.
[10] 'Introduction: Containing an Historical Account of the Origin and Establishment of the Society of Antiquaries', Archaeologia (London, 1770), p. xxxviii.
[11] Rousseau and Haycock, 'Voices Calling for Reform', p. 381. See also David Boyd Haycock, '"The Cabal of a Few Designing Members": The Presidency of Martin Folkes, PRS, and the Society's First Charter', Antiquaries Journal 80 (2000), pp. 273–84.
[12] Joan Evans, A History of the Society of Antiquaries (Oxford: Oxford University Press, 1956), p. 83, as quoted in Rousseau and Haycock, 'Voices Calling for Reform', p. 382.

philosophy and antiquarianism beautifully. At the time, Folkes may have also thought a bloc of Antiquaries supportive of his aims could have served as a counterweight to Sloane's presidential power.

When Folkes became President of the Society of Antiquaries in 1750, he thus faced opposition to obtaining a Royal Charter for Antiquaries from older members of the Society, who remembered how they were almost subsumed into the Royal Society. That year, William Stukeley did not attend Antiquaries in silent protest, writing rather disparagingly to Maurice Johnson, 'I come home directly from Crane Court and taking my contemplative pipe, I minute down, what I remember, of all that passes. This is one reason, why I never go to the antiquarians, who have foolishly altered their meeting, to the same night. So that by mixing two entertainments, they remember nothing distinctly of either'.[13]

Stukeley was keeping busy enough without attending the Society of Antiquaries. He wrote to Maurice Johnson on 15 May 1750:

> We have here 3 friendly conversations in the week where a collection of us meet at 5 in the afternoon; over a dish of thea, & nothing else. One at Dr Hills, son to Dr Hill of Peterborough; Another at Dr [James] Parsons in redlyon square. Another at Mr Sherwoods, Devonshire street, surgeon; beside Dr Mortimer on a fryday, in queen square, which is a sort of echo to the Royal Society, so we have literal conversation enough.[14]

Stukeley was referring to Dr Theophilus Hill, one of his mentors, who was the father of John Hill, who had disparaged Folkes's Presidency of the Royal Society; surgeon James Parsons lived just south of Stukeley's rectory in Queen Square; Sherwood was Dr Noah Sherwood FRS, and Cromwell Mortimer, of course, was secretary to the Society. It is not particularly surprising, due to Stukeley's enmity towards Folkes for his freethinking beliefs, that he was meeting regularly with John Hill and his father. Even by April 1751, Stukeley wrote again to Johnson, 'the antiquaries are in high spirits in hope of obtaining a charter. I never go to their meetings. tis absurd to run from Royal Society to a new kind of entertainment, whereby both are tumbled out of our mind'.[15]

Folkes had other sources of opposition to the Charter. Numismatists George North and Andrew Coltée Ducarel, who became keeper of the Lambeth Library, and George Vertue also opposed incorporation. (North's opposition must have caused Folkes particular personal pain, as Folkes had considered him his friend and given him open access to the Royal Society Library, presumably for research

---

[13] Stukeley to Johnson, 16 January 1750, SGS/Stukeley/23, in Honeybone and Honeybone, *Correspondence*, p. 141.

[14] Stukeley to Johnson, 15 May 1750, SGS/Stukeley/25, reprinted in Honeybone and Honeybone, *Correspondence*, p. 150.

[15] Stukeley to Johnson, 13 April 1751, SGS/Stukeley/27, reprinted in Honeybone and Honeybone, *Correspondence*, p. 167.

for his monograph on Arabic numerals.[16]) Antiquary and Cambridge clergyman William Cole (1714–82) recorded Ducarel's remarks about the matter: 'The Patrons of the Charter were possibly Mr Folkes, but I am certain Mssrs Theobolds, Cesteret, the York Family, and Lord Willoughby of Parham…were the chief movers in it…the Charter was obtained contrary to the Sense of the Majority of the Members, and more to please…[Folkes's] Vanity, than to answer any other purpose'.[17] Cole continued:

How that may be I won't pretend to say: Dr Ducharel I believe was no Favourite of Mr Folkes; and so Allowances are to be made. I well know, that the older members of the Society, as Dr Rawlinson, Mr Willis, Mr Vertue, and among them also myself, tho' a young Member of it,[18] were very averse to be incorporated, and thought the way they were in then, preferable to what was designed. Dr Rawlinson, in particular, a most useful member, was so disgusted at their new Proceedings that he gave to the University a noble legacy he had willed to the Society.[19]

In 1741, Rawlinson had been excused by Folkes for not paying his Royal Society dues, as he faced difficulties settling his brother Thomas's indebted estates, and they both moved in Masonic circles. Folkes also helped broker a parcel of medals that Rawlinson ordered from Italy, so he could have seen Rawlinson's decision as a decided snub.[20] Cole, however, may have been overstating his case. It is true that

to the Bodleian, that 'sanctuary for use and curiosity' as he described it, Rawlinson bequeathed all his manuscripts, charters and seals, as well as a selection of his printed books. To St John's College, Oxford he bequeathed the bulk of his landed property, some of his books, and his coins and medals. In addition to his earlier endowment of a chair in Anglo-Saxon at Oxford, he made provision for the salary of the keeper of the Ashmolean Museum.[21]

However, when the Society of Antiquaries, 'having elected Rawlinson vice-president in 1753, in 1754 removed him from the council on account of his Jacobite

[16] MS Y.c. 946 (1), Martin Folkes to Francis Hauksbee, 10 July 1746, Folger Library, Washington DC.
[17] BL Add MS/5833, f. 158v., British Library, London.
[18] Cole was elected to the Society of Antiquaries in 1747.
[19] BL Add MS/5833, f. 158v., British Library, London.
[20] CMO/3/89/1, 14 January 1740/1, List of members of the Royal Society who have been excused from payments to the Society, Royal Society Library, London; Bodl. MS Raw. N.6. Numismatic Collections, Letter of Martin Folkes to Richard Rawlinson, 2 April 1743, Bodleian Library, Oxford.
[21] Mary Clapinson, 'Rawlinson, Richard (1690–1755), Topographer and Bishop of the Nonjuring Church of England', Oxford Dictionary of National Biography. 23 Sep. 2004; https://doi.org/10.1093/ref:odnb/23192 [Accessed 1 December 2019]. Rawlinson was made a Freemason at Lodge No. 37, meeting at the Sash and Cocoa Tree, eventually becoming Worshipful Master of Lodge No. 43, which met at the Rose, Cheapside, London. See John Cherry, Richard Rawlinson and his Seal Matrices (Oxford: Ashmolean, 2006), p. 37.

loyalties,…Rawlinson revoked his bequest to the antiquaries'.[22] Political belief, rather than the Antiquaries Charter itself, seemed to be behind Rawlinson's decision in making his decisions about his estate.

George Vertue's letter of 15 March 1751 to Ducarel also tells a slightly different story from Cole about the Charter. It seems that the objections among the Antiquaries were mostly financial, without a larger understanding of the benefits the Charter would bring. Vertue wrote:

> I wish, amongst the numerous assembly last night you had been at the Society, to have heard and seen the debates and motions upon the reading of the draughts of the Charter for incorporating our Society, and appointing a Committee to sign a petition to be delivered to his Majesty as soon as possible, perhaps in a week or less time. The principal promoters or movers in this affair, besides the President, was Mr. Theobald, Mr. Webb, Sir Joseph Ayloffe, Mr. Wray, Dr. Milles, &c. Little or nothing was offered or said, in opposition to these expensive schemes, by the members of the first institution, and not of the Royal Society schemers. But, when the great question was put, some disputes arose about voting by proxy. At length that was carried, being supposed that they were mostly procured to increase the number of votes one way; as when the ballot was proposed and agreed to be whether a petition should be presented to his Majesty, as was read, in order to obtain the confirmation of the charter, at the conclusion of the ballot stood 59 for the question, and 9 dissentient—when a Committee was immediately appointed, to withdraw for half an hour, to consider and agree to name 20 persons, including Officers of the Society, to be of Council to the Society, and to be named in the Charter personally and the petition that is to be presented. After these names were settled in a room below, they returned to the company, and reported what had been resolved, and the names of the 20 members. Then ordered to proceed. All this while little regard had been made of the charge and necessary expence of these extraordinaries, till Mr. Webb mentioned that all care should be taken to make the expence as moderate as could be, but money must be ready ; upon which the Treasurer, Mr. Compton, being desired to mention what money in hand that could be spared on this occasion, said one hundred pounds then could be advanced ; but, as that was not near what was necessary, Mr. Theobald rose up, and said it would be convenient to think of a method of raising more, by each member paying down two guineas apiece ; and that might for the present, unless more should be absolutely wanted, be passed as other affairs, *nem. con.*; nobody being willing to take upon them to contest or represent publicly what they privately said or thought; that had done without such expensive costs, and

---

[22] Mary Clapinson, 'Rawlinson, Richard (1690–1755)'.

could not foresee how they could do better ; nor to what benefit it would end to the advantage or knowledge of Antiquities.[23]

As Haycock and Rousseau have noted, on 24 May 1750 North wrote to Ducarel with fears that incorporation would cost the Antiquaries £200 to £300 in fees, swallowing up 'much of the annual income of the Society...From the first mention of incorporation in his correspondence, North interpreted such amalgamation as plot either to line the pockets of the officers or to subsume it into the Royal Society'.[24]

In the 1740s, Folkes had been charged with restoring the Royal Society to financial health by collecting dues in arrears. It seemed like in this case, he was trying to do the same for Antiquaries to set it on a sound institutional basis. Risking great unpopularity, Folkes used his amiability and skills as a committee-man to push the Charter through. From his work at the Foundling Hospital to incorporate that institution, Folkes had seen the advantages a Royal Charter might bring to the Antiquaries. Although it ultimately proved to be expensive, the bill of incorporation totalling £326.12.6, the Antiquaries could then hold property perpetually to the value of £1,000 and receive bequests.[25] The advantages of the Charter particularly became an issue when, in 1749, Lord Coleraine made a bequest of his collection of drawings and engravings to the Society, but because at the time the 'society was not a chartered body, it could not hold property by law. The bequest was therefore invalid'.[26] Folkes may simply have wished to avoid similar events in future. Even Stukeley admitted to Johnson that after incorporation, 'our Antiquarys, to whom I never go, have after a huge struggle got out of a tavern, & fix'd their tabernacle in a great upper room over the Master of the Rolls's gate-way'.[27] Indeed, in 1754, the Society of Antiquaries moved out of the Mitre Tavern into rooms in Chancery Lane, London, which had previously belonged to the Master of the Rolls in the seventeenth century. Their move, along with the Charter, 'signalled the rise in status and aspirations of the Society'.[28]

We also have to remember that in 1751 the incorporation of learned societies was novel, and 'apart from the Royal Society, no such incorporation had occurred'.[29] There were in fact also close resemblances between the two Charters; each Charter had a preamble, and 'the ascription of royal foundation and patronage was borrowed by the Antiquaries from the Royal Society's second charter of 1663'.[30] Folkes, who had written in manuscript a history of the Royal Society, had

---

[23] Nichols, John, *Literary Anecdotes of the Eighteenth Century*, 6 vols (London: Nichols, Son and Bentley, 1812), vol. 2, pp. 712–13.

[24] Rousseau and Haycock, 'Voices Calling for Reform', p. 383.

[25] Pugh, 'Our First Charter', p. 350.   [26] Sweet, *Antiquaries*, p. 87.

[27] Stukeley to Johnson, 9 January 1753, SGS/T3, reprinted in Honeybone and Honeybone, *Correspondence*, p. 185.

[28] Stukeley to Johnson, 9 January 1753, SGS/T3, reprinted in Honeybone and Honeybone, *Correspondence*, p. 186, footnote 5.

[29] Pugh, 'Our First Charter', p. 353.   [30] Pugh, 'Our First Charter', p. 354.

made careful notes about their creation, so he was, in that sense, an ideal person to see the matter through.[31] For instance, Folkes had kept careful note of when the Royal Society was granted a Statute of Mortmain on 17 December 1736 which let it hold properties in perpetuity up to £1,000.[32] The similarities between the two Charters may have been interpreted by his fellow antiquaries as a nefarious plot to unite them to the Royal Society, but it may have just been pragmatism on Folkes's part.

Stukeley commented, however, that as a result of incorporation there were 'great bickerings and ballotings among' the Antiquaries, and Cole also confirmed the subsequent divisions in the Society:

It occasion'd a great deal of ill blood among the discrepant Members, and it is not improbable the Mr Folkes might not have escaped some Censures from the Opposition, which might disgust him, and determine his Favour more towards the Royal Society.[33]

Folkes's will of 12 September 1751 demonstrated his attitudes towards the two organizations of which he was President. He bequeathed both money and mementos to the Royal Society which were important to its heritage, including £200 and a painting of the Society's intellectual father, the Lord Chancellor Francis Bacon. The painting was by the studio of Paulus van Somer (1576–1621) in which Bacon's left arm is leaning on a table upon which rests a bag holding the Great Seal of England, signifying his office as its keeper (1617).[34] Folkes also bequeathed his own presidential portrait painted by Hogarth, showing him with his hand pugnaciously extended in lively debate (see cover image) and, on an unknown date, had presented another portrait to the Society of Isaac Newton by Vanderbank featuring an *ouroboros* [a snake holding its own tail in its mouth]. The ouroboros was the 'alchemical tail-devouring symbol implying eternity because of constant regeneration, and with eternity signifying wisdom; The ouroboros also correlates alchemy with his interest in the revelations of antiquity', appropriate iconography as Folkes edited

[31] RS/MS/702, Martin Folkes, 'An Account of the Royal Society from Its First Institution', Royal Society Library, London.

[32] RS/MS/702, Martin Folkes, 'An Account of the Royal Society from Its First Institution', Royal Society Library, London; Journals of the House of Commons (London: H. M. Stationery Office, 1803), vol. 22, p. 710.

[33] Stukeley to Johnson, 9 January 1753, SGS/T3, reprinted in Honeybone and Honeybone, *Correspondence*, p. 185; BL Add MS/5833, f. 158v.

[34] The bequest from Martin Folkes's will is recorded in The Royal Society Council Minutes: 'he [the President] further acquainted them, that the said Mr William Folkes and Mrs Lucretia Folkes, the Executors of the said Will had also Sent to the Society The Lord Bacon's Picture bequeathed to them by the Said Will for which Mr Hauksbee had given a Receipt'. (RS/CMO/4, 21 November 1754). The will stated: 'I do give to the Royal Society...my Picture of the Great Lord Chancellor Bacon'. (PROB/11/809, Will of Martin Folkes of Hillingdon, Norfolk, 3 July 1754, The National Archives). The painting is still within the Royal Society's collection, archive number P/4, Royal Society Picture Library, London. https://pictures.royalsociety.org/image-rs-9655 [Accessed 9 January 2019].

Newton's works on biblical chronology (see chapter four).[35] Folkes also gave a carnelian seal ring of the Royal Society's arms and motto 'Nullius en verba' set in gold and cut by Seaton, 'intended by the donor to be worn by the present and future Presidents'.[36] Peter Davall, secretary of the Royal Society, received £100 and in the first Codicil of Folkes's will he was given a silver repeating watch for serving as an executor of his estate. Folkes's dear friend, George Graham, received £50, and his friend and fellow antiquary Henry Stuart Stevens was bequeathed a reflecting telescope made by renowned instrument maker James Short.

The fact that Folkes, one of its foremost antiquaries of the age, left nothing to the foremost antiquarian society of the age speaks volumes.[37] Ultimately, although intellectually he was in sympathy with the amalgamation of the two organizations to continue their research program of 'scientific antiquarianism', emotionally and practically it was not to be. The reasons for the unification not taking place seemed more to do with personal acrimony than with disciplinal distinctions. Instead, he promoted a Charter to permit the Society of Antiquaries to survive as its own organization and promote the study of material culture. Indeed, on 23 April 1752, with Folkes incapacitated by illness (see section 9.2), Stukeley began attending Antiquaries again, and read his 'Memoirs towards a History of the Antiquarian Society', dedicated to the Earl of Macclesfield. Stukeley's memoirs included a proposal to 'move the furtherance of the Corporation' that Folkes created.[38] Stukeley advocated for the Antiquaries to have formal involvement in the preservation of public monuments, as well as to create an academy of antiquaries to do research in particular 'heads of inquiry', including coats of arms, funerals, parishes, and 'measures of land among the English'. No mention was made by Stukeley of Folkes's efforts towards the charter which made the entertainment of such proposals possible.

9.2 Last Years

On 26 September 1751, only five weeks before the Antiquaries Charter was approved, what Madame Geoffrin feared most occurred. No doubt due to his

[35] John Vanderbank, *Portrait of Isaac Newton*, 1727, oil on canvas, 1270 mm x 1016 mm, P/92, Royal Society of London; Milo Keynes, *The Iconography of Sir Isaac Newton to 1800* (Woodbridge: The Boydell Press, 2005), p. 39.
[36] PROB/11/809/301, The National Archives, Kew. https://discovery.nationalarchives.gov.uk/details/r/D553698 [Accessed 29 November 2019]; *The London Gazette*, July 2, 1754 –July 4, 1754; Issue 1269. 17th–18th Century Burney Collection Newspapers, British Library, London. My thanks to Christian Dekesel for an annotated copy of the will. The ring Folkes left the Royal Society is now lost.
[37] A point also made by Haycock, 'The Cabal of a Few Designing Members', p. 282.
[38] SAL/MS/266, William Stukeley, 'Memoirs towards an History of the Antiquarian Society' read by Stukeley on 23 April 1752, 'Copyed by Mr Ames from mine, with Additions', Society of Antiquaries of London Library, London.

girth and overindulgence, and perhaps the stress of dealing with Antiquaries, Folkes had a stroke. Stukeley wrote to Maurice Johnson of the Spalding Gentlemen's Society rather gleefully and cruelly, 'our President Folks has had a paralytic stroke: but was an infidel even in philosophy. For we have good reason to believe, that electrifying is highly useful in the case. He is in a very poor condition.'[39] Henry Baker addressed the Royal Society in 1747 about medical electricity, publishing a number of articles about it in the *Philosophical Transactions*, as the 'conception of the *electric* fluid as a permeating ether lent itself to the conclusion that it had an effect on the human body.'[40] In query 24 of his *Opticks*, Newton had offered an alternative to the older model of nervous action caused by animal spirits travelling through hollow fibres, suggesting that nerves were solid filaments infused with aether which excited nervous impulse.[41] The aether was thought to oscillate in the nerves, transmitting nervous impulses to the brain, and electrotherapy was thus thought to have benefits in paralysis.

Ironically, several years earlier on 20 January 1746/7, Folkes abstracted a letter he received from James Simon of Dublin, claiming 'Electricity if properly applied, might produce some good effects in Nervous complaints.'[42] The letter described how a stroke victim with a paralysed right side named Henry McComock was taken to a Mr Booth, 'a young Gentleman who reads lectures here in Experimental Philosophy where he was strongly Electrified', whereupon 'he began to move his arm, and can now walk with a stick'. It is unknown if Folkes agreed to be medically electrified as was the fashion.

Thomas Birch, who was serving as one of the Society's auditors of the Treasurer's Accounts, was a little kinder than Stukeley in his account, recording merely that Folkes was 'seiz'd with a Palsy', which paralysed his left side.[43] Birch and Folkes were close colleagues, having exchanged notes previously on several antiquarian matters; in 1737, Folkes passed along a request to Birch to borrow some letters of 'MyLord Russel and that of his Lady to the King' from Birch's collection that the Duke of Montagu wished to see.[44] In 1739, Folkes asked Birch for his opinion about a paper he had read to the Society of Antiquaries about the Roman foot; Folkes worried its conclusion needed improving, as he 'had been ill in the forepart of the week and was forc'd to finish it in a hurry to

[39] Stukeley to Johnson, 24 April 1752 SGS/T3, in Honeybone and Honeybone, *Correspondence*, p. 178.

[40] Henry Baker, 'A letter from Mr. Henry Baker F. R. S. to the President, concerning several medical experiments of electricity', *Philosophical Transactions* 45, 486 (1748), pp. 270–75. Paola Bertucci, 'The Shocking Bag: Medical Electricity in mid 18th-Century London', *Nuova Voltiana* 5 (2003), pp. 31–42, on p. 35.

[41] William Clower, 'The Transition from Animal Spirits to Animal Electricity: A Neuroscience Paradigm Shift', *Journal of the History of the Neurosciences* 7, 3 (1998), pp. 201–18, on p. 209.

[42] MS 5403/57, Memoirs, Letters and Papers sent to The Royal Society of London, and mostly published in the *Philosophical Transactions*, Wellcome Library, London.

[43] BL Add MS/4222, Thomas Birch, 'Memoirs of the life of Martin Folkes', ff. 30r–31v, British Library, London.

[44] BL Add MS/4307, Folkes to Birch, 9 August 1737, f. 94, British Library, London.

be read'.[45] Birch also compiled an unpublished memoir of Folkes's life.[46] By the end of 1751, it was clear to Birch and others that Folkes was rapidly declining. By 28 November, Folkes had resigned his chair in his beloved Royal Society Dining Club, as his illness that prevented him from 'coming abroad, was not likely to be removed'.[47] Although Folkes was still performing his duties as Royal Society President and recommending others to membership, a note to Birch of 20 February shows a shaky, weak hand.[48] Thomas Birch wrote to Philip York on 14 October 1752,

> Our President whom I visited yesterday, complains greatly of a Cold, which affected him in such a manner yesterday Sennight with a convulsive cough, that he was thought actually expiring; & this cough still returns frequently, tho' in a less Degree. He gives no intimation of any Desire to be discharg'd from the Office, which he is never likely to be again capable of executing; & we are unwilling to take any step this year, that may possibly hurt him. But when we proceed to a new choice, Lord Macclesfield will have no Competitor, for Lord Charles Cavendish has lately declar'd to Mr. Watson his Resolution not to accept of the Presidentship.[49]

A month later, Folkes indicated he would resign, and Macclesfield accepted the offer to be President of the Royal Society.[50] On 25 June 1754, Folkes had a second stroke and died on Friday, 28 June at 4 a.m.

At his request, he was buried in the chancel of the parish church of Hillington, Norfolk 'aforesaid near unto the bodies of my most honour'd parents only I desire that a plain flat Stone of black Marble may be laid over my body with my name and the day and Year of my death engraved upon it'[51] (see figure 9.1). Unlike the monument for Montagu, his own memorial would be simple. The *Public Advertiser* of Wednesday, 3 July 1754 indicated that 'the remains of Martin Folkes, Equ. Will on Saturday morning next be carried out of Town to be interred in the Family vault in Norfolk'.[52] It was reported that his corpse was 'met by upwards of 40 of his principal Tenants, about ten Miles from Hillington, who attended it in Funeral Procession to the Place of Interment. There was the greatest

---

[45] BL Add MS/4307, Folkes to Birch, 7 December 1739, f. 96, British Library, London.

[46] BL Add MS/4222, Thomas Birch, 'Memoirs of the life of Martin Folkes', British Library, London.

[47] Geikie, *Annals of the Royal Society Club*, p. 41.

[48] BL Add MS/4307, f. 97, Folkes to Birch, 20 February 1751/2, British Library, London.

[49] BL Add MS/35398, Hardwicke Papers, Birch to Philip Yorke (Lord Royston), 14 October 1752, ff. 104–5. British Library, London.

[50] David P. Miller, 'The 'Hardwicke Circle': The Whig Supremacy and Its Demise in the 18th-Century Royal Society', *Notes and Records of the Royal Society of London* 52, 1 (1998), pp. 73–91, on p. 77.

[51] PROB/11/809/301, Will of Martin Folkes, 3 July 1754, The National Archives, Kew.

[52] *Public Advertiser*, Wednesday, July 3, 1754; Issue 6139. 17th–18th Century Burney Collection Newspapers, British Library, London.

Fig. 9.1 Tomb of Martin Folkes, Church of St Mary, Hillington, Norfolk, Photograph by the Author.

Concourse of Persons upon this Occasion that has been known at that Village in the Memory of Man'.[53]

Folkes was nursed in his last two years by Helen Betenson (d. 1788), his neighbour and friend in Queen Square. Helen was the sister of Sir Richard Betenson, 4th Baronet of Wimbledon, Surrey (d. 1786) and Sheriff of Kent, who married Folkes's youngest daughter Lucretia in 1756.[54] At the time of her marriage, Lucretia was a thirty-six-year-old spinster, and she served as Folkes's executrix, with William Folkes as executor; Folkes left Lucretia his house in Queen Square, his library, which the newspapers reported as 'said to equal, if not exceed', that of bibliophile Richard Mead's, as well as his plate, coin collection, together with a sum of £12,000 and also £200 in stocks of South Sea Annuities.[55] To his older daughter, Dorothy, widowed by her marriage to bookseller William Rishton and left with three children, he was not as generous. In his original will, she would have received £12,000 and an extra £2,000, together with a gilt silver salver, his portrait painted by Thomas Gibson (a founding member of Godfrey Kneller's St Martin's Lane Academy and friend of Hogarth) and her own portrait in crayon by George Knapton (1697–1778), an oil and pastellist apprenticed to Folkes's friend Jonathan Richardson.[56] The second Codicil, added to his will on 3 November 1751, changed these terms completely. The executors of his will had to invest these two sums in securities, and only the dividends and interests of these sums could be paid to his

---

[53] *Whitehall Evening Post or London Intelligencer*, July 11, 1754–July 13, 1754; Issue 1273. 17th–18th Century Burney Collection Newspapers, British Library, London.

[54] George Edward Cokayne, ed., *The Complete Baronetage*, 5 vols (no date, c.1900); reprint, (Gloucester: Alan Sutton Publishing, 1983), vol. III, p. 278.

[55] *Public Advertiser*, Wednesday, July 3, 1754; Issue 6139. 17th–18th Century Burney Collection Newspapers, British Library, London.

[56] PROB/11/809/301, Will of Martin Folkes, 3 July 1754, The National Archives, Kew. My thanks to Dr C. E. Dekesel F.S.A. for this information. Dorothy's Portrait is listed in Neil Jeffares, 'George Knapton', *Pastels and Pastellists*, portrait J.432.181. http://www.pastellists.com [Accessed 15 January 2020].

daughter Dorothy to be used for her upkeep and the education of her three children. The invested sums had to be divided between these children when they reached the age of twenty-one.[57] This may have been done to protect his daughter against a fortune hunter or perhaps her late husband's relatives. The portrait of Dorothy may have been later sold at the auction of Folkes's effects.[58]

These changes that Folkes made were not without controversy. In an undated letter to his brother William, Folkes wrote, 'now the case is I am without suffering a recovery in no capacity to secure any fortunes to my girls or to make any provision for any younger children I may happen to have, the concern and uneasiness for this was very great upon me when I thought I should die in my late illness, and indeed bitterer to me than Death itself. I am therefore preparing to suffer such a recovery to enable me to secure fortunes to my girls and younger children.'[59] He assured his brother he would not 'alienate any part of the estate' upon doing so, putting 'all remainders again on the same foot my father left them'. The comment about his younger children is interesting; we recall that in 1743, Folkes received a letter from one Louisa Edwards advising her cousin had two children 'very much like yourself', so perhaps he was by private means providing for their welfare, as they are not acknowledged in his will.

Lucretia sadly only outlived her father by two years, dying on 6 June 1756, most likely because of complications in childbirth.[60] Lucretia's memorial in St George's Church, Wrotham was by Nicholas Read, but it was linked with a design in Roubiliac's catalogue, so one wonders if Folkes or his brother William had a hand in its commission before he died[61] (see image 9.2).

[57] My thanks to Dr C. E. Dekesel F.S.A. for this information.
[58] 'A lady's head, ¾ crayons, Lot 92, £2/1', in *A Catalogue of the Genuine and Curious Collection of Mathematical Instrument, Gems, Pictures....of Martin Folkes, Esq.*
[59] MS/MC/783–1, Undated letter from Martin Folkes to William Folkes, Norfolk Record Office, Norwich.
[60] *Gentleman's Magazine* 27 (1758), p. 292. Her memorial is in St George, Wrotham, Kent in the Betenson Chapel. The inscription reads, 'Sacred to the memory of Mrs Lucretia Betenson, the beloved wife of Richard Betenson, esq; only son of Sir Edward Betenson, of this county, bart. Her early death fixed deep in the breasts of her disconsolate friends and inexpressible and lasting sorrow, as she was an affectionate wife, a sincere and steady friend, ever compassionate to the sorrows and bountifull to the wants of her fellow creature; in a word, an amiable pattern of every Christian virtue. She was daughter and coheiress of Martin Folkes, of Hillington in Norfolk, esq, who was president of the Royal Society, and distinguished by his extensive learning amongst the brightest ornaments of the age. This monument was erected by the care and direction of the aforesaid Richard Betenson, esq.' See John Thorpe, *Registrum Roffense: Or, A Collection of Antient Records, Charters and Instruments of Divers Kinds, Necessary for Illustrating the Ecclesiastical History and Antiquities of the Diocese and Cathedral Church of Rochester* (London: W. and J. Richardson, 1769), p. 832. Betenson had no children, and the baronetcy became extinct.
[61] David Bindman and Malcolm Baker, *Roubiliac and the Eighteenth Century: Monument: Sculpture as Theatre* (New Haven: Yale University Press, 1995), p. 375, note 12. Folkes's oldest daughter Dorothy, who had married William Rishton on 28 April 1742, was given interests and dividends on a set of securities for her upkeep and the education of her three children, Lucretia Eleanor Rishton (1742–1801), Ellen Rishton (1744–1802), and Martin Folkes William Rishton (b. 1747). Stukeley claimed William Rishton was an indigent bookkeeper who 'used her very ill', and indeed she was not given as much out of Folkes's estate. See Stukeley, *The Family Memoirs*, p. 99.

Fig. 9.2 Monument, Lucretia Folkes Betenson, St George's Church, Wrotham, Kent, Photograph by the Author.

In the Codicil to his will dated 17 October 1751, just after he had his first major stroke on 26 September, Folkes stated:

> I do hereby give devise and bequeath unto/ Mrs. Helena Bettenson Daughter of Sr. Edward Bettenson Bart./ the Sum of two hundred Pounds as a small token of my thankfullness/ to her for her great kindness to my Daughter and in gratitude for/ her kind regard to my self in my present Illness.[62]

This may have been simple friendship, if not for the subsequent and elaborate bequest that Helen made for him, which may have indicated a deeper connection. Dekesel has surmised that Stukeley's growing disapproval of Folkes may even have stemmed from disapproval of what he thought was an amorous relationship between Martin and Helen.[63] Suffice it to say, the evidence we have is suggestive of a close bond.

[62] PROB/11/809/301, Will of Martin Folkes, 3 July 1754, The National Archives, Kew.
[63] Christian Deckesel, *Salon: The Newsletter of the Society of Antiquaries* 226 (2010). https://www.sal.org.uk/2020/02/sal226/ [Accessed 1 September 2020].

In her own will, probated on 26 November 1788, she indicated that

I give and bequeath to my said Trustees and Executors named and appointed and to their Executors and Administrators the Sum of One thousand Pounds of lawful British Money In Trust nevertheless that they and the Survivor of them his Executors or Administrators shall and do lay out the same in the purchase and erection of a Handsome Marble Monument in Westminster Abbey with a proper Inscription to be thereon Engraved and approved of by the President and Council of the Royal Society or the Major part of them present at any Meeting for that purpose to the Memory of Martin Foulkes Esquire sometime president of the said Royal Society and which Monument I desire may be Erected as soon as convenient after my decease and placed as near to that of Sir Isaac Newton as may be.[64]

On 17 March 1790, the Dean and Chapter of Westminster Abbey gave the executors the permission for the erection of a monument in the north aisle of the Cathedral and a "fine" of 50 Guineas was paid by Lord Amherst on 31 March 1791 (See figure 9.3).[65] The Royal Society, who had previously been given a copy of the will, met on 31 March 1791 to discuss 'An inscription intended for the Monument of the late Martin Folkes Esq. was in consequence of the Will of the late Mrs Helen Bettenson laid before the President and Council for his Approbation. The said Inscription was read and the Council gave their consent to its being engraven on the Monument'.[66]

The monument, executed by William Tyler and Robert Ashton, shows a seated figure of Folkes, eyes closed in contemplation. He is in typically neoclassical Roman dress, a reference to the connections between the Roman collegia and Freemasonry. A stone mason, Ashton himself may have had connections to Freemasonry. In 1788, he 'tendered for building the Freemasons' Tavern; although he did not get the commission, he was employed on decorative carving for the building from 1791 until 1792'.[67] In Tyler and Ashton's design, Folkes leans on two books, one a treatise on medals labelled S.A.L. (Society of Antiquaries of

---

[64] PROB/11/1171/319, Will and Testament of Helen Betenson, The National Archives, Kew.
[65] My thanks to Dr C. E. Dekesel F.S.A. for this information.
[66] RS/JBO/34/47, Minutes of a meeting, 21 March 1791. My thanks to Dr C. E. Dekesel F.S.A. for this information. The inscription read (translated from the Latin): Sacred to the memory of MARTIN FOLKES Esq. of Hillington in the County of Norfolk, who, under the auspices of Newton happily devoted his mind, his energies, his life, to the study of philosophy, that lofty subject. He was for long a Fellow of the Royal Society, and in 1741 was deservedly elected its President. He peacefully submitted to our common fate on 28th June 1754 aged 63. This marble was erected according to the will and testament of Helen, only sister of Richard Betenson Kt. who took as wife Lucretia, younger daughter of Martin Folkes. 1788. W. Tyler Invt. R. Ashton Sculpt.
[67] 'Robert Ashton', in Ingrid Roscoe, Emma Hardy, and M. G. Sullivan, eds, *A Biographical Dictionary of Sculptors in Britain, 1660–1851* (New Haven: Yale University Press, 2009), online version: https://www.henry-moore.org/archives-and-library/sculpture-research-library/biographical-dictionary-of-sculptors# [Accessed 12 April 2019].

London), the other the *Philosophical Transactions of the Royal Society*, a reference to his dual presidencies of these scholarly societies. An adjacent cherub carefully measures a globe with callipers and another is peering into a microscope with close attention. This iconography was an allusion to Folkes's lifelong passions for metrology and astronomy, to his work with the hydra and Trembley, and to his treatise describing the instruments of Leeuwenhoek. Above is an urn covered with drapery that another cherub holds aloft. It was an elaborate gift and a magnificent monument, done out of honour and love.

9.3 The Legacy of Martin Folkes

Folkes is a complex and enigmatic figure. The scattering of his archives, his character assassination by John Hill and William Stukeley, and the unfortunate timing of the paralytic strokes he suffered at the very end of his presidencies, led historians to have a low opinion about the state of the Royal Society during his tenure. As a result, although he was famous in his lifetime, there has been a type of collective amnesia about his historical significance, which this reappraisal of his life and letters has tried to rectify. Why does he need to be remembered?

It is evident that Folkes's promotion of the setting of standards of measurement via precision instruments witnessed the transition of instruments from trading tools to machines that described the universe, making them symbols of not only natural philosophy but also trade and power. As Lugli reminds us, quantification, scale, and magnitude also played a role in the construction of early modern knowledge.[68] Folkes well understood that the use of separate national measures made the construction of hypotheses in natural philosophy needlessly complicated, frustrating reproducibility and international collaborations. In his support of metrological research programmes and instrument makers in the Royal Society, Folkes helped 'standardize' science in a very real way. His own collection of instruments was serious and well chosen, complete with a Francis Hauksbee air pump, a Graham sidereal and mean-time clock, quadrants by Jonathan Sisson and John Bird, and a Richard Cushee pocket globe, probably bought from Sisson's shop.[69] The author has little doubt that he would have loved the metric system, had he been able to see its creation in France's revolutionary government, though he may have not been as happy seeing the height of the Trajan column he obsessively observed and measured during his Grand Tour transformed from the satisfyingly round figure of one hundred Roman feet into 29.635 metres.[70]

---

[68] Emanuele Lugli, *The Making of Measure and the Promise of Sameness* (Chicago: University of Chicago Press, 2019), pp. 9–10.

[69] My thanks to Jim Bennett for these points.   [70] Lugli, *The Making of Measure*, p. 11.

Fig. 9.3 Monument, Martin Folkes, Westminster Abbey, Image © 2019 Dean and
Chapter of Westminster.

Matter-of-fact empiricism, archaeo-astronomical techniques such as those he used with the Farnese Globe, and interest in the cultural context and manufacture of artefacts such as the coinage all played a role in his work, but all of this work depended on an understanding of precise weights, measures, and instruments. Even Folkes's practice of Freemasonry and support of giving the craft a historical grounding with James Anderson's *Constitutions* can be seen in this light. Freemasonry was a practice of belief, but it was also based on past stone cutters using engineering principles that required some degree of metrological understanding.

It was Folkes's dedication to metrology, combined with an interest in the material culture of the past, that also anticipated the development of antiquarianism into what we know as archaeology. But it was antiquarianism that was uniquely grounded in the eighteenth century; while Folkes saw to it that objects and papers from the Society's early history were preserved during his Presidency, this preservation could be so that those objects could be employed for experimental use and reuse. What we now consider totemic books, such as Robert Hooke's *Micrographia*, had their copperplates modified for 'improvement' and republication.

At the same time, Folkes well knew that instruments were particularly important, because mere sense perception can thoroughly trick us. He loved playing with optical illusion and imitation in his scientific work and in his art connoisseurship, which may have been behind his love of the theatre early in his life. He even married an actress, a mistress of illusion, who ironically in the throes of mental illness became unable to distinguish fact from fiction herself. However, though Folkes in his portraits imitated his mentor Newton's visage and iconography as he assumed his role of Royal Society President, he was not play-acting, but using the power of illusion for his own serious aims. Folkes's dedication to internationalizing Newtonianism made him a statesman of natural philosophy on the Continent, but he went beyond being merely well connected in the Republic of Letters to being a participant and demonstrator of the primacy of English optics, the precision instrumentation in geodesy, and the predictive validity of the laws of English physics. And part of that ambassadorship was about associating his name and his very appearance with his great mentor.

Our own perception concerning one of Folkes's key rivals, Sir Hans Sloane, may also have been faulty and has recently been changing. Sloane has in the past been glorified as a President of the College of Physicians, purveyor of chocolate, founder of the Chelsea Physic Garden, and the founding father of the British Museum. As we have seen in this book, he was President of the Royal Society, triumphing over the upstart Folkes in his election. But now, thanks to the work of James Delbourgo, there is the disquieting realization that Sloane's collections were partially funded by enslaved labour on Jamaican sugar plantations which the family of his wife Elizabeth Langley Rose owned and from which they both immensely profited.[71] While Sloane has not been pushed off his pedestal completely, his statue in the British Museum's Enlightenment Gallery was recently moved in August 2020 to a display case to explain more clearly the role of his work and the implications of his collecting in the British Empire.[72] He is not known just for Sloane Square anymore, but for more troubling associations.

And of the upstart Folkes, the 'infidel' Folkes, the incapacitated Folkes paralysed by two strokes who was satirized by Hill and disparaged by Stukeley for his supposedly poor administration of the Royal Society? That Folkes was surely not perfect, overeating and overdrinking, unfaithful to his wife, his materialist beliefs clinically cold when describing parasitic twinning. That Folkes, like Sloane, had stock in the slave-trading South Sea Trading Company, as did his predecessor Newton. But the Folkes in this book was also elected FRS at the age of twenty-three and subsequently made a Vice President by Newton due to his mathematical gifts. This Folkes argued for the liberty of discussion within the Royal Society Council and for foreign Fellows to have a vote, even if it cost him the presidential election against Sloane. This Folkes instituted the Copley Medal to recognize scientific genius, and promoted the work of those from lower social stations to further natural philosophy. This Folkes also questioned racial prejudice. We recall in 1747 that Folkes explained to his Jewish friend Emanuel Mendes da Costa, 'but we are all citizens of the world, and see different customs and tastes without dislike or prejudice, as we do different names and colours'.[73] In his assessment of Folkes's life, Thomas Birch wrote 'The Generosity of his Temper was no less remarkable than the Civility & Vivacity of his Conversation'.[74] This Folkes deserves more than our collective amnesia.

---

[71] See James Delbourgo, *Collecting the World: The Life and Curiosity of Hans Sloane* (Cambridge, MA: Harvard University Press, 2018).

[72] David Olusoga, 'It is not Hans Sloane who has been erased from history, but his slaves', *The Guardian*, 30 August 2020, https://www.theguardian.com/commentisfree/2020/aug/30/it-is-not-hans-sloane-who-has-been-erased-from-history-but-his-slaves [Accessed 1 September 2020].

[73] John Nichols, ed., *Illustrations of the Literary History of the Eighteenth Century*, 8 vols (London: for the Author, 1817–58), vol. 4, pp. 635–7.

[74] BL Add MS/4222, f. 32r.

Afterword: Folkes and Voltaire

There may be another important reason to remember Martin Folkes. In 1733, Voltaire's *Letters concerning the English nation* was published, appearing in English before being published in French the following year. Voltaire wrote his classic work after his visit to England in 1726–8, and in his Letter 24 complained that the Royal Society 'mixes indiscriminately literature with physics', a pointed criticism of Folkes's 'scientific antiquarianism'.[1] Voltaire objected that 'a dissertation on the Head-dresses of the *Roman* Ladies' would be in the same publication 'with an hundred or more new [mathematical] Curves'.[2] He rather petulantly 'wished to see a distinction as existed in France between the Académie Française, concerned with French language and letters, and the Académie des Sciences'.[3] Ultimately Voltaire got his wish, as despite any efforts Folkes made to bear, the Society of Antiquaries of London and the Royal Society never did unite.

One wonders if Voltaire previously discussed his concerns about interdisciplinarity with Folkes, as the two natural philosophers were correspondents throughout their lives.[4] On 10 October 1739, Voltaire wrote to Folkes from Paris in reference to his *Réponse aux objections principales qu'on a faites en France contre la philosophe de Newton*, a tract he wrote in support of his *Éléments de la Philosophie de Newton* (1738).[5] Voltaire conceived of the *Éléments* as a '*machine de guerre*' directed against the Cartesian establishment, which he believed was holding France back from the modern light of scientific truth. Vociferous criticism of Voltaire and his work quickly erupted, with some critics emphasising his rebellious and immoral proclivities while others focused on his precise scientific views'.[6] According to Shank, Voltaire and Émilie De Châtelet engaged in

---

[1] Voltaire [François-Marie Arouet], *Letters Concerning the English Nation* (London: C. Davis, 1741), p. 184. I am using the second edition. Another revised edition in 1778 appeared as *Lettres philosophiques sur les Anglais* (*Philosophical Letters on the English*).

[2] Voltaire, *Letters Concerning the English Nation*, p. 185.

[3] 'Letters Concerning the English Nation', Voltaire Foundation, University of Oxford, http://www.voltaire.ox.ac.uk/news/lettres-sur-les-anglais/lettres-sur-les-anglais-royal-society-letter-24 [Accessed 29 November 2019].

[4] It is acknowledged that Folkes and Voltaire were infrequent correspondents, but Voltaire's letters to English philosophers were also few overall, his frequency of letters directly related to the physical proximity to France. See Dan Edelstein and Biliana Kassabova, 'How England Fell Off the Map of Voltaire's Enlightenment', *Modern Intellectual History* (2018). doi:10.1017/S147924431800015X.

[5] Voltaire also sent another copy of his *Réponse* to Robert Smith, Professor of Astronomy at Cambridge. See Harcourt Brown, *Science and the Human Comedy: Natural Philosophy in French Literature from Rabelais to Maupertuis* (Toronto: University of Toronto Press, 1976), e-book.

[6] J. B. Shank., 'Voltaire', in Edward N. Zalta, ed., *The Stanford Encyclopedia of Philosophy* (Fall 2015 Edition). https://plato.stanford.edu/archives/fall2015/entries/voltaire/ [Accessed 29 November 2019]. See also J. B. Shank, *The Newton Wars and the Beginning of the French Enlightenment* (Chicago: University of Chicago Press, 2008).

a campaign on behalf of Newtonianism, putting in their sights 'an imagined monolith called French Academic Cartesianism as the enemy against which they in the name of Newtonianism were fighting', the main artillery of their battle Voltaire's *Éléments de la Philosophie de Newton*.[7] Voltaire wrote to Folkes in a fit of pique:

> Sir, I Do my self the honour to send you this little answer. I was oblig'd to write against our antineutonian cavillers.
>
> I am but a man blind of one eye expostulating with stark blind people who deny, there is such Thing as a sun.
>
> I'll be very happy if this conflict with ignorant philosophers may ingratiate my self with a such a true philosopher as you are.[8]

In 1743, upon his election to the Royal Society, three years before he was elected to the Académie Française, Voltaire wrote to Folkes, again in some frustration with his continued fight for Newtonianism and against those irritatingly persistent Cartesian vortices.[9]

> One of my strongest desires was to be naturaliz'd in England; the royal society, prompted be you vouchsafes to honour me with the best letters of naturalisation. My first masters in your frée and learned country, were Shakespear, Adisson, Dryden, Pope; I made some steps afterwards in the temple of philosophy towards the altar of Newton. I was even so bold as to introduce into France some of his discoveries; but j was not only a confessor to his faith, I became a martir. I could never obtain the privilege of saying in print, that light comes from the sun and stars, and is not waiting in the air for the sun's impulsion; that vortices cannot be intirely reconcil'd with mathematics; that there is an evident attraction between the heavenly bodies, and such trash.
>
> But the liberty of the press was fully granted to all the witty gentlemen who teach'd us that attraction is a chimera, and vortices are demonstrated, who printed that a mobile lanch'd out from on high describes a parabola because of the resistance from the air below, that t'is false and impious to sai, light comes from the sun. Even some of them printed, *col la licenza dei superiori*, that Newton ridiculously mistook, when he learn'd from experience the smaller are the pores of transparent bodies, the more pellucid they are; they alleg'd very wisely that the

[7] J. B. Shank, 'Voltaire', *The Stanford Encyclopedia of Philosophy*.
[8] Voltaire [François-Marie Arouet], 'Voltaire [François Marie Arouet] to Martin Folkes [ffolkes]: Saturday, 10 October 1739—[letter]'. letter nº: D2088 *Electronic Enlightenment*, ed. Robert McNamee et al. Vers. 2.4. University of Oxford. http://www.e-enlightenment.com/item/voltfrVF0910014b1key001cor/ [Accessed 15 May 2014].
[9] See G. R. de Beer, 'Voltaire, F.R.S.', *Notes and Records of the Royal Society of London* 8, 2 (1951), pp. 247–52.

widest windows give the greatest admittance to light in a bed chamber. These things I have seen in our booksellers shops, and at their shops only.

You reward me sir for my sufferings. The tittle of brother you honour me with is the dearest to me of all titles. I want now to cross the sea to return you my hearthy thanks, and to show my gratitude and my veneration for the illustrious society of which you are the chief member.

Be pleas'd sir to be so Kind as to present your worthy bretheren with my most humble respects. I hope you will sait to mylord duke of Richemont [Richmond], to mr Jurin, mr Turner etc. how deeply I am sensible of their favours. I am with the greatest esteem, and the most sincere respect

Sir Your most humble and faithfull servant, I dare not say brother Voltaire.[10]

As Cronk has indicated, Voltaire referred to himself as a 'martir' for the treatment he received following his publication of his *Lettres philosophiques*; 'in May 1734, a *letter de cachet* was issued against Voltaire, making him liable for arrest', and a court of law ordered that a copy of his book be 'ceremoniously burned in the courtyard of the *parlement*'.[11] Voltaire also reminded Folkes of his visit to England fifteen years earlier and his acquaintance with Charles Lennox, the Duke of Richmond, James Jurin, and 'Mr Turner', who was Shallet Turner, Professor of Modern Languages at Cambridge. All three had nominated Voltaire as Fellow of the Royal Society as a 'gentlemen well known by Several curious and valuable Works...well skill'd in Philosophical Learning'.[12] For all his support of Newton, and his comments about Newton's funeral and monument in Westminster Abbey, Newton and Voltaire had not met before Sir Isaac died in March 1727. During his stay in England from May 1726 until the autumn of 1728, Voltaire did, however, meet Newton's niece Catherine Barton Conduitt, who told him the apple story, a story that Folkes also related (see chapter two), and Voltaire related twice in his writings.

The correspondence between Voltaire and Folkes, Newtonian to Newtonian, thus suggests a long acquaintance. Was Voltaire introduced to Folkes as well during his visit to London in 1726–8? It is possible. As Shank has indicated, 'given his other activities, it is also likely that Voltaire frequented the coffeehouses of London even if no firm evidence survives confirming that he did'.[13] Voltaire did live at the White Wig (known also as the White Peruke) on Maiden Lane,[14] and was said to have dined at the Bedford Head Tavern, one of the places in the 1720s

[10] RS/MM/16/41, Letter from François-Marie Arouet de Voltaire, Paris, to Martin Folkes, 29 March 1743. A copy can be found at MM/15/10.
[11] Nicholas Cronk, *Voltaire: A Very Short Introduction* (Oxford: Oxford University Press, 2017), p. 47.
[12] RS/EC/1743/07, Royal Society Library, London.
[13] J. B. Shank, 'Voltaire', *The Stanford Encyclopedia of Philosophy*.
[14] Voltaire, *Select Letters of Voltaire*, ed. and trans. Theodore Besterman (London: Thomas Nelson and Sons, 1963), p. 29; Norma Perry, 'The Rainbow, the White Peruke and the Bedford Head: Voltaire's London Haunts', *Voltaire and the English*, Studies on Voltaire and the Eighteenth century, vol. 179. The Voltaire Foundation at the Taylor Institution, Oxford (Oxford: Voltaire Foundation, 1979), pp. 203–20.

in which Folkes attended Masonic meetings and met with John Byrom. And at one of the coffee-houses, called Buttons, which was near Covent Garden Piazza on Russell Street, we may have some firmer evidence that Voltaire met Folkes.

A sketch attributed to Hogarth *c.*1720 at Buttons depicts Martin Folkes examining a watch with an unknown gentleman sitting beside him, handing him an obscure object, perhaps a knife to pry open the watch, a coin, or another timepiece.[15] Buttons was a famous gathering place for the literati, with Addison, Steele, Pope, Swift, and Arbuthnot assembling each evening, evaluating their own work and that of others over a dish of coffee and a pipe. In 1786, Samuel Ireland did an aquatint of Hogarth's work, where he identifies the figures as Martin Folkes and playwright, author, and journalist Joseph Addison. Folkes is easily discernible, but the latter identification is impossible, as Addison died in 1719.[16] It is possible that Ireland thought the portrait was of Addison, as Addison owned Buttons coffee-house, but artist Michael Dahl's portrait of Addison, painted in 1719, shows only vague similarity to the Hogarthian sketch, as does Godfrey Kneller's portrayal engraved by John Faber the Younger (see figures 4.3, 9.4a, and 9.4b).

---

[15] William Hogarth, *Examining a watch; two men seated at a table, the older (Martin Folkes) looking through his eyeglasses at a watch, a paper headed 'Votes of the Commons' (?) on the table.* Pen and brown (?) ink and wash, over graphite, *c.*1720, Image 1861,0413.508, ©The Trustees of the British Museum. This drawing is part of a set of four owned by engraver and prints dealer Samuel Ireland, portrayed in his *Graphic Illustrations of Hogarth* (1794–9) as a series of characters in Button's coffee-house. Although Ireland is known for spurious attributions of characters portrayed in Hogarth's works, Lawrence Binyon thought 'the most plausible of Ireland's identifications is that of Martin Folkes', due to its similarity with the later Hogarth oil portrait; Binyon also firmly considered the drawings by Hogarth (Lawrence Binyon, *Catalogue of Drawings by British Artists and Artists of Foreign Origin working in Great Britain preserved in the Department of Prints and Drawings in the British Museum*, 4 vols (London: Order of the Trustees, 1898–1907), vol. 2, p. 321). In the catalogue raisonné of Hogarth's drawings, A. P. Oppé also mentions Ireland's problematic attributions, but Hogarth is still identified by him as the artist due to the 'careful, sensitive treatment of the faces' and the clumsy bodies typical of Hogarth's other works done at the time. He does note, however, that the drawing style and use of media is different from Hogarth's early drawing style (A. P. Oppé, *The Drawings of William Hogarth* (New York: Phaidon, 1948), pp. 30–1). On the other hand, Sheila O'Connell, retired assistant keeper of Prints and Drawings, British Museum believes the set of drawings suspicious due to the Hogarthomania of the later eighteenth century (email of 15 August 2020). See also Sheila O'Connell, 'Appendix: Hogarthomania and the Collecting of Hogarth', in David Bindman, ed., *Hogarth and his Times: Serious Comedy* (Berkeley and Los Angeles: University of California Press, 1997), pp. 58–61, on p. 59. However, if the drawing is not by Hogarth, that does not mean it is not Folkes and Voltaire by a contemporary. My thanks to Sheila O'Connell and Elizabeth Einberg for discussing the drawing with me.

[16] The British Museum Collection Online, https://www.britishmuseum.org/collection/object/P_1861-0413-508 [Accessed 29 November 2019]. Samuel Ireland, *Graphic Illustrations of Hogarth* (London, 1794), facing p. 31. Ireland claims that he bought this sketch from a 'Mr Brent', an old gentleman who was an intimate of Hogarth. Although 'Mr Brent' might be apocryphal, he appears again in relation to a painting in a nineteenth-century edition of Hogarth's works, *Anecdotes of William Hogarth, Written by Himself* (London: J. B. Nichols and Son, 1833), p. 376. The author claims that Hogarth presented to his friend 'Brent' a painting called *St. James Day, or First Day of Oysters*, owned by G. Weller and exhibited at 'Mr. Forest's Piccadilly'. The painting was possibly apocryphal, but it is however mentioned in the *Catalogue of Prints and Drawings in the British Museum, Division 1, Political and Personal Satires* (London: Trustees of the British Museum, 1873), vol. 2, p. 678.

Fig. 9.4a Michael Dahl, *Joseph Addison*, oil on canvas, 1719, 40 1/2 in. × 31 1/4 in. (1029 mm × 794 mm), © National Portrait Gallery, London.

Fig. 9.4b Godfrey Kneller, *Joseph Addison*, engraved by John Faber the Younger, *c.* 1733, Image: 12 9/16 × 9 11/16 inches (31.9 × 24.6 cm), Yale Centre for British Art, Paul Mellon Collection.

The sketch attributed of the unknown man sitting with Folkes does, however, have remarkable similarity to an oil portrait of the young Voltaire painted by Nicolas de Largillière done immediately before Voltaire's visit to England. As the work of Samuel Taylor has shown, there were two portraits done of Voltaire by Nicolas de Largillière, one c.1718, now in the Musée Carnavalet, and a 1740 copy, now in the Musée National du Château de Versailles[17] (see figures 9.5a, 9.5b, and 9.5c). While likeness is not proof, the sketch and Largillière's portraits both portray a heart-shaped face with defined cheekbones, straight eyebrows, a dimpled chin, and pronounced nose, with the same facial proportions. Voltaire's high forehead in the Hogarthian sketch was compensated by a wig that sat lower on the brow. The artist was also known for his character studies, in which he skillfully delineated the salient features of the figure. The sketch attributed to Hogarth also shows a young man of very slender body, a physiognomy borne out by Voltaire's acquaintances when he was in London. As Ballantyne remarked, Voltaire 'seems undoubtedly to have been in a sickly state of body during the whole period of his residence in England'; in a letter to Nicolas-Claude Thieriot of 1729, Voltaire proclaimed, 'I have been very ill, I was very weak when I arrived'.[18] At the Palladian mansion of Eastbury in Dorset, Voltaire had met Edward Young, the author of *Night Thoughts*, who wrote the famous description of him after a discussion of Milton's *Paradise Lost*: 'You are so witty, profligate and thin, At once we think thee Milton, Death and Sin'.[19]

As Voltaire did not speak English when he came to England, he spent a large portion of his time with the London Huguenot refugee community, with whom Folkes was acquainted through Abraham de Moivre and Desaguliers. Folkes also spoke fluent French and was intimately familiar with French natural philosophy. As Voltaire wished to publish his *La Henriade*, he also sought out Huguenot printers, who ultimately published it.[20] Voltaire had presented a copy of his *Essay upon the Civil Wars of France* (1727) to Sir Hans Sloane, inscribing it in his own

[17] *Portrait de Voltaire (1694-1778) en 1718*, Paris Musées: Les Musées de la Ville de Paris, http://parismuseescollections.paris.fr/fr/musee-carnavalet/oeuvres/portrait-de-voltaire-1694-1778-en-1718#infos-principales [Accessed 30 November 2019]. 'L'existence de la version de Versailles, de plus belle qualité, a fait douter de l'authenticité du portrait du musée Carnavalet, dont l'historique est pourtant assez sûr. D'après les recherches de Samuel Taylor, il semble que le tableau de Carnavalet soit bien l'original et que le tableau de Versailles soit une réplique autographe, demandée par Voltaire à l'artiste vers 1740. La différence de qualité peut s'expliquer par la plus grande notoriété du modèle en 1740 : Largillierre aurait alors pris plus de soin à représenter le plus célèbre écrivain de toute l'Europe qu'il n'en avait pris, vers 1720, à représenter un auteur encore à ses débuts'.

[18] Archibald Ballantyne, *Voltaire's Visit to England 1726-1729* (London: Smith, Elder and Co, 1898), pp. 37–8.

[19] Anthony Netboy, 'Voltaire's English Years', *VQR: A National Journal of Literature and Discussion* (Spring 1977), https://www.vqronline.org/essay/voltaire%E2%80%99s-english-years-17261728 [Accessed 29 November 2019].

[20] Norma Perry, 'Voltaire's View of England', *Journal of European Studies* 7, 26 (1977), pp. 77–94, on pp. 81–4; Edelstein and Kassabova, 'How England Fell Off the Map of Voltaire's Enlightenment', p. 17. Charles Lennox and his wife were subscribers for five copies, and Folkes had three copies of the 1728 edition in his own library.

Fig. 9.5a Nicolas de Largillière
(1656-1746), *Voltaire,* oil on canvas,
*c.*1718–24, 80 cm × 65 cm, Musée
Carnavalet, Histoire de Paris, CC Open
Access license.

Fig. 9.5b Nicolas de Largillière,
Voltaire, oil on canvas, *c.*1724–5,
© Château de Versailles, Dist.
RMN-Grand Palais/
Christophe Fouin.

Fig. 9.5c Close-up of Hogarthian Sketch in Figure 4.3,
reversed horizontally.

handwriting, indicating they had been acquainted; Folkes and Sloane, of course, knew each other intimately.[21] The evidence is certainly suggestive that Voltaire and Folkes met well before Folkes's Grand Tour and subsequent visit to Paris in 1739. If so, Folkes would have been pleased that the relatively unknown young man he may have met in the 1720s in London had so distinguished himself to be admitted to the Royal Society two decades later.

The Hogarthian sketch is also more largely symbolic for the character of natural philosophy and antiquarianism manifested during Folkes's presidencies. Folkes is examining a watch, a symbol for the growing precision of timepieces, such as John Harrison's clocks for calculating longitude, or Graham Graham's horological achievements such as the dead-beat escapement and mercury pendulum. It was not surprising Folkes loved watches. Time derivatives permeate all of Newtonian mechanics, the ultimate variable on which all others depend, a replacement for the Cartesian vortices that Voltaire ultimately rejects. Even the Bible was subjected by Newton, and then Folkes, to chronologies created from astronomical calculations, Cotes's least squares method employed (it was hoped) to create a newly precise timeline of divinity. The probable length of human life itself could be calculated with De Moivre's new statistical methods, and Folkes and his brother William had Roubiliac visually represent the moment when the Duchess of Montagu's life was cut short by the Fates on her monument.

In pen and ink, the sketch too captured that precise moment in time between Folkes and his companion with that 'indefatigable watchfulness' considered by the founding members of the Royal Society as the essential quality of the 'experimental philosopher'.[22] Both natural philosophers and antiquaries in the eighteenth century were concerned with the 'visual, resonant, tactile materiality of reality', as Robert Hooke advocated in his *Micrographia*, the use of a 'sincere Hand and a faithful Eye, to examine and to record, the things themselves as they appear'.[23] Folkes examines the watch with magnifying eyeglasses, much like those he used to examine his coins or medals: as a numismatist and antiquary assessing and comparing their merits, fine details, their specific gravities determined with precision for his *Tables of English Silver and Gold Coins*, written to honour Newton. Folkes also thoroughly understood Newton's *Opticks* and the nature of optical illusion and human perception, directing Vanderbank to employ it on his own portraiture slyly to honour Newton in a unique manner.

---

[21] Ballantyne, *Voltaire's Visit to England*, p. 114. The inscribed copy is in the British Library, shelfmark C.60.g.11.

[22] Thomas Sprat, *The history of the Royal Society* (London: T. R. for J. Martyn, 1667), p. 337; Frédéric Ogée, 'Je-sais-quoi, William Hogarth and the representations of the forms of life', in Frederic Ogée, Peter Wagner, and David Bindman, eds, *Hogarth: Representing Nature's Machines* (Manchester: Manchester University Press, 2001), pp. 71–81, on p. 72.

[23] Ogée, 'Je-sais-quoi', p. 72.

The artist of the sketch too is sly. In this conversation piece, one is confronted by a daily environment of functional objects, furniture, a pipe, and a copy of 'Votes for the Commons', its truncated voting record. After the Exclusion Crisis, the House permitted a daily printing and publication of the vote, because 'it was impossible to Keep the Votes out of the Coffee-Houses'.[24] It seemed that parliament wanted to counter any false information being circulated in these venues: 'fake news'. In the sketch, there is the spectacle of a living world before one perceives the satirization of the characters that whiled away their time in the coffeehouse. For both Folkes and the Hogarthian artist, 'the work of art is an optical instrument which intensifies reality, opens it out to the eye and allows for a true experience of it'.[25]

Despite his emulation of Newton, Folkes was not always slavish in his admiration of Newton, not sharing his religious devotion. When taking Newtonianism on tour, Folkes would verify Newton's anti-Trinitarianism by repeatedly examining the ancient biblical codex in Rome to see for himself, no doubt with his eyeglasses. *Nullius in verba*. Newton, by contrast, had never left England, and relied on a second-hand account. Neither for Folkes nor for Voltaire were the chimeras of speculative philosophy or religious obscurantism convincing. Hooke stated, 'The truth is, the Science of Nature has been too long made only a work of the Brain and the Fancy: It is now high time that it should return to the plainness and soundness of Observations on material and obvious things'.[26] Folkes would have agreed.

---

[24] Martyn Atkins, 'Persuading the House: The Use of the Commons Journals as a Source of Precedent' in Paul Evans, eds, *Essays on the History of Parliamentary Procedure: In Honour of Thomas Erskine May* (London: Bloomsbury, 2017), pp. 69–86, on p. 77.

[25] Ogée, 'Je-sais-quoi', p. 81.

[26] Robert Hooke, 'Preface' to *Micrographia* (London: Jo. Martyn and Ja. Allestry, 1665), p. 5.

Bibliography

MANUSCRIPTS

Académie des Sciences, Paris

L'archive un dossier, Martin Folkes
Fonds Maupertuis

Beinecke Rare Book and Manuscript Library, Yale

The Osborn Collection
  Henry Baker Correspondence Box 19572
  Martin Folkes to Reverend Charles Morgan, 23 June 1724

Biblioteca dell'Archiginnasio, Bologna

Fondo Leprotti

Biblioteca Lancisiana di Roma

MS 281 and 282, Letters of Leprotti

Bodleian Library, Oxford

Arch.Num. X. 28,28*, [44 plates illustrating English gold and silver coins, circulated for observations to the committee of the Society of Antiquaries, in preparation for the edition of 1763, Tables of English silver and gold coins]
MS Ashmole 1816, Correspondence of Edward Lhwyd
MS Eng misc. *c.* 113, William Stukeley, Letter Book
MS Eng misc. *c.* 323, William Stukeley, The History of the Temples of the Ancient Celts, 1722
MS Eng. misc. *c.* 444, Martin Folkes, 'Journey from Venice to Rome'
MS Raw. N.6. Numismatic Collections
  Martin Folkes to Richard Rawlinson, 2 April 1743

British Library, London

Add MS 4222, 'Memoir of Martin Folkes', Birch's Biographical Collection, Collection of biographical notices and memoranda of literary, scientific and other illustrious persons, made by or for Thomas Birch; circa 1730–66, Vol II, D-L

Add MS 4307, Correspondence, Letters addressed to Thomas Birch, Volume 8, 1728–66

Add MS 4391, Papers Related to Coinage and Minting, ff. 31, 32 Martin Folkes, President of the Royal Society: Table of the standard of English silver coins, 1066–1601. Engr.: 1763

Add MS 4432–48, Collections relating to the Royal Society (incorporated in 1662) made by Thomas Birch while Secretary of the Society (1752–65), consisting of letters, articles or communications sent for consideration, many of which were subsequently printed in Philosophical Transactions of the Royal Society, draft minutes, etc. 1660–1765. Seventeen volumes

Add MS 4456, Verses, etc., almost entirely by Thomas Birch or addressed to him

Add MS 5833, Volume XXXII of Rev. William Cole's manuscripts, containing various miscellaneous collections, copies of correspondence, etc

Add MS 8834, Alexander Gordon, 'An Essay towards Illustrating the History, Chronology, and Mythology of the Ancient Egyptians' (1741)

Add MS 19939, Original letters of Dr Richard Pococke [Bishop successively of Ossory and Meath] to his mother, written during his travels in Italy, Germany, etc.; 30 Nov/10 Dec: 1733–Aug. 1737; and containing minute accounts of the places visited by him

Add MS 22067, Miscellaneous letters and papers, mostly addressed to, or concerning, Martin Folkes, Attorney-General to Katherine Queen Dowager, 1639–1704

Add MS 22978, Tour through France and North Italy, including Rome, in the form of letters to his mother; 28 Aug./8 Sept. 1733–30 June/11 July, 1734; with corrections in Pococke's hand, Italy: Travels in, by Dr Pococke: 1733–7

Add MS 23096, Volume 5 of Miscellaneous notes and extracts, chiefly relating to art and antiquities; with autobiographical and other memoranda, items of expenses, etc., by George Vertue; 1715–52

Add MS 32699, Correspondence of the Duke of Newcastle f. 24 Martin Folkes, Letters to the Duke of Newcastle: 1742–50

Add MS 32723, Correspondence of the Duke of Newcastle, f. 361 Martin Folkes, Letters to the Duke of Newcastle: 1742–50

Add MS 35397, Hardwicke Papers, Thomas Birch to Lord Hardwicke, 8 September 1750

Add MS 35398, Hardwicke Papers, Thomas Birch to Philip Yorke (Lord Royston), 14 October 1752, ff. 104–5

Add MS 52362, Minutes of the Egyptian Society, 1741–3

Add MS 58318, Diary of Andrew Mitchell, Travels in Italy, Rome, 1733

Add MS 71593, Dropmore Papers III, Vol. VII, ff. 97–9 Martin Folkes, President of the Royal Society: Coinage: Table of relative value of silver at different times, by M. Folkes; with anon. explanatory note: [aft. 1754]: Partly printed

MS Edgerton 1041–1042, Minutes of the Society of Antiquaries, from its formation in 1717 to the end of 1751, with a list of the members to Jany. 1758, and of the purchases and donations to June 1753

MS Facsimile 589, Letters to Dr Antonio Cocchi. Photostats (40) of letters to Dr Antonio Cocchi, mainly from scholars and scientists, including M. Folkes, J. Spence, R. Mead, Lord John Hervey, R. Hervey(?), P Coste, La Condamine, Alan Ramsey, T. Birch, Abbé Nollet, G. van Swieten, and from the family of Conte Enrico Baldasseroni of Florence; 1730–5. Enlarged from microfilms in the University of London library. Presented by Dr. Nicholas Adolph Hans

Leghorn MS Facs 589, photographs from manuscripts in the family archive of Conte Enrico Baldasseroni in Florence, a descendant of Antonio Cocchi

MSS Sloane, Correspondence to Hans Sloane from Martin Folkes

Cambridge University Library

Keynes MS 130, f. 5, Martin Folkes's recollections of Newton
MS Add 3989, Observations upon the Prophecies of Daniel and the Apocalypse of St John

Christ Church Archives, Oxford

MS. 541a, Archbishop Wake's autobiography
William Wake Letters

Clare College, Cambridge

Charles Morgan Papers, NRA 33326
Clare College Letterbook

C.P. Hoare & Co. Archives

Ledger: 36/357/429, Martin Folkes

Electronic Enlightenment

Letters of Maupertuis and Voltaire

Folger Shakespeare Library, Washington DC

MS Y.c. 946 (1), Letter from Martin Folkes to Francis Hawksbee, 10 July 1746
MS Folger W.b. 111, f. 10a, 23 January 1714, Costume Bill for Lucretia Bradshaw, Drury Lane
MS Folger W.b. 111, f. 13(a), 26 April 1714, Costume Bill for Lucretia Bradshaw, Drury Lane

Freemasons Hall Library, London

William Stukeley Manuscripts
  Ashtaroth, ABRAHAM
  On Egyptian Antiquitys. 1742

Guildhall Library, London

MS 6832, Diocesan Record Offices, p. 26, Marriage Entry for Martin Folkes and Lucretia Bradshaw, 18 October 1714

Huntingdonshire Library and Archives, Huntingdon, Cambridgeshire

HINCH 2/69, Deed of Indemnity. 1. Rt. Hon. John, 5th Earl of Sandwich. 2. William Palmer of Brampton, co. Huntingdon, Esq.

Kent Record Offices, Canterbury

MS U1590/C21, 2nd Earl Lord Stanhope/Martin Folkes letters

John Rylands Library, Manchester

Eng MSS 19, Henry Baker Correspondence

London Metropolitan Archives

A/CSC/1514, Charter of the Corporation of the Sons of the Clergy, 1 July 1678
A/FH/A09/001, Billet Book, March-May 1741
A/FH/K02/001–023 (1739–1800) (X041/014) General Committee Minutes, Foundling Hospital

The National Archives, Kew

MINT 19/6/vol. 4, 127a, 'Newton manuscripts in the library of the Royal Mint'
PROB/682/5, Will of William Wake, Lord Archbishop of Canterbury, 1 March 1737
PROB/11/809/301, Will of Martin Folkes, 3 July 1754
PROB/31/50/413, Will of Solomon Negri, 26 July 1727
SP 36/19/54–58 'An account of a highway robbery', 19 June 1730

New College, Oxford

NCL MSS 361.1–4, Four Newton Autograph Volumes, containing drafts related to Newton's chronological work

Norfolk Record Office, Norwich

NRS 20658, Martin Folkes Esq. *Memoranda in case he should go abroad again, respecting his Estates, foreighn money & Places*
MC 50-1-4, Martin Folkes's copy of will of Martin Browne
MC 50-1-6, Will of Dorothy Hovell Folkes, 28 May 1719
MC 50-1-8, Genealogy of Martin Folkes Senior and his parents in the Norwich Archives: 'An Extract, taken out of the Registry book, of Births, Marriages, and burials, kept in the parish of Rushbrooke, in the county of Suffolk'
MC 50-2-114, Letter from Charles Lennox to William Folkes, 2 August 1740
MC 50-2-115, Letter from Charles Lennox to William Folkes, 3 August 1740

MC 50-2-116, Letter from Charles Lennox to William Folkes, 8 August 1740

MC 50-3-5, Martin Folkes, Caen to William Folkes, Chancery Lane, London, 23 April 1739

MC 50-6, Martin Folkes to William Folkes regarding boundaries of Hillington Estate fields, 24 March 1746, with hand-drawn maps by Folkes

MC 342-1, Statement to pay interest in South Sea Stock (£1285) jointly owned by Martin Folkes and William Churchill to Churchill, 20 July 1730

MC 783-1, Undated letter from Martin Folkes to William Folkes

NRS 11526, Manuscript will of Dorothy Folkes, mother of Martin Folkes; codicil regarding Lucretia Folkes, wife of Martin Folkes

NRS 11567, Manuscript will of Martin Folkes

Nottingham University Library, Department of Manuscripts and Special Collections

Ga 12885, Notebook containing Extracts taken from amongst Sir Isaac Newton's papers relating to directions about the trail of the monies of gold and silver in the pix and other technical notes relating to the Mint

Parish of St Giles-in-the-Fields, London

Christening Book

Pierpont Morgan Library, New York

MS MA 176E, Deed of Conveyance, House of Nell Gwyn

MS MA 179 A, Letters patent under the Great Seal, 1676 Dec. 1, granting to William Chaffinch and Martin Folkes (agents for Nell Gwyn and Henry Jermyn, Earl of St Albans) reversion and inheritance of the property then occupied by Nell Gwyn

Royal Society Library, London

Mezzotints of Presidents, Martin Folkes

RSC, Royal Society Club Archive

RS/Cl.P/2/17, 'Some Account of Mr Leeuwenhoek's curious microscopes lately presented to the Royal Society' by Martin Folkes'

RS/Cl.P/8i/67, Observations of Jupiter's satellites and Saturn's rings with reflecting telescope by John Hadley, 1721

RS/CMB/63, Repository Committee of the Royal Society, 15 January 1730–30 October 1733

RS/CMO/1, Minutes of meetings of the Council of the Royal Society, 13 May 1663–8 February 1682/83

RS CMO/2, Minutes of meetings of the Council of the Royal Society, 15 February 1682-9 to November 1727

RS/CMO/3, Minutes of meetings of the Council of the Royal Society, 4 January 1727/28–30 November 1747

RS/CMO/4, Minutes of meetings of the Council of the Royal Society, 6 January 1748–27 January 1763

RS/DM/5/8/2, Chronological list dating from 1633–1722, of donations to the Royal Society, 'extracted out of Journal Books of the weekly meetings'

RS/EC series, Certificates of Election and Candidature for Fellowship of the Royal Society, 1731–

RS/EL/M3/32, Pietro Antonio Michelotti to John Machin, 6 October 1733

RS/JBC/12, Minutes of ordinary meetings of the Society, 1720–6

RS/JBO/10, Journal Book of the Royal Society Volume 10, 1696–1702

RS/JBO/11, Journal Book of the Royal Society Volume 11, 1702–14

RS/JBO/12, Journal Book of the Royal Society Volume 12, 1714–20

RS/JBO/13, Journal Book of the Royal Society Volume 13, 1720–6

RS/JBO/14, Journal Book of the Royal Society Volume 14, 1726–31

RS/JBO/15, Journal Book of the Royal Society Volume 15, 1731–4

RS/JBO/16, Journal Book of the Royal Society Volume 16, 1734–6

RS/JBO/17, Journal Book of the Royal Society Volume 17, 1736–9

RS/JBO/18, Journal Book of the Royal Society Volume 18, 1739–42

RS/JBO/19, Journal Book of the Royal Society Volume 19, 1742–5

RS/JBO/20, Journal Book of the Royal Society Volume 20, 1745–8

RS/JBO/21, Journal Book of the Royal Society Volume 21, 1748–51

RS/JBO/22, Journal Book of the Royal Society Volume 22, 1751–3

RS/JBO/34/, Journal Book of the Royal Society Volume 34, 1789–92

RS/LBO/16/Copies of letters sent to the Royal Society—Volume XVI, 1712–23

RS/L&P/1/174, 6 February 1742, Letter, 'Note on Réaumur's history of insects, 6th volume, being an extract from a letter referring to the account of Trembley's freshwater polyp' from Pierre Louis Moreau de Maupertuis to unknown recipient [Folkes]

RS/L&P/1/360, Copy of Letter of Gasparo Cerati to Cromwell Mortimer, 1 January 1745

RS/L&P/1/479, Paper, 'Journal of observations made in Peru; and account by Martin Folkes' by Antonio de Ulloa, 1746

RS/L&P/2/10D/60, 62, 63, Accounts of the 1749 London Earthquake

RS/L&P/9/385, 'An Account of a Print showed to the Royal Society, 25 April 1745'

RS/L&P/12, Decade 12 of scientific letters and papers sent to the Royal Society

RS/MM/20, Miscellaneous Manuscripts

RS/MS/130/11/11, Drawing of a ballistic pendulum by Benjamin Robins

RS/MS/213, Table of Contents of Journal Book, 1741–8, compiled by Martin Folkes

RS/MS/250, Papers and Correspondence of Martin Folkes, 1710–54

RS/MS/645, Minutes of Meetings of the Council of the Royal Society

RS/MS/702, An Account of the Royal Society from it's [sic] first institution by Martin Folkes

RS/MS/704, Index of Papers read before the Royal Society, 1716–1738, compiled by Martin Folkes

RS/MS/865, Letters of Charles Lennox, 2nd Duke of Richmond and Duke of Aubigny, to Martin Folkes, 1725–1744

RS/MS/790, Correspondence of Martin Folkes

RS/MM/16/41, letter from François Marie Arouet de Voltaire, Paris, to Martin Folkes, 29 March 1743. A copy can be found at RS/MM/15/10

RS/RBO/13/5, 'An account of a Tract de hyrometris et corum defectibus' by Martin Folkes

RS/RBO/19/52, 'Of the Runic Characters of Helsingland (northern Sweden)' by Andreas Celsius

RS/RBO/19/57, Observations of the Aurora Borealis witnessed in London from September 1735 to April 1736 by Andreas Celsius

Society of Antiquaries of London Library

SAL/MS/264 B, Register transcribed by Joseph Ames of papers read or submitted to
    the Society
SAL/MS/265 Minute Books, vols 1–33, 1717–1817
SAL/MS/266, William Stukeley, 'Memoirs towards an History of the Antiquarian Society'
    read by Stukeley on 23 April 1752, 'Copyed by Mr Ames from mine, with Additions'
SAL/MS/657, papers relating to the publishing of Folkes's posthumous numismatic works

Spalding Gentlemen's Society, Spalding, Lincolnshire

William Stukeley, Memoirs of the Royal Society, 4 vols, 1740–1750

Saint-Omer Bibliothèque d'agglomération du Pays, France

MS 786, 'An abstract of Chronology by Sir Isaac Newton'

Trembley Family Archives, Geneva

Folkes-Trembley Correspondence

University Library Erlangen-Nürnberg

MS 1471, Friendship album for Christoph Jacob Trew (1695–1769)

University of Virginia, Charlottesville

MS 4530; Cromwell Mortimer Royal Society letters, 1746–9

Victoria & Albert, National Art Library, London

Autobiographical memoranda, correspondence with William Arderon in 4 vols, GB 072
    Forster Collection

Wellcome Library, London

MS 1302, Charles Bonnet, Recherches sur la respiration des chenilles. Author's holograph
    MS. Holograph Abstract of the same in French by Abraham Trembley. Holograph
    translation of the Abstract by Martin Folkes. Fragments of the author's holograph notes
    on 'Salamandre', 'Limaçon', 'Memory', etc. Endorsed by Martin Folkes in respect of
    the 'Abstract' as read before the Royal Society, and printed in the 'Transactions',
    No. 487

MS 2391, Memoirs and Papers sent to The Royal Society of London, Collection I. Eleven short Memoirs and Papers sent to the Royal Society of London, including four holograph items by Folkes: all are initialed, signed, or endorsed by him

MS 2392, Memoirs and Papers sent to The Royal Society of London, Collection II. Fourteen short Memoirs and Papers sent to the Royal Society of London, initialed, signed, or endorsed by M. Folkes

MS 4726, William Stukeley, The pictures in the Cottonian MS. of Genesis [Otho B. vi]. Author's holograph MS. Illustrated with 20 plates pasted down, and cut up from the original two engraved copper-plates issued by the Antiquaries Society in 1749, Signature of Stukeley inside the upper cover, and a note by him on the first (unnumbered) leaf '13. Aug. 1749. Martin Folkes Esq. LL.D. gave me the set of prints done by the Antiquarian Society'

MS 5403, Memoirs, Letters and Papers sent to The Royal Society of London, and mostly published in the *Philosophical Transactions*. Most are initialled by Martin Folkes and endorsed with the dates when they were read to the Society

MS 6143, Letters from Jurin to Antonie van Leeuwenhoek (1632–1723), 1722–24

West Sussex Record Office, Chichester

Goodwood MS 110, Correspondence between Charles Lennox, 2nd Duke of Richmond and Martin Folkes

Goodwood MS 112, Correspondence between William Folkes and Martin Folkes, 8 September 1751

PRINTED PRIMARY SOURCES AND PRIMARY SOURCE EDITIONS

Aga, Cassem, *An account of the success of inoculating the small-pox in Great Britain, for the years 1727 and 1728. With a comparison between the mortality of the natural small-pox, and the miscarriages in that practice; as also some general remarks on its progress and success, since its first introduction. To which are subjoined, I. An account of the success of inoculation in foreign parts. II. A relation of the like method of giving the small-pox, as it is practised in the kingdoms of Tunis, Tripoli, and Algier. Written in Arabic by his excellency Cassem Aga, ambassador from Tripoli. Done into English from the French of M. Dadichi, His Majesty's interpreter for the eastern languages.* London: J. Peele, 1729.

Ahlers, Cyriacus, *Some Observations Concerning the Woman of Godlyman in Surrey. Made at Guilford on Sunday, Nov. 20. 1726: Tending to Prove Her Extraordinary Deliveries to be a Cheat and an Imposture.* London: J. Roberts, 1726.

Algarotti, Francesco, *Oeuvres de comte Algarotti traduit de l'italien*, 7 vols. Berlin: G. J. Decker, 1772.

Algarotti, Francesco, *Opere Varie Del Conte Francesco Algarotti*. Venice: Giambattista Pasquali, 1757.

Ames, Joseph, *Numismata Antiqua in tres partes divisa. Collegit olim et aeri incidi vivens curavit Thomas Pembrochiae et Montis Gomerici Comes.* London: n. p., 1747.

'An Account of a comparison lately made by some gentleman of the Royal Society, of the standard of a yard, and the several weights lately made for their use; with the original standards of measure and weights in the exchequer, and some others kept for public use, at Guild-hall, Founders hall, the tower &c', *Philosophical Transactions* 42, 470 (1743), pp. 541–56.

An Account of the Foundling Hospital in London for the Maintenance and Education of Exposed and Deserted Young Children. London: Luke Hansard, 1817.

Anderson, James, *The New Book of Constitutions of the Antient and Honourable Fraternity of Free and Accepted Masons*. London: Caesar Ward and Richard Chandler, 1738.

[Anonymous], 'A Geometrical Use Proposed from the Fire-Works', *Gentleman's Magazine* 18 (November 1748), p. 488.

Baker, Henry, 'An Account of Mr Leeuwenhoek's Microscopes', *Philosophical Transactions* 41, 458 (1739–41), pp. 503–19.

Baker, Henry, 'An easy method of procuring the true impression or figure of medals, coins, &c. humbly addressed to the Royal Society', *Philosophical Transactions* 43, 172 (1744), pp. 135–7.

Baker, Henry, *Employment for the Microscope*. London: R. Dodsley, 1753.

Baker, Henry, 'A letter from Mr. Henry Baker F. R. S. to the President, concerning several medical experiments of electricity', *Philosophical Transactions* 45, 486 (1748), pp. 270–75.

Baker, Henry, *Micrographia Restaurata*. London: John Bowles, 1745.

Baker, Henry, *The Microscope Made Easy*. London: R. Dodsley, 1743.

[Bellers, Fettiplace], *Injur'd Innocence: A Tragedy. As it is Acted at the Theatre-Royal in Drury Lane*. London: J. Brindley, 1735.

Berkeley, George, *Saggio d'una nuova teoria sopra la visione... ed un discorso preliminare al Trattato della cognizione*. trans. Giovanni Pisenti. Venice: Francesco Storti, 1732.

Berward, Christian, *Interpres phraseologiae metallurgicae*. Frankfurt: Johann David Zunner, 1684.

Besterman, Theodore, ed. and trans., *Select Letters of Voltaire*. London: Thomas Nelson and Sons, 1963.

Betterton, Thomas, *The history of the English stage from the Restauration to the Present Time*, London: E. Curll, 1741.

Bond, Donald F., ed., *The Tatler*, 3 vols. Oxford: Clarendon Press, 1987.

Bradley, James, 'A Letter from the Reverend Mr. James Bradley Savilian Professor of Astronomy at Oxford, and F.R.S. to Dr Edmond Halley Astonom. Reg. &c. giving an Account of a new discovered Motion of the Fix'd Stars', *Philosophical Transactions* 35, 406 (1727–8), pp. 637–61.

[Byrom, John,] *A Full and True Account of an Horrid and Barbarous Robbery, Committed on Epping-Forest, upon the Body of the Cambridge Coach, in a letter to M.F. Esq*. London: J. Roberts, 1728.

Byrom, John, *The Private Journal and Literary Remains of John Byrom*. 2 parts in 4. Manchester: Chetham Society, 1854–7.

A Catalogue of the Genuine and Curious Collection of Mathematical Instruments, Gems, Pictures, Bronzes, busts, Urns, Cabinets, curious Clocks, Book Cases, etc of Martin Folkes, Esq. London: Samuel Langford, 1755.

A catalogue of the... library of Martin Folkes, esq., president of the Royal society... lately deceased; which will be sold by auction by Samuel Baker... To begin... February 2, 1756, and to continue for forty days successively (Sundays excepted). London: Samuel Baker, 1756.

Celsius, Anders, 'Observations of the Aurora Borealis Made in England by Andr. Celsius, F. R. S. and Secr. R. S. of Upsal in Sweden', *Philosophical Transactions* 39, 441 (1735), pp. 241–4.

Celsius, Anders, 'An Explanation of the Runic Characters of Helsingland', *Philosophical Transactions* 40, 445 (1737/8), pp. 7–13.

Cibber, Colley, *An Apology for the Life of Colley Cibber: Comedian and Late Patentee of the Theatre Royal*. London: John Watts, 1740.

Clarke, John and Samuel, eds, *Jacobi Rohaulti Physica*. London: Jacobi Knapton, 1697; trans. John Clarke, London, 1723.

Cohen, I. B. ed., *A Treatise of the System of the World by Isaac Newton*. Philadelphia: American Philosophical Society, 2004.

Conti, Antonio, *Lettere scelte di celebri autori all' Ab. Antonio Conti*. Venice: Domenico Fracasso, 1812.

Cooper, Anthony Ashley, Earl of Shaftesbury, *Characteristicks of Men, Manners, Opinions, Times*, 3 vols. London: s.n., 1711.

A copy of the Royal Charter, Establishing an Hospital for the Maintenance and Education of Exposed and Deserted Young Children. London: J. Osborn, 1739.

The Daily Journal. London, 1721–37.

Defoe, Daniel, 'On the Playhouse in the Haymarket', *A Review of the Affairs of France* (3 May 1705), vol. 2., pp. 103–4.

De Moivre, Abraham, *The Doctrine of Chances*. London: W. Pearson, 1718.

Desaguliers, J. T., 'An Account of Experiments made on 27 April 1719 to find how much Resistance of Air retards falling Bodies', and 'Further Account of Experiments made on 27 April last', *Philosophical Transactions* 30, 362 (1717–19), pp. 1071–8.

Desaguliers, J. T., 'Optical Experiments made in the beginning of August 1728, before the President and Several Members of the Royal Society, and Other Gentlemen of Several Nations, upon Occasion of Signior Rizzetti's Opticks...', *Philosophical Transactions* 35, 406 (1727–8), pp. 596–629.

Doble, C. E., ed., *Remarks and Collections of Thomas Hearne*, 11 vols. Oxford: Clarendon Press, 1885–1921.

Dürer, Albrecht, *Divae Parthenices Mariae Historiam ab Alberto Durero [Life of the Virgin]*. Nürnberg: Hieron. Hölzel, 1511.

Edwards, George, *Natural History of Uncommon Birds and of Some Other Rare and Undescribed Animals*. London: For the author, 1743–51.

Eisenschmid, Johann Caspar, *De ponderibus et mensuris...De valore pecuniae veteris*. Strasbourg: Henr. Leo Stein, 1708.

The Evening Post. London: E. Berington, 1710–32.

Fabretti, Raphaelis, *De Columna Traiani syntagma*. Rome: Nicolai Angeli Tinassij, 1693.

Finberg, Alexander J., ed., '[George] Vertue's Note Book B.4 [British Museum. Add. MS. 23, 704]', *The Volume of the Walpole Society* 22 (1933–4), pp. 143–62.

Folkes, Martin, 'An Account of the Aurora Borealis, Seen at London, on the 30th of March Last, as It Was Curiously Observ'd by Martin Folkes', *Philosophical Transactions* 30, 352 (1719), pp. 586–8.

Folkes, Martin, 'An Account of some human Bones incrusted with Stone, now in the Villa Ludovisia at Rome: communicated to the Royal Society by the President, with a Drawing of the same', *Philosophical Transactions* 43, 477 (1744), pp. 557–60.

Folkes, Martin, 'An Account of the Standard Measures Preserved in the Capitol at Rome', *Philosophical Transactions* 39, 442 (1735–6), pp. 262–6.

Folkes, Martin, 'An observation of three mock-suns seen in London, Friday, Sept. 17, 1736', *Philosophical Transactions* 40, 445 (1736), pp. 59–61.

Folkes, Martin, 'On the Trajan and Antonine Pillars at Rome: Read 5 February 1735–6', *Archaeologia* 1, 2 (1779), pp. 117–21.

Folkes, Martin, 'A remark on Father Hardouin's amendment of a passage in Pliny's natural history, lib. 11. LXXIV Edt. Paris. Folio, 1723', *Philosophical Transactions* 44, 482 (1746), pp. 365–70.

[Folkes, Martin,] 'Some Account of the Insect called the Fresh-water Polypus, before-mentioned in these Transactions, as the same was delivered at a Meeting of the Royal Society by the President, on Thursday, March 24. 1742–3', *Philosophical Transactions* 42, 469 (1742–3, pub. 1744), pp. 422–36.

Folkes, Martin, 'Some account of Mr. Leeuwenhoek's curious microscopes lately presented to the Royal Society', *Philosophical Transactions* 32, 380 (1723), pp. 446–53.

Folkes, Martin, *A table of English gold coins: from the eighteenth year of King Edward III, when gold was first coined in England: with their several weights and present intrinsic values*. London: Society of Antiquaries, 1736.

Folkes, Martin, *A Table of English silver coins from the Norman conquest to the present time: with their weights, intrinsic values, and some remarks upon the several pieces*. London: Society of Antiquaries, 1745.

Folkes, Martin, *Tables of English Silver and Gold Coins: First published by Martin Folkes, Esq; and now Re-printed, With Plates and Explanations*. London: Society of Antiquaries, 1763.

[Folkes, William,] 'Part of a letter from William Folkes, Esq; F.R.S. to his brother the President, concerning a shock of an earthquake felt at Newton in Northamptonshire, on Sunday. September 30, 1750', *Philosophical Transactions* 46, 497 (1750), pp. 701–2.

Franklin, Benjamin, *Experiments and Observations on Electricity, Made at Philadelphia in America, by Mr. Benjamin Franklin, and Communicated in several Letters to Mr. P. Collinson, of London, F.R.S.* London: Printed and sold by E. Cave, at St John's Gate, 1751.

Gauger, Nicolas, 'Lettre à M. l'abbé Conti, juillet 1727', *Continuation des Mémoires de Littérature et d'Histoire* 5, 1 (1728), pp. 10–51.

Gebelin, François and M. André Morize, eds, *Correspondance de Montesquieu*, 2 vols. Bordeaux: Gounouilhou, 1914.

Gentleman's Magazine, London: 1731–1922.

Ghobrial, John-Paul A., 'The Life and Hard times of Solomon Negri: An Arabic Teacher in Early Modern Europe'. In Jan Loop, Alastair Hamilton, and Charles Burnett, eds, *The Teaching and Learning of Arabic in Early Modern Europe*. Leiden: Brill, 2017, pp. 310–31.

Gildon, Charles, *The Life of Mr. Thomas Betterton: The Late Eminent Tragedian*. London: Robert Gosling, 1710.

Gordon, Alexander, *An Essay towards explaining the hieroglyphical figures on the coffin of the ancient mummy belonging to Captain William Lethieullier* (London: For the author, 1737.

Gordon, Alexander, *An Essay Towards Explaining the Antient Hieroglyphical Figures on the Egyptian Mummy, in the Museum of Doctor Mead, Physician in Ordinary to his Majesty*. London: For the author, 1737.

'Graham's brass standard yard', Science Museum Group Collection, London. https://collection.sciencemuseumgroup.org.uk/objects/co8028873/grahams-brass-standard-yard-measuring-yard [Accessed 4 August 2020].

'Graham's diagonal scale', Science Museum Group Collection, London. https://collection.sciencemuseumgroup.org.uk/objects/co60115/diagonal-scale-screwed-to-mahogany-base-drawing-instruments-diagonal-scales [Accessed 4 August 2020].

Greaves, John, *A Discourse of the Romane Foot, and Denarius: from when, as from two principles, the Measures, and Weights, used by the Ancients, may be deduced*. London: M. F., 1647.

Greene, Robert, *ΕΓΚΥΚΛΟΠΑΙΔΕΙΑ (Encyclopaedia) or method of instructing pupils*. Cambridge: n. p., 1707.

Greene, Robert, *The Principles of the Philosophy of the Expansive and Contractive Force, or an Enquiry into the principles of the Modern Philosophy, that is into the General Chief rational Sciences*. Cambridge: Cornelius Crawfield, 1727.

Greenhill, Thomas, *[Nekrokedeia]: Or, the Art of Embalming*. London: For the author, 1705.

Hadley, John, 'An Account of a Catadioptrick Telescope, Made by John Hadley, Esq; F. R. S. With the Description of a Machine Contriv'd by Him for the Applying It to Use', *Philosophical Transactions* 32, 376 (1722–3), pp. 303–12.

Hadley, John, 'The Description of a new Instrument for taking Angles', *Philosophical Transactions* 37, 420 (1731), pp. 147–57.

Hales, Stephen, 'Some Considerations on the causes of earthquakes', *Philosophical Transactions* 46, 497 (1750), pp. 669–81.

Hall, A. Rupert, J. F. Scott, Laura Tilling, and H.W. Turnbull, eds, *The Correspondence of Isaac Newton*, 7 vols. Cambridge: Cambridge University Press, 1959–81.

Halley, Edmond, 'Observations of the late total eclipse of the sun on the 22d of April last past, made before the Royal Society at the house in Crane Court in Fleet-Street London', *Philosophical Transactions* 29, 343 (1714), pp. 245–62.

[Hill, John,] 'Art. CXXXIX. Philosophical Transactions, Number 488. For the Month of June, 1748', *Monthly Review* 2 (1749–50), pp. 466–75.

Hill, John, *A review of the works of the Royal Society*. London: R. Griffiths, 1751.

Honeybone, Diana and Michael, eds, *The Correspondence of William Stukeley and Maurice Johnson, 1714–54*. Woodbridge: Boydell Press, 2014.

Hooke, Robert, *Micrographia*. London: Jo. Martyn and Ja. Allestry, 1665.

Hooke, Robert, *The Posthumous Works of Robert Hooke; containing his Cutlerian Lectures, and other discourses read at the meetings of the illustrious Royal Society…published by R. Waller*. London: Samuel Smith and Benjamin Walford, 1705.

Horace Walpole's Correspondence. New Haven: Yale University Press, online edition. http://images.library.yale.edu/hwcorrespondence [Accessed 18 January 2020].

Hutton, Charles, *Tracts on Mathematical and Philosophical Subjects*. London: F. C. and J. Rivington, 1812.

'Introduction: containing an historical account of the origin and establishment of the Society of Antiquaries', *Archaeologia* 1 (1770), pp. i–xxxix.

Ireland, Samuel, *Graphic Illustrations of Hogarth*. London: R. Faulder, 1794.

Jessop, T. E. and A. A. Luce, eds, *The Works of George Berkeley, Bishop of Cloyne*, 9 vols. London: Nelson, 1948–64.

J.G.D.M.F.M., *Relation Apologique et Historique de la Société des Franc-Maçons*. Dublin: False imprint, 1738.

Juan, Jorge and Antonio de Ulloa, *Relación Histórica del Viage a la América Meridional*. 4 vols. Madrid: Antonio Marin, 1748.

Jurin, James, 'Dedication to Martin Folkes, Esq; Vice-President of the Royal Society', *Philosophical Transactions* 34, 392 (1727), p. i.

Knight, Gowin, 'A Description of a Mariner's Compass contrived by Gowin Knight, M.B. F.R.S', *Philosophical Transactions* 46, 495 (1750), pp. 505–12.

Knight, Gowin, 'A letter to the President, concerning the poles of magnets being variously placed', *Philosophical Transactions* 43, 476 (1745), pp. 361–3.

Labaree, Leonard W. and Barbara B. Oberg, eds, *The Papers of Benjamin Franklin*, 28 vols. New Haven: Yale University Press, 1961–90.

Le Blanc, Jean-Bernard, *Letters on the English and French Nations*, 2 vols. London: J. Brindley, 1747.

Lémery, Nicolas, *Course of chymistry: containing an easie method of preparing those chymical medicines which are used in physick; with curious remarks and useful discourses upon each preparation, for the benefit of such a desire to be instructed in the knowledge of this art. - The 3rd. ed., transl. from the 8th ed. in the French*. London: Walter Kettilby, 1698.

Leti, Gregorio, *Critique historique, politique, morale, économique & comique, sur les loteries, anciennes & modernes, spirituelles & temporelles*. Amsterdam: Chez les Amis de l'Auteur, 1697.

Lister, Martin, *De Fontibus medicates Angliae*. London: Walter Kettilby, 1684.

Lister, Martin, *Exercitatio anatomica In qua de Cochleis, Maximè Terrestribus & Limacibus, agitur*. London: Samuel Smith and Benjamin Walford, London, 1694.

London Evening Post. London.

The London Gazette. London.

Lyster [Lister], Martin, 'The Third Paper of the Same Person, Concerning Thunder and Lightning being from the Pyrites', *Philosophical Transactions* 14, 157 (1684), pp. 517-19.

McKenzie, D. F. and C. Y. Ferdinand, ed. William Congreve, *Works*, 3 vols. Oxford: Oxford University Press, 2011.

MacPike, Eugene Fairfield, ed., *Correspondence and Papers of Edmond Halley*. London: Taylor and Francis, Ltd., 1937.

Magnus, Olaus, *Historia Olai Magni Gothi Archiepiscopi Upsalensis…* Basel: Henric Petrina, 1567.

Manley, Mrs [Delariviere], *The Royal Mischief; A Tragedy. As it is Acted by His Majesties Servants*. London: R. Bentley, S. Saunders, and J. Knapton, 1696.

Martyn, John, *The Philosophical Transactions from the Year 1743, to the Year 1750, Abridged and Disposed under General Heads*. London: Lockyer Davis and Charles Reymers, 1750.

Millan, J., *Coins, Weights and Measures. Ancient and Modern. Of all Nations. Reduced into English on above 100 Tables. Collected and Methodiz'd from Newton, Folkes, Arbuthnot, Fleetwood…* London: J. Millan, 1749.

Montagu, John, 4th Earl of Sandwich, *A Voyage Performed by the Late Earl of Sandwich Round the Mediterranean in the Years 1738 and 1739 Written by Himself*. London: T. Cadell, 1799.

Monthly magazine, or British register (1796–1822).

Morgan, Charles, *Six philosophical dissertations…published by Dr. Samuel Clarke in his notes upon Rohault's Physics*. Cambridge: Fletcher and Hodson, 1770.

[Mottley, John,] *A Complete List of All the English Dramatic Poets and of all the Plays ever Printed in the English Language, to the Present Year M, DCC, XLVII*, appended to Thomas Whincop, *Scanderbeg: or, Love or Liberty. A Tragedy*. London: W. Reeve, 1747.

Murdoch, Patrick, *An Account of Sir Isaac Newton's Philosophical Discoveries in Four Books by Colin Maclaurin*. London: For the author's children, 1748.

Newton, Isaac, *The mathematical principles of natural philosophy. By Sir Isaac Newton. Translated into English by Andrew Motte…in two volumes*. London: Benjamin Motte, 1729.

Newton, Isaac, *Opticks*. New York: Dover, 1952.

Nichols, John, *Biographical and Literary Anecdotes of William Bowyer: Printer, F.S.A*. London: For the author, 1782.

Nichols, John, *Literary Anecdotes of the Eighteenth Century*, 6 vols. London: Nichols, Son, and Bentley, 1812.

Parsons, James, 'An Account of a Very Small Monkey, Communicated to Martin Folkes Esq; LL. D. and President of the Royal and Antiquarian Societies, London; By James Parsons M. D. F. R. S', *Philosophical Transactions* 47 (1751-20), pp. 146-50.

Pearce, Zachary, *A Commentary with Notes on the Four Evangelists and the Acts of the Apostles*. London: E. Cox, 1778.

Perrault, Charles, *Parallèle Des Anciens et Les Modernes En Ce Qui Regarde L'Éloquence*, 4 vols. Paris: J. B. Coignard, 1688–97.

Perry, Charles, *A View of the Levant: Particularly of Constantinople, Syria, Egypt and Greece*. London: T. Woodward, 1743.

Péter, Róbert, Jan A. M. Snoek, and Cécile Révauger, eds, *British Freemasonry, 1717–1813*, 5 vols. New York: Routledge, 2016.

Piles, Robert de, *The Art of Painting: and the Lives of Painters: Containing, a Compleat Treatise of Painting, Designing, and the Use of Prints*. London: J. Nutt, 1706.

Pine, John, *The Procession and Ceremonies Observed at the Time of the Installation of the Knights Companions of the Most Honourable Military Order of the Bath*. London: S. Palmer and J. Huggonson, 1730.

Pix, Mary, *The Deceiver Deceiv'd*. London: R. Basset, 1698.

Plan of the cities of London and Westminster and borough of Southwark, with the contiguous buildings is humbly inscribed taken by John Rocque, Land-Surveyor, and engraved by John Pine. London, 1749.

Pococke, Richard, *A Description of the East and Some Other Countries*. London: W. Bowyer, 1743.

Polignac, Cardinal Melchior de, *Anti-Lucretius sive de Deo et Natura*. Paris: H. -L. and J. Guérin, 1747; Amsterdam: M. M. Rey, 1748.

Public Advertiser, London 1744–94.

Registres de l'Academie, 1746–86, Berlin-Brandenburgische Akademie der Wissenschaften. https://akademieregistres.bbaw.de/index.html [Accessed 30 November 2019].

Reland, Adrian, *Palaestina ex monumentis veteribus illustrate*. Utrecht: Willem Broedelet, 1714.

Reynardson, Samuel, 'A State of the English Weights and Measures of Capacity', *Philosophical Transactions* 46 (1749), pp. 54–71.

Richardson, Jonathan, *An essay on the whole art of criticism as it relates to painting and an argument in behalf of the science of the connoisseur*. London: W. Churchill, 1719.

Robins, Benjamin, 'Demonstration of the Eleventh Proposition of Sir I. Newton's Treatise of Quadratures', *Philosophical Transactions* 34, 397 (1727), pp. 230–6.

Robinson, Bryan, *Isaac Newton's Account of the Aether*. Dublin: G. and A. Ewing, 1745.

Robinson, Bryan, *A Treatise of the Animal Oeconomy*. Dublin: George Grierson, 1732.

Roe, Shirley A., 'John Turberville Needham and the Generation of Living Organisms', *Isis* 74, 2 (1983), pp. 158–84.

Rusnock, Andrea, ed., *The Correspondence of James Jurin (1684–1750): Physician and Secretary of the Royal Society*. Amsterdam and Atlanta, GA: Rodopi, 1996.

Rymsdyk, John and Andrew van, *Museum Britannicum, Being an Exhibition of a Great Variety of Antiquities*. London: I. Moore, 1778.

Sansovino, Francesco, *Venetia, città nobilissima, et singolare: descritta in XIIII. Libri*. Venice: Steffano Curti, 1663.

Saunderson, Nicholas, *The Elements of Algebra, in Ten Books*. Cambridge: Cambridge University Press, 1741.

Shakespeare, William, *The works of Shakespeare: in seven volumes. Collated with the oldest copies, and corrected; with notes, explanatory and critical by Mr. Theobald*, 7 vols. London: A. Bettesworth and C. Hitch, 1733.

Shaw, Thomas, *Travels or Observations relating to several parts of Barbary and the Levant*. Oxford: Printed at the Theatre, 1738.

Shaw, William A., ed., *Calendar of Treasury Books, Volume 19, 1704–1705*. London: His Majesty's Stationery Office, 1938. *British History Online*, http://www.british-history.ac.uk/cal-treasury-books/vol19/pp461-468 [Accessed 29 May 2019].

Sheppard, F. H. W., ed., *Survey of London: Volumes 29 and 30, St James Westminster, Part 1.* London: London County Council, 1960. *British History Online*, http://www.british-history.ac.uk/survey-london/vols29-30/pt1/pp115-118 [Accessed 28 May 2019].

Sheppard, F. H. W., ed., *Survey of London: Volumes 31 and 32, St James Westminster, Part 2.* London: London County Council, 1963. *British History Online*, http://www.british-history.ac.uk/survey-london/vols31-2/pt2/pp167-173 [Accessed 28 May 2019].

Sheppard, F. H. W., ed., *Survey of London: Volume 36, Covent Garden.* London: London County Council, 1970. *British History Online*, http://www.british-history.ac.uk/survey-london/vol36/pp207-218 [Accessed 21 November 2019].

A short Account of the Rise and Establishment of the general Fund of Charity for the Relief of distressed Masons. London: J. Scott, 1754.

Sprat, Thomas, *The history of the Royal Society.* London: T. R. for J. Martyn, 1667.

Stanhope, Philip Dormer, Earl of Chesterfield. *Letters written by the late Right Honourable Philip Dormer Stanhope, Earl of Chesterfield, to his son, Philip Stanhope, Esq.* London: E. Lynch, 1774.

Stearns, Raymond Phineas, ed., *Martin Lister, A Journey to Paris in the Year 1698.* Urbana, Chicago and London: University of Illinois Press, 1967.

Stewart, Philip and Catherine Volpilhac-Auger, eds, *Oeuvres Complètes de Montesquieu.* Sociéte Montesquieu. Paris: ENS Éditions and Classique Garnier, 2014-.

Stukeley, William, *The Family Memoirs of the Rev. William Stukeley M.D.*, 2 vols, Vol. LXXIII, Publications of the Surtees Society. London: Surtees Society, 1882-7.

Stukeley, William, 'On the Causes of Earthquakes', *Philosophical Transactions* 46, 497 (1749-50), pp. 641-6.

Stukeley, William, *The Medallic History of Marcus Aurelius Valerius Carausius*, 2 vols in 1. London: Charles Corbet, 1757.

Stukeley, William, *The Philosophy of Earthquakes, Natural and Religious.* London: Charles Corbet, 1750.

Summers, Montague, ed., *Dryden the Dramatic Works*, 6 vols. London: Nonesuch Press, 1931.

Thorpe, John, *Registrum Roffense: Or, A Collection of Antient Records, Charters and Instruments of Divers Kinds, Necessary for Illustrating the Ecclesiastical History and Antiquities of the Diocese and Cathedral Church of Rochester.* London: W. and J. Richardson, 1769.

Torkos, Justo Johanne and William Burnett, 'Observations anatomico-medicœ, de monstro bicorporeo virgineo a. 1701. Die 26 Oct. In Pannonia, infra comaromium, in possessione szony, quondam quiritum bregetione, in lucem edito, atque A. 1723. Die 23 Febr. Posonii in Cœnobio Monialium S. Ursulæ morte functo ibidemque sepulto. Authore Justo Johanne Torkos, M. D. Soc. Regalis Socio', *Philosophical Transactions* 50 (1757), pp. 311-22.

Tressan, Comte De, *Essai sur le fluide électrique*, 2 vols. Paris: Chez Buisson, 1786.

Tyson, Edward, 'Carigueya seu Marsupiale Americanum Masculum, or the Anatomy of a Male Opossum: In a letter to Dr. Edward Tyson, from Mr William Cowper', *Philosophical Transactions* 24, 290 (1704), pp. 1565-90.

Vanbrugh, Sir John, *The Confederacy: A Comedy.* New York and Boston: Wells and Lilly, 1823.

Vertue, George, *Medals, coins, great seals, and other works of Thomas Simon, engraved and described by George Vertue.* London: J. Nichols, 1753, 2nd edn, 1780.

Voltaire [François-Marie Arouet], *Letters Concerning the English Nation.* London: C. Davis, 1741.

Wallis, John, 'A Note of Dr Wallis, Sent in a Letter of Febr 17 1672/3 Upon Mr Lister's Observation Concerning the Veins in Plants, Published in Number 90 of these Tracts', *Philosophical Transactions* 8, 95 (1672), p. 6060.

Waterland, Daniel, *Advice to a Young Student. With a Method of Study for the Four First Years*. London: John Crownfield, 1740.

Watkins, Francis, *A Particular Account of the Electrical Experiments Hitherto made publick*. London: For the author, 1747.

Watson, William, 'An Account of the Experiments made by some Gentlemen in the Royal Society, in order to measure the absolute Velocity of Electricity', *Philosophical Transactions* 45, 489 (1748), pp. 491–6.

Watson, William, 'A Collection of the Electrical Experiments Communicated to the Royal Society', *Philosophical Transactions* 45, 485 (1748), pp. 49–120.

Weekly Journal or British Gazetteer. London: J. Read, 1715–25.

Whitehall Evening Post or London Intelligencer. London: Charles Corbet, 1746–59.

Whiston, William, *Memoirs of the Life and writings of Mr William Whiston*, 3 vols. London: For the authors, 1753.

Wilson, Benjamin, *An Essay Towards an Explication of the Phaenomena of Electricity, Deduced from the Aether of Sir Isaac Newton*. London: C. Davis, 1746.

Winter, E., *Die Registres der Berliner Akademie der Wissenschaften 1746–1766*. Berlin: Akademie Verlag, 1757.

Vitruvius, *On Architecture, in two volumes*. trans. Frank Granger. London: William Heinemann Ltd, 1934.

Young, Edward, *Love of Fame, the Universal Passion. In Seven Characteristical Satires*. 2nd edn. Dublin: Sarah Powell, for George Ewing, 1728.

Wilson, James, *Mathematical Tracts of the late Benjamin Robins esq.*, 2 vols. London: J. Nourse, 1761.

SECONDARY SOURCES

Adams, Frank Dawson, *The Birth and Development of the Geological Sciences*. New York: Dover Publications, 1938.

Adolph, Anthony, *Full of Soup and Gold: The Life of Henry Jermyn*. London: For the author, 2006.

Aldridge, Alfred Owen, 'Benjamin Franklin and Jonathan Edwards on Lightning and Earthquakes', *Isis* 41 (1950), pp. 162–4.

Allen, Louise, *St. Edmund's Church and the Montagu Monuments*. Oxford: Shire Publications, 2016.

Allibone, T. E., *The Royal Society and Its Dining Clubs*. Oxford: Pergamon Press, 1976.

Ament, Ernest J., 'The Anti-Lucretius of Cardinal Polignac', *Transactions and Proceedings of the American Philological Association* 101 (1970), pp. 29–49.

Andrews, Jonathan and Andrew Scull, *Customers and Patrons of the Mad-Trade: The Management of Lunacy in Eighteenth-Century England*. Berkeley: University of California Press, 2003.

Andrews, Jonathan and Andrew Scull, *Undertaker of the Mind: John Monro and Mad-doctoring in Eighteenth-century England*. Berkeley: University of California Press, 2001.

Anis, M., 'The First Egyptian Society in London (1741–3)', *Bulletin de l'Institut français d'archéologie Orientale* 50 (1950), pp. 99–105.

'The Architecture of Opera Houses', Victoria & Albert Museum, London. https://www.vam.ac.uk/articles/opera-architecture [Accessed 17 November 2019].

Armitage, Angus, *Edmond Halley*. London: Thomas Nelson, 1966.

Armitage, Geoff, *The Shadow of the Moon: British Solar Eclipse Mapping in the Eighteenth Century*. Tring: Map Collector Publications Ltd, 1997.

Atherton, Margaret, *Berkeley's Revolution in Vision*. Ithaca: Cornell University Press, 1990.

Atkins, Martyn, 'Persuading the House: The Use of the Commons Journals as a Source of Precedent'. In Paul Evans, ed., *Essays on the History of Parliamentary Procedure: In Honour of Thomas Erskine May*. London: Bloomsbury, 2017, pp. 69–86.

Baillon, Jean-François, 'Early Eighteenth-century Newtonianism: The Huguenot Contribution', *Studies in History and Philosophy of Science* 25 (2004), pp. 533–48.

Baird, Rosemary, 'Richmond House in London: Its History, Part I', *British Art Journal* 8, 2 (Autumn 2007), pp. 3–15.

Baker, C. H. Collins, 'Sir James Thornhill as Bible Illustrator', *Huntington Library Quarterly* 10, 3 (May, 1947), pp. 323–7.

Baker, Malcolm, 'Attending to the Veristic Sculptural Portrait in the Eighteenth Century'. In Malcolm Baker and Andrew Hemingway, eds, *Art as Worldmaking: Critical Essays on Realism and Naturalism*. Manchester: Manchester University Press, 2018, pp. 53–69.

Baker, Malcolm, ' "For Pembroke Statues, Dirty Gods and Coins": The Collecting, Display, and Uses of Sculpture at Wilton House', *Studies in the History of Art* 70 (2008), pp. 378–95.

Baker, Malcolm, 'Making the Portrait Bust Modern: Tradition and Innovation in Eighteenth-century Sculptural Portraiture'. In Jeanette Kohl and Rebecca Müller, eds, *Kopf/Bild Die Büste in Mittelalter und Früher Neuzeit*. Munich and Berlin: Deutscher Kunstverlag, 2007, pp. 347–66.

Baldwin, Olive and Thelma Wilson, 'The Subscription Musick of 1703-4', *The Musical Times* 153, 1921 (Winter 2012), pp. 29–44.

Ballantyne, Archibald, *Voltaire's Visit to England 1726–1729*. London: Smith, Elder, and Co, 1898.

Ball, W. W. Rouse, *A History of the Study of Mathematics at Cambridge*. Cambridge: Cambridge University Press, 1889.

Barlow, Graham F., 'Vanbrugh's Queen's Theatre in the Haymarket, 1703-9', *Early Music* xvii, 4 (November 1989), pp. 515–22.

Barnard, John, D. F. McKenzie, and Maureen Bell, eds, *The Cambridge History of the Book in Britain, Volume IV 1557–1695*. Cambridge: Cambridge University Press, 2008.

Barrett, Katy, 'Mr Whiston's Project for Finding the Longitude', Board of Longitude. https://cudl.lib.cam.ac.uk/view/MS-MSS-00079-00130-00002/2 [Accessed 18 June 2019].

Barzazi, Antonella, *Gli affanni dell'erudizione: studi e organizzazione culturale degli ordini religiosi a Venezia tra Sei e Settecento*. Venice: Istituto Veneto di Scienze, Lettere ed Arti, 2004.

Battin, Jacques, 'Montesquieu les sciences et la médecine en Europe', *Histoire des Sciences Médicales* xli, 3 (2007), pp. 243–54.

Bawden, Tina, Dominik Bonatz, Nikolaus Dietrich, Johanna Fabricius, Karin Gludovatz, Susanne Muth, Thomas Poiss, and Daniel A. Werning, 'Early Visual Cultures and Panofsky's *Perspektive als 'symbolische Form'*, *eTopoi: Journal for Ancient Studies* 6 (2016), pp. 525–70.

Beach, Matthew J. and David Pascoe, 'The Role of Hydra Vulgaris (Pallas) in Assessing the Toxicity of Freshwater Pollutants', *Water Research* 32, 1 (1998), pp. 101–6.

Beech, Martin, 'The Makings of Meteor Astronomy: Part VI', WGN, *The Journal of the International Meteor Organization* 22, 2 (1994), pp. 52–4.

Beer, G. R. de, 'Voltaire, F.R.S.', *Notes and Records of the Royal Society of London* 8, 2 (1951), pp. 247–52.

Bellhouse, David R., *Abraham de Moivre: Setting the Stage for Classical Probability and its Applications*. Boca Raton: CRC Press, 2011.

Bellhouse, David R., Elizabeth Renouf, Rajeev Raut, and Michael A. Bauer, 'De Moivre's Knowledge Community: An Analysis of the Subscription List to the Miscellanea Analytica', *Notes and Records of the Royal Society* 63 (2009), pp. 137–62.

Bellhouse, David R., 'Lord Stanhope's Papers on the Doctrine of Chances', *Historia Mathematica* 34 (2007b), pp. 173–86.

Bellhouse, David R. and Christian Genest, 'Maty's Biography of Abraham De Moivre, Translated, Annotated and Augmented', *Statistical Science* 22, 1 (2007), pp. 109–36.

Benedict, Barbara, 'Collecting Trouble: Sir Hans Sloane's Literary Reputation in Eighteenth-century Britain', *Eighteenth-century Life* 36, 2 (Spring 2012), pp. 111–42.

Bennett, J. A., Catadioptrics and Commerce in Eighteenth-century London', *History of Science* 44 (2006), pp. 247–77.

Bennett, J. A., 'James Short and John Harrison: Personal Genius and Public Knowledge', *Science Museum Journal* 2 (Autumn 2014), http://dx.doi.org/10.15180/140209.

Bennett, J. A., 'Obituary R. E. W. Maddison (1901–93)', *Annals of Science* 52, 3 (1995), p. 306.

Bensaude-Vincent, Bernadette and Christine Blondel, eds, *Science and Spectacle in the European Enlightenment*. Aldershot: Ashgate, 2008.

Beresiner, Yasha, 'Masonic Caricatures: For Fun or Malice – 300 Years of English Satirical Prints', *Ars Quatuor Coronatorum* 129 (September 2016), pp. 1–38.

Berman, Ric, 'The Architects of Eighteenth-Century English Freemasonry, 1720–1740'. PhD Dissertation. University of Sussex, 2010.

Berman, Ric, *The Foundations of Modern Freemasonry*. Brighton: Sussex University Press, 2012.

Berry, Helen, *Orphans of Empire: The Fate of London's Foundlings*. Oxford: Oxford University Press, 2019.

Bianchi, Paola and Karin Wolfe, eds, *Turin and the British in the Age of the Grand Tour*. Cambridge: Cambridge University Press, 2017.

Biermann, K. R. and G. Dunken, eds, *Deutsche Akademie der Wissenschaften zu Berlin. Biographischer Index der Mitglieder*. Berlin: Akademie Verlag, 1960.

Bindman, David and Malcolm Baker, *Roubiliac and the Eighteenth-century Monument: Sculpture as Theatre*. London: Paul Mellon, 1995.

Binyon, Lawrence, *Catalogue of Drawings by British Artists and Artists of Foreign Origin working in Great Britain preserved in the Department of Prints and Drawings in the British Museum*, 4 vols. London: Order of the Trustees, 1898–1907.

Black, Jeremy, *France and the Grand Tour*. Houndmills: Palgrave, 2003.

Boardman, J., J. Kagan, C. Wagner, and C. Phillips, *Natter's Museum Britannicum. British Gem Collections and Collectors in the Mid-eighteenth Century*. Oxford: Archaeopress in Association with the Hermitage Foundation UK and the Classical Art Research Centre, 2017.

Boni, G., 'Trajan's Column', *Proceedings of the British Academy* 3, 1–6 (1907), pp. 93–8.

Boran, Elizabethanne and Mordechai Feingold, eds, *Reading Newton in Early Modern Europe*. Leiden: Brill, 2017.

Bossut, Charles, *A General History of Mathematics from Earliest Times to the Middle of the Eighteenth Century*. London: J. Johnson, 1803.

Bowater, Laura and Kay Yeoman, 'Case Study 5.2: Evaluating an activity for the "Norfolk Science Past, Present and Future" Event'. In *Science Communication: A Practical Guide for Scientists* (New York: John Wiley and Sons, 2012), pp. 110–13.

Brennan, T. Corey, 'The 1644 Visit of the Englishman John Evelyn to the Villa Ludovisi', Archivio Digitale Boncompagni Ludovisi. https://villaludovisi.org/2012/12/03/1644-english-diarist-john-evelyn-visits-the-villa-ludovisi [Accessed 25 June 2019].

Bridgewater, David, '18th-Century Portrait Sculpture'. http://english18thcenturypor
traitsculpture.blogspot.com/2016/05/bust-of-isaac-newton-by-roubiliac-at.html [Accessed
12 January 2020].

Bridgewater, David, 'Portraits of Martin Folkes'. http://bathartandarchitecture.blogspot.
com/20150521archive.html. [Accessed 12 January 2020].

Brooks, Helen E. M., 'Theorizing the Woman Performer'. In Julia Swindells and David
Francis Taylor, eds, *The Oxford Handbook of the Georgian Theatre 1737–1832*. Oxford:
Oxford University Press, 2014, pp. 551–67.

Brown, Harcourt, 'Madame Geoffrin and Martin Folkes: Six New Letters', *Modern
Language Quarterly* 1, 2 (1940), pp. 215–21.

Brown, Harcourt, *Science and the Human Comedy: Natural Philosophy in French Literature
from Rabelais to Maupertuis*. Toronto: University of Toronto Press, 1976.

Brown, Stephen W. and Warren McDougall, eds, *Edinburgh History of the Book in Scotland:
Volume II, Enlightenment and Expansion: 1707–1800*. Edinburgh: Edinburgh University
Press, 2011.

Browne, C. L. and L. E. Davis, 'Cellular Mechanisms of Stimulation of Bud Production in
Hydra by Low Levels of Inorganic Lead Compounds', *Cell and Tissue Research* 177, 4
(February 1977), pp. 555–70.

Browne, Theodore, 'The Mechanical Philosophy and the "Animal Oeconomy": A Study in
the Development of English Physiology in the Seventeenth and Early Eighteenth
Century', PhD Dissertation. Princeton: Princeton University, 1968.

Brundtland, Terje, 'Francis Hauksbee and His Air Pump', *Notes and Records of the Royal
Society* 66 (2012), pp. 1–20.

Buccleuch and Queensberry, Richard, Duke of, *Boughton: The House, Its People and Its
Collections*, ed. John Montagu Douglas Scot. Hawick: Caique Publishing, 2006.

Buchwald, Jed Z. and Mordechai Feingold, *Newton and the Origin of Civilization*.
Princeton: Princeton University Press, 2012.

Buhl, Marie-Louise, Erik Dal, and Torben Holck Colding, *The Danish Naval Officer:
Frederik Ludvig Norden*. Copenhagen: The Royal Danish Academy of Sciences and
Letters, 1986.

Burke, Peter, 'Images as Evidence in Seventeenth-century Europe', *Journal of the History of
Ideas* 64, 2 (2003), pp. 273–96.

Bertucci, Paola, 'The Shocking Bag: Medical Electricity in mid-18th-Century London',
Nuova Voltiana 5 (2003), pp. 31–42.

Cailhon, François, 'Secondat, Jean-Baptiste de'. trans. Philip Stewart, in *Dictionnaire
Montesquieu* [online]. Directed by Catherine Volpilhac-Auger, ENS Lyon, September
2013. http://dictionnaire-montesquieu.ens-lyon.fr/fr/article/1376477218/en [Accessed
23 September 2019].

Cameron, J. K., 'Leaves from the Lost Album amicorum of Sir John Scot of Scotstarvit',
Scottish Studies 28 (1987), pp. 35–48.

Cantor, Geoffrey, *Optics after Newton: Theories of Light in Britain and Ireland, 1704–1840*.
Manchester: Manchester University Press, 1983.

Cantor, Geoffrey, *Quakers, Jews, and Science: Religious Responses to Modernity and the
Sciences in Britain, 1650–1900*. Oxford: Oxford University Press, 2005.

Cantor, Geoffrey, 'Quakers in the Royal Society, 1660–1750', *Notes and Records of the Royal
Society* 51, 2 (1997), pp. 175–93.

Cantor, Geoffrey, 'The Rise and Fall of Emanuel Mendes da Costa', *English Historical
Review* 116 (2001), pp. 584–603.

Carpenter, Audrey T., *John Theophilus Desaguliers: A Natural Philosopher, Engineer and
Freemason*. London: Bloomsbury, 2011.

Carter, Harry, ed., *A History of the Oxford University Press, Volume 1: to 1780*. Oxford: Oxford University Press, 1975.

Casini, Paolo, *Hypotheses non fingo: Tra Newton e Kant*. Rome: Edizioni di Storia e Letteratura, 2006.

Casini, Paolo, 'The Reception of Newton's Opticks in Italy'. In J. V. Field and Frank A. J. L. James, eds, *Renaissance and Revolution: Humanists, Scholars, Craftsmen and Natural Philosophers in Early Modern Europe*. Cambridge: Cambridge University Press, 1997, pp. 215–29.

Chaiklin, Martha, 'Ivory in World History: Early Modern Trade in Context', *History Compass* 8, 6 (June 2010), pp. 530–42.

Chaney, Edward, *The Evolution of the Grand Tour: Anglo-Italian Cultural Relations Since the Renaissance*. London and Portland: Frank Cass, 1998.

'The Changing Properties of Smalt Over Time', Tate Britain, January 2007, https://www.tate.org.uk/about-us/projects/changing-properties-smalt-over-time [Accessed 26 August 2018].

Chapman, Robert A., 'The Manuscript Letters of Stephen Gray, FRS (1666/7–1736)', *Isis* 49, 4 (1958), pp. 414–23.

Cherry, John, *Richard Rawlinson and his Seal Matrices*. Oxford: Ashmolean, 2006.

Crichton-Miller, Emma, 'Jonathan Richardson by himself', *Apollo*, 27 July 2015, https://www.apollo-magazine.com/jonathan-richardson-by-himself [Accessed 20 March 2020].

Christie, J. R. R., 'Ether and the Science of Chemistry: 1740–1790'. In G. N. Cantor and M. J. S. Hodge, eds, *Conceptions of Ether: Studies in the History of Ether Theories, 1740–1900*. Cambridge: Cambridge University Press, 1981, pp. 86–110.

Clark, Peter, *British Clubs and Societies 1580–1800: The Origins of an Associational World*. Oxford: Oxford University Press, 2000.

Clower, William, 'The Transition from Animal Spirits to Animal Electricity: A Neuroscience Paradigm Shift', *Journal of the History of the Neurosciences* 7, 3 (1998), pp. 201–18.

Coates, Clive, 'History in a Bottle', *Decanter*, 1 April 2002. https://www.decanter.com/features/history-in-a-botttle-248958/ [Accessed 29 December 2019].

Cohen, I. Bernard, 'Neglected Sources for the Life of Stephen Gray (1666 or 1667–1736)', *Isis* 45, 1 (1954), pp. 41–50.

Cokayne, George Edward, ed., *The Complete Baronetage*, 5 vols. No date, *c.*1900. Reprint. Gloucester: Alan Sutton Publishing, 1983.

Collins, John Churton, *Voltaire, Montesquieu and Rousseau in England*. London: Eveleigh Nash, 1908.

Cope, Kevin L., 'Notes from Many Hands: Pierre Lyonnet's Redesign of Friedrich Christian Lesser's Insecto-Theology'. In Brett C. McInelly and Paul E. Kerry, eds, *New Approaches to Religion and Enlightenment*. Vancouver and Madison: Fairleigh Dickinson University Press, 2018, pp. 1–34.

Craske, Matthew, *The Silent Rhetoric of the Body: A History of Monumental Sculpture and Commemorative Art in England, 1720–70*. New Haven: Yale University Press, 2007.

Cronk, Nicholas, *Voltaire: A Very Short Introduction*. Oxford: Oxford University Press, 2017.

Crosland, Maurice, 'Explicit Qualifications as a Criterion for Membership of the Royal Society: A Historical Review', *Notes and Records of the Royal Society of London* 37, 2 (1983), pp. 167–87.

'Cuff Microscope', Science Museum, http://www.sciencemuseum.org.uk/broughttolife/objects/display?id=93111 [Accessed 15 October 2019].

Curl, James Stevens, *The Egyptian Revival: Ancient Egypt as the Inspiration for Design Motifs in the West*. Abingdon and New York: Routledge, 2005.

Curran, Brian, *The Egyptian Renaissance: The Afterlife of Ancient Egypt in Early Modern Italy*. Chicago: University of Chicago Press, 2007.

Daston, Lorraine, *Classical Probability in the Enlightenment*. Princeton: Princeton University Press, 1995.

Davis, Ernest, 'Review of Newton and the Origin of Civilization, by Jed Z. Buchwald and Mordechai Feingold', https://cs.nyu.edu/davise/papers/Newton.pdf [Accessed 4 September 2018].

Davis, John W., 'The Molyneux Problem', *Journal of the History of Ideas* 21, 3 (July–September 1960), pp. 392–408.

Dawson, Warren R., 'The First Egyptian Society', *The Journal of Egyptian Archaeology* 23, 2 (Dec. 1937), pp. 259–60.

Dawson, Warren R., 'Pettigrew's Demonstrations Upon Mummies: A Chapter in the History of Egyptology', *The Journal of Egyptian Archaeology* 20/3–4 (November 1934), pp. 170–82.

Dearnley, John, 'Patronage and Sinecure: Examples of the Practice of Bishop Hoadly at Winchester (1734–61)', *Proceedings of the Hampshire Field Club of Archaeology* 65 (2010), pp. 191–201.

DeJean, Joan, *Ancients against Moderns: Culture Wars and the Making of a Fin de Siècle*. Chicago: University of Chicago Press, 1997.

Delbourgo, James, *Collecting the World: The Life and Curiosity of Hans Sloane*. Cambridge, MA: Harvard University Press, 2018.

Dickie, Simon, *Cruelty and Laughter: Forgotten Comic Literature and the Unsentimental Eighteenth Century*. Chicago: University of Chicago Press, 2011.

Dictionary of Irish Biography Online. https://dib.cambridge.org/.

Dobbs, Betty Jo Teeter and Margaret C. Jacob, *Newton and the Culture of Newtonianism*. Amherst: Humanity Books, 1995.

Dobrée, Bonamy and Geoffrey Webb, eds, *The Complete Works of Sir John Vanbrugh*. 4 vols. London: Nonesuch Press, 1927–8.

Doran, John, *'Their Majesties' Servants', or Annals of the English Stage*. London: Wm. H. Allen and Co, 1865.

Drewitt, Frederic Dawtrey, *Bombay in the days of George IV: memoirs of Sir Edward West, chief justice of the King's court during its conflict with the East India company, with hitherto unpublished documents*. London: Longmans, Green, and Co., 1907.

Eaglen, Robin J., 'The Illustration of Coins: An Historical Survey, Part I', Presidential Address 2009', *British Numismatic Journal* (2010), pp. 140–50. https://www.britnumsoc. org/publications/Digital%20BNJ/pdfs/2010_BNJ_80_7.pdf [Accessed 7 September 2019].

Edelstein, Dan, 'The Egyptian French Revolution: Antiquarianism, Freemasonry and the Mythology of Nature'. In Dan Edelstein, ed., *The Super Enlightenment: Daring to Know Too Much*. Oxford: Voltaire Foundation, 2010, pp. 215–41.

Edelstein, Dan and Biliana Kassabova, 'How England Fell Off the Map of Voltaire's Enlightenment', *Modern Intellectual History* (2018). https://doi.org/10.1017/S147924431800015X.

Eisler, William, 'The Construction of the Image of Martin Folkes (1690–1754) Part I', *The Medal* 58 (Spring 2011), pp. 4–29.

Eisler, William, 'The Construction of the Image of Martin Folkes (1690–1754) Art, Science and Masonic Sociability in the Age of the Grand Tour Part II', *The Medal* 59 (Autumn 2011), pp. 4–16.

Eisler, William, *Lustrous Images from the Enlightenment: The Medals of the Dassiers* of Geneva. ed. Matteo Campagnolo. Geneva: Skira, 2010.

Eisler, William, 'Paul Mellon Centre Rome Fellowship: The Medals of Martin Folkes: Art, Newtonian Science and Masonic Sociability in the Age of the Grand Tour', *Papers of the British School at Rome* 78 (November 2010), pp. 301–2.

Elliot, J. Keith, 'The Text of the New Testament'. In Allan J. Houser and Duane F. Watson, eds, *A History of Biblical Interpretation, Vol. 2: The Medieval Through the Reformation Periods*. Grand Rapids, Michigan and Cambridge: William B. Eerdmans Publishing, 2003.

Elliot, Paul and Stephen Daniels, 'The "school of true, useful and universal science"? Freemasonry, Natural Philosophy, and Scientific Culture in Eighteenth-century England', *The British Journal for the History of Science* 39 (2006), pp. 207–29.

Ellis, Heather, *Generational Conflict and University Reform: Oxford in the Age of Revolution*. Leiden: Brill, 2018.

Evans, Joan, *A History of the Society of Antiquaries*. Oxford: Oxford University Press, 1956.

Fara, Patricia, *An Entertainment for Angels: Electricity in the Enlightenment*. London: Icon Books, 2017.

Fara, Patricia, '"Master of Practical Magnetics": The Construction of an Eighteenth-century Natural Philosopher', *Enlightenment and Dissent* 14 (1995), pp. 52–87.

Fara, Patricia, *Sympathetic Attractions: Magnetic Practices, beliefs, and Symbolism in Eighteenth-century France*. Princeton: Princeton University Press, 1996.

Farrer, Edmund, *Portraits in Suffolk Houses (West)*. London: Bernard Quaritch, 1908.

Feingold, Mordechai, 'Confabulatory life'. In P. D. Omodeo and K. Freidrich, eds, *Duncan Liddel (1561–1613): Networks of polymathy and the Northern European Renaissance*. Leiden: Brill, 2016, pp. 22–34.

Feingold, Mordechai, 'Learning Arabic in Early Modern England'. In Jan Loop, Alastair Hamilton, and Charles Burnett, eds, *The Teaching and Learning of Arabic in Early Modern Europe*. Boston and Leiden: Brill, 2017, pp. 33–56.

Ferguson, J. P., An *Eighteenth-century Heretic: Dr Samuel Clarke*. Kineton: Roundwood Press, 1976.

Ferreiro, Larrie D., *Measure of the Earth: The Enlightenment Expedition that Reshaped Our World*. New York: Basic Books, 2013.

Ferrone, Vincenzo, *The Intellectual Roots of the Italian Enlightenment: Newtonian science, religion, and politics in the early eighteenth century*. Atlantic Highlands, New Jersey: Humanities Press, 1995.

Finnegan, Rachel, 'The Divan Club, 1744–46', *EJOS* IX (2006), pp. 1–86.

Finnegan, Rachel, *English Explorers in the East (1738–45): The Travels of Thomas Shaw, Charles Perry and Richard Pococke*. Leiden: Brill, 2019.

Finnegan, Rachel, 'The Travels and Curious Collections of Richard Pococke, Bishop of Meath', *Journal of the History of Collections* 27, 1 (2015), pp. 33–48.

Fisher, John, 'Conjectures and Reputations: The Composition and Reception of James Bradley's Paper on the Aberration of Light with Some Reference to a Third Unpublished Version', *British Journal of the History of Science* 43, 1 (2010), pp. 19–48.

Flage, Daniel E., 'George Berkeley (1685-1753)', *Internet Encyclopedia of Philosophy*. http://www.iep.utm.edu/berkeley [Accessed 3 December 2016].

Fontes da Costa, Palmira, 'The Culture of Curiosity at the Royal Society in the First Half of the 18th Century', *Notes and Records of the Royal Society of London* 56 (2002), pp. 147–66.

Fontes da Costa, Palmira, *The Singular and the Making of Knowledge at the Royal Society of London in the 18th Century*. Newcastle-upon-Tyne: Cambridge Scholars Publishing, 2009.

Force, James E., *William Whiston: Honest Newtonian*. Cambridge: Cambridge University Press, 1985.

Fordham, Douglas, *British Art and the Seven Years' War: Allegiance and Autonomy*. Philadelphia: University of Pennsylvania Press, 2010.

Foyster, Elizabeth, 'At the Limits of Liberty: Married Women and Confinement in Eighteenth-century England', *Continuity and Change* 17, 1 (2002), pp. 39–62.

Frandsen, Paul John, *Let Greece and Rome Be Silent: Frederick Ludvig Norden's Travels in Egypt and Nubia, 1737–1738*. Copenhagen: Museum Tusculanum Press, 2019.

Fyfe, Aileen, Julie McDougall-Waters, and Noah Moxham, '350 Years of Scientific Periodicals', *Notes and Records: The Royal Society Journal of the History of Science* 69 (2015), pp. 227–39.

Garrett, Brian, 'Vitalism and Teleology in the Natural Philosophy of Nehemiah Grew (1641–1712)', *British Journal of the History of Science* 36, 1 (2003), pp. 63–81.

Gascoigne, John, *Cambridge in the Age of the Enlightenment: Science, Religion and Politics from the Restoration to the French Revolution*. Cambridge: Cambridge University Press, 1989.

Gee, Brian, Anita McConnell, and A. D. Morrison-Low, *Francis Watkins and the Dollond Telescope Patent Controversy*. Abingdon: Ashgate, 2004.

Geikie, Sir Archibald, *Annals of the Royal Society Club; the record of a London dining-club in the eighteenth and nineteenth centuries* (London: Macmillan and co., 1917).

Gerrard, Christine, *Aaron Hill: The Muses' Projector, 1685–1750*. Oxford: Oxford University Press, 2003.

Gibson, Elizabeth, 'Owen Swiney and the Italian Opera in London', *The Musical Times* 125, 1692 (February 1984), pp. 82–96.

Gibson, Susannah, *Animal, Mineral, Vegetable: How Eighteenth-century Science Disrupted the Natural World*. Oxford: Oxford University Press, 2015.

Gibson-Wood, Carol, *Jonathan Richardson, Art Theorist of the English Enlightenment*. New Haven and London: Yale University Press, 2000.

Gibson-Wood, Carol, 'Jonathan Richardson as Draftsman', *Master Drawings* 32, 3 (Autumn 1994), pp. 203–29.

Gillis, John R., *For Better, for Worse: British Marriages, 1600 to the Present*. Oxford: Oxford University Press, 1985.

Gingerich, Owen, 'Eighteenth-century Eclipse Paths'. In Owen Gingerich, ed., *The Great Copernicus Chase and Other Adventures in Astronomical History*. Cambridge: Cambridge University Press, 1992, pp. 152–9.

Goff, Moira, *The Incomparable Hester Santlow*. Farnham and Burlington, VT: Ashgate, 2007.

Gömöri George, and Stephen D. Snobelen, 'What He May Seem to the World: Newton's Autograph Book Epigrams', *Notes and Records: The Royal Society Journal of the History of Science* 74, 3 (2020), pp. 409–52.

Goodman, Dena, *The Republic of Letters: A Cultural History of the French Enlightenment*. New York: Cornell University Press, 1994.

Gordon-Lennox, Charles Henry, 8th Duke of Richmond and Earl of March, *A Duke and His Friends: The Life and Letters of the Second Duke of Richmond*. London: Hutchinson and Company, 1911.

Gowing, Ronald, *Roger Cotes: Natural Philosopher*. Cambridge: Cambridge University Press, 1983.

Graciano, Andrew, 'The Memoir of Benjamin Wilson FRS (1721–88): Painter and Electrical Scientist', *Walpole Society* 74 (2012), pp. 165–243.

Graziosi, Elisabetta, 'Women and Academies in Eighteenth-century Italy'. In Paula Findlen, Wendy Wassyng Roworth, Catherine M. Siena, eds, *Italy's Eighteenth Century: Gender and Culture in the Age of the Grand Tour*. Stanford: Stanford University Press, 2009, pp. 103–19.

Greengrass, Mark, Daisy Hildyard, Christopher D. Preston, and Paul J. Smith, 'Science on the Move: Francis Willughby's Expeditions'. In Tim Birkhead, ed., *Virtuoso by Nature: The Scientific World of Francis Willughby FRS* (1635–1672). Leiden: Brill, 2016, pp. 142–226.

Greig, Martin, 'The Reasonableness of Christianity? Gilbert Burnet and the Trinitarian Controversy of the 1690s', *Journal of Ecclesiastical History* 44, 4 (October 1993), pp. 631–51.

Grell, Ole Peter and Andrew Cunningham, eds, *Centres of Medical Excellence? Medical Travel and Education in Europe, 1500–1789*. Aldershot: Ashgate, 2010.

Griffiths, Antony, *The Print in Stuart Britain, 1603–1689*. London: British Museum Press, 1998.

Guerlac, Henry, *Newton on the Continent*. Ithaca: Cornell University Press, 1981.

Guerrini, Anita, 'Advertising Monstrosity: Broadsides and Human Exhibition in Early Eighteenth-century London'. In Patricia Fumerton and Anita Guerrini, eds, *Ballads and Broadsides in Britain, 1500–1800*. Aldershot: Ashgate, 2010, pp. 109–30.

Guerrini, Anita, 'Anatomists and Entrepreneurs in Early 18th-century London', *Journal of the History of Medicine and Allied Sciences* 59, 2 (2004), pp. 219–39.

Guerrini, Anita, *Experimenting with Humans and Animals: From Galen to Animal Rights*. Baltimore: Johns Hopkins University Press, 2003.

Guerrini, Anita, 'The Tory Newtonians: Gregory, Pitcairne and Their Circle', *Journal of British Studies* xxv (1986), pp. 288–311.

Guerrini, Luigi, *Antonio Cocchi naturalista e filosofo*. Florence: Polistampa, 2002.

Guertin, S., and G. Kass-Simon, 'Extraocular Spectral Photosensitivity in the Tentacles of *Hydra vulgaris*', *Comparative Biochemistry and Physiology Part A: Molecular & Integrative Physiology* 185 (June 2015), pp. 163–70.

Guthrie, Neil, *The Material Culture of the Jacobites*. Cambridge: Cambridge University Press, 2013.

Hall, A. Rupert, *All Was Light: An Introduction to Newton's Opticks*. Oxford: Clarendon Press, 1993.

Hall, Marie Boas, *The Library and Archives of the Royal Society: 1660–1990*. London: The Royal Society, 1992.

Hamill, John and Pierre Mollier, 'Rebuilding the Sanctuaries of Memphis: Egypt in Masonic Iconography and Architecture'. In Jean-Marcel Humbert and Clifford Price, eds, *Imhotep Today: Egyptianizing Architecture*. London: UCL Press, 2016, pp. 207–20.

Hancox, Joy, *The Queen's Chameleon: The Life of John Byrom. A Study of Conflicting Loyalties*. London: Jonathan Cape, 1994.

Hanham, Andrew, 'The Politics of Chivalry: Sir Robert Walpole, the Duke of Montagu and the Order of the Bath', *Parliamentary History* 35, 3 (2016), pp. 262–97.

Hans, Nicholas, 'The Masonic Lodge in Florence in the Eighteenth Century', *Ars Quatuor Coronatorum* 61 (1958), pp. 109–12.

Hanson, Craig Ashley, *The English Virtuoso: Art, Medicine, and Antiquarianism in the Age of Empiricism*. Chicago: University of Chicago Press, 2009.

Harland-Jacobs, Jessica, *Builders of Empire: Freemasonry and British Imperialism, 1717–1927*. Chapel Hill: University of North Carolina Press, 2007.

Harskamp, Jaap, *Streetwise: Art at Heart of the City, Streetscapes from Lorenzetti to Mondrian*. Armorica Editions, 2015. https://issuu.com/bookhistory/docs/streetwise_f53b04df393fff [Accessed 28 August 2018].

Harter, H. Leon, 'The Method of Least Squares and Some Alternatives—Part I', *International Statistical Review* 42, 2 (August 1974), pp. 147–74.

Hartman, Lee, '"Que sera sera": The English Roots of a Pseudo-Spanish Proverb', *Proverbium* 30 (2013), pp. 51–104.

Hartwig, Melinda, 'Method in Ancient Egyptian Painting'. In Valérie Angenot and Francesco Tiradritti, eds, *Artists and Colour in Ancient Egypt: Proceedings of the Colloquium Held in Montepulciano, August 22nd–24th, 2008*. Montepulciano: Missione Archaeologica Italiana a Luxor, 2016, pp. 28–56.

Hartwig, Melinda, *Tomb Painting and Identity in Ancient Thebes, 1419–1372 BCE*. Turnhout: Brepols, 2004.

Haycock, David Boyd, 'Ancient Egypt in 17th and 18th Century England'. In Peter Ucko and Timothy Champion, eds, *The Wisdom of Egypt: Changing Visions Through the Ages*. London: UCL Press, 2003, pp. 133–61.

Haycock, David Boyd, '"The Cabal of a Few Designing Members": The Presidency of Martin Folkes, PRS, and the Society's First Charter', *Antiquaries Journal* 80 (2000), pp. 273–84.

Haycock, David Boyd, *William Stukeley: Science, Religion and Archaeology in Eighteenth-century England*. Woodbridge: Boydell Press, 2002.

Heilbron, John, *Electricity in the 17th and 18th Centuries: A Study of Early Modern Physics*. Berkeley: University of California Press, 1997.

Heilbron, John, *Elements of Early Modern Physics*. Berkeley, Los Angeles and London: University of California Press, 1982.

Heilbron, John, *Physics at the Royal Society during Newton's Presidency*. Los Angeles: William Andrews Clark Memorial Library, 1983.

Held, Anders, *A History of Probability and Statistics and Their Applications before 1750*. Hoboken, NJ: John Wiley and Sons, 2003.

Henry, John, 'Gravity and *De gravitatione*: The Development of Newton's Ideas on Action at a Distance', *Studies in History and Philosophy of Science Part A* 42, 1 (March 2011), pp. 11–27.

'Her Majesty's', *Theatre Trust*. https://database.theatrestrust.org.uk/resources/theatres/show/1993-her-majesty-s-london [Accessed 15 October 2019].

Hervey, S. H. A. H., *Biographical list of boys educated at King Edward IV's Grammar School (Bury St. Edmunds England)*. Bury St Edmunds: Paul and Mathew, 1908.

Herzfeld, Chris, *The Great Apes: A Short History*. New Haven: Yale University Press, 2017.

Hibbert, Christopher, Ben Weinreb, John Keay, and Julia Keay, eds, *The London Encyclopaedia*. 3rd edn. New York: Pan–Macmillan, 2011.

Higgitt, Rebekah, 'Equipping Expeditionary Astronomers: Nevil Maskelyne and the Development of "Precision Exploration"'. In F. MacDonald and C. W. J. Withers, eds, *Geography, Technology and Instruments of Exploration*. Basingstoke: Ashgate, 2015, pp. 15–36.

Higgitt, Rebekah, '"In the Society's Strong Box": A Visual and Material History of the Royal Society's Copley Medal, c.1736–1760', *Nuncius: Journal of the Material and Visual History of Science* 34, 3 (2019), pp. 284–316.

Highfill, Jr, Philip, Kalman A. Burnim, and Edward A. Langhans, *A Biographical Dictionary of Actors, Actresses, Musicians, Dancers, Managers and Other Stage Personnel in London, 1660–1880*, 12 vols. Carbondale: Southern Illinois University Press, 1973–1993.

[Hire, Philippe de la,] *New Elements of Conic Sections*. trans. Bryan Robinson. London: Dan Midwinter, 1704.

Hogan, Charles Beecher, *The London Stage 1776–1800: A Critical Introduction*. Carbondale: Southern Illinois University Press, 1968.

Holloway, Sally, *The Game of Love in Georgian England: Courtship, Love and Material Culture*. Oxford: Oxford University Press, 2019.

Home, R., 'Force, Electricity, and the Powers of Living Matter in Newton's Mature Philosophy of Nature'. In M. J. Osler and P. L. Farber, eds, *Religion, Science, and Worldview: Essays in Honor of Richard S. Westfall*. Cambridge: Cambridge University Press, 1985, pp. 95–117.

Home, R. W., 'The Royal Society and the Empire: The Colonial and Commonwealth Fellowship Part 1: 1731–1847', *Notes and Records of the Royal Society* 56, 3 (2002), pp. 307–32.

Horwood, Alfred J., 'The Manuscripts of Sir Wm. Hovell Browne Ffolkes, Bart., at Hillington Hall, co. Norfolk', in *The Third report of the Royal Commission on Historical Manuscripts*. London, HMSO, 1872.

Hoskin, Michael A., '"Mining all within": Clarke's Notes to Rohault's *Traité de Physique*', *The Thomist* 24 (1961), pp. 353–63.

Houston, R. A., 'Madness and Gender in the Long Eighteenth Century', *Social History* 27, 3 (2002), pp. 309–26.

Hume, Robert D and Harold Love, eds, *Plays, Poems, and Miscellaneous Writings Associated with George Villiers, Second Duke of Buckingham*. Oxford: Oxford University Press, 2007.

Hurley, David Ross, 'Dejanira, Omphale, and the Emasculation of Hercules: Allusion and Ambiguity in Handel', *Cambridge Opera Journal* 11, 3 (1999), pp. 199–214.

Hyslop, James, 'John Mayall and Reproductions of Early Microscopes', Explore Whipple Collections, Whipple Museum of the History of Science, University of Cambridge, 2008. https://www.whipplemuseum.cam.ac.uk/explore-whipple-collections/microscopes/dutch-pioneer-antoni-van-leeuwenhoek/mayall-reproductions [Accessed 30 September 2019].

Iliffe, Rob, 'Aplatisseur du Monde et de Cassini': Maupertuis, Precision Measurement, and the Shape of the Earth in the 1730s', *History of Science* xxxi (1993), pp. 335–75.

Iliffe, Rob, *Priest of Nature: The Religious Worlds of Isaac Newton*. Oxford: Oxford University Press, 2017.

Iltis, Carolyn, 'Leibniz and the Vis Viva Controversy', *Isis* 62, 1 (1971), pp. 21–35.

Ingamells, John, *A Dictionary of British and Irish Travellers in Italy, 1701–1800: Compiled from the Brinsley Ford Archive*. New Haven: Yale University Press, 1997.

'Isaac Newton (1642–1727), Object ID ZBA1640, Royal Museums Greenwich'. https://collections.rmg.co.uk/collections/objects/220530.html [Accessed 7 January 2020].

Jacob, Margaret, *The Radical Enlightenment: Pantheists, Freemasons and Republicans*. London: George Allen and Unwin, 1981.

James, Douglas, 'Portraits in Medical Biography: Alexander Pope (1688–1744), Poet, Patient, Celebrity', *Journal of Medical Biography* 21, 4 (2013), pp. 200–8.

James, T. G. H., *The British Museum and Ancient Egypt*. London: British Museum Press, 1981.

Jankovic, Vladimir, *Reading the Skies: A Cultural History of English Weather, 1650–1820*. Manchester: Manchester University Press, 2000.

Janssen, Lydia, 'Antiquarianism and National History. The Emergence of a New Scholarly Paradigm in Early Modern Historical Studies', *History of European Ideas* 43, 8 (2017), pp. 843–56.

Jansson, Sven B. F., *Runes in Sweden*. London: Phoenix House, 1962.

Jessop, T. E., *A Bibliography of George Berkeley*. 2nd edn. International Archives of the History of Ideas, 66. New York: Springer, 1973.

'John Byrom Collection', Chetham's Library, Manchester. https://library.chethams.com/collections/printed-books-ephemera/john-byrom-collection/ [Accessed 10 August 2019].

Johnson, W., 'Aspects of the Life and Works of Martin Folkes (1690–1754)', *International Journal of Impact Engineering* 21, 8 (1998), pp. 695–705.

Johnson, W. 'Early Bridge Consultants, Benjamin Robins, F.R.S. and Charles Hutton, F.R.S. and Mis-judged Bridge Designer, Thomas Paine', *International Journal of Mechanical Sciences* 41 (1999), pp. 741–8.

Johnstone, James Fowler Kellas, *The Alba Amicorum of George Strachan, George Craig, Thomas Cumming*. Aberdeen University Studies, 95. Aberdeen: University of Aberdeen, 1924.

Jungnickel, Christa and Russell McCormmach, *Cavendish: The Experimental Life*. Lewisburg, PA: Bucknell University Press, 1999.

Keller, Vera, Anna Marie Roos, and Elizabeth Yale, eds, *Archival Afterlives: Life, Death and Knowledge-Making in Early Modern British Scientific and Medical Archives*. Leiden: Brill, 2017.

Kennedy, James, *A Description of the Antiquities and Curiosities in Wilton House, Salisbury*. London: E. Easton, 1769.

Keynes, Milo, *The Iconography of Sir Isaac Newton to 1800*. Woodbridge: Boydell and Brewer, 2005.

King, Bob, 'Happy Nights with the Hyades', *Sky and Telescope* (30 January 2019). https://www.skyandtelescope.com/observing/happy-nights-with-the-hyades [Accessed 14 December 2019].

Kisby, Fiona, *Music and Musicians in Renaissance Cities and Towns*. Cambridge: Cambridge University Press, 2005.

Kolbe, George, 'Godfather to all Monkeys: Martin Folkes and his 1756 Library Sale', *The Asylum* 32, 2 (April–June 2014), pp. 38–92.

Lake, Crystal B., 'Plate 1.20: Medals of Henry VIII, Edward VI, Elizabeth I, and James I', *Vetusta Monumenta: Ancient Monuments*, https://scalar.missouri.edu/vm/vol1plate20-sixteenth-century-english-medals [Accessed 2 January 2019].

Lake, Crystal B. and David Shields, 'Plates 1.37–1.38: Tables of English Coins', *Vetusta Monumenta: Ancient Monuments*, https://scalar.missouri.edu/vm/vol1plates37-38-table-of-english-coins [Accessed 2 January 2020].

Latour, Bruno, *Science in Action: How to Follow Scientists and Engineers through Society*. Cambridge, MA: Harvard University Press, 1987.

Laudan, Larry, *Science and Hypothesis: Historical Essays on Scientific Methodology*. Dordrecht: Springer, 1981.

Lenhoff, Howard M. and Sylvia G. Lenhoff, 'Abraham Trembley and his Polyps, 1744: The Unique Biology of Hydra and Trembley's Correspondence with Martin Folkes', *Eighteenth-century Thought* 1 (2003), pp. 255–80.

Lenhoff, Sylvia G. and Howard M. Lenhoff, and Abraham Trembley, *Hydra and the Birth of Experimental Biology, 1744: Abraham Trembley's Mémoires Concerning the Polyps*. Pacific Grove, CA: Boxwood Press, 1986.

Leonarducci, Don Gasparo, *La Provvidenza. Cantica seconda. I primi quattro canti inedita del P. don Gasparo Leonarducci della Congregazione di Somasca*. Venice: Dalla Tipografia Di Alvisopoli, 1827.

Levine, Joseph M., *The Battle of the Books: History and Literature in the Augustan Age*. Ithaca: Cornell University Press, 1991.

Levitin, Dmitri, 'Egyptology, the Limits of Antiquarianism, and the Origins of Conjectural History, c. 1680–1740: New Sources and Perspectives', *History of European Ideas* 41, 6 (2015), pp. 699–727.

Lieberkühn, Johann Nathanael, (1711–1756) Molecular Expressions, https://micro.magnet.fsu.edu/optics/timeline/people/lieberkuhn.html [Accessed 25 September 2019].

Lilti, Antoine, *The World of the Salons: Sociability and Worldliness in Eighteenth-century Paris*. Oxford: Oxford University Press, 2015.

Lippincott, Kristen, 'A Chapter in the "Nachleben" of the Farnese Atlas: Martin Folkes's Globe', *Journal of the Warburg and Courtauld Institutes* 74 (2011), pp. 281–99.

Lippincott, Louise, 'Arthur Pond's Journal of Receipts and Expenses, 1734–50', *The Volume of the Walpole Society* 54 (1988), pp. 220–333.

Lock, Frederick Peter, 'The Dramatic Art of Susanna Centlivre', PhD Dissertation. Ontario: McMaster University, 1974.

Lockwood, Mike and Luke Barnard, 'An Arch in the UK: Aurora Catalogue', *A & G: News and Reviews in Astronomy and Geophysics* 56 (August 2015), pp. 4.25–4.30.

Lowerre, Kathryn, *Music and Musicians on the London Stage, 1695–1705*. London and New York: Routledge, 2016.

Lux, David S. and Harold J. Cook, 'Closed Circles or Open Networks: Communicating at a Distance During the Scientific Revolution', *History of Science* 36, 2 (1998), pp. 179–211.

Lynall, Gregory, *Swift and Science: The Satire, Politics, and Theology of Natural Knowledge, 1690–1730*. New York: Palgrave, 2012.

Lyons, Henry G., 'The Officers of the Society (1662–1860)', *Notes and Records: The Royal Society Journal of the History of Science* 3 (1940–1), pp. 116–40.

Lyons, Henry G., *The Royal Society, 1660–1940*. Cambridge: Cambridge University Press, 1949.

Lyons, Henry G., 'Two Hundred Years Ago: 1739', *Notes and Records: The Royal Society Journal of the History of Science* 2, 1 (1939), pp. 34–42.

Lyte, Sir H. C. Maxwell, *A History of Eton College*. London: Macmillan, 1911.

McCann, Timothy J., 'Two of the Stoutest Legs in England: The 2nd Duke of Richmond's Leg Break in 1732', *Sussex Archaeological Collections* 150 (2012), pp. 139–41.

McClellan III, James, *Science Reorganized: Scientific Societies in the Eighteenth Century*. New York: Columbia University Press, 1985.

McConnell, A., 'L. F. Marsigli's visit to London in 1721, and His Report on the Royal Society', *Notes and Records of the Royal Society of London* 47 (1993), pp. 179–204.

MacGregor, Arthur, 'Patrons and Collectors: Contributors of Zoological Subjects to the Works of George Edwards (1694–1773)', *Journal of the History of Collections* 26, 1 (2014), pp. 35–44.

McKie, Douglas and Gavin Rylands de Beer, 'Newton's Apple', *Notes and Records of the Royal Society of London* 9, 1 (1951), pp. 46–54.

Maddison, R.E.W., 'A Note on the Correspondence of Martin Folkes, P. R. S', *Notes and Records of the Royal Society of London* 11, 1 (1954), pp. 100–9.

Mandelbrote, Scott, 'Eighteenth-century Reactions to Newton's Anti-Trinitarianism'. In James E. Force and Sarah Hutton, eds, *Newton and Newtonianism: New Studies*. Dordrecht and Boston: Kluwer, 2004, pp. 93–111.

Mandelbrote, Scott and Helmut Putte, eds, *The Reception of Isaac Newton in Europe*, 3 vols. London: Bloomsbury Academic, 2019.

Mandelkern, India Aurora, 'The Politics of the Palate: Taste and Knowledge in Early Modern England', PhD Dissertation. Berkeley: University of California, Spring 2015.

Marland, Hilary, 'Women and Madness', Centre for the History of Medicine, University of Warwick. Marland, Hilary, 'Women and Madness', [Accessed 28 December 2019].

Marples, Alice, 'Scientific Administration in the Early Eighteenth Century: Reinterpreting the Royal Society's Repository', *Historical Research* 92, 255 (February 2019), pp. 183–214.

Martin, Susan, 'Actresses on the London Stage: A Prosopographical Study', PhD Dissertation. University Park, PA: Pennsylvania State University, 2008.

Massai, Sonia, 'Nahum Tate's Revision of Shakespeare's *King Lears*', *Studies in English Literature* 40, 3 (Summer 2000), pp. 435–50.

Matikkala, Antti, *The Orders of Knighthood and the Formation of the British Honours System*. Cambridge: Cambridge University Press, 2008.

Mauelshagen, Franz, 'Networks of Trust: Scholarly Correspondence and Scientific Exchange in Early Modern Europe', *The Medieval History Journal* 6 (2003), pp. 1–32.

Mazzotti, Massimo, 'Newton for Ladies: Gentility, Gender and Radical Culture', *British Journal of the History of Science* 37, 2 (2004), pp. 119–46.

Mendelson, Sara and Patricia Crawford, *Women in Early Modern England, 1550–1720*. Oxford: Clarendon Press, 1998.

Milhous, Judith and Robert D. Hume, 'A Letter to Sir John Stanley: A New Theatrical Document of 1712', *Theatre Notebook* 43, 2 (1989), pp. 71–80.

Milhous, Judith and Robert D. Hume, *The London Stage 1660–1800, Part 2: 1700–1729: Draft of the Calendar for Volume I, 1700–1711*. http://www.personal.psu.edu/hb1/London%20Stage%202001/lond1707.pdf [Accessed 17 October 2019].

Milhous, Judith and Robert D. Hume, 'The Silencing of Drury Lane in 1709', *Theatre Journal* 32, 4 (Dec. 1980), pp. 427–47.

Milhous, Judith and Robert D. Hume, 'Theatrical Politics at Drury Lane, New Light on Letitia Cross, Jane Rogers and Anne Oldfield', *Bulletin of Research in the Humanities* (Winter 1982), pp. 412–29.

Milhous, Judith and Robert D. Hume, *Vice Chamberlain Coke's Theatrical Papers, 1706–1715*. Carbondale and Edwardsville: Southern Illinois University Press, 1982.

Miller, David, 'The "Hardwicke Circle": The Whig Supremacy and Its Demise in the 18th-Century Royal Society', *Notes and Records of the Royal Society of London* 52, 1 (1998), pp. 73–91.

Miller, David, '"Into the Valley of Darkness": Reflections on the Royal Society in the Eighteenth Century', *History of Science* 27, 2 (1 June 1989), pp. 155–66.

Money, John, *Experience and Identity: Birmingham and the West Midlands 1760–1800* Manchester: Manchester University Press, 1977.

Moore, Keith, 'Chimes at Midnight', Repository Blog, The Royal Society, 22 December 2014, https://blogs.royalsociety.org/history-of-science/2014/12/22/chimes-at-midnight/ [Accessed 19 November 2019].

Morgan, Augustus du, *A Budget of Paradoxes*. London: Longmans, Green, and Co., 1872.

Morgan, M. J., *Molyneux's Question: vision, touch, and the philosophy of perception*. Cambridge: Cambridge University Press, 1977.

Morris, S. B., 'New Light on the Gormogons and Other Imitative Societies', *Ars Quatuor Coronatorum* 126 (2013), pp. 15–70.

Moschini, Giovanni Antonio, *Della letteratura veneziana del secolo XVIII fino a' nostri giorni*, 4 vols. Venice: Dalla Stamperia Palese, 1806.

Moser, Stephanie, *Wondrous Curiosities: Ancient Egypt at the British Museum*. Chicago: University of Chicago Press, 2006.

Mulvey-Roberts, Marie, 'Hogarth on the Square: Framing the Freemasons', *British Journal for Eighteenth-century Studies* 26 (2003), pp. 251–70.

Munby, A. N. L., 'The Distribution of the First Edition of Newton's *Principia*', *Notes and Records of the Royal Society of London* 10, 1 (1952), pp. 28–39.

Murdoch, Tessa, 'Roubiliac as an Architect? The Bill for the Warkton Monuments', *Burlington Magazine* 122, 922 (January 1980), pp. 40–6.

Murdoch, Tessa, 'Spinning the Thread of Life: The Three Fates, Time and Eternity'. In Diana Dethloff, Tessa Murdoch, Kim Sloan and Caroline Elam, eds, *Burning Bright: Essays in Honour of David Bindman*. London: UCL Press, 2015, pp. 47–54.

Murdoch, Tessa, 'Roubiliac as an Architect? The Bill for the Warkton Monuments', *Burlington Magazine* 122, 922 (January 1980), pp. 40–6.

Murphy, Kathleen Susan, 'James Petiver's "Kind Friends" and "Curious Persons" in the Atlantic World: Commerce, Colonialism and Collecting', *Notes and Records: the Royal Society Journal of the History of Science* 74, 2 (2020), pp. 259–74.

Nau, Elizabeth, *Lorenz Natter (1705–1763): Gemmenschneider und Medailleur*. Biberach an der Riß: Biberacher Verlagsdruckerei, 1966.

Navari, Leonora, *Greece and the Levant, the Catalogue of the Henry Myron Blackmer Collection of Books and Manuscripts*. London: Maggs Bros, 1989.

Neher, Alistair, 'The Truth about Our Bones: William Cheselden's Osteographia', *Medical History* 54 (2010), pp. 517–28.

Netboy, Anthony, 'Voltaire's English Years', *VQR: A National Journal of Literature and Discussion* (Spring 1977), https://www.vqronline.org/essay/voltaire%E2%80%99s-english-years-17261728 [Accessed 29 November 2019].

Nichols, John, ed., *Bibliotheca topographica Britannica*, 8 vols. London: J. Nichols, 1780–90.

Nichols, John. ed., *Illustrations of the Literary History of the Eighteenth Century*, 8 vols. London: For the Author, 1817–58.

Nickson, M. A. E., *Early Autograph Albums in the British Museum*. London: Trustees of the British Museum, 1970.

Nordenmark, N. V. E., *Anders Celsius: Professor i Uppsala 1701–44*. Luleå: Almqvist & Wiksell, 1936.

Norman, Larry F., *The Shock of the Ancient: Literature and History in Early Modern France*. Chicago: University of Chicago Press, 2011.

North, John, 'The Satellites of Jupiter: From Galileo to Bradley.' In Alwyn van der Merwe, ed., *Old and New Questions in Physics, Cosmology, Philosophy, and Theoretical Biology*. Boston: Springer, 1983, pp. 689–718.

O'Connell, Sheila, 'Appendix: Hogarthomania and the Collecting of Hogarth'. In David Bindman, ed., *Hogarth and his Times: Serious Comedy*. Berkeley and Los Angeles: University of California Press, 1997, pp. 58–61.

Ogée, Frédéric, 'Je-sais-quoi, William Hogarth and the representations of the forms of life'. In Frederic Ogée, Peter Wagner, and David Bindman, eds, *Hogarth: Representing Nature's Machines*. Manchester: Manchester University Press, 2001, pp. 71–81.

Önnerfos, Andreas, 'The Earliest Account of Swedish Freemasonry? *Relation apologique* (1738) Revisited', *Ars Quatuor Coronatorum* 127 (2014), pp. 1–34.

Önnerfos, Andreas, 'Secret Savants, Savant Secrets: The Concept of Science in the Imagination of European Freemasonry'. In Martin Stuber, André Holenstein, Hubert Steinke, and Philippe Rogger, eds, *Scholars in Action: The Practice of Knowledge and the Figure of the Savant in the 18th Century*, 2 vols. Leiden: Brill, 2013, pp. 433–57.

Oppé, A. P., *The Drawings of William Hogarth*. New York: Phaidon, 1948.

Ottaway, Susannah R., *The Decline of Life: Old Age in Eighteenth-century England*. Cambridge: Cambridge University Press, 2004.

'Oughtred, William', MacTutor, University of St Andrews, http://mathshistory.st-andrews.ac.uk/Biographies/Oughtred.html [Accessed 8 December 2019].

Oxford Dictionary of National Biography. https://www.oxforddnb.com/.

Pagan, Hugh, 'Martin Folkes and the Study of English Coinage in the Eighteenth Century'. In R. G. W. Anderson, M. L. Caygill, A. G. MacGregor and L. Syson, eds, *Enlightening the British: Knowledge, discovery and the museum in the eighteenth century*. London: British Museum Press, 2003, pp. 158–63.

Papy, Jan, 'The Scottish Doctor William Barclay, his Album Amicorum, and His Correspondence with Justus Lipsius'. In Dirk Sacré and Gilbert Tournoy, eds, *Myricae: Essays on neo-Latin Literature in Memory of Jozef Ijsewijn*. Supplementa humanistica Lovaniensia, 16. Leuven: Leuven University Press, 2000, pp. 333–96.

Pasachoff, Jay M., 'Halley as an Eclipse Pioneer: His Maps and Observations of the Total Solar Eclipses of 1715 and 1724', *Journal of Astronomical History and Heritage* 2, 1 (1999), pp. 39–54.

Pastorino, Cesare, 'Measuring the Past: Quantification and the Study of Antiquity in the Early Modern Period', unpublished paper.

Paulson, Ronald, *Hogarth: The Modern Moral Subject, 1697–1732*. Cambridge: Lutterworth Press, 1991.

Peck, Linda, 'Uncovering the Arundel Library at the Royal Society: Changing Meanings of Science and the Fate of the Norfolk Donation', *Notes and Records of the Royal Society of London* 52, 1 (1998), pp. 3–24.

'Pembroke Book 1747'. Holabird-Kagin Americana, 22 February 2014 Auction, Lot 1362, https://www.liveauctioneers.com/item/24254201_1747-pembroke-book [Accessed 4 September 2019].

Perry, Norma, 'The Rainbow, the White Peruke and the Bedford Head: Voltaire's London Haunts'. In *Voltaire and the English*, Studies on Voltaire and the Eighteenth Century, 179. The Voltaire Foundation at the Taylor Institution, Oxford. Oxford: Voltaire Foundation, 1979, pp. 203–20.

Perry, Norma, 'Voltaire's View of England', *Journal of European Studies* 7, 26 (1977), pp. 77–94.

Piggott, Stuart, *William Stukeley: An Eighteenth-Century Antiquary*. New York: Thames and Hudson, 1995.

Pope, Alexander, *Works of Alexander Pope*, ed. William Roscoe, 10 vols. London: J. Rivington, 1824.

Prescott, Andrew, 'John Pine: A Remarkable 17th-century Engraver and Freemason', https://freemasonrymatters.co.uk/latest-news-freemasonry/john-pine-a-remarkable-17th-century-engraver-and-freemason/ [Accessed 12 January 2020].

Prescott, Andrew, 'John Pine: A Sociable Craftsman', *MQ Magazine* 10 (July 2004), p. 8. http://www.mqmagazine.co.uk/issue-10/p-08.php [Accessed 15 August 2019].

Pugh, R. B., 'Our First Charter', *The Antiquaries Journal* 62 (March 1982), pp. 347–55.

Radcliffe, David Hill, 'Spencer and the Tradition: English Poetry 1579–1830'. http://spenserians.cath.vt.edu/TextRecord.php?action=GET&textsid=34514 [Accessed 4 October 2019].

Ratcliff, Marc J., 'Abraham Trembley's Strategy of Generosity and the Scope of Celebrity in the Mid-Eighteenth Century', *Isis* 95, 4 (2004), pp. 555–75.

Ratcliff, Mark J., *The Quest for the Invisible: Microscopy in the Enlightenment*. Aldershot: Ashgate, 2009.

Read, Sara, *Menstruation and the Female Body in Early Modern England*. London: Palgrave Macmillan, 2013.

Rée, Jonathan, 'I tooke a bodkine', *London Review of Books* 35, 19 (10 October 2013). https://www.lrb.co.uk/the-paper/v35/n19/jonathan-ree/i-tooke-a-bodkine [Accessed 22 December 2019].

Rees, Abraham, 'Stenography'. In *The Cyclopaedia: Or Universal Dictionary of Arts, Sciences, and Literature*, 41 vols. London: Longman, Hurst, Rees, Orme, and Brown, 1802–19, vol. 34.

The Renaissance Duke. Catalogue to 2018 exhibition, Goodwood House, Sussex. https://www.goodwood.com/globalassets/venues/goodwood-house/summer-exhibition-2017.pdf [Accessed 9 October 2019].

Review of the 'Royal Mischief', The Reviews Hub. Rose Playhouse, London, 16 September 2016. https://www.thereviewshub.com/southwest-template-draft-3-2-5/ [Accessed 15 October 2018].

Riggs, Christina, *Egypt: Lost Civilizations*. London: Reaktion Books, 2017.

Rimbault, Edward F., ed., *Memoirs of Musick by the Hon. Roger North*. London: George Bell, 1846.

Robert, G. C. B., 'Historical Argument in the Writings of the English Deists', DPhil Dissertation. Oxford: Worcester College, 2014.

Roos, Anna Marie, *Goldfish*. London: Reaktion Books, 2019.

Roos, Anna Marie, 'Luminaries in Medicine: Richard Mead, James Gibbs, and Solar and Lunar Effects on the Human Body in Early Modern England', *Bulletin of the History of Medicine* 74, 3 (2000), pp. 433–57.

Roos, Anna Marie, *Martin Lister and his Remarkable Daughters: The Art of Science in the Seventeenth Century*. Oxford: Bodleian Library Publishing, 2018.

Roos, Anna Marie, 'Taking Newton on Tour: The Scientific Travels of Martin Folkes, 1733-35', *The British Journal for the History of Science* 50, 4 (2017), pp. 569–601.

Rosenheim, Max, 'The album amicorum', *Archaeologia* 62 (1910), pp. 251–308.

Rosenmeyer, Patricia A., *The Language of Ruins: Greek and Latin Inscriptions on the Memnon Colossus*. Oxford: Oxford University Press, 2018.

Rousseau, George, 'The Eighteenth Century: Science'. In Pat Rogers, ed., *The Eighteenth Century*, The Context of English Literature Series. London: Holmes & Meier, 1978.

Rousseau, George S. and David Haycock, 'Voices Calling for Reform: The Royal Society in the Mid-eighteenth Century: Martin Folkes, John Hill and William Stukeley', *History of Science* 37, 118 (1999), pp. 377–406.

Rowland, Richard, *Killing Hercules: Deianira and the Politics of Domestic Violence, from Sophocles to the War on Terror*. New York: Routledge, 2016.

Rowlands, Peter, *Newton and Modern Physics*. London: World Scientific, 2018.

Ruding, Roger, *Annals of the Coinage of Great Britain and its Dependencies: From the Earliest Period of Authentic History to the Reign of Victoria*, 3 vols. London: For the author, 1812.

Ruestow, Edward, *The Microscope in the Dutch Republic: The Shaping of Discovery*. Cambridge: Cambridge University Press, 2004.

Rumbold, Margaret E. *Traducteur Huguenot: Pierre Coste*. New York: Peter Lang, 1991.

Rusnock, Andrea, 'Correspondence networks and the Royal Society, 1700–50', *British Journal of the History of Science* 32 (1999), pp. 155–69.

Sadie, Julie Anne, ed., *Companion to Baroque Music*. Berkeley: University of California Press, 1998.

Sakaguchi, Tatsuma, Yoshinori Hamada, Yusuke Nakamura, Yuki Hashimoto, and Hiroshi Hamada A-Hon Kwon, 'Epigastric Heteropagus Associated with an Omphalocele and Double Outlet Right Ventricle', *Journal of Paediatric Surgery Case Reports* 3, 10 (October 2015), pp. 469–72.

Salmon, Frank, 'British Architects and the Florentine Academy, 1753-94', *Mitteilungen des Kunsthistorischen Instituts in Florenz*, 34, 1/2 (1990), pp. 199–214.

Schaefer, B. E., 'The Epoch of the Constellations on the Farnese Atlas and Their Origin in Hipparchus' Lost Catalogue', *Journal for the History of Astronomy* 36 (2005), pp. 167–96.

Schaffer, Simon, 'Fontanelle's Newton and the Uses of Genius', *L'Esprit Createur* 55, 2 (Summer 2015), pp. 48–61.

Schaffer, Simon, 'Glass Works: Newton's Prisms and the Uses of Experiment'. In David Gooding, Trevor Pinch, and Simon Schaffer, eds, *The Uses of Experiment: Studies in the Natural Sciences*. Cambridge: Cambridge University Press, 1989, pp. 67–104.

Schaffer, Simon and Stephen Shapin, *Leviathan and Air Pump: Hobbes, Boyle, and the Experimental Life*. Princeton: Princeton University Press, 1985.

Schilt, Cornelis J., 'Of Manuscripts and Men: The Editorial History of Isaac Newton's Chronology and Observations', *Notes and Records: The Royal Society Journal of the History of Science* 74, 3 (2020), pp. 387–403.

Schliesser, Eric, 'Newton and Newtonianism in Eighteenth-Century British Thought'. In James A. Harris, ed., *The Oxford Handbook of British Philosophy in the Eighteenth Century*. Oxford: Oxford University Press, 2013. doi: 10.1093/oxfordhb/9780199549023.013.003.

Schüller, Volkmar, 'Samuel Clarke's Annotations in Jacques Rohault's *Traité de Physique*, and How They Contributed to Popularising Newton's Physics'. In Wolfgang Lefèèvre, ed., *Between Leibniz, Newton and Kant: Philosophy and Science in the Eighteenth Century*. Dordrecht: Springer, 2001, pp. 95–110.

Scouten, Arthur H. and Robert D. Hume, 'Additional Players' Lists in the Lord Chamberlain's Registers, 1708–1710', *Theatre Notebook* 37 (1983), pp. 77–9.

Seifert, Christian Tico, ed., *Rembrandt: Britain's Discovery of the Master*. Edinburgh: National Galleries of Scotland, 2018.

Seising, Rudolf, Menso Folkerts, and Ulf Hashagen, eds, *Form, Zahl, Ordnung. Studien zur Wissenschafts-und Technikgeschichte. Festschrift für Ivo Schneider zum 65. Geburtstag*. Stuttgart: Franz Steiner, 2004.

Shalev, Zur, 'Measurer of All Things: John Greaves (1602–1652), the Great Pyramid, and Early Modern Metrology', *Journal of the History of Ideas* 63, 4 (2002), pp. 555–75.

Shalev, Zur, 'The Travel Notebooks of John Greaves'. In Alistair Hamilton, Maurits H. Van Den Boogert, and Bart Westerweel, eds, *The Republic of Letters and the Levant*. Leiden: Brill, 2005, pp. 77–103.

Shank, J. B., 'Voltaire'. *The Stanford Encyclopedia of Philosophy*. Fall 2015 Edition. Edward N. Zalta, ed. https://plato.stanford.edu/archives/fall2015/entries/voltaire/ [Accessed 29 November 2019].

Shields, David, 'Plate 1.69, Standard of Weights and Measures', *Vetusta Monumenta*. A Digital Edition, https://scalar.missouri.edu/vm/vol1plate69-weights-and-measures-1497 [Accessed 28 October 2019].

Shilliam, Nicola J., 'Mummies and Museums: Egyptology in Britain in the Early Eighteenth Century', Princeton University Library. https://library.princeton.edu/news/marquand/2019-03-18/mummies-and-museums-egyptology-britain-early-eighteenth-century [Accessed 14 January 2020].

Shorvon, Simon and Alastair Compston, *Queen Square: A History of the National Hospital and its Institute of Neurology*. Cambridge: Cambridge University Press, 2018.

Simon, Robin, *Hogarth, France and British Art*. London: Hogarth Arts, 2007.

Singh, Prince Frederick Duleep, *Portraits in Country Houses of Norfolk*. Norwich: Jarrod and Sons, 1928.

Small, Alastair and Carola Small, 'South Italy, England and Elysium in the Eighteenth Century', *Antiquaries Journal* 79 (1999), pp. 333–42.

Smith, John, Mezzotint Print Maker, National Portrait Gallery, London. https://www.npg.org.uk/research/programmes/early-history-of-mezzotint/john-smith-mezzotint-printmaker-biography [Accessed 20 June 2019].

Solano, Francisco de, *Don Antonio de Ulloa, Paradigma del Marino Científico de la Ilustración Española*. Coimbra: Universidade de Coimbra, 1990.

Soren, David and Noelle Soren, *A Roman Villa and a Late Roman Infant Cemetery: Excavation at Poggio Gramignano Lugnano in Teverina*. Rome: L' Erma di Bretschneider, 1999.

Sorrenson, Richard, 'George Graham, visible technician', *British Journal for the History of Science* 32 (1999), pp. 203–21.

Sorrenson, Richard, 'Towards a History of the Royal Society in the Eighteenth Century', *Notes and Records of the Royal Society of London* 50, 1 (1996), pp. 29–46.

Sotheby and Company, London, sale catalogue 27 June 1932, lots 111–123A.

Sotheby's, *The Library of a Greek Bibliophile, Aldines and an important Qur'an*, 28 July 2020, Lot 85, Perry: *A View of the Levant*. https://www.sothebys.com/en/buy/auction/2020/the-library-of-a-greek-bibliophile-travel-books-aldines-and-an-important-quran/perry-a-view-of-the-levant-particularly-of [Accessed 28 July 2020]

Steele, Brett D., 'Muskets and Pendulums: Benjamin Robins, Leonhard Euler, and the Ballistics Revolution', *Technology and Culture* 35, 2 (April 1994), pp. 348–82.

Stolzenberg, Daniel, *Egyptian Oedipus: Athanasius Kircher and the Secrets of Antiquity*. Chicago: University of Chicago Press, 2013.

Sweet, Rosemary, *Antiquaries: The Discovery of the Past in Eighteenth-Century Britain*. London: Hambleton, 2004.

Summers, Montague. *The Restoration Theatre*. New York: Macmillan, 1933.

Tarek, Ahmed, Mohamed Abdel-Rahman, Nesma Mohamed, and Ahmed Abedallatif, 'Study of Some Types of Different Wrappings on Ibis Mummies from Catacombs of Tuna-el-Gebel, Hermopolis', 2016, Poster session, https://isaae2016.sciencesconf.org/89426 [Accessed 29 July 2019].

Terrall, Mary, *Catching Nature in the Act: Réaumur and the Practice of Natural History in the Eighteenth Century*. Chicago and London: University of Chicago Press, 2015.

Terrall, Mary, *The Man who Flattened the Earth: Maupertuis and the Sciences in the Enlightenment*. Chicago: University of Chicago Press, 2002.

Te Velde, H., 'A Few Remarks Upon the Religious Significance of Animals in Ancient Egypt' *Numen* 27 (1980), pp. 76–82.

Thackray, Arnold, *Atoms and Powers: An Essay on Newtonian Matter Theory and the Development of Chemistry*. Cambridge, MA: Harvard University Press, 1970.

Thomas, David, 'The 1737 Licensing Act and its Impact'. In Julia Swindells and David Francis Taylor, eds, *The Oxford Handbook of the Georgian Theatre 1737–1832*. Oxford: Oxford University Press, 2014, pp. 91–106.

Thomas, Jennifer M., 'A "Philosophical Storehouse": The Life and Afterlife of the Royal Society's Repository', PhD Thesis. London: Queen Mary, University of London, 2009.

Thomas, Keith, 'Gentle Boyle', review of Steven Shapin's *Social History of Truth: Civility and Science in 17th-Century England*, London Review of Books 16, 18 (22 September 1994), pp. 14–16. https://www.lrb.co.uk/v16/n18/keith-thomas/gentle-boyle [Accessed 10 March 2019].

Thompson, C. J. S., *The Mystery and Lore of Monsters: With Accounts of Some Giants, Dwarfs and Prodigies*. London: Williams and Norgate Ltd, 1930.

Touber, Jetze, 'Applying the Right Measure: Architecture and Philology in Biblical Scholarship in the Dutch Early Enlightenment', *The Historical Journal* 58, 4 (2015), pp. 959–85.

Trompf, Garry W., 'On Newtonian History'. In Stephen Gaukroger, ed., *The Uses of Antiquity: The Scientific Revolution and the Classical Tradition*. New York: Springer, 2013, pp. 213–49.

Turner, Dawson, *Catalogue of the Manuscript Library of the Late Dawson Turner*. London: Puttick and Simpson, 1859.

Twyman, Michael, ed., *The Encyclopedia of Ephemera: A Guide to the Fragmentary Documents of Everyday Life for the Collector, Curator, and Historian*. New York: Routledge, 2018.

Ukers, William H., *All About Coffee*. New York: Tea and Coffee Trade Journal Company, 1922. http://www.web-books.com/Classics/ON/B0/B701/15MB701.html [Accessed 2 June 2019].

Van Blanken, Kerrewin, 'Earthquake Observations in the Age before Lisbon: Eyewitness Observation and Earthquake Philosophy in the Royal Society, 1665–1755', *Notes and Records: The Royal Society Journal of the History of Science*, https://doi.org/10.1098/rsnr.2020.0005

Venn, J. and J. A. Venn, *Alumni Cantabrigienses, A Biographical List of all Known Students, Graduates and Holders of Office at the University of Cambridge, from the Earliest Times to 1900, Part I: From the Earliest Times to 1751, Volume II: Dabbs-Juxton*. Cambridge: Cambridge University Press, 1922.

Venn, J., *Biographical History of Gonville and Caius College, 1349–1713*. Cambridge: Cambridge University Press, 1897.

Vine, Angus, *In Defiance of Time: Antiquarian Writing in Early Modern England*. Oxford: Oxford University Press, 2010.

Vingopoulou, Ioli, 'Pococke, Richard. A Description of the East', Travelogues, Aikaterini Laskaridis Foundation. http://eng.travelogues.gr/collection.php?view=219 [Accessed 26 August 2018].

Wagner, Gillian, *Thomas Coram, Gent (1668–1751)*. Woodbridge: Boydell Press, 2004.

Walford, Edward, 'Queen Square and Great Ormond Street'. In *Old and New London: volume 4* (London: Cassell, Petter and Galpin, 1878), pp. 553–64, *British History Online*, http://www.british-history.ac.uk/old-new-london/vol4/pp553-564 [Accessed 26 April 2019].

Wallace, Daniel B., 'The Textual Problem in 1 John 5:7–8', Bible.org. https://bible.org/article/textual-problem-1-john-57-8 [Accessed 20 December 2019].

Walsh, Kirsten, 'Newton's Epistemic Triad', PhD dissertation. Dunedin: University of Otago, 2014.

Wardhaugh, Benjamin, *Gunpowder and Geometry: The Life of Charles Hutton: Pit Boy, Mathematician, and Scientific Rebel*. London: William Collins, 2019.

Warneke, Sara, *Images of the Educational Traveller in Early Modern England*. Leiden: Brill, 1995.

Warner, Deborah, 'The Oldest Microscope in the Museum', Smithsonian, 13 May 2015. Smithsonian/https://americanhistory.si.edu/blog/microscope [Accessed 1 October 2019].

Waters, Alice N., 'Ephemeral Events: English broadsides of Early Eighteenth-century Solar Eclipses', *History of Science* 37 (1999), pp. 1–43.

Weinshenker, Anne Betty, *Falconet, His Writings and His Friend Diderot*. Geneva: Librairie Droz, 1966.

Weitzmann, Kurt, ed., *Age of spirituality: late antique and early Christian art, third to seventh century; Catalogue of the exhibition at the Metropolitan Museum of Art, November 19, 1977, Through February 12, 1978*. New York: Metropolitan Museum of Art, 1979.

Weitzmann, Kurt and Herbert Kessler, *The Cotton Genesis: The Illustrations in the Manuscript of the Septuagint, Volume 1. British Library, Codex Cotton Otho B. Vi*. Princeton: Princeton University Press, 1987.

Weld, Charles, *A History of the Royal Society: with memoirs of the Presidents*, 2 vols. Cambridge: Cambridge University Press, 2011, reprint of 1848 edition.

Werrett, Simon, *Fireworks: Pyrotechnic Arts and Sciences in European History*. Chicago: University of Chicago Press, 2010.

Werrett, Simon, *Thrifty Science: Making the Most of Materials in the History of Experiment*. Chicago: University of Chicago Press. 2019.

Werrett, Simon, 'Watching the Fireworks: Early Modern Observation of Natural and Artificial Spectacles', *Science in Context* 24, 2 (2011), pp. 167–82.

Westfall, Richard, *Never at Rest: A Biography of Isaac Newton*. Cambridge: Cambridge University Press, 1983.

Whalley, Paul E. S., 'The Authorship and Date of the *Micrographia restaurata*, with a Note on the Scientific Name of the Silverfish (Insecta, *Thysanura*)', *Journal of the Society of the Bibliography of Natural History* 6, 3 (1972), pp. 171–3.

Wheatley, Henry Benjamin and Peter Cunningham, *London Past and Present: Its History, Associations, and Traditions*. Cambridge: Cambridge University Press, 2011.

Whiteside, D. T., 'Kepler, Newton and Flamsteed on Refraction Through a "Regular Aire"', *Centaurus* 24 (1980), pp. 288–315.

Wilson, Bronwen, 'Social Networking: the "album amicorum" and Early Modern Public Making'. In Massimo Rospocher, ed., *Beyond the Public Sphere: Opinions, Publics, Spaces in Early Modern Europe*. Bologna: Società editrice il Mulino, 2012, pp. 205–23.

Wilson, John Harold, *All the King's Ladies: Actresses of the Restoration*. Chicago: University of Chicago Press, 1958.

de Winkel, Marieke, *Fashion and Fancy: Dress and Meaning in Rembrandt's Paintings*. Amsterdam: Amsterdam University Press, 2006, p. 189.

Wright, Rebecca, 'The Georgian Theatre Audience: Manners and Mores in the Age of Politeness, 1737–1810'. MA Thesis. University of London, 2014.

Young, Mark Thomas, 'Nature as Spectacle: Experience and Empiricism in Early Modern Experimental Practice', *Centaurus* 59 (2017), pp. 72–96.

Younger, Fletcher William, *Bookbinding in England and France*. London: Seeley and Company, 1897.

Zwierlein, Cornel, *Imperial Unknowns: French and British in the Mediterranean, 1650–1750*. Cambridge: Cambridge University Press, 2016.

Index

Abgali, Mohammed Ben Ali 102–3, 213
Academia Scientiarum Germanica
   Berolinensis 247
Académie Française 353–4
Académie Royale des Sciences 1, 26, 96, 247, 353
Academy of Saumur 26
Addison, Joseph 184, 356–7
aether 291–2, 294–6, 298, 342
Aga, Cassem Algaida 104
Ahlers, Cyriacus 117–20
Alexander the Great 130
Albani, Alessandro 147
Algarotti, Francesco 149 159, 162, 169, 321
   his Newtonianism 167
   metrology and aesthetics of Trajan's pillar
      154–6, 161, 303
Allamand, Johannes Nicolaas Sebastiaan 260
Ames, Joseph 200–1, 206
Amherst, Lord 347
Amyand, Claude 320
Anderson, Dr James 194
   Book of Constitutions (1723) 76, 83–4, 86,
      209, 350
Anstis, John, Garter King at Arms 90
D'Anteney 114, 120
antiquarian science 2–6, 149, 176
   see also scientific antiquarianism
anti-Trinitarianism 15, 127–8, 361
Antonio da Ponte's Bridge 151
Apollonius 102
Aquinas, Thomas 282
Arbuthnot, Sir John 36, 79, 356
archaeology, early development of 3, 5, 11, 215,
      236, 350
Archaeologia 6
Archbishop of Canterbury 121
Arderon FRS, William 274, 275–7
Aristotle 42
Aristotelian Chain of Being 13, 257, 282
Artur, Jacques François 274
Arundel, 2nd Earl of 219
Arundel Collection 107, 187, 196
Ashburnham 183
Ashmolean Museum 14, 43, 209, 258, 337
Ashton, Robert 347
Askew, Anthony 243

Asmasaeus of Udine, Romulus 243
Aston, Francis 108
Athenian Mercury 66
d'Aubignac, François-Hédelin 328
Aubigny, Duke of 13
aurora borealis 11, 22, 42–3, 168, 170
Ayscough, James 277
azimuth compass 249–50

Babbage, Charles 7, 11
Bacon, Francis 2, 277, 340
Bacon, Montagu 318
Baker, Henry 13, 14, 110, 247, 266, 267, 269,
      274–287, 290, 342
   Employment for the Microscope (1753) 206,
      207, 258, 275, 280
   parasitic twins 13, 257
   publishing of Trembley 273–4
   The Microscope Made Easy (1742) 275–6, 278
   re-engraving of plates for Hooke's
      Micrographia 16, 276–78
Baker, Malcolm 179, 204, 329
Ballantyne 358
ballistics 8, 246, 247, 251–2, 317
Baltimore, Lord, governor of Maryland 202
Bank of Charity 80
Banks, Mr 256
Banks, Sir Joseph 7
Barkway 256
Barlow, Thomas 39
Barrow, Isaac 38
Barry, Elizabeth 50, 54, 57, 62, 65, 68
Bartholin, Thomas 148, 287
Barton Conduitt, Catherine 355
Bayfield, Josiah 305
Bayle, Pierre 39
Beacon Hill, Essex 256
Beale, John 14
Bedford, Duchess of 290
Bedford Head Tavern 39, 78–9, 355–6
Beech 43
Belchier FRS, John 179, 242
Bellers, Fettiplace 49–50
Bellers, John 49
Bentinck, Charles 264, 304
Bentley, Richard 11, 134, 174, 239–40

Berain, Jean 282
Berdmore, Samuel 292
Berkeley, George 148, 160–1, 162–3, 252
Bernard, Edward 102
Bernoulli, Jean 251
Bernoulli, Nicolas 157
Berry, Helen 309, 311
Betenson, Helen 344, 346–7
Betenson, Richard, 4th Baronet of
    Wimbledon 18, 344
Bethlam Royal Hospital (Bedlam) 314, 315
Betterton, Thomas 51, 57, 65
Bickham, George 217
biological vitalism 246, 247, 289
    see also vitalism
Birch, Thomas 15, 39, 138, 243, 251, 269, 299,
    326, 342–3, 351
Bird, John 348
Le Blanc, Jean-Bernard 31
Board of Longitude 1, 9, 246
Boerhaave, Herman 294
Bombasius, Paulus 127
Bonnet, Charles 264
Book of Wisdom 138
Booth, Barton 66, 67
Borelli 38
Borlase, William 251
Bos, Lambert 189
Bossuet, Jacques 129
Boughton House 90, 210, 324, 326, 327
Boulle, André-Charles 330
Bowen, Jemmy 56
Bower, Archibald 70
Boyle Lectures 125
Boyle, Robert 15, 36, 43, 294
Bracegirdle, Anne 57, 71
Bradley, James 9, 15, 97, 172, 175, 246, 260,
    298, 299
    Astronomer Royal 12
    Savilian Professor of Astronomy 42
    discovery of the aberration of light 97, 170
Bradshaw, Lucretia 1, 145–7, 190, 313–5
Breslau 31
Brett, Colonel Henry 62
Briggs 36
The British Apollo 66, 67
British Museum 190, 250, 310, 325, 350
    Enlightenment Gallery 351
Broughton, Michael 293
Lord Brouncker 277
Brown, Hugh 252
Browne, Sir Thomas 316
Browne, Sir William 115
Bruno, Giordano 209
Budé, Guillaume 3

Van der Bucht, Gerard 163
de Buffon, Comte Georges Louis Leclerc 13,
    265, 298
Burges, Colonel Elizeus 147
Burnaby 59
Burnet, Gilbert 102, 129
Burnet, William 287
Burrell, Peter 310
Burroughs, Mr 308
Busby, Richard 102
Butler, Samuel 86
Butler, William, LLB, Prebendary of St Paul's 71
Buttons, Russell Street 356
Byrom, John 72, 80–1, 102, 114–5, 125, 170,
    239, 316, 356
    Jacobitism 117–20, 121–2
    inventor of shorthand 80, 114, 116, 120, 319

Cadogan, Lord 121
Caen Military Academy 252–3, 316–7
Calvin, John 204
Campanile, Venice 150
Camus 169
Canaletto 152, 325
Canton FRS, John 256
Caponi, the Marquess 153
Cappel, James 26–7, 189
Cappel, Louis 26
Capon, William 61
Princess Caroline 131
Queen Caroline 309
Carteret Webb, Philip 335
Cartwright, Dr. 296
Cassini, Jacques 169, 172–3
de Castro Sarmento, Dr Jacob 324
Caswell 36
Queen Catharine Braganza 19
Catullus 270
Cavendish, Charles 28, 343
Cavendish-Watt-Lavoisier controversy 298
Celsius, Anders 12, 149, 153, 168–71
Centlivre, Susanna 64, 66
Cerati of Pisa, Count Gaspero 262, 266
Chaffinch, William 20–21
Challis, Martin 330–1
Challis, Thomas 330
Chambers, Ephraim 32
Chancery Lane 339
Chardin, Jean-Siméon 282–3
Charles II of England 19, 65, 117, 196, 277, 333
Charles Emmanuel III of Sardinia 254
Charles, Earl of Burford 21
The Charterhouse 292–3
De Châtelet, Émilie 353–4
Cheale, John 228

Chelsea Physic Garden 350
Chéron 211
Cheselden, William 39, 72, 320–2
Lord Chesterfield 300
Chetham's Library, Manchester 116
Cheyne, George 157
Chichele, William 23
Chichele, Henry, Archbishop of Canterbury 23
Chichester Cathedral 333
Chinoiserie 324–5
Chiron the centaur 130
'Chivalric Enlightenment' 18, 73, 90
Christian VI of Denmark 211
Churchill, Charles, Brigadier 69
Cibber, Colley 49, 60, 63, 65, 67–9, 125
Clairaut 169
Clare Hall, Cambridge 33, 316
Clarke, John 34–35
Clarke, Samuel 34, 37–8, 128, 129, 130, 148
Clayton, Thomas 72
Clüver, Philipp 26
Cocchi, Antonio 12, 131, 145, 166–67, 190, 234, 313
Codex Vaticanus 126–7, 189, 361
coffee-houses 361
Cohen, I.B. 135
Cole, William 72, 337
Coleraine, Lord 74, 339
College of Arms 92
Collier, William 66, 68
Collinson, Peter 244, 306
Committee of Common Council of the City of London 253
Conduitt, John 131, 179–80, 194–5, 196
Congreve, William 59
Connoisseur 329
Conti, Abbé Antonio Schinella 131, 149, 157, 160, 163–6, 167
Cook, Hal 237
Cook, James 310
'Cooke' manuscript 75
Cooper, Anthony Ashley, 3rd Earl of Shaftesbury 77–8
Copley Medal, see Royal Society
Copley, Sir Geoffrey 241–2
Coram Foundling Hospital 1, 309, 316, 329, 339
Coram, 'Captain' Thomas 309–10
Corelli, Arcangelo 56
Corneille 329
Corporation of the Sons of the Clergy 71
da Costa, Emanuel Mendes 251, 256, 322–4, 351
da Costa, Joseph Salvador 323–4
da Costa, Samuel 6–7
de la Costa 308

Coste, Pierre 29
Cotes, Roger 34, 36, 100–1, 106, 200, 263, 360
    Plumian Professor of Astronomy 136
Cotton Genesis 188
Cotton, Robert 183
Cottrell, Sir Clement 104
Cowper, Willam 225
Crane Court see Royal Society
Crivelli, Father Giovanni 164–5
Cromwell, Oliver 201
Cronk 355
Cross, Letitia 56
Crowne, John 63
Crudeli, Tommaso 313–4
Cuff, John 110, 266, 271, 272, 277
    Folkes' patronage of 246–7, 258,
    rejection by Royal Society 278–80
Culpeper 279
Cushee, Richard 348

Dahl, Michael 356–7
Dale, Samuel 125
Earl of Dalkeith, Francis Scott 79, 80, 84
Dampier, Thomas 216–7, 219–20
Danneskiold-Samsøe, Ulrich, Count of 211
Dassier, Jacques Antoine 184, 204, 206
Davall, Peter 28, 298, 341
Davenant, Sir William 50
Deacon, Dr 120
Defoe, Daniel 57, 91, 274
Delbourgo, James 350
Deloraine, Earl of 93
Denne, Archbishop 245
Derbyshire 323
Dereham, Sir Thomas 12, 157, 159
Derham, William 125
Descartes 36, 38, 160–1
Desaguliers, John Theophilus 25–6, 46, 78, 84, 86, 87, 89, 110, 158, 159, 163, 223, 241, 242, 248, 267, 289, 358
Devonshire, 2nd Duke of 248
Diderot 162
Dilettanti 213
Diodorus Siculus 230
Ditton Park, Buckinghamshire 324, 333
Divan Club 213
Doggett, Thomas 68, 69
Donthorne, W. J. 22
Doppelmayr, Johann Gabriel 191
Doran, John 69
Douglas, James 114
Drury Lane 63, 66, 72, 315
Ducarel, Andrew Coltée 336–9
Dublin 294, 296, 297, 342
Dublin Courant 327

Dufay, Charles 18, 290
Duffield, Michael 315
Dugdale, William 196
Dughet, Gaspard 300
Dugood FRS, William 248
Dunton, John 66
Dürer, Albrecht 108, 303

East India Company 256, 323
Eastbury, Dorset 358
Eden, Abraham 110
Edward I 196
Edwards, George 267–8, 282
Edwards, Louisa 312, 345
Edwards, Thomas 270
Egyptian Society 6, 208, 239, 243, 269
Eisenschmid, Johann Caspar 150
Eisler, William 184, 234, 313
electricity 292
Ellicot, John 253
Ent, George 108
Epicureanism 284–5
Erasmus, Desiderius 127
Estienne, Gommar 243
Etheredge, George 63
Evening Journal 124
Evening Post 80
Euclid 35, 76
Euler, Leonhard 252, 253
Evelyn, John 105, 107, 184, 277
exchequer annuities 31
Eyles, Sir Joseph 310

Faber, John 177–8
Faber the Younger, John 333, 356–7
Fabretti 153
Fairchild Lecture 245
Fane, Charles 167
Fara 248
Farnese Globe 11, 27, 132, 173–4, 350
Farnese, Cardinal Alessandro 243
Fauquier, W. F. 213
Fawkener, William 310
Ferchault de Réaumur, René Antoine 13
Ferguson, James 9
Fermor, Sir William, Baron Lempter 219
Ferrone, Vincenzo 164
Fielding, Henry 145
Filason 279
Flamsteed, John 99, 112
Fleet Street 266, 335
Fleetwood, Charles 54
Fleetwood, Bishop 186
Fleming, Richard 186
Fletcher, John 63

Florence 313
Florentine Academy 12, 167
Florentine Masonic Lodge 1, 234, 314
Folkes, Dorothy 72, 312, 315, 344–5
Folkes, Lucretia, daughter of Martin 72, 269,
    315, 322, 325, 344–6
Folkes, Lucretia, wife of Martin, see Bradley,
    Lucretia
Folkes Sr., Martin 19, 25
Folkes, Thomas 29
Folkes, William 29, 72, 114, 115, 316–8, 330
    as a lawyer 25, 325, 327, 334, 344–5
    failed political run as Whig MP 12–13
    experience of earthquake 308–9
Fontanelle 259
Foundling Hospital, see Coram Foundling
    Hospital
Fountaine, Sir Andrew 202–4
Fowke (Folkes) Elizabeth 20
Franklin, Benjamin 298, 305, 306–7, 308
Franks, Naphtali 324
Franz Stephan of Lorraine 314
Freman FRS, William 176–7
Friend FRS, John 83, 103
Fyfe 241

Gale, Roger 112, 186
Gale, Samuel 242
Galileo 108
Gamma Draconis 169–170
Garfoot, William 331
Garrick, David 54
Gauger, Nicolas 164
Gellée, Claude (Lorrain) 300
Gentleman's Magazine 256, 309
geodesy 149, 168–9, 174, 291, 350
Geoffrin, Madame Marie Thérèse Rodet 13, 237,
    257, 341
George I 34, 62, 90, 93, 102, 116–7
George II 94, 104, 194, 293, 309
Gibson, Thomas 344
Gifford, Andrew 200
Gildon, Charles 50–1
Giustiniani, Girolamo Ascanio 160
Giustiniani Palace 165
Glover, Philips FRS 81, 118, 121–2, 127–8, 129
Goddard, Jonathan 70
Godmarsham, Kent 256
Goodman, Dena 259
Goodwood estate 93, 102, 320–23, 333
Gordon, Alexander 232–3
Gormogons 85–6
Graham FRS, George 331, 348, 360
    instrument maker and horologist 9, 16, 46,
    111, 169, 171–2, 331

Folkes patronage and gifting of his devices
    170, 258
friendship with Folkes 72, 80–1, 115, 121,
    247, 285, 341
as FRS 297
trials, demonstrations and experiments
    197–8, 200, 298–9
Graham, Richard 298
Grand Lodge of Freemasons 73–9, 85, 91, 94
Granger, James 190
Graunt, John 31
Gravesande, Willem 25, 163
Gray, Stephen 8, 41, 194, 289
Great Ormond Street 243
Greaves, John 138, 149, 152–3, 211, 228
Green Park 255
Greene, Robert 35, 36, 38, 111–2
Greenhill, Thomas 230
Gregorian Calendar 12
Gregory, David 38, 83
Gresham, Sir Thomas 70
Grew, Nehemiah 38, 108, 109, 225, 284
Grey, Henry, Lord Chamberlain, Earl of
    Kent 62, 65
Grishow, Mr 299
Grollier, Jean 243
Gronovius 265
Guerlac, Henry 176
Guerrini 285
Gwyn, Nell 20

Hadley, John 96, 98, 238
Hales FRS, Stephen 8, 228, 242, 297, 305, 308
Halley, Edmond 97, 102, 108, 133–4, 247
    aurora borealis 42–3
    life tables 31
    'mathematical party' in the Royal Society 245
    solar eclipse 99–101, 256
    trials, demonstrations and experiments 46
Hamerani, Ottone 184–5, 261
Handel, George Frideric 62, 94, 227, 329
Hanover, George 311
Hardinge, Nicholas 269
Harley, Edward, 2nd Earl of Oxford and
    Mortimer 198, 334
Harris, Joseph 197
Harris, William 320
Harrison, John 9–10, 174, 246, 360
Hartsoeker, Nicolaas 280
Harvey, William 38
Hassell, Richard, FRS 121
Hatton Garden 310, 311
Hauksbee, Francis 106, 115, 240, 290, 348
Hawksmoor, William 58
Haycock 219, 244, 245, 339

Hayman, Francis 292
Haymarket Theatre 94
Hearne, Thomas 123–4
Heathcote, Gilbert 115
Heidegger, Johann Jacob 94
Heilbron 8, 295
hell-fire club, infidel club 124
Helsham, Richard 295
Henry IV 190
Henry V 190
Henry VI 190
Henry VII 197, 201
Henry VIII 187, 196, 197
Henry, Lord Jermyn 19
Herbert, Henry, 9th Earl of Pembroke 79
Herbert, Thomas, 8th Earl of Pembroke
    45, 46, 334
Hermann, Jakob 157
Herodotus 214–5, 218, 223, 224, 230, 233, 236
Hervey, Lord John 89
Hickman, Dr 245–6
Higgitt, Rebekah 194
Highgate 303–4
Highmore, Joseph 92, 222
Hill, Aaron 66, 67, 68
Hill, John 6, 7, 200, 241, 281, 351
    friendship with Stukeley 245, 336, 348
Hill, Theophilus 245, 336
Hill, Thomas 44, 93, 95
Hill, Tom 146
Hillington Hall, Norfolk 16, 22, 23, 26, 39, 42,
    78, 100, 168, 330
    Folkes' burial 343–4
Hipparchus 11,130
Hippocrates 295
de la Hire, Pierre 36, 296
Hoadly, Benjamin, Bishop of Winchester 24,
    114, 115
Hodgson, Mary 59
Hogarth, William 1, 58, 282, 300, 356
    as a Freemason and satirist of
        Freemasonry 85–91
    portraits of Folkes 86, 340, 356, 358–61
    social links with Folkes 91, 260, 269, 292,
        300, 344
Holder, William 274
Hollar, Wenceslaus 277
Hooke, Robert 4, 70, 276–8, 361
    Micrographia (1665) 16, 38, 276–8, 350, 360
de l'Hôpital 36
Horace 233
Hovell, Dorothy 19, 22–23, 71
Hovell, Ethelreda 71
Howard, Henry, 6th Duke of Norfolk and 22nd
    Earl of Arundel 107

Howard, Hugh 190
Hudson, Thomas 292, 300
Huggonson, John 91
Huguenot community 25, 29, 79, 358
Humber Castle 10
Hume-Campbell, 3rd Earl of Marchmont,
    Hugh 269, 274
Hutton, Charles 252
Huygens, Christian 15, 38, 111
Hyde, John 293
hydra, *see* polyp
hydrostatic balance 196, 263

Inchiquin, Earl of 93
Ireland, Samuel 356
Isle of Wight 325
Itinerarium Curiosum 91

Jackson, John 130
Jacob, Margaret 78
Jacobite revolt 74
Jallabert, Jean 260
James I 186, 188, 192, 195
James, Lord Beauclaire (Beauclerk) 20
James, Lord Cavendish 248
Jason the Argonaut 130
Johnson, Maurice 214, 339
    Friendship with Stukeley 201, 245, 308, 325,
        336, 342
    President and founder of the Spalding
        Gentlemen's Society 183, 186
Jones, Jezreel 104
Jones, William 35, 36, 249, 253
Juan, Don Jorge 260
Jurin, James 113, 118, 122, 253, 355
    correspondence with Leeuwenhoek 110
    Friendship with and support of Folkes 105–6,
        121, 123, 147–8
    FRS and Secretary to the Royal Society 106,
        121, 123
    trials, demonstrations and experiments 46, 96

Keen, Theophilus 67
Kentish Town 325
Kepler, Johannes 38
King of Portugal 248
King, William 113
Kircher, Athanasius 105, 208, 212
Kit-Cat Club 57, 178
Kisby, Fiona 151
Knapton, George 344
Kneller, Sir Godfrey 39–40, 44, 45, 125, 210,
    300, 344, 356–7
Knight, Gowin 246–251

Lake, Crystal B. 187, 193, 201
Lamb, Mr 310
Lambeck, Peter 189
Lambert, George 292
Lambeth Library 336
Langley Rose, Elizabeth 350
Lapland 12, 169
de Largillière, Nicolas 358, 359
Larouche, James 56
Lassells, Richard 105
Latour, Bruno 237
Laughton, Richard 33, 34, 35, 144
Leake, Stephen Martin 187
Lebeck 210
Leclerc, Georges-Louis 13, 26, 280
van Leeuwenhoek, Antonie 38, 106, 278–80, 348
    correspondence with James Jurin 110
    donation of microscopes to the Royal Society
        13, 16, 110–11, 258–9, 278
    Folkes' treatise on his instruments 110–1,
        258–9, 348
Leibniz, Gottfried Wilhelm 71, 116, 252, 298
Leigh, Anthony 125
Leigh, Francis 67, 68
Leland, John 3
Lely, Sir Peter 333
Lemery, Nicolas 38
Lenhoff, Harold and Sylvia G. 18, 267
Lennox, Charles, 2nd Duke of Richmond and
        Goodwood 25, 146, 151, 229, 254, 260, 262,
        271–2, 293, 312, 314, 317–8, 325, 329, 355
    annuities and lotteries 32
    Coram Foundling Hospital 309
    envoy to the Hague 271
    freemasonry 44, 79–81, 84, 124, 313
    Folkes as a Knight's Companion to
        Lennox 49, 90, 93–4, 333
    firework display at Richmond House 227–8
    friendship for and patronage of Folkes 13,
        17–18, 126, 147, 317–8, 319–324
    Lady Sarah 'Taw' 268, 309, 333–4
    oriental antiquarianism 102–4, 227–9
    patronage of Abraham Trembley 257,
        258, 266–9
    patronage of George Edwards 267–8
    President of the Society of Antiquaries 334
    royal connections 116–7, 312
    support of Folkes for President of Royal
        Society 237–8
Lens, Bernard 70, 97, 98
Lesser, Friedrich Christian 273
Lethieullier, Smart 213, 233–4
Lethieullier, William 226
Leti, Gregorio 32
Levitin 214

Lewis, Samuel 22
Leycester, Sir Peter 121
Leyden Jar 260, 298
Lhwyd, Edward 258
Lieberkühn, Johann Nathanael 266, 267
Lincoln's Inn Fields 19
Lincoln's Inn Fields Theatre 51
Lindsey 279
Linnaeus 323
Lister FRS, Martin 149, 258, 275, 283
    Historiae Conchyliorum (1685–92) 206
    sulphur 43, 297, 305
Locke, John 128, 162, 167, 180–1, 321
Lombard, Jean-Louis 252
London Bridge 253
London Evening Post 321
London-Field, Hackney 256
Longitude Prize 9
Longleat House 243
Longueville, French artist 292
Longueville, Grey, Bath King of Arms
    90, 92–3
Lowther, James 238
Lucian 223
Lucretia 315–6
Lucretius 133
Lugli 348
Lully, Jean-Baptiste 56
Lux, David 237
Lyonnet, Pierre 273

Macclesfield Library at Shirburn Castle 16
McComock, Henry 342
McDougall-Waters 241
MacGregor, Arthur 268
Machin, John, Gresham Professor of Astronomy
    79, 120, 166, 237
Maclaurin, Colin 9, 10, 296
Maddison, R. E. W. 17
magnetism 247
Magnus effect 254–5
Mainwaring, Arthur, MP 69
Malebranche 160–1
Malpighi 278
Manfredi, Eustachio 157, 160
de Mangueville, Peter 28
Manilius 11
Manley, Mary de la Rivière 51–2
Mann, James 277
Mann, Sir Horace, 1st Baronet
Mann, Nicholas 243
Maple, William 296
Marivaux 259
Marples 223
Marsigli, Luigi Ferdinando 110, 223

material culture 2, 6, 75, 116, 150, 215, 218, 233,
    341, 350,
Martinelli, Cristino 32
Mary I 190
Matikkala 18, 73, 91
Maunder minimum 42
Maupertuis, Pierre Louise 170, 172, 174, 247,
    257, 264
    'capotes anglaises' or condoms 262, 312
    expedition to measure the Earth's
        circumference 9, 12
    expedition to measure the Earth's
        shape 168–9, 176
    materialism 281
Mead, Dr Richard 103, 112, 243, 315, 344
    Mead's support for Folkes 121, 231
Melanchthon, Philip 204
Mercator 36
metrology 149, 168, 196, 208, 211, 236, 246, 248,
    291, 348, 350
Michelotti, Pietro 166
microscopy 13, 247, 274–6
Miles, Reverend Dr Henry 14, 274, 280
Miller, Andrew 210
Milles, Jeremiah 212, 217
Milnes 36
Mitchell, Sir Andrew 153, 213
Mitchell, Dr John 306
Mitre Tavern 260, 316, 335, 336
Modena, Duke of 227
de Moivre, Abraham 35, 204, 247
    Annuities upon Lives (1725, 1743) 29,
        30–31, 360
    as Folkes' tutor and link between Newton and
        Folkes 28–30, 32, 33, 157, 160
    The Doctrine of Chances (1718, 1738) 28, 29
    as a Huguenot 26, 358
    as a mathematician 11, 27
Molyneux, Samuel 170
Molyneux, William 162
Le Monnier 169
de Mons, Marie Catherine 263
Churchill, Lady Mary, Duchess of Montagu
    325–9, 360
Montagu House, Whitehall 310, 325, 330
Montagu, Charles, 1st Earl of Halifax 45
Montagu, John, 2nd Duke of Montagu 124, 227,
    254, 262, 266, 324, 330, 342, 343
    connection with Folkes via De Moivre 28, 29
    Coram Foundling Hospital 309–10, 311
    family sepulchre at Warkton,
        Northamptonshire 1, 327–30, 333, 360
    freemasonry 79, 80–1, 83–4, 93, 125
    friendship with Folkes 325, 334
    interest in the hydra 266–7

Montagu, John, 2nd Duke of Montagu (*cont.*)
  Lady Mary Churchill, Duchess of
    Montagu 267, 290, 325, 327–9, 360
  medal portrait by Dassier 204
  Order of the Bath 90
  oriental antiquarianism and the Egyptian
    Society 103, 210–12, 214, 217, 222,
  Royal Military Academy, Woolwich 252–3
Montagu, John, 4th Earl of Sandwich 211,
  222–3
Montagu, Edward 29
de Montaudouin, Nicholas 260–1
Montesquieu 212, 247, 261–3, 312, 317
  Montesquieu's son, Jean-Baptiste de Secondat,
    Baron de La Brède 261, 263, 317
Monthly Magazine 78
Morelluis, Andreas 185
de Morgan, Augustus 111–2
Morgan, Charles 97–100, 134, 136
  Newtonianism at Clare College, Cambridge
    34–5, 37, 111
  oriental antiquarianism 101–2, 104
Mortimer, Cromwell 228, 277, 336
  earthquake 304
  Repository of the Royal Society 108
  secretary of the Royal Society 40, 237, 336
Mosca, Felice 157
de la Motte 329
Motteux, Peter Anthony 55
Mottley, John 64
Moxham 241
Muller, John 253
Muller, J.S. 300
Munro, John 315
Murdoch, Tessa 330
Murphy, Kathleen 261
Musée Carnavalet 358–9
Musée National du Château de Versailles 358–9

Nahum, Tate 65
Natter, Lorenz 184
Neale, Thomas 195
Needham FRS, Joseph Turberville 280–1
Negri, Solomon 103
Neher, Allister 321
Nelme, Anthony 72
Nesbit, Robert 119
de Neve, Peter 186, 195
Newton, Sir Isaac 42, 87, 102, 113, 116, 131–5,
  148, 153, 168, 169, 179–181, 243, 245, 251,
  255, 257, 289, 347, 351, 353
  Abregé de la Chronologie (1726) 131
  A Course of Experimental Philosophy
    (1734-44) 25
  Arithmetica Universalis 30
Chronology of Ancient Kingdoms Amended
    (1729) 11, 15, 26, 74, 79, 124, 128, 130,
    136–8, 150, 170–71, 196, 200, 218, 294
  electricity 289, 291–2, 294–5, 297, 299, 342
  Folkes as devoted disciple, editor and
    disseminator 1, 11, 12, 15, 74–5, 111, 124,
    130, 132–3, 136–9, 149, 183, 189, 208, 235,
    347, 350, 361
  freemasonry 78, 81–2, 91
  heterodoxy 128
  international Newtonianism 25, 26, 37, 124,
    149, 154, 156–167, 176, 324, 354
  Newton-Leibniz calculus dispute 28, 79, 116,
    157, 252, 298
  Newtonian optics 12, 38, 157–60, 162–7, 360
  Newtonianism 2, 9, 10, 12, 96, 155, 176, 181,
    194, 253, 350, 353
  Newtonianism in the Royal Society 105, 111,
    157, 194
  *Observations Upon the Prophecies of Daniel
    and the Apocalypse of St John* (1733) 74,
    124, 128, 137
  Opticks 8, 26, 29, 37, 158–9, 162–4, 176,
    294–5, 342, 360
  patronage of Folkes 47, 87, 123, 148, 351
  portraiture 5, 44–6, 97, 132, 176–81, 295–6,
    300, 303, 340, 350, 360
  presidency of the Royal Society 44, 47, 112, 114
  Principia 7, 37, 47, 131–132, 134–5, 136, 169,
    175, 245, 251, 263, 296, 299
  Sir Isaac Newton's Head Tavern 290
  shape of the Earth 172–5
  theories of fluid mechanics 253, 255
  Treatise of Quadratures 251
  Treatise of the System of the World 134
  Voltaire 354–5
  Warden of the Mint 191, 194–6, 296
  Whig Member of Parliament 45, 83
Nicholson, Bishop 186
Nine Years' War 31
Nisbett, Dr 255
Norden, Captain Frederick Lewis 211–12, 228
Norfolk Library 107
North, George 201, 336
North, Roger 56

Oldenburg, Henry 14, 275, 279
Oldfield, Anne 49, 54, 63, 65, 68–9
Ord, Robert 119
Order of the Bath 83, 90–96, 124, 187, 204,
  324, 333
Order of the Garter 90, 333
Lord Orrery 300
d'Ortous de Mairan, J. J. 247
Ossorio, Cavaliere Giuseppe 254–5, 256, 260

Oughtred, William 35
Ouguela *see* Rassem
ouroboros 340–1
Outhier, Abbé 169
Oxford, Earl of 121

Pack, George 59, 62, 67
Paetus, Lucas 152
Pagan, Hugh 194, 197
Palazzo Giustinani 12
Palazzo della Zecca 191
Palmer, Samuel 91
Pamphili, Teresa Grillo 147
Panini, Giovanni Paolo 300
Paracelsus 248
Parker, George, 2nd Earl of Macclesfield 28, 304,
    308, 341, 343
Parmentier, Jacques 292
Parsons, James 267, 282, 304, 336
Pastorino 150
Pausanias 243
Payne, George 76, 80
Pearce, Zachary 131, 136
Pelham, Henry 317
Pellet, Thomas 74, 79, 124, 130, 131, 136–7
Pemberton, Henry 91, 115, 251
Pembroke, Thomas, 8th Earl of 189, 190, 191,
    201–3, 205
Pembroke, Henry, 9th Earl of 203, 205
Pennsylvania Gazette 305
Percival, Sir John 148, 160
Perrault, Charles 155–6, 213
Perry, Charles 211–2, 231, 233–5
Perry, Francis 200
Perry, Micajah 310
Perry, William, FRS 107
Pesselier, Charles-Etienne 282
Petiver, James 261
Petty, William 31
Philip of Spain 190
Philippe II, Duke of Orléans 198
physiognomy 184
Piazza San Marco 152
de Piles, Roger 301
Pine, John 91, 94, 95
Pisenti, Giovan Bernardo 160
Pitcairne, Archibald 83
Pitt, Thomas 198, 200
du Plessis, James Paris 287
Pliny 42, 223
Pococke, Reverend Richard 225–7, 229–30,
    233–4, 239
    Egyptian travels 209
    member of the Egyptian Society 211, 212,
    239, 243

Poleni, Giovanni 147–8
Cardinal Polignac 284
polyp 13, 18, 244, 246, 257–8, 264–7, 269–73,
    281, 324
Pomfret, Lord 187
Pond, Arthur 300
Pontac's Head 122
Pope, Alexander 28, 39, 258, 356
Pope Clement XII 314
Porcacchi, Tommaso 329
Poro, Jacomo 285–7
Porter, Mary 62
Portsmouth, Earls of 195
Postlethwaite, John 103
Pound, Reverend James 9, 15, 41–2,
Poussin, Nicolas 300
Powell, George 66, 67, 68
Power Henry 278
Priestley, Joseph 307
Prince of Wales 117
Procurator of Venice 191
Public Advertiser 343
Pugh 334
Purcell, Henry 56

Quare, Daniel 331
Queen Square, London 22, 39, 72, 121, 128, 138,
    187, 196, 198, 205, 243, 245, 249, 267, 288,
    290, 303–4, 310, 336, 344
Queen's Theatre, Haymarket 57, 60–2, 66

Rassem 104, 111
Ratcliff 257, 265, 272
Rawlinson, Richard 207, 337–8
Ray, John 149, 277
Read, Nicholas 345
de Réaumur, Antoine Ferchault 264–5, 267, 273–4
Rée, Jonathan 130
Rees, Abraham 114
Regius Chairs of Modern History 91
de Requeleyne, Hilaire-Bernard, baron de
    Longpierre 87
Rembrandt 300–3
Republic of Letters 350
Revocation of the Edict of Nantes 25
Reynardson, Samuel 198–9
Rhodes, Mr 167
van Rhoon, Count Willem Bentinck 246
Rialto Bridge, Venice 150
Riccoboni, Luigi 328
Rich, Christopher 51, 62, 65, 68
Richard II 190
Richard III 192
Richardson the Elder, Jonathan 39, 92, 124, 142,
    178, 301–3, 344

Richardson the Younger, Jonathan 128, 129–30
Richmond House 227–8
Rishton, William 312, 344
Riva, Ludovico 166
Rizzetti, Giovanni 158–9, 160, 163–5
Robins, Benjamin 8, 246–7, 251–257
Robinson, Dr Bryan 294–6, 299
Roderick, Richard 217
Roe, Shirley 280
Rohault, Jacques 36–8
Rolli, Paolo 329
Roman foot 149–153, 197
Rooke, Henry 335
Roque, John 91
Roubiliac, Louis-François 177, 179, 203–4, 205,
    292, 295, 321, 345
    Montagu sepulchre at Warkton 1, 325, 328–9,
    333, 360
Rousseau 7
Rousseau, George 244, 339
Royal Academy of Sciences at Berlin 299
Royal College of Physicians 268, 350
Royal Dublin Society 296
Royal Military Academy, Woolwich 252, 256
Royal Society 14, 241, 248
    Copley Medal 10, 110, 194, 207, 241, 244,
        246, 255, 264, 268, 273, 275, 351
    Crane Court 44, 266, 335, 336
    Dining Club 210–11, 260, 335, 343
    Library 15, 237, 242–3, 336
    Library and Repository Committee 108–109,
        242, 279
    Philosophical Transactions 4, 6, 17, 38, 93, 96,
        97, 105, 106, 110, 113, 123, 146, 153, 157,
        159, 168, 170, 200, 207, 225, 229, 241, 251,
        256, 263, 265, 274, 275, 276, 278, 287, 304,
        308, 342, 348
    Repository Museum Collection 13, 241, 244
    Royal Society Archive 14
    Royal Society Council 40, 251, 335
    Folkes as President 174, 178, 181
    Thursday club, see Dining club
Royal Commission on Historical
    Manuscripts 16
Royal Observatory at Greenwich 180
Ruding 201
Rushbrooke Hall, Suffolk 19
Rutty, William 79, 123
Rysbrack, John Michael 179

St Geminianos Church, Venice 151
St George's Church, Wrotham 345
St Helen's Church, Bishopsgate 70
St James's Square 19
St John's College, Oxford 337

St Luke's Hospital for Incurable Lunatics 314
St Martin's Lane Academy 292, 344
St Thomas's Hospital 315
Salisbury, Earl of 310
Sanderson, Mr 310
Sandwich, Lord 231, 233, 235
    Egyptian Society 211, 213, 216, 220–1, 223–5
    travels in Egypt 212–3, 221
Sansovino, Francesco 151
Sansovino, Jacopo 3, 151, 191
Santlow, Hester 66–7
Saunderson, Nicholas 162–3
scientific antiquarianism 183, 207–8, 236,
    341, 353
scientific peregrination 148–9, 176
Scott, George Lewis 32, 253, 298
Seaby 201
de Secondat, Charles, Baron de Montesquieu,
    see Montesquieu
seismology 303
Shadwell, Charles 66, 67
Shadwell, Thomas 63–4
Shank 353–4
Shapin, Steven 93
Sharp, Abraham 112
Sharp, John, Archbishop of York 186, 190
Shaw, Thomas 104–5, 209
Sherwin, Parson 319–20
Sherwood, Noah 336
Shields, David 193, 197, 198, 201
Shooter's Hill 298
Short, James 341
Simon, James 342
Simon, John 44
Simon, Thomas 189
Sisson, Jonathan 348
Slaughter's coffee-house 31
Sloane, Sir Hans 5, 7, 39, 108, 112–4, 115–21,
    123, 148, 159, 176, 191, 237–9, 242, 268,
    286, 287, 350–1, 358, 360
    British Museum 190, 310, 351
    competition for presidency of Royal Society
        with Folkes 74, 112–124, 178, 237,
        335–6, 350
    collections 186, 207, 215, 224, 244, 250, 261,
        282, 310, 351
    rapprochment with Folkes 144, 160, 238
    royal patronage of the Royal Society and free
        speech 116–7, 119–20
    satirization 113
Sloane, William FRS 81, 123
Smith, Charles John 61
Smith, John, printmaker 44, 45, 143
Smith, John, uncle to Roger Cotes 100
Smith, Robert 34, 106, 161

Smith, Thomas 147
Snead 115
Snelling, Thomas 201
Snow, Matthew 95
Society of Antiquaries 1, 2, 4, 6, 149, 168, 170, 173
    Archaeologia 6
    establishment in 1717 4, 73
    Farnese Globe 12, 173
    Folkes' numismatics 186–8, 191,
        193–4, 195–7
    Folkes as President 1, 174, 189, 333–341
    freemasonry 73–4, 90
    links and mooted merger with Royal
        Society 18, 168, 170, 275, 335, 353
    metrology 149, 342
    Newtonianism 171
    Royal Charter 334–40
    Vetusta Monumenta 187–88, 198–99
Society of Isaac Newton 340
Solomon's Temple 76, 78, 134, 137–8, 150, 234
van Somer, Paulus 340
Sorghvliet 246, 271, 288
Sorrell, Francis 80
South Sea Trading Company 351
Southampton Street, Covent Garden 39
Spalding Gentleman's Society 183, 188, 193
    Maurice Johnson 184, 186, 245, 325, 342
    William Stukeley 217, 245, 325, 342
The Spectator 68
Sprat, Thomas 277
Squire, Dr 308
Stanhope FRS, Charles 226, 298
Stanhope, Philip Dormer, 4th Earl of
        Chesterfield 31, 32, 78, 166, 254, 261,
        298, 299
Stanhope, William 297
Statute of Mortmain 340
Steele, Richard 8, 64, 356
von Steinberg, George Friedrich 30
Sterrop, George 277
Stewart 285
Stirling, James 78
Stonecastle, Henry see Baker, Henry
Stonehenge 10, 39
von Stosch, Baron Philipp 184, 211
Strabo 224
Stuart, Alexander 242
Stuart Stevens, Henry 341
Stukeley, William 100–1, 115–6, 125, 145, 183,
        186, 201, 214, 227, 244–6, 272, 283, 325, 334
    correspondence with Folkes 2
    censure of Folkes' irreligiousness 6, 124, 283
    Egyptian Society 212, 213, 215, 217–9, 222,
        223, 225, 231–3
    freemasonry 73, 91, 125

hostility towards Folkes 71, 144, 189, 193,
        245–6, 336, 342, 346
John Hill 245, 336, 348, 351
as member and first Secretary of Society of
        Antiquaries 73, 186, 339–41
as member of Royal Society 244–5,
        272, 305–6
Memoirs of Sir Isaac Newton's Life (1752) 112
patronage of John Harrison 10
rector at St George the Martyr, Queen
        Square 39
The Philosophy of Earthquakes (1750) 306–8
Sturmius 35, 36
Surgeon General of Ireland 294
Surugue, Pierre-Louis 282–3
Swift 7, 356
Swiney, Owen 49, 62

Tatler 62, 63, 64, 66
taverns
Taylor, Brook 79
Taylor, Samuel 358
Tencin, Madame 259
Tenison, Thomas 34
Terrasson, Jean 209
Theobald, James 242
Theobald, Lewis 1
Thieriot, Nicolas-Claude 358
Thomas, Lord Jermyn 19
Thornhill, James 72, 91, 292
Thornton, Connell 329
Thucydides 270
Tillotson, John 34
Toft, Mary 117, 287
Tonson, Jacob 131
Tradescant Collection, Lambeth 209
Trajan Pillar 153–6, 161, 303, 348
Trembley, Abraham 13, 18, 110, 246–7, 249,
        257–274, 271–4, 279, 281, 284, 287–8, 318,
        324, 348
    Henry Baker 273
    Copley Medal 264, 273
    Cuff's microscope 272
    Folkes' first encounter in France 13, 257
    Folkes' mentorship and patronage of 263,
        265, 267, 272, 273, 274
Trew, Christoph Jacob 192
Trinity College, Cambridge 190, 256
Trompf, Gary 78
Turner, Dawson 275
Turner, Sir John 12–13
Turner, Shallet 355
Tuscher, Carl Marcus 212, 234
Tyler, William 347
Tyson, Edward 225

de Ulloa, Antonio 174–6
Universal Spectator 274
University of Coimbra 324

Valla, Lorenzo 3
Valoue, James 242
Vanbrugh, John 57, 59–62
Vanderbank, John 5, 132, 177–8, 179, 180–1,
    303, 340, 360
de Varenne, Pierre 116
Varignon, Pierre 26
Vaughan, John, 3rd Earl of Carbery 44
Verbruggen, John 62
Lord Vere Beauclerk 310
de la Vergne, Louis Elisabeth, Comte de
    Tressan 289
Versailles 324
Vertue, George 45, 200–2, 204, 242, 334–9
    engraver to the Society of Antiquaries 187–9,
    190, 198
Vienna Genesis 189
Villa Ludovisi 105
'Vinegar Bible' 71
Virgil 234, 319
vis viva controversy 148, 252
vitalism 13, 246, 257, 280–1
Vitruvius 151
Vivares 300
Voltaire 1, 257, 353–361
    *Letters concerning the English Nation/Lettres
    philosophique*, 112, 353, 355
Vossius 218

Wagner, Gillian 310, 311–3
Wake, William, Archbishop of Canterbury 9,
    23–4,34, 42, 71, 204
Wallis, John 35, 36, 274, 283–4
Walpole, Horace 114, 227, 228
Walpole, Robert 83, 90
Walsh, William 59
Wandelaar, Jan 273
Wanley, Humphrey 103, 186, 334
War of the Austrian Succession 176, 227, 271, 273
Ward, John 200
Ward, Seth 35
Waring, Edward 34

Warner 272
Waterland, Daniel 35, 37, 128
Watkins, Francis 290–1
Watson FRS, William 8, 290, 293, 297–8,
    299, 343
Weekly Journal 274
Weld, Charles 147
Weldon, John 56, 59
Wellcome Trust 17
Wesley, Charles and John 114
West, James 194, 237, 238–9, 242
Westminster Abbey 349, 355
Westminster Bridge 298, 299
Wharton, Philip, 1st Duke of Wharton 81, 82,
    83, 84, 86, 124
Whiston, William 34, 40, 97–100, 128, 130, 139
White Wig or White Peruke on Maiden
    Lane 355
White, Gilbert 275
White, John 119, 121–2
White, Taylor 267
Whiteside, John 43
Wilks, Robert 49, 63, 68, 69
William III 194
Willis, Browne 187–88, 190, 191–2, 195,
    243, 337
Willoughby, Lord 200, 337
Willughby, Francis 277
Wilson, Benjamin 8, 292–300
Wilton 202–3
Wissing, Willem 45
de Witte 36
Wolleston, Francis 293
Wolters 266
Wood, Sarah 313
Woodward, John 43
Wray, Daniel 217, 269, 297
Wren, Sir Christopher 19, 45

York Buildings 56
Yorke, Philip, 2nd Earl of Hardwicke 251,
    269, 343
Young, Edward 113, 358

Zeeman, Enoch 44
Zollman, Philip Henry 267